Artur Braun

Quantum Electrodynamics of Photosynthesis

Mathematical Description of Light, Life, and Matter

DE GRUYTER

Author
Dr. Artur Braun
Empa
Swiss Federal Laboratories for
Materials Science & Technology
Ueberlandstr. 129
8600 Duebendorf
Switzerland
Artur.Braun@alumni.ethz.ch

ISBN 978-3-11-062692-6
e-ISBN (PDF) 978-3-11-062994-1
e-ISBN (EPUB) 978-3-11-062700-8

Library of Congress Control Number: 2020944017

Bibliographic information published by the Deutsche Nationalbibliothek
The Deutsche Nationalbibliothek lists this publication in the Deutsche Nationalbibliografie; detailed bibliographic data are available on the Internet at http://dnb.dnb.de.

Cover image: Artur Braun
Typesetting: le-tex publishing services GmbH, Leipzig
Printing and binding: CPI books GmbH, Leck

www.degruyter.com

Dedicated to the German Physicist Gerald Hegedus

Preface

This is the third book that I am publishing with De Gruyter. The first one dealt with the use of x-ray methods for the study of electrochemical systems and, specifically, the use of synchrotron methods for energy materials. The focus was on the microscopic level of energy storage and conversion. The cover photo shows seaweed that I picked up from the shallow water at a beach in Jeju Island in South Korea, which I used to indicate that biological systems were also electrochemical systems. It was an inspiration for how to look into the physics and chemistry of batteries, fuel cells and photoelectrochemical cells, and biological motifs with synchrotron radiation.

In my second book with De Gruyter, I focused on the systems aspects of electrochemical energy converters and storage devices, and only to a lesser extent on the physics and chemistry of the underlying materials using radiation-based analytical methods. That book dealt with the application aspects, where suddenly size mattered, and microscopic methods were no longer very helpful. In that book, I already drew connections between energy, environment, and economy.

This third book is about primary production, which is a term used in biology and means the synthesis, assembly, and production of biomass from water and carbon dioxide, powered by sunlight. We know this process as photosynthesis. However, the term primary production already indicates the link to working life, economy, sustainability, and many other relevant sectors.

Primary production is measured as gram carbon per square meter per day. Carbon comes from the carbon dioxide. The conversion of carbon dioxide to carbon, or carbohydrates, by photosynthesis is called carbon-dioxide assimilation. Primary production on Earth amounts to 100 gigatons of carbon per year. A similar number holds for the primary production of biomass in the oceans. Around 1% of the energy from the Sun that arrives on our globe is used for this primary production.

We get our food from primary production, and to some extent also our energy from biomass burning and, increasingly, from direct conversion of sunlight to heat and electricity. This book will show the microscopic molecular mechanisms of the conversion of light to chemical energy in the photosynthesis process, and also to some extent the nuclear mechanisms of light generation in the Sun, which is the primary energy source for primary production and, thus, the basis for life.

120 years ago it was discovered that light is an effect of electrodynamics. At around the same time, quantum mechanics was developed, which, with greater success, would account for discrepancies between experimental observation of the light-electric effect and classical light theory. The remaining discrepancies between experiment and theory were further resolved by what is known as second quantization, that is, the quantization of the electromagnetic field and what, in fact, is quantum electrodynamics (QED).

https://doi.org/10.1515/9783110629941-201

This book shall not replace specialized literature on quantum electrodynamics and that on photosynthesis. The purpose is to have a book that combines both fields and extends the idea of cause and chance, and path integrals, to the domain of finance and economy. Life is based on science and on economy. Further, the limited size and resources of our planet are a driving force for mankind to look further ahead to space exploration.

QED sticks with the wave nature of many physical phenomena and the Lagrange density, the integration of which amounts to actual physical observations. I use this book as an opportunity to point the reader to other fields in the sciences where the wave nature is observed, and these are, specifically, the fields of finance and economy. Finance and economy fit in the scope of the book because of their mathematical parallelism in quantitative treatment and because of their relevance for the use and distribution of the produce and fruits of primary production on the stock market.

Most of this book was written in the years 2018–2020, which were years of climate change rallying and climate change politics. In Germany, parliament decided on a new CO_2-tax. Sixty years ago, photosynthesis pioneer Melvin Calvin from Berkeley concluded his studies, which today are listed as 23 key papers in *The path of carbon in photosynthesis* (Calvin and Benson 1948), which culminated in a *Lebenswerk* for which he received the 1961 Nobel Prize in Chemistry "for his research on the carbon-dioxide assimilation in plants" (Kauffmann and Mayo 1996; Calvin 1961a). Plants and algae assimilate the carbon dioxide in the atmosphere and produce oxygen, and planting trees has been suggested as one solution to reduce the concentration of CO_2 in the atmosphere.

My previous two lecture and reading books with De Gruyter (Braun 2017, 2019a) were written with Microsoft Word and a specifically designed book template from De Gruyter. The relatively large number of mathematical equations made me consider LaTeX for this third book. The relatively larger number of references made me reconsider MS Word again, and in the end, I opted for LaTeX. Personally, I prefer drawing the Feynman diagram with a pen and paper or, in the worst case, with MS Power Point. However, LaTeX provides very nice packages[1] that allow you to draw Feynman diagrams with ease.

I provide many graphics for the reader, which I have taken from references, with the necessary reuse licenses. I have included many photographs, my own data, and sketches drawn with computer software, and some just drawn with colored markers and pencil on paper. Nowadays, I believe, drawing and sketching is no longer part of the curriculum, but in my time at medical school at RWTH Aachen, drawing was part of the curriculum. In your studies, drawing helps you learn about form and function. Consider drawing with pen and paper as an exercise. Richard Feynman started drawing when he was 44.

1 I used the package TikZ-Feynman from Joshua Ellis (Ellis 2017).

When I was student at RWTH Aachen, I had very good professors and enjoyed their lectures, particularly when they strayed away from their written manuscript and started facing the students in the audience – RWTH is a very large polytech, and we often had 100 and more students attending the lectures – and adding anecdotes to the dry matter of science. In their spirit, I have spiced up the book with a number of personal anecdotes and hope that the reader will find them useful for their further growth and understanding of life and science.

Artur Braun
Zürich, May 2020

Acknowledgements

My gratitude goes to Elton Cairns, Professor of Chemical Engineering at UC Berkeley and Principal Investigator at Lawrence Berkeley National Laboratory (LBNL). As my supervisor at Berkeley, he gave me the opportunity to join the group of UC Davis Biophysics Professor Stephen "Steve" P. Cramer, the Physical Biosciences group at LBNL, along with Drs. Hongxin Wang, Uwe Bergmann, Pieter Glatzel, Tobias Funk, Weiwei Gu, and Daulat Patil.

There I learnt how to access functional components in photosynthesis apparatus with protein x-ray spectroscopy and the necessary method development. Elton promoted the use of all these methods for battery research and development. It was mind blowing when I found that the redox processes in the manganese in lithium batteries were not fundamentally different from the cycle of manganese in the oxygen evolving complex in photosystem II.

It would take me another 6 years from then to get an opportunity to join photosynthesis research by ways of semiconductor photoelectrochemistry – because I inherited, meanwhile at Empa in Switzerland, a project on photoelectrochemistry from Andri Vital, who had decided to leave Empa for a position in industry. This project was an internal Empa project, funded by the Empa Board of Directors' 6th competition for internal funding. This project's money brought Renata Solarska (then from the Jan Augustynski school at the University of Geneva, now a professor in Warsaw) in my group as a postdoctoral researcher, an expert in photoelectrochemistry to whom I owe my first insight in PEC plus the shopping list from my first SNF R'equip grant for instrumentation.

Having her in my group provided me with more research funds, mediated by solar energy Professor Andreas Luzzi, science diplomat with the Swiss Federal Office of Energy (BfE). With this BfE money, we hired Debajeet K. Bora from India, a PhD student with the University of Basel Chemistry Professor Edwin "Ed" C. Constable. I had not known Ed until I heard his talk on photosynthesis and coordination compounds at a workshop at Professor Michael Grätzel's lab at EPFL in Lausanne.

This workshop in Lausanne was a key event and motivator for my future work on solar energy conversion. There was a gentleman from Hawai'i, Eric L. Miller (I didn't know him then and later learnt that he was a science diplomat like Andreas Luzzi), who gave a 360° overview of PEC and solar hydrogen production and showed work from another gentlemen, Clemens Heske from University of Nevada Las Vegas (UNLV) with all the fancy soft x-ray spectroscopy methods that I had been using for ceramic fuel cells, batteries and fossil fuel combustion aerosols, and I then knew I was in the right community.

Moreover, Ed Constable gave a talk on his field of expertise, dye sensitizers and coordination compounds, and he linked it to the hydrogenases, nitrogenases and photosystem II, topics, which I remembered from my time at Berkeley. I was excited to win

https://doi.org/10.1515/9783110629941-202

him as thesis advisor for Mr. Bora, who is now heading a young research group as Professor in Bangalore. With respect to Professor Grätzel, I had the privilege of joining him as co-PI for the European project NanoPEC and got to collaborate with Scott Warren and Kevin Sivula, back then EPFL postdocs and group leaders, now professors. Such large EU projects help me increase the radius of my international collaborations.

I owe Dr Rita Toth – she was in my group for 8 years – more project funding with Ed Constable, based on which we could hire two PhD students, Florent Boudoire at Empa and Roche Walliser at Basel, while Rita could secure extra funding for herself with a Marie Heim-Vögtlin Fellowship. In this setting, along with Jakob Heier in the Functional Polymer Lab at Empa, we published a landmark paper on moth-eye microstructure photoelectrodes, which in 2014 would bring us high-profile press coverage from La Liberte, Le Monde and The Economist, and an invitation to Washington D.C. where we met U.S. Secretary of State, John Kerry. I think here I have to be grateful not only to my associates but also to David Rothkopf from Foreign Policy magazine, and his creative team who spotted our work and lifted us onto their podium.

By suggestion from Prof. Louis Schlapbach in 2008, back then Empa CEO, I was inspired to propose a Symposium at the Materials Research Society (MRS), and I decided to try one on photoelectrochemistry and photocatalysis. Alan Hurd from LANSCE in Los Alamos was MRS President back then, and he responded to my email inquiry less than 24 hours later with "Great idea, Artur! Do this, [...], and do that [...]". This became the first PEC symposium at the MRS, with Jinhua Ye from NIMS, Paul Alivisatos from LBNL, John Turner from NREL, and Egbert Figgemeier from Bayer AG. Since, the MRS Spring and Fall Meetings have had symposia on PEC, solar fuels and artificial photosynthesis every year, sometimes several in parallel. There I got to meet my future collaborators Elena Rozhkova from ANL, and Lionel Vayssieres.

I was inspired by the PEC research at University of Hawai'i at Manoa, and decided to make a sabbatical in Eric Miller's Lab at Hawai'i Natural Energy Institute (HNEI) in the School of Ocean and Earth Sciences and Technology (SOEST). Arriving in Honolulu in November 2010, I found Eric Miller was no longer there, as he had just assumed a position with the U.S. Department of Energy in the Hydrogen Program. Thus, Nicolas Gaillard was now in charge and leading the research and operations of the HNEI Thin Film Lab. I had the privilege and pleasure of moving into Eric's office, which allowed me a great open-door view over the Manoa Valley with the virtually daily rainbow.

Nicolas, Alexander de Angelis and Yangzen Chen showed me how to make magnetron sputtered photoelectrodes and take calibrated PEC and solar cell measurements. At a local meeting in downtown Honolulu, we also got privileged information about the just established Joint Center for Artificial Photosynthesis (JCAP) from Berkeley's Peidong Yang.

At the Dean's Reception by the International House of UH Manoa, at College Hill, I got to know Sam Wilson from Center for Microbial Oceanography: Research and Education (C-MORE), who studies cyanobacteria. He frequently does campaigns with the ship in the Pacific Ocean and measures the hydrogen production of cyanobacteria, which certainly was very interesting for me. Eventually, we co-authored a joint paper on bioelectric interfaces for artificial photosynthesis.

In 2017, I spent my sabbatical at Yonsei University in Seoul, Korea in the Department of Mechanical Engineering, with Professor Wonhyoung Ryu. We have a similar scientific and technological view; our background is the inorganic energy conversion, and we look forward to biological systems performing the future energy storage and conversion job. Along with his colleagues in Stanford, he works on biological energy conversion and studies thylakoids and their membranes with nano tips and other fancy methods. In Professor Ryu's group, I learnt about the extraction of thylakoids and how to measure their photocurrents in photoelectrochemical cells. I am grateful to him, his students and his school.

Finally, I want to express my gratitude to Professor Krisztina Schrantz from Szeged University, who for 4 years worked in my group on PEC and conducted our collaboration with Professor Linda Thöny-Meyer and Julian Ihssen and Michael Richter in Empa Sankt Gallen – on genetically modified cyanobacteria. This research was funded by the Swiss Federation and the VELUX Foundation.

Contents

List of Figures

https://doi.org/10.1515/9783110629941-203

List of Tables

https://doi.org/10.1515/9783110629941-204

Prologue

U.S. PRESIDENT: You don't think I should take the chance?
BENJAMIN RAND: Absolutely not.
U.S. PRESIDENT: Mr. Gardiner, do you agree with Ben, or do you think we can stimulate growth through temporary incentives?
CHAUNCEY GARDINER: As long as the roots are not severe, all is well and all will be well in the garden.
U.S. PRESIDENT: In the garden?
CHAUNCEY GARDINER: Yes. In the garden, growth has its season. First comes spring and summer. But then we have fall and winter. And then we get spring and summer again.
U.S. PRESIDENT: Spring and summer?
CHAUNCEY GARDINER: Yes.
U.S. PRESIDENT: And fall and winter?
CHAUNCEY GARDINER: Yes.
BENJAMIN RAND: I think what our insightful young friend is saying is that we welcome the inevitable seasons of nature, but we're upset by the seasons of our economy.
CHAUNCEY GARDINER: Yes. There will be growth in the spring.
U.S. PRESIDENT: Well Mr. Gardiner, I must admit that it's one of the most refreshing and optimistic statement I've heard in a very long time. I admire your good, solid sense. That's precisely what we lack on Capitol Hill.

From the movie "Being there" (Ashby 1979; Baum 2016).

Chauncey Gardiner is an illiterate private gardener without any birth record or identification but well behaved, with the manners and clothing of a butler. By accident, literally, he makes it to the home of Benjamin Rand, Chairman of the First American Financial Corporation, who is advisor to and friend of the US President. The story is from a novel by Jerzy Kosinski (Being There, (Kosinski 1970)) and was made into a movie in 1979 (Ashby 1979; Baum 2016). In the above scene, the President visits the home of Mr. Rand in order to receive advice on how to get out of the recession, a national economic crisis. Clueless Chauncey is triggered by the word *growth* and, thus, naively talks about the seasons and growth in the garden. The movie stars Peter Sellers, Shirley MacLaine, and Melvyn Douglas. While the author is finishing this book (Braun 2020), the world is going through an economic crisis, triggered and welcomed by world leaders in response to the Covid-19 crisis (Zhang and Liu 2020; Cohen and Kupferschmidt 2020).

https://doi.org/10.1515/9783110629941-205

1 Introduction

This book promises the treatment of two fields of science. One is quantum electrodynamics *QED*[1], the science of the interaction of electric and magnetic fields, and electromagnetic waves, with space and with matter and with energy, as it originates from fields and waves. Therefore, the book will contain a substantial introduction to classical and modern electrodynamics, but with a view on the second field: photosynthesis.

Photosynthesis is the process by which plants and some primitive animals, cyanobacteria, absorb light and convert its radiative or photonic energy into chemical energy for the buildup of organic matter, which is known as primary production, and for the process of life.

Many books, lectures, classes, and courses on quantum electrodynamics deal exclusively, and only exclusively, with the mathematical formalism of quantum electrodynamics but hardly show any applications in real life. This is a pity because the value of a theory comes with its validity for an application. Also, there are countless such applications, the most wonderful of which, in my opinion, is the one being showcased here in this book ...

While photosynthesis is known as the process that occurs in the biosphere only, by default and definition, we might anticipate that I treat *QED* only for this purpose and region, for the light absorption and potential re-emission in the biosphere. However, I also dedicate one large chapter (Chapter 5) to the generation of photons and where they come from: the sun.

All energy that we know, at large, originally comes in one form or another form from the sun. Therefore, nuclear fusion processes, as they occur everywhere in the sun and in the stars are treated here as well. Hydrogen is being burnt to helium while it releases cosmic γ-radiation in the 4π-direction, i.e., in all directions. Further, some of the radiation generated arrives on Earth in wavelengths that we know, which is direct visible sunlight, scattered visible sunlight, and infrared radiation and UV radiation. Visible light is that which is used by plants for photosynthesis. So, photons are generated in and on the sun, they travel through space to Earth and are absorbed and converted in the biosphere as the primary production of biomass.

My publisher requested that I write this book according to the state-of-the-art of the field. I have problems with this because the state-of-the-art implies that something is a finished product. My aim is certainly to provide readers with a finished book. However, I do not consider the contents finished, and this for a particular pedagogical reason. The theories that I present in this book are not God given. They were created by researchers a hundred years ago and then developed further. The early forms of the theory are not taught in school, and, therefore, some of their originality goes missing.

1 QED also stands for the Latin words *quod erat demonstrandum*, in German *was zu beweisen war*, and means "which was to be proven" and was in older times used as the final words in a mathematical proof or philosophical arguments.

https://doi.org/10.1515/9783110629941-001

2 Nature and natures of light

2.1 Personal perception of light

In Figure 2.1, we see a realistic nature scene[1] in a rural part of Switzerland, within walking distance from the geographical center of gravity of the Canton Zürich. In the center is a paved asphalt road. On the right is a crop field with densely planted corn crops as high as 150 cm. The corn plants are still green. On the left we see the outskirts of a field with wheat corn. The plants are yellow and ripe, with a height of 50 cm. Two weeks later it will have been harvested already. Next to that field is a patch with two hands full of large trees such as fir, oak, and nut, and some bushes. The height of the tallest trees is around 15 meters. These trees cast a shadow on the road where curves to the right.

Behind that patch of trees on the left there is an extension of the wheat field. Far way ahead of you there is a forest with mixed trees like fir and oak and beech with a height of around 15 to 20 meters. The trees have all a green "skirt"; this is why the forest appears green – at least at this time of the year.

The right top of Figure 2.1 shows a blue sky with some white clouds. The photo was obviously taken in daytime – in sunshine – because the trees cast a shadow and because the sky is blue. The sun is to the left shoulder of the photographer. The sun would impress us with its bright white yellow shine – if it were just recorded in the photo.

I am, therefore, adding a second photo in Figure 2.2, which shows the sun just in the center, partially covered by the tops of three tall, slim palm trees. Here, the sky is blue, and there are no clouds. The photo was taken in the parking lot of Murray's Family Farms, 5 km east of Bakersfield in the Mojave desert. While I was there, the temperature was 32 °C after lunch, according to the records available on the Internet. This is way above the average temperatures there in April, which is typically between 11 °C and 25°. At the time of writing these lines, December 22, 2019, the temperature is 20 °C, and it is raining. The land there is desert land, but agricultural technology makes that from the airplane you can see large green patches where crops and fruit are grown – and sold in the Red Barn of Murray's Family Farms, for example. While it is very hot and sunny outside, inside the grocery store, you freeze because of the air conditioning.

How do we interpret what we see in the Figure 2.1? We see that the plants at large are green. Therefore, the plants emit green light. Our eyes are the light detectors (for

1 Is it really nature? Is it nature because it is green, with plants? I do not think so. This is all cultivated land, roads, meadows and crop fields, and woods and trees. Nature certainly has its peculiarities, but the landscaping is manmade and not natural, with the exception of the general geological formation of the area.

https://doi.org/10.1515/9783110629941-002

Fig. 2.1: Nature scene with road, crops, trees, and forest in rural Switzerland near the center of Canton Zurich. 14 July 2018. Photo taken by Artur Braun.

Fig. 2.2: Murray's Family Farms at Highway 58 near Bakersfield CA, 13 April 2015. The white–yellow bright sun shines through a cloudless blue sky over the Mojave Desert. The brownish haze over the horizon is from air pollution in nearby Bakersfield. Photo taken by Artur Braun.

the physiology of the eye, see the very early book by Hermann von Helmholtz, 1867, (von Helmholtz 1867)), but also our skin can detect heat or temperature as the infrared radiation that comes directly from the sun or indirectly from the ambient air.

The sky is blue to our eyes. This is because the sky reflects blue light. Where does blue light come from, if it does not come from the sun? Blue light does come from the sun, as all light comes from the sun, unless it comes from a synthetic source like a fire, a lightning bug, electric light, a lightning flash, aurora borealis, or the light from the stars. Moonshine is sunshine that is reflected by the moon. In rare cases, the light that comes from the moon to our eyes may be light reflected by the Earth to the moon and then reflected back to our eyes, but in the first place, Earth received that light from the sun. Therefore, the sun is our primary light source and – thus – also our primary energy source. Therefore, when we deal with photosynthesis we also must deal, at least partially, with astronomy.

We also must deal with meteorology because the weather is part of our observations. The clouds can cast shadows over the plants, but the clouds also contain the necessary water for rain, which plants need to function and survive. However, also the atmosphere as such typically has minimum humidity even when no clouds are visible. Also the soil on which the plants grow has some humidity, which is why geography and geology are important fields for photosynthesis. The Atacama Desert in Chile is the driest nonpolar desert, but it still has some minimum humidity, which allows for the growth of some flora. With respect to any flora, botany is certainly a core field for photosynthesis.

We see that very humid and warm places typically have flourishing flora, whereas dry deserts hardly have any. However, when there is plenty of water, at some desert places, such as in an oasis or a river, plants will grow green. Murray's Family Farms is in Bakersfield, a very dry area, but with artificial watering it is possible to grow 25% of all produce exported by the United States to other countries in the St. Joaquin Valley in California.

The solar power that arrives on the Earth's surface – on average – is defined as the solar constant (Johnson 1954). This is a mere technical quantity and cannot be considered a nature constant. It is roughly around a power density of $1.367\ \mathrm{kW/m^2}$. The sun has a huge power of 3.8×10^{26} Watt. Such numerals and numbers are important for technologists, engineers, and economists. As an exercise, find the statistics that tell us how much power or energy your country needs per year. This is the power for all individual households, mobility and transportation, and for industry, the manufacturing for your own country's goods, and for goods that your country exports. Then look up the area (in square kilometers) of your country and calculate, how much of the area needs to be used to collect sunlight and convert it with solar cells into electricity.

As a further exercise you can compare countries' "performance" and sustainability. A sunny country with a large area and a small population and little industrial activity can sustain itself well with solar power. A small and heavily industrialized country with manufacturing that needs a lot of energy for production will possibly not be able to power all its businesses with solar and wind power alone. On 5 November 2019, the German government held a press conference on Germany's future with respect to hydrogen technology and economy (phoenix 2019). They said Germany was too small to produce all the necessary solar fuels and would, therefore, explore opportunities elsewhere with large sunny countries, like Africa and Australia. This fits with the words of my colleague at the University of Pretoria in South Africa, Professor Egmont Rohwer,

Fig. 2.3: University of Pretoria Chemistry Professor Egmont Rohwer speaking at a diplomatic meeting at the Future Africa Campus on renewable energy, CO_2 reduction, and solar fuels. He explicitly mentioned that South Africa could become a solar fuel exporting country. Photo taken 30 July 2019.

who at a diplomatic meeting in July 2019 (Figure 2.3) on his campus said that South Africa would have great opportunities to export solar fuels in the future[2].

Both photos shown in Figures 2.1 and 2.2 were taken during daytime, in the afternoon. At night, there is no light shining on the plants. There is still a number of photons going down to Earth in this region, but it is negligibly small. When you launch a rocket and move yourself away from Earth, you will pass the different layers, the different spheres that span around the globe. You will leave the biosphere where the plants grow and the birds fly after a few hundreds of meters, or when measured from the sea level to the highest mountains where biological systems are still present.

Then you will soon also leave the atmosphere where humans and animals cannot breathe anymore because there is no sufficient oxygen concentration for survival anymore. Look at Figure 2.4, where all the spheres, biosphere, atmosphere, stratosphere, and so on are sketched until heights of 1500 km. Then you move into the stratosphere and, finally, the ionosphere, where the remaining gas atoms and molecules are ionized from the ionizing cosmic radiation. I have been told that cosmic radiation is not the correct term. The ionizing radiation actually comes just from the sun, including protons. This by far overshadows the actual cosmic radiation coming from extrastellar regions. From here, you will see the Earth as the blue planet, largely because of the

2 As a matter of fact, in 2013, I said the same at a diplomatic mission in Pretoria, when we were setting up the extension of the Swiss South African Joint Research Programme SSAJRP. As South Africa is a wine producing and wine exporting country, with so much sun on its land, it could be the perfect place for producing solar fuels and exporting them as well.

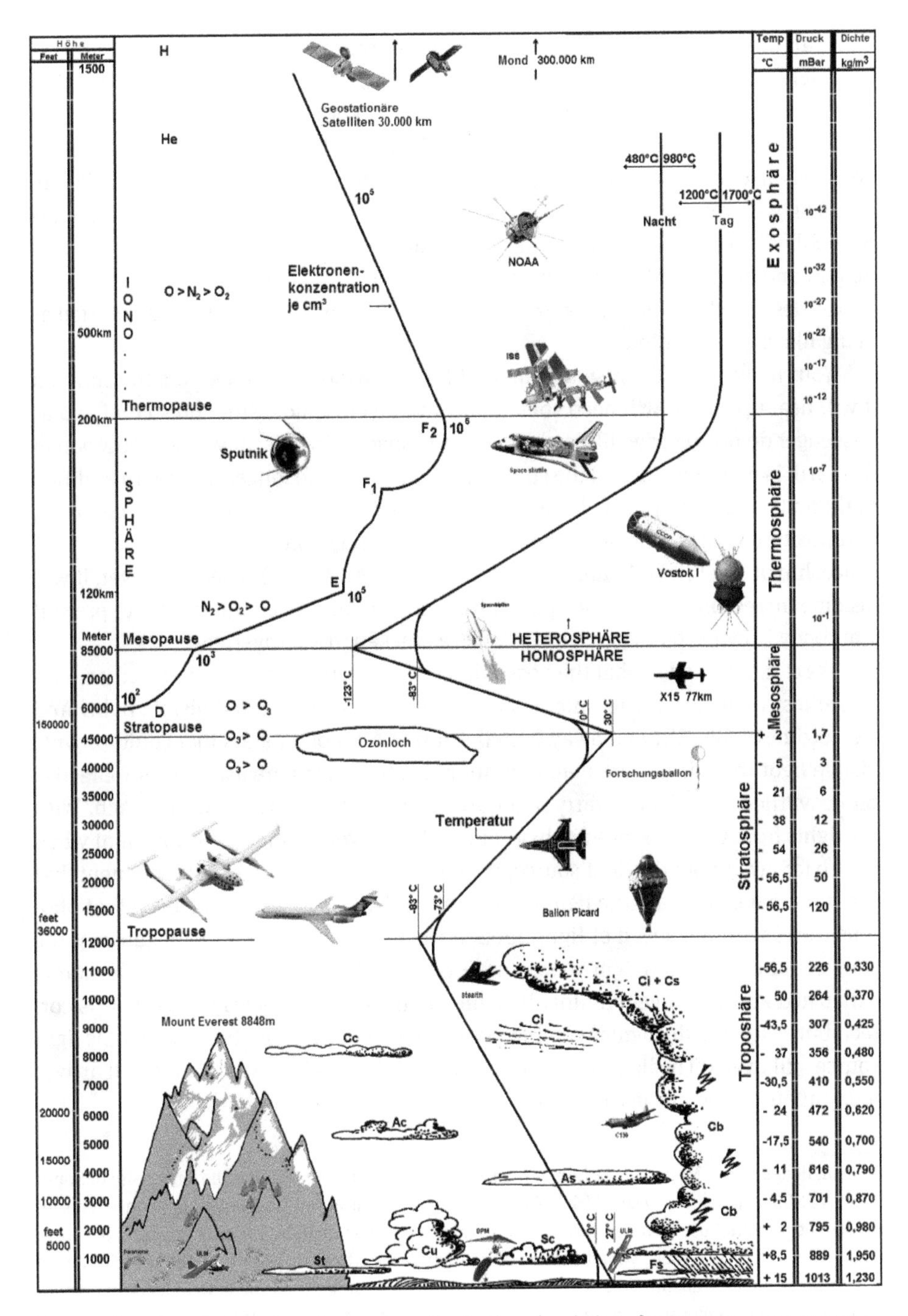

Fig. 2.4: Overview of atmosphere layers above the Earth and variation of temperature, gas pressure, and gas density as a function of height above Earth. From Wikimedia Commons, free media repository. This file is licensed under the Creative Commons Attribution-Share Alike 3.0 Unported license. Courtesy Hans Cools, Cdt v/h Vlw i.r.

blue part of the sunlight scattered back from the atmosphere. You may see the sun as the bright shining star that you are known to see from the Earth's surface. In all other directions you will see no light but only the black $\kappa o \sigma \mu o \sigma$. The first flights into the stratosphere with balloons provided new insight into the coloring of the skies. In his book, Niklitschek (1949) explains how Russian stratosphere pilots reported (12 February 1934, with reference to the Ossoaviachim program and a mission where all pilots were killed) that the color of the sky was blue at 8.5 km height, dark blue at 11 km height, and deep violet at 13 km height. At 19 km, it was a deep violet-blue. For 21 km it, was described as black violet-gray. Finally, at 22 km height, the sky was reported as being black-gray in color.

Your rocket capsule or spaceship is still in such close distance from the sun that it will heat up, and it will need some protective layer, some insulation so that you as passenger do not get fried like a chicken in the spaceship. It is only when the Moon or the Earth is between you and the sun that the sun's rays no longer hit you. In addition to thermal radiation, you will also be hit by cosmic radiation from γ-rays, neutrons, and positrons. They will cause damage to all materials, including the surface of the spaceship or satellite[3] (Madiba 2017; Madiba et al. 2018, 2017). Then, however, it will become incredibly cold for the spaceship, and you need a heating system to prevent your body from becoming an ice block like a chicken in a refrigerator.

We can reflect over what the possibilities are that life, as far as it is based on photosynthesis, can sustain in outer space. We need CO_2, water and visible light. And we need nutrients. Very far away from the sun or from other stars that could provide visible light there is the problem that not enough photons are available for effective photosynthesis. Look at a starry night and think that there is no sun in reach. With the lights from the stars, even if they count billions, you cannot grow any plant when it is so far away from the light source, the energy source. One may now comment that the stars that send that little light to us would maybe no longer exist, given that photons travel with the speed of light through the universe and that the distance of the stars could be a million light years away from us, which means that when the star is exhausted and no longer shining, its light could still be on its way to us. This is correct, but as long as the light is coming, it's coming. It does not matter whether the light source still exists. The light from extinct stars millions of light years away still arrives today at our observatories on Earth.

3 My colleague Prof. Malik Maaza from iThemba LABS in Somerset West and the University of South Africa (UNISA) in Pretoria, South Africa, works in various scientific fields of materials science. This includes the potential use of vanadium dioxide as thermochromic material for heat shielding purposes in so-called cube satellites. Vanadium dioxide undergoes a phase transition at about 30 °C, above which it can block transmission of heat radiation and thus protect, for example, electronic equipment behind the shielding. However, when vanadium oxide comes under irradiation from γ-rays and neutrons, defects may be created in the material, and this may potentially alter the Curie points of the material. His PhD student Itani Madiba worked in my group for 1 year in order to analyze the electronic structure of this material.

Moreover, when you get so close to a star that you can harvest sufficient energy, you have to shield your distance to the sun properly to avoid exposure to the dangerous cosmic radiation, which can harm not only cube satellites but also the DNA of plant cells. The necessary shielding for that on Earth is the atmosphere and the magnetic field of our Earth.

We should also be aware that the light that arrives from the sun on Earth has traveled over 8 minutes through our solar system – with the speed of light. It is also noteworthy that the light produced in the center of the sun takes many million years before it can arrive at the surface of the sun. These are the astronomical dimensions around the photosynthesis that we observe on our Earth.

2.2 History

In old times men believed that seeing was an active act done by the eyes. Our eyes would spot actively all visible objects with some resolving power, like throwing spears to the objects, and when we hit them, we see them. At some point, in later ages, it was understood that there must be a light source before we can see anything. We either see the light source directly, or we see objects indirectly when they are illuminated by the light source. Our eyes were understood then as passive devices that would receive the light. Making this transition in understanding was an important step in increasing the consciousness about the human vision apparatus.

2.3 Mystics and metaphysics of light; popular cultures

Without light there is darkness, and we cannot see and can no longer move around safely. A lack of light means danger, unless we sleep in a safe place. In older times, man could move around only during daylight – when the sun was shining or at night in the moonlight. With the discovery of fire, man had an artificial light source and, thus, extended the range of his activity also into dark caves. Fire gave man power.

Still, the sun was and is the most important light source. It is, thus, not surprising that the sun has become an icon in mankind's culture. The Old Testament is the entrance to the three religions of Judaism, Christianity, and Islam. In the Jewish Thora you can find the book of Genesis , which explains how the world was created by God. I show below the 31 statements of Genesis. Genesis means creation. Specifically, it means the creation of Heaven and Earth, of man, beast, and plants, and, finally, the book tells us "Be fruitful, and multiply, and replenish the Earth, and subdue it." This is a well-known command. The other well-known command in the book of Genesis is *fiat lux*, which is taken from the third sentence of the book and means "Let there be light". For the interested reader, I have copied the book of Genesis in the Appendix 1.

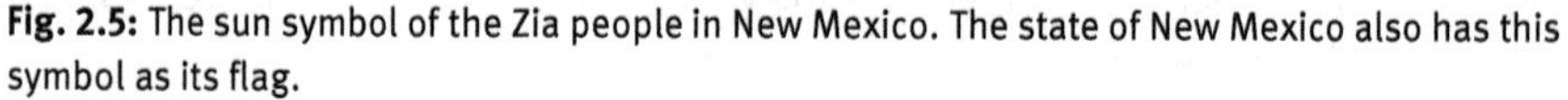

Fig. 2.5: The sun symbol of the Zia people in New Mexico. The state of New Mexico also has this symbol as its flag.

Thus, light is deeply rooted in our religion. There are many ancient people for whom the sun had the position of a god. For the Zia people, the sun is a divine god. The Zia people in New Mexico, a tribe of now only 850 Pueblo Indians, claims ownership of the Zia symbol, which is a red stylized sun on a yellow background, as shown in Figure 2.5. The red symbol is a circle, at the boundary of which four sets of ray lines point in all four directions: four to the North, four to the East, four to the South, and four to the West. Meanwhile, the Zia symbol has become the symbol of the State of New Mexico. Said Amadeo Shije, Governor of the Zia at a public hearing in Albuquerque on 8 July 1999: "The Zia sun symbol was and is a collective representation of the Zia Pueblo. It was and is central to the Pueblo's religion. It was and is a most sacred symbol. It represents the tribe itself." Tribal Administrator Peter Pino said "For the Zia people, the sun symbol is an exceptionally significant religious and cultural symbol." (Shije 1999; Turner 2012)

In middle school we learnt how to understand and treat light as a bundle of rays, which we could symbolize with straight lines. As light was implicitly considered as having a direction, we use arrows instead of just straight lines. The arrows point away from the light source. Often, the petals of a flower, like the yellow ones of a sunflower, are implied to be sun rays that point away radially from the plant, see, for example, Figure 2.6.

Fig. 2.6: Sunflowers in a sunflower field, photographed at 18:55 on 14 July 2018, not long before sunset. With the sun now on their back, the flowers are still pointing in a south-eastern direction, where they received most radiation from around noon. The yellow leaves are often pictured as the rays of the sun.

Common daily life observations such as the occurrence of an image of an object, the mirror image in a mirror, the magnification of an object by a lens, or the projection of an object on a screen, can all be mathematically described by what we call geometrical optics (Bethe 1938; Born 1933).

I have family ties to Korea and spend a lot of time there, and I frequently travel there also for research projects. In Korea, you will notice the swastika symbol, which is a cross with right angles at the tip, which also resembles the symbol of national socialism (nazism). You can find the swastika in Korea, Japan, and India on many ornaments (Park et al. 2019) in restaurants and also on many Buddhist temples, such as the one shown in Figure 2.7. The symbol resembles the sun and the eternal return of the sun, the cycle of the seasons and that of day and night, and the hope that after a dark night, there will be a sunny day, and after a cold winter with no food, there will be summer again with fruit we can harvest. In short, the sun is the symbol for life.

The sun is also present in popular cultures. Countless songs and poems deal with the sun. The Beatles' *"Good day, sunshine"*, The Velvet Underground's *"Ride into the sun"*, Boney M.'s *Sunny*, or *Sonne und Liebe, soleil et l'amour* by Sacha Distel, to name only a few well-known songs. Some of these songs have a light, jolly spirit, whereas

Fig. 2.7: This Buddhist temple in Jong-no, in Seoul, Korea, is one of the countless examples where swastikas, you can see it in white on the gable below the roof, are used as ancient symbols for the sun. Photo taken by Artur Braun.

others reflect a deep philosophy, existentialism even, such as *Der Rat der Motten* by Witthüser and Westrupp, where the council of the moths is interested in revealing the secret of the candlelight. A young team of moths flies towards the candle, and only the moth that gets burnt in the flames gets to know the secret, but it keeps it for itself because it dies in the flame:

> Im Osten dämmert der Morgen schon,
> da fliegt eine junge Motte davon,
> stürzt in die Flamme, deren Kraft sie nicht kennt,
> glüht kurz auf und verbrennt!

2.3.1 The Sun Chariot of Trundholm

In 1902, near the city of Trundholm in Denmark, a farmer found a bronze sculpture that resembles a horse carriage. Its age is estimated at 3500 years, which dates it back to the Bronze Age. The sculpture is known to experts as the Sun Chariot from Trundholm (Bradley 2008). A photo is shown in Figure 2.8. It resembles a horse carriage with

Fig. 2.8: The Sun Chariot from Trundholm is a bronze sculpture discovered by a farmer in Denmark in 1902. Photo taken by Malene Thyssen, https://creativecommons.org/licenses/by-sa/3.0/deed.en.

a sun on it, and the wheels of the carriage are sun wheels, i.e., wheels with a fourfold cross as the spines. The wagon carries a large bronze disk that is partially still plated with gold. The disk has two sides: a golden bright one resembling sun and day, and on the back side a dark side that resembles night. The bright side is filled with ornaments that present a geometrical regularity, which can be an interesting puzzle for intelligent people with creativity and imagination. An elaborate quantitative study on the relationship between geometrical structures in the disk and the structure of a calendar was done 10 years ago by Christoph Sommerfeld, who finds that the creator of bronze was aware of the Metonic cycle[4] (Sommerfeld 2010). More recently, an Italian physicist came up with the idea that the sun wheel of the bronze sculpture found in Trundholm could actually be a calendar because the regular pattern drawn on the wheel allows for the interpretation of a week with 8 days and a year with 360 days (Sparavigna 2012).

Light also stands for enlightenment, for intellectual insight into how world affairs run. Johannes Kepler was a mathematician born in the late fifteenth century. He also occupied himself with astrology and astronomy. I do not know whether these two fields were distinguished in the fifteenth century. He served, for example, as fortune teller to Wallenstein and also collaborated with Tycho Brahe, the best astronomer of

4 The Metonic cycle means that 19 years equal to 235 synodic months and 6940 full days.

the time. Eventually, Kepler, a mathematician, could bring sense into the many empiric observations made by himself and other astronomers, and found that the planets travel around the sun and that the trajectories form what is known as the five platonic bodies. He considered this discovery of the geometry and architecture of the solar systems as the enlightenment. A true enlightenment, however, was the mathematical rationale of the heliocentric architecture (postulated by Nikolaus Kopernikus before) of the building of planets, their moons, and the sun. We know that Galileo Galilei struggled with the Church about the claim that the sun was the center of the system, and not the Earth. This is not withstanding that Kopernikus was a man of the Church and that Kepler was a religious man. Further, this is not withstanding that the religious astronomers were tough men, egomanes who would engage in intrigues. Read, for example, the review on the competition between Kepler and Brahe (Voelkel 2008) and part of Kepler's biography (Graneau and Graneau 2006):

> Throughout his life Kepler was at odds with the intellectual establishment represented at his time by scholars and leaders of the Christian Church. The church hierarchy could have easily ruined his career as a mathematician and astronomer. He did not have the means to practise science without financial support from others and therefore had to disguise his scientific convictions best he could.

A hundred years later, scientific progress mounted also in social and political progress, and that age was called the Age of Enlightenment. In previous ages, men had been the slaves of mysticism, and only a small elite had the insight or an educated guess of the true forces that keep the world running. The laws of mechanics were known not only for the bodies on Earth but also for celestial bodies. The mystic field of astrology became the scientific field of astronomy. In 1955, Richard Feynman delivered an address (Feynman 1955) at Caltech on the value of science and how the freedom to doubt was the result of a struggle against authority in the Middle Ages. Galileo Galilei was one scientist who fell victim to authoritarianism, which back then was exercised by clericalism.

In the New Testament (testamentum 1994), in John's gospel, the term light is used with respect to spiritual enlightenment. I have reproduced part of John's gospel in the Appendix 2. There it says "In him was life; and the life was the **light** of men." Typically, light stands for the sun because the sun is the major lighting source. There are countless other examples for the relevance of the sun in human history and culture. However, not everybody associates light with the sun only. I am one of the founding members of the Daylight Academy. The DLA is an off-spring of the VELUX Foundation in Switzerland (Stiftung et al. 2015). VELUX is a Danish window manufacturing company known for its roof windows and skylights.

There, daylight has a different meaning than sunlight. Daylight is a term of its own, possibly because in architecture, you play with windows and shading and directing light to particular places by architectural tricks. You do not want to be directly exposed to sunlight in your home maybe. You want to be able to read a book, and that requires the proper lighting, and for other situations you want different lighting.

Lighting can be controlled with windows and with shades and with reflecting surfaces. Human well-being is one interest of the VELUX Foundation and the DLA. A couple of year ago, we wrote a paper on the aspects of lights, the VELUX Lightbook (Norton et al. 2017).

2.3.2 Stone monuments in the Sahara Desert as geographical markers

When I did my physics thesis for my diploma studies at KFA Jülich, I was given the opportunity to spend 2 weeks at the Institute of Crystallography at CNRS in Grenoble, France. I think that was in the summer of 1995. My host was well known for his computer simulations for so-called LEED IV studies. LEED stands for low-energy electron diffraction, and IV stands for IV curves. When you record the electric current as a function of electric voltage, then you obtain an IV curve. The LEED method is used for crystallography with very high surface sensitivity. It helps to determine the crystal lattice of the first few atomic layers on a solid surface. Because electrons from the electron gun of the LEED instrument system interact very strongly with the electrons from the probed surface, the collisions can no longer be modeled with kinematic theory. Therefore, dynamic scattering theory must be applied, which is very costly in terms of computation time. Back then, we had to do the simulations on a CRAY supercomputer, which was hosted in Paris.

My host has a hobby that he has developed to professionalism. The French have a very long summer vacation, and he used his vacations together with his family to travel to the Sahara and discover engravings and stone sculptures. Back then, he told me he had published almost as many papers in this field of archeology as he had in surface science. At that time, in the mid 1990s, he was also authoring a book on this topic. Anyway, along his ways through the Sahara desert he also noticed stone monuments, actually stones and rocks laid in patterns that served as local markers for directions. He found that they typically were directed to the east and pointed to the direction of the sunrise (Gauthier 2009; Gauthier and Gauthier 1996).

Apparently, in very old ages, humans who traveled across the Sahara region arranged these stones as marks for traveling. The monuments would guide the way for other travellers so that they would get not lost in the wide desert. He found that such monuments were not used throughout the entire North Africa. In Niger, for example, no such monuments for orientation were found; if they were found, they were directed towards the sunrise in the east.

2.3.3 From astrology to metaphysics

The reliable occurrence of sunrise in the morning in the east and sunset in the west, the change of day and night, summer and winter, the regularities in the seasons and the regularities in the observations with stars in the dark night – these were the scientific facts the early humans noticed. They combined their observations with mystic beliefs about the gods, which would move as the sun over the sky or as planets through the night. They would connect the dots in the starry sky, the visible zodiac, into patterns that would make sense to them, such as the animals and hybrids that we still know as the 12 star signs (Tierkreiszeichen). In the past, this was astronomy, the naming of the celestial space and giving meaning into the observations, until the fifteenth century. Until then, astrology and theology were considered exact sciences (Feyerabend 1975). With the Enlightenment in the age of reason in the eighteenth century, societal progress followed scientific progress, which had begun with the discoveries of Keppler and Galilei, the celestial mechanics that is part of astronomy. Mankind has, thus, witnessed the decline of the science of astrology, and the rise of astronomy; see also Chapter 6. Paul Feyerband suggested that we should not necessarily forget and discard a declining science but allow for its potential resurgence when it would be better in solving problems (Feyerabend 1994).

2.4 Astronomy

The discrete spectral analysis of the light from the sun and the stars provided great insight into their elemental composition. However, fundamental insight into the statistical nature of light was gained from the studies of the light that was emitted from them, considered as just glowing objects. The spectral decomposition (spectroscopy) of the light that is emitted from such glowing matter allows us to not only know its elemental composition but also gives some fundamental insight into the statistical nature of light.

Thermodynamic considerations done by Kirchhoff have shown that once the emitter is absolutely black, then the spectrum of this emitter depends only on the temperature and not on its elemental composition. This sounds confusing because we can hardly imagine an emitter with a black color. Even worse, black means having no color at all, because all radiation with its characteristic colors are absorbed.

It is important to note the distance between Earth and the sun. As the trajectories of the planets around the sun are not circles but ellipses, we cannot simply tell one radius, but two radii, which are called aphelion and perihelion. Roughly, Earth is 95 million miles or 150 million kilometers away from the sun. This distance is defined as the astronomical unit of length AU (Huang et al. 1995).

The speed of light is very fast but not infinitely fast. It was Olaf Römer who in 1676 published a method for the proof that the speed of light is finite and not infinite (Romer 1676). He based his conclusions on observations that he had made with the moons of the planet Jupiter.

2.5 Remote sensing with satellites

For the quantitative assessment of photosynthesis and agriculture (primary production), it helps when you can have a look at Earth from an airplane or even from outer space (anonymous, cia 1969). This can be done with satellites at a very large distance (large "height" no longer applies in astronomy. Can you tell what I mean, as an exercise?) from Earth. A very recently launched such satellite is the Deep Space Climate Observatory, in brief DSCOVR (formerly called TRIANA, but DSCOVR has a different scientific nature (Valero 2006)). Have a look at the simple sketch in Figure 2.9. This satellite flies at a distance of around 1.5 million km from Earth. This is 1/100 of an astronomical unit. There is a particular reason for that. When we consider the sun with large mass M and Earth with smaller mass m in the distance of 1 AU, the center of gravity of this two-body system can be mathematically calculated. Let us now put a satellite with mass m_S for which the condition should hold that the satellite's centrifugal force is equal and compensated by the gravitational force of the center of gravity of M and m.

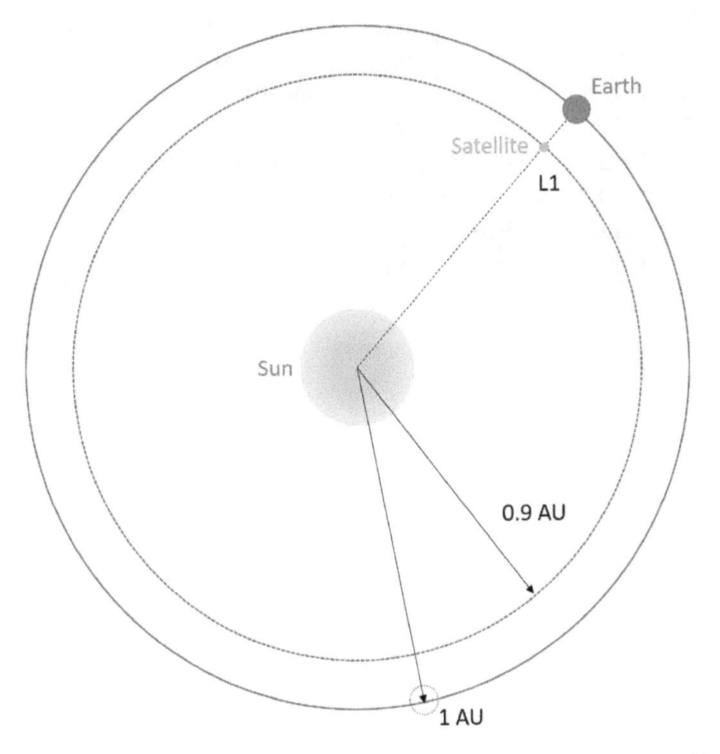

Fig. 2.9: Construction of the first Lagrangian point L1. The sun is the central body and the Earth is the second large body on its trajectory around the sun. The combined gravitational forces of the sun and the Earth acting on the satellite are equal to the centripetal force of the satellite on its trajectory around the sun. This condition is valid for the trajectory around the sun in around 0.9 AU and is called the first Lagrangian point.

Keep in mind that DSCOVR will be flying at the first Lagrange Point, or L1, which is about 1/100 of an astronomical unit (AU). Since 1 AU is approximately 150 million km or 93 million miles, DSCOVR will be roughly 1.5 million km or 0.93 million miles from Earth. The combined gravitational forces of the sun and Earth acting on the satellite is equal to the centripetal force of the satellite on its trajectory around the sun. This condition is valid for the trajectory around the sun in around 0.9 AU and is called the first Lagrangian point. The Lagrange points are actually solutions to the three-body problem, which were found first by Leonhard Euler (Euler 1767) and later supplemented by Lagrange. It is a fundamental mathematical problem in physics to find the exact equations of motion for a system of more than two moving bodies. Euler was maybe the first to discover that this is possible when the three bodies are positioned on a straight line, or when they are on the points of a triangle with three identical sides (Fieber 1960; Euler 1767). Everything else is more difficult and not exactly solvable.

This is comparable to the geostationary satellites, whose trajectory physics is based on the equilibrium of the gravitational force between Earth and satellite, and the centripetal force of the satellite. This condition is given at a satellite trajectory of 38000 km above Earth. The Lagrange 1 satellites are 40 times that distance away from Earth. The mathematical description of the Lagrange points is explained in a brief paper by Cornish (Cornish 1998). The Lagrange conditions are valid also for other trajectories: L2, L3, L4, and L5. On 20 July 2015, NASA released to the world the first image of the sunlit side of Earth captured by its EPIC camera on NOAA's DSCOVR satellite. You can view the movie on NASA's website, see Sharghi (2015). A comprehensive book about the mechanics and its mathematical description for satellites is found in the book by Koon, Lo, Marsden, and Ross, which is available for free on the Internet (Koon et al. 2011).

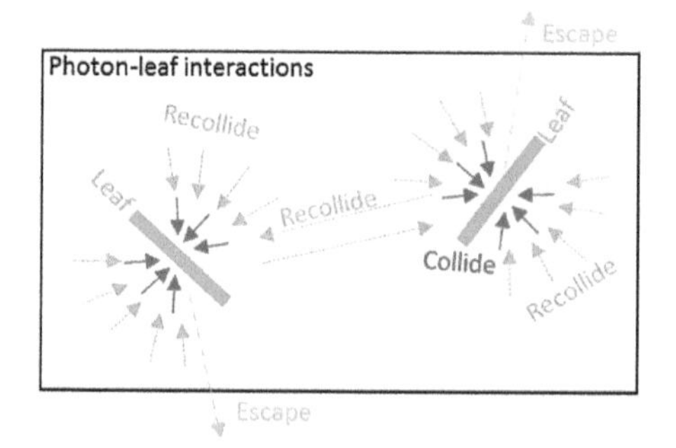

Fig. 2.10: Recollision and escape probabilities. Solid arrows depict photons incident on leaf surfaces from different directions. Reprinted from Remote Sensing of Environment, 198, Bin Yang, Yuri Knyazikhin, Matti Mõttus, Miina Rautiainen, Pauline Stenberg, Lei Yan, Chi Chen, Kai Yan, Sungho Choi, Taejin Park, Ranga B. Myneni, Estimation of leaf area index and its sunlit portion from DSCOVR EPIC data: theoretical basis, 69–84, Copyright (2017), with permission from Elsevier (Yang et al. 2017).

Figure 8.21 shows a photo taken from DSCOVR (Deep Space Climate Observatory) (Burt and Smith 2012; Frey and Davis 2016), launched 11 February 2015) on August 15. The wildfires not only destroy the vegetation that performs photosynthesis. Its clouds can also block the sunlight to some extent. The latter holds also, for example, for volcano eruptions. This satellite-based DSCOVR data can also be used to determine the portion of leaves that is exposed to sunlight (Yang et al. 2017), see Figure 2.10. Data from this satellite are also shown for wildfires in Canada in 2018, see Section 8.3. It is a special field of mathematical physics that deals with the interaction of photons with vegetation. This does not primarily concern photosynthesis, but the development of concepts for the determination of the reflectivity function of vegetation canopies for remote sensing with visible light and the near infrared and ultraviolet, see Figures 2.11 and 2.12. The field is, therefore, termed photon–vegetation interaction (Myneni and Ross 1991).

Remote sensing is not only used for land observations (Landsat satellites), but also for maritime observations, because the sea is full of life, too. This includes plankton, of course. NASA has operated a programme for that purpose called Plankton, Aerosol, Cloud and Ocean Ecosystem, the PACE programme (Omar et al. 2018; Gorman et al. 2019; Werdell et al. 2019; Uz et al. 2019)[5].

The Central Intelligence Agency had shown an interest in remote sensing and signal intelligence with respect to the assessment of the photosynthesis and primary production capacity of countries, see the reference from CIA PRD, also the comment by rocket and space flight pioneer Wernher von Braun (Stuhlinger and Ordway 1994).

The (anonymous) author of the Intelligence School Training Manual Number 4 (anonymous, cia 1969) quotes rocket and spaceflight pioneer Wernher von Braun, who apparently gave an interview in the U.S. News and World Report of 12 December 1966, page 66, on the question of to what extent spaceflight could help fight hunger in the world. Said Wernher von Braun:

> It has been demonstrated with airplane flights, using some sophisticated photographic equipment and remote sensors, that from high altitudes you can distinguish very clearly rye from barley, soybeans from oats. Moreover, you can distinguish healthy crops from sick ones. You can, for example, distinguish corn afflicted by black stain rust from healthy corn. You can also find out whether the proper fertilizer has been applied, whether there is too much salinity in the soil.
> By continuously surveying and re-surveying the tilled areas of the world-by keeping track of each patch of land as it develops from the planting season in the spring to the harvesting season in the autumn you can predict very well the crop expectations on a global scale. When drought hits an area, you will find a local setback. If some crop has been damaged or destroyed by hail, your satellite-mounted remote sensors will find it.
> As you get closer to the harvesting period you can, by feeding all that information into a computer, predict just how much of a crop to expect, and what kind, and when and where.
> Of course, you would need plenty of correlation data before the data produced by such a satellite system would be reliable. You get this correlation simply by comparing the "ground truth", or the facts determined by a man walking through a field, with what the satellite equipment sees in that same field.

5 https://pace.oceansciences.org/documents.htm.

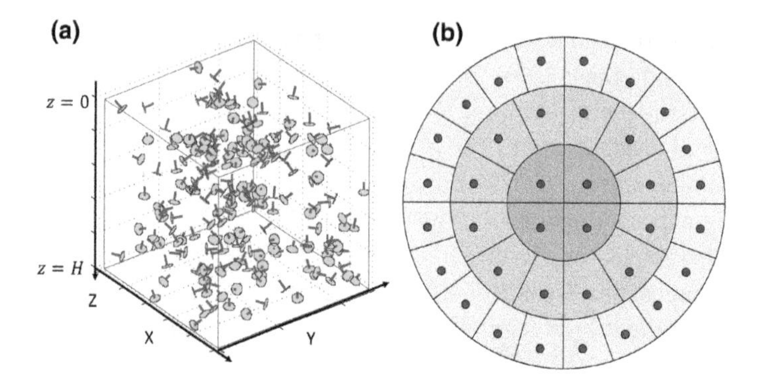

Fig. 2.11: Stochastic model of the canopy structure. Points are scattered in a volume V according to a stationary Poisson point process of intensity d (panel a). On each of these points, a disc of radius r (panel b) is placed. Their random orientation is generated with a leaf normal distribution function. The discs represent bi-Lambertian leaves, i.e., the incident photons are reflected from, or transmitted though, the disc in a cosine distribution about its upward normal. The disc is divided into n equal areas, which represent the smallest resolvable scale. Panel (a) shows a realization of canopy structure with 207 leaves, each containing $n = 36$ equal areas. The leaf radius to canopy height ratio, r/H, and the mean leaf area volume density, uL, are 0.03 and 0.5, respectively. Leaf normals are shown as blue bars. More details can be found in Appendix A. Reprinted from Remote Sensing of Environment, 198, Bin Yang, Yuri Knyazikhin, Matti Mõttus, Miina Rautiainen, Pauline Stenberg, Lei Yan, Chi Chen, Kai Yan, Sungho Choi, Taejin Park, Ranga B. Myneni, Estimation of leaf area index and its sunlit portion from DSCOVR EPIC data: Theoretical basis, 69–84, Copyright (2017), with permission from Elsevier (Yang et al. 2017).

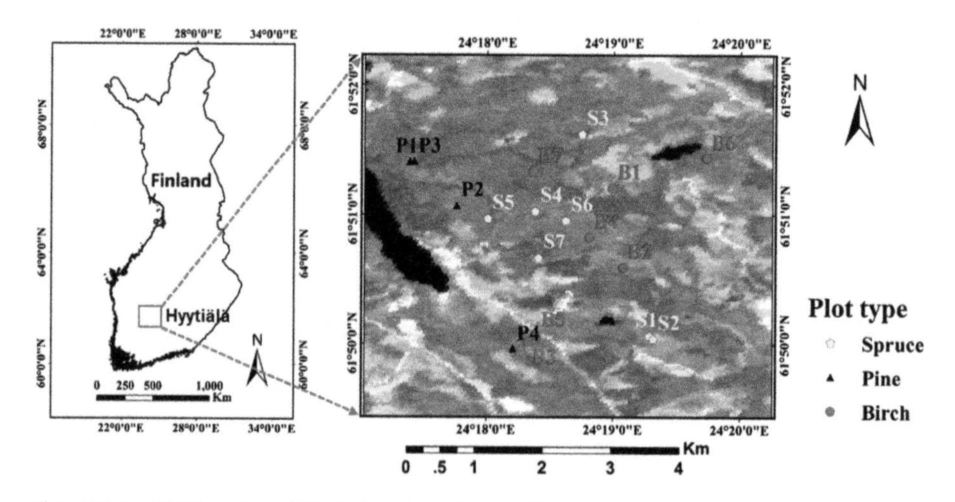

Fig. 2.12: Hyytiälä forest and distribution of study sites. The true color composite image is from Hyperion hyperspectral cube acquired on 3 July 2010. Reprinted from Remote Sensing of Environment, 198, Bin Yang, Yuri Knyazikhin, Matti Mõttus, Miina Rautiainen, Pauline Stenberg, Lei Yan, Chi Chen, Kai Yan, Sungho Choi, Taejin Park, Ranga B. Myneni, Estimation of leaf area index and its sunlit portion from DSCOVR EPIC data: Theoretical basis, 69–84, Copyright (2017), with permission from Elsevier (Yang et al. 2017).

Astronomy can tell the elemental composition of stars and solar systems, galaxies, and nebulae which are millions of light years away from Earth. It is, thus, not surprising that remote sensing based on photo images from airplanes and satellites can produce quantitative information on the status of the primary production and thus agricultural performance and thus economical performance anywhere on the globe – unless the plants are hidden in greenhouses with artificial lighting.

2.6 Remote sensing with airplanes and drones

Not everybody can afford a satellite for counting the crops on the fields. How is crop counting done anyway? What you need is spatial information and species information. You get the species information by the spectroscopic signature of the plants. The absorption spectra and emission spectra of plants are rich in signatures and can be used as fingerprint information. So you need a spectroscopy detector for this task. For the spatial information, it is sufficient to have a two-dimensional detector. Practically, this requirement amounts to the need for a camera with a high spectral and reasonable spatial resolution. Such a camera can be mounted on a satellite, on a high altitude airplane, an air surveillance plane, a helicopter, and today even drones become suitable vehicles to port a camera over agricultural areas.

Note, however, that it is hard, even impossible to acquire an absorption spectrum of a plant from a very remote satellite. Even fluorescence spectra may be difficult to obtain; usually you are in close contact with the plant with a fluorescence spectrometer, which are available today as handheld devices. What is utilized is the reflectance spectra. However, there is a range between in close contact with the plant and being in space in a satellite, far away from the plant. Airplanes and helicopters, and nowadays also drones (to me they are still model airplanes and helicopters), are being used to record images from plants.

The Swiss start-up company Gamaya SA, a spin-off from Ecole Polytechnique Federal de Lausanne (EPFL), develops technology for precision farming and builds cameras for surveillance and recognition of crop fields from above. The firm was established in the year 2015. So it is a young firm. They show a so-called white paper on their website, which was published in the Bornimer Agrartechnische Berichte (Constantin et al. 2015). The paper shows data which were recorded at the experimental agriculture test station Strickhof in Lindau, a village halfway between Zürich and Winterthur. Strickhof is a farm that used to be a "farming college for the poor" in the year 1818. Today, it is a research center (Wuersten 2016)[6].

In that paper (Constantin et al. 2015), you see examples where a drone with two Gamaya cameras was used to record images of a local experimental agricultural crop area with subsequent data analysis of the images and identification of the spectro-

6 https://www.agrovet-strickhof.ch/standorte/lindau.

scopic signatures for the various crops. When you look for further publications of Gamaya SA in the web of knowledge data base[7], you find four more papers. The most recent (status January 2020) was published in *Astronomy & Astrophysics*, where infrared and microwave data from planet Pluto were used (Meza et al. 2019). One of the almost 100 authors is from Gamaya. The first paper with affiliation Gamaya SA is from 2016 and deals with the retrieval of biomass information from the Siberian forest. It deals with image and data analyses, and the data were from the Landsat satellite (Stelmaszczuk-Górska et al. 2015).

An example where hyperspectral imaging can be used is demonstrated by a paper on the fruit and leaves of oranges. For a citrus farmer to be sure of how much of the oranges are already orange in color and how many are not ripe yet but green in color, an image analysis can be useful (Torres et al. 2019). Figure 2.13 shows a comparison of the reflectance spectra of green leaves and orange fruit (I have been to Jeju island several times and I am pretty sure the authors mean mandarins or clementines, and not oranges) obtained with two different methods, this is, the shortwave infrared (SWIR) method and the visible near infrared method (Vis/NIR).

The upper panel in Figure 2.13 shows the reflectance spectra of leaves and oranges obtained with Vis/NIR. There is a small reflectance peak at 540 nm and a large reflectance plateau onset at 700 nm and beyond. There are some differences in the intensity from both species. By the way, the fruit was measured on Jeju island in South Korea, which is well known for its tangerine industry. In a later chapter you will learn again about seaweed food production on Jeju island.

In the bottom panel in Figure 2.13 you see the reflectance spectra recorded with the SWIR. Here, the spectral characteristics are considerably different than in the Vis/NIR spectra. At 1150 nm there is a gross contrast between green leaves and orange fruit. The difference in reflectance amplitude between leaves and oranges is also substantial and allows for better spectral discrimination of both. Figure 2.14a shows an image, a photograph of two pieces of fruit along with the leaves, and it is not printed in true colors in the original publication (Torres et al. 2019). The researchers had to find a way from the spectra to get the best discrimination of fruit against leaf, and they did so by applying a principal component analysis (PCA) to the spectra. The chlorophyll a signature at 698.2 nm is a very strong marker. Another marker was taken as 951.2 nm, where we noticed the flank with decreasing reflectance – in Figure 2.13.

Figure 2.13b shows the mask binary image which only allows discrimination of the plants (white) against a background (black). Figure 2.13 shows the chemical contrast image, which is based on a subtraction of images recorded with (actually, filtered for) 951 and 658 nm light. This band ratio $R_{\lambda 698}/R_{\lambda 951}$ provided the highest "classification accuracy" (Torres et al. 2019). In hyperspectral imaging it is possible to make out spots on the fruit and check them for degradation and poor quality.

7 www.webofknowledge.com.

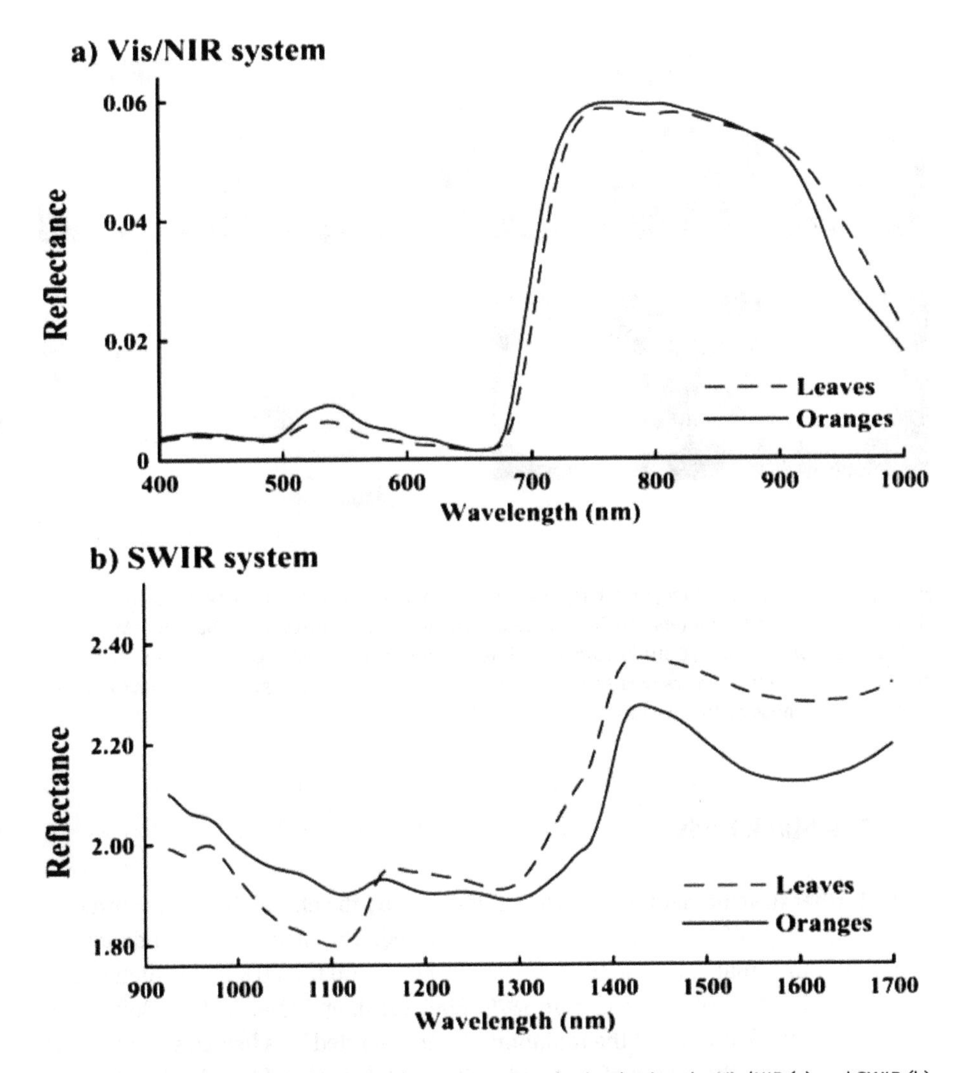

Fig. 2.13: spectral features for leaves and green oranges obtained using the Vis/NIR (a) and SWIR (b) hyperspectral imaging systems. Reprinted from Computers and Electronics in Agriculture, 167, Torres, Irina; Sánchez, María-Teresa; Cho, Byoung-Kwan; Garrido-Varo, Ana; Pérez-Marín, Dolores, Setting up a methodology to distinguish between green oranges and leaves using hyperspectral imaging, 105070, Copyright (2019), with permission from Elsevier (Torres et al. 2019).

Hence, such camera systems can be used in fruit packing factories to select the high quality fruit from the bulk of the produce. This is certainly not remote sensing, but in-factory or on-plant sensing.

Remote sensing is a field of its own for various applications, for example for hyperspectral imaging in forestry (King 2000). A survey on the various remote sensing platforms and sensors can be found in a paper by Toth and Jóźków (2016).

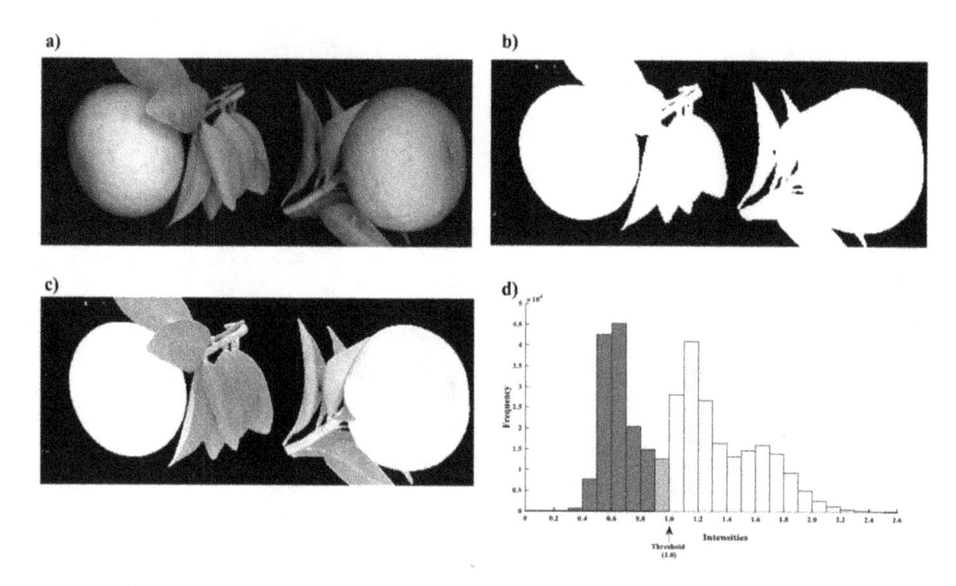

Fig. 2.14: (a) reflectance image, (b) binary image (mask), (c) ratio image, (d) histogram of frequencies. Reprinted from Computers and Electronics in Agriculture, 167, Torres, Irina; Sánchez, María-Teresa; Cho, Byoung-Kwan; Garrido-Varo, Ana; Pérez-Marín, Dolores, Setting up a methodology to distinguish between green oranges and leaves using hyperspectral imaging, 105070, Copyright (2019), with permission from Elsevier (Torres et al. 2019).

2.7 The black body

In the further development of the science of optics by the end of the nineteenth century, the concept of the black body was introduced. The black body stands for the radiation source that emits radiation[8] once it is heated up from 0 K to any higher temperature T. First, Josef Stefan (Stefan 1879; Ebel 2014), and later Ludwig Boltzmann (Boltzmann 1884) found that the radiation power P emitted by a heated source would scale linearly with the area A of that body with the 4th power of the absolute temperature T (in Kelvin; K) of that body:

$$\frac{P(T)}{A} = \sigma \cdot T^4 \tag{2.1}$$

In short, Equation 2.1 means that when you double the temperature T of the light source, this will quadruple the light power P emitted. Maybe even more important is that this law does not depend on the kind of material used as the light emitter. The Stefan–Boltzmann law is a universal law. Several scientists were involved in the

8 Note: at that early stage, the concept of photons, i.e., light particulates did not yet exist. Light was made by rays that point like vectors from the source.

discovery of the law (Crepeau and Asme 2009). The proportionality constant σ is the Stefan–Boltzmann constant[9]. The Stefan–Boltzmann constant σ is a nature constant, as it is composed by a number of other well-known constants:

$$\sigma = \frac{2 \cdot \pi^5 \cdot k_B^4}{15 \cdot h^3 \cdot c^2} \sim 5.7 \cdot 10^{-8} \frac{\mathrm{W}}{\mathrm{m^2 K^4}} \tag{2.2}$$

An application of the Stefan–Boltzmann equation can be found, for example, in a paper on the occultation of the planet Pluto with nitrogen gas, see Meza et al. (2019). The equation is used to determine the equilibrium temperature at Pluto.

The emission spectrum of the sun resembles, to some extent, the theoretical spectrum of the black body, which was derived by Max Planck from fundamental statistical ideas (Planck 1901a). The distribution of the electromagnetic radiation with frequencies ν of a black body with temperature T is described by following relation:

$$U_\nu^0(\nu, T)\,\mathrm{d}\nu = \frac{8\pi h \nu^3}{c^3} \frac{1}{\mathrm{e}^{(h\nu/kT)} - 1} \mathrm{d}\nu \tag{2.3}$$

The black-body radiation law has a functional relationship between the temperature of the black body and its emission spectrum; the well-known laws of Rayleigh and Jeans, Stefan–Boltzmann, and Wien can be derived from it. From the position of the maximum of the emitted intensity, we can derive that temperature. As we cannot peek so easily into the inner part, into the center of the sun, the temperature determined represents the conditions on the sun's surface.

You can construct a black emitter by making a furnace that is completely closed, except for a peek hole that you use for detecting the radiation inside. Now you heat the furnace up from ambient temperature, and it will start producing radiation. However, even below ambient temperature, it will emit radiation because it has a temperature. This condition holds even down to just above 0 K, right before it is absolutely cold and would no longer emit any radiation. Now let us make a Gedankenexperiment, where we allow white light from outside to enter the closed furnace through the tiny peek hole. The light rays will distribute inside the furnace, but the chance, i.e. the probability, that all light will find its way back by reflection through the same peek hole to the outside is extremely small

I recommend here a conference presentation that Walther Gerlach delivered on 22 September in the year 1936 in Dresden at the assembly of German nature scientists and physicians (Gerlach 1936). He refers there to a paper, which, in turn, refers to a paper by Abu Muhammad Ali Ibn Hazm al-Andalusî, a Spanish Arab (or an Arab Spaniard),

---

9 The Stefan–Boltzmann constant σ is not to be mistaken for the Boltzmann constant k_B, which was recently determined with an accuracy that allows us to define the international norming system SI: C. Gaiser, B. Fellmuth, N. Haft, A. Kuhn, B. Thiele-Krivoi, T. Zandt, J. Fischer, O. Jusko, W. Sabuga: Final determination of the Boltzmann constant by dielectric constant gas thermometry. Metrologia 54, 280–289 (2017).

who wrote an essay about the colors, similar to Johann Wolfgang von Goethe, who wrote a book in several volumes (von Goethe 1810) on the scholarship and science of the colors. Gerlach cites Abu Muhammad Ali Ibn Hazm al-Andalusî in German:

> Wir finden, wenn an der Wand eines geschlossenen Hauses zwei Löcher geöffnet werden und hinter das eine ein schwarzer Vorhang gehängt und das andere offen gelassen wird, daß der von der Ferne Hinblickende zwischen beiden keinen Unterschied machen würde.

Which means: "We find when the wall of a closed building has two holes, and when one of the holes has a black curtain behind it, then from far away an observer will not find any difference between both holes." Gerlach's paper (Gerlach 1936) also contains very good references with remarks that are worthwhile looking into. See here also the paper by Bergdolt (1922).

The idea of the black body has been associated with Otto Lummer and Willy Wien (Nobel Prize in Physics 1911), who published a paper in 1895 (Wien and Lummer 1895) with title *Method on the examination of the radiation laws of absolutely black bodies.* However, it was Ludwig Boltzmann already 1 year before (Boltzmann 1884) who derived Stefan's law of the dependence of the radiation depending on the temperature by electromagnetic light theory.

This probability is even smaller when the walls of the furnace interior are sufficiently irregular.

The radiation of this tiny peek hole is then virtually the same as the radiation of an absolutely black plate (body, area). This radiation is, therefore, called black-body radiation.

Max Born refers the construction of this oven in his book (Born 1933) to Wien and Lummer (Wien and Lummer 1895), but according to Walter Gerlach (Gerlach 1936), it was nobody lesser than Ludwig Boltzmann who first figured out the idea of the black body.

The empiric observation that blue light comes not directly from the direction of the sun but from virtually everywhere else, isotropic from all other directions, means that blue light is scattered by the atmosphere and then reflected to our eyes. Let us consider the extreme case where the sun at sunset can have a dark reddish color, and we see the sun round like a disk with a very sharp boundary against the horizon, and where the sky is blue without any sharp boundary. This observation is a manifestation of the law by, Equation 2.4, which states that the "intensity" of the scattered light scales reciprocally with the 4th power of the wavelength of the light:

$$B(\lambda, T) = \frac{2ck_B T}{\lambda^4} \tag{2.4}$$

The "intensity" is here the spectral radiance B, which is measured in the power of light (Watt) per steradian (sr; this is a solid angle in a three-dimensional geometry) per square meter (m^2). For this metric of radiance, we have to consider the light source in the center of a sphere, and the light source emits its radiation homogeneously in all directions (we physicists say 4π). This is the same as the sun in the universe, the rays of which point radially into the cosmos. Now let us consider Earth, which moves around the sun in 1 year. The cross section of Earth can roughly be determined by the Earth's diameter. With this cross section, we collect the light from the sun on our trajectory around the sun in 1 year. From Kepler (Holder 2011, 2015), we know the following:

1. The orbits of the planets are ellipses with the sun in one of the focal points.
2. The radius vector of the planet covers equal areas in equal times.
3. The squares of the revolution times are in the same relation as the cubes of the average distances.

I can recommend the book by Niklitschek (1949) *Ausflug ins Sonnensystem* for an entertaining and clarifying explanation of the astronomy relevant for this book. Kepler worked as a mathematician and as an astronomer. He exchanged letters with Galileo Galilei and he went to study and work with Tycho Brahe, who was the unchallenged astronomer of the time. However, in comparison with the analyst Kepler, Brahe's work focused on the collection of astronomic observations and precise collection of data. Kepler's mathematical mind allowed him to derive quantitative models for the motion of the planets that could accurately predict the trajectories and positions of the planets as a function of time[10].

Since the Earth's trajectory around the sun is an ellipse, the distance between Earth and the sun is not a constant number. At some time, we are closer to the sun, at some time we are more remote from the sun. When we are closer to a radiation source with our body area (cross section), we capture more radiation than when we are far away from that radiation source. This is the consequence of a simple geometrical relation, which is that the area of a sphere scales with the square of the radius of the sphere: $A = 4\pi r^2$. The same holds for the segment of that sphere. People who work with x-rays and neutrons or at nuclear facilities, and who are subject to and protected by radiation regulations (Environmental Safety and Health, EHS) learn that they should always try to increase the distance from the radiation source. Their body cross section will absorb the less harmful high-energy radiation (x-rays and γ-rays) the further away they are from the radiation source. The angle, the solid angle that

10 What counts in science and in many fields of life is that we are able to connect data to relations and results. Even a large number of data by itself does not yet constitute a model or a theory about mechanisms that produce these data. Richard Feynman, in his Douglas Robb Memorial Lectures – Part 1: Photons – Corpuscles of Light, refers to Mayan Indian priests and scholars who were able to predict astronomical events based on their calendar system, which they could calculate with the laws of arithmetics with high accuracy and precision. "The more accurately they can do it, the fact that they know that they have to change it by six days, so far added nothing to the understanding of it." (Feynman 1979).

captures their body cross section, measured at the position of the radiation source becomes smaller and smaller, the further away they move from that source.

Therefore, while the cross-sectional area of Earth is certainly the same on its trajectory around the sun, its distance may be different, and, therefore, the "number of rays" that arrive on the cross section are different. In order to account for this change, the metric of the steradian, which accounts for the solid angle, is introduced in the spectral radiance $B(\lambda, T)$. It is an exercise for the reader to plot $B(\lambda, T)$ in the range 0–2500 nm for the wavelength for several temperatures like 50, 250, 500, 1500, 2500, 5000, and 10000 K.

The constant c in the above formula 2.4 is the speed of light, k_B is Boltzmann's constant, T is the temperature of the radiation source, actually the black-body radiation temperature, and λ is the wavelength of the radiation[11]. In this formula, the wavelength is a parameter and not just a single constant number. The black body irradiates an entire spectrum in the ideal case, with wavelengths including very small and very long wavelengths[12]. The photons have an energy $E = h\nu = hc/\lambda$.

In his article, Lord Rayleigh (Rayleigh 1900) explained the dependency from the 4th power of the wavelength but was not sure whether this would also represent the experimental observations. A small error in his derivation was corrected later by Jeans:

> It seems to me that Lord Rayleigh has introduced an unnecessary factor 8 by counting negative as well as positive values of his integers ξ, η, ζ.

Anyway, the law of Rayleigh and Jeans describes the spectral distribution of the radiation emitted by the black body. There is, however, an inconsistency with this law, which you become aware of at second glance, in that the intensity will become infinitely high when the wavelength of the radiation becomes very small. For ultraviolet light, which has a short wavelength, the intensity would literally explode to infinity, the effect of which is, thus, termed ultraviolet catastrophe. Because this effect has not been experimentally observed, the law of Raleigh and Jeans is not entirely correct, specifically not for the short wavelengths.

---

11 In theoretical physics, c is sometimes taken as 1, so their actual speed is then given in multiples of c. In condensed matter physics, the temperature T is sometimes divided by a characteristic temperature T_C, for example a Curie point or some other phase transition temperature, so that the data are reported versus the ratio T/T_C, which is then called *reduced temperature*.

12 When you enter into a scientific career and get exposed to scattering and spectroscopy in various communities, you may experience that some communities look at photons with respect to their energy E, typically measured in electron Volt eV. When I do x-ray spectroscopy, I use this unit. Other communities would rather use the wavelength in nm; this holds, for example, for those who use optical spectroscopy. In vibration spectroscopy such as infrared (IR) and Raman spectroscopy, people use wavenumbers 1/cm. When you employ impedance spectroscopy, you use the quantity Hertz (Hz) for the frequency f.

In 1896, Wilhelm Wien published (Wien 1896) his derivation of a radiation law that would fix the problems of the Rayleigh–Jeans law with the short wavelengths:

$$\phi(\lambda) = \frac{C}{\lambda^5} \cdot \frac{1}{\exp\left(\frac{C}{\lambda T}\right)} \tag{2.5}$$

Max Planck and Friedrich Paschen had doubts whether Wien's radiation law was fully accurate, which is notwithstanding that Planck noted that he felt Wien's radiation law was valid at large. By looking into interpolations of the laws of Rayleigh and Jeans, and Wien, and by considering the variation of entropy of resonators that would emit radiation, he eventually arrived at a relation between wavelength and spectral distribution, which formally does not differ considerably from Wien's radiation law (Planck 1900a, b, 1901a, b):

$$\phi(\lambda) = \frac{C}{\lambda^5} \cdot \frac{1}{\exp\left(\frac{C}{\lambda T}\right) - 1} \tag{2.6}$$

It is an exercise for the reader to derive the laws of Wien and Rayleigh and Jeans by looking for approximations for very small and very large wavelengths. You can also try to find the maximum of the distribution of intensity as a function of T; by doing so you will derive Wien's displacement law[13]. The effect of this law – which is a consequence of Planck's radiation law – is that the wavelength at which the black body emits the maximum intensity scales with the reciprocal of the absolute temperature of the black body:

$$\lambda_{\text{max}} = \frac{2897.8\,\mu\text{m K}}{T} \tag{2.7}$$

This means that when we are able to determine the spectral intensity of the emitter, then we can determine the temperature of the emitter by the location of the maximum of the spectral intensity. Figure 2.15 shows a solar emission spectrum, i.e., the photon flux versus the wavelength of the photons. The maximum intensity is found at around 600 nm. Figure 2.15 shows an experimentally determined spectrum of the solar flux.

---

13 Here is another anecdote, which I owe to my friend Klaus Kurre (Boeker et al. 1997). At RWTH Aachen, Professor Heinz Genzel (Brauer 2003) was in charge of the physics practical laboratory courses. He also took the exams. When it was Klaus' turn, the Professor asked him what the temperature of the sun's surface is Klaus told me, he struggled a little, but said something like 4000 to 5000 Kelvin (information he just had read in *Scientific American* a week before), and then the professor asked what color is the highest sensitivity of the *human eye*. Here, I should add that Klaus was originally a *medical student*, and had got a glimpse about the evolution of life in general and the specular apparatus of the human eye, and could thus could make the connection between photosynthesis of plants with green leaves, the color of the sun, and the adaptation of the eye during evolution, and so he mentioned Wien's displacement law. The moment when Professor Genzel spotted that glimpse on Klaus' mind, Klaus passed the exam.

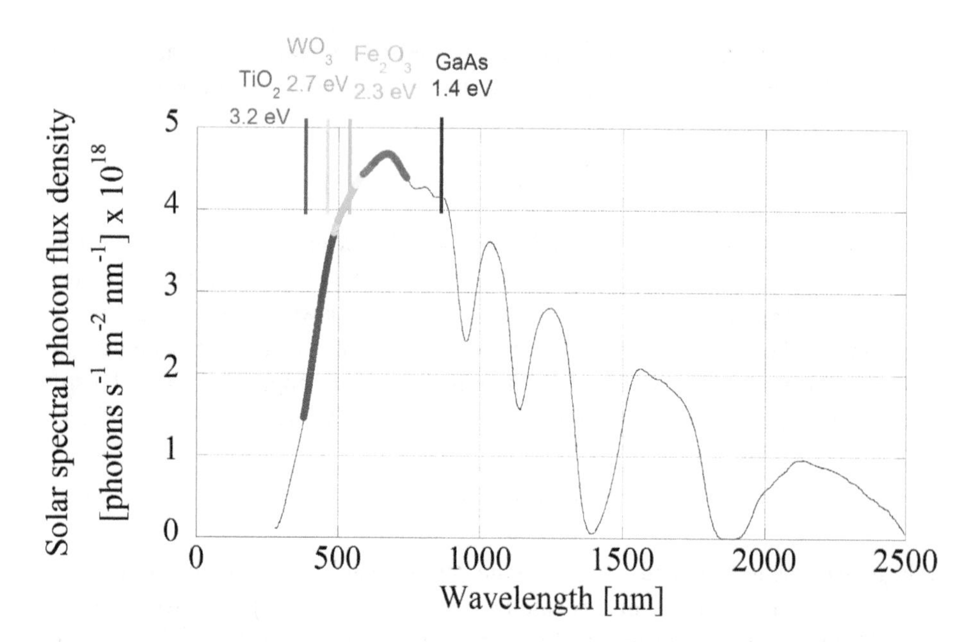

Fig. 2.15: The solar spectrum. The thick colored lines denote the actual color of the light in the particular wavelength range.

We know now that light can be considered as the radiation from a heated body. When the body is absolutely cold, with its temperature at $T = 0$ Kelvin, it should not irradiate at all. Heisenberg's uncertainty principle forbids ($\Delta p \cdot \Delta q \geq \hbar/2$) an exact temperature of $T = 0$ Kelvin, because as such, the atoms would no longer move or vibrate, as their positions would be fixed. Astronomical studies show that the universe contains a homogeneous and isotropic background radiation with a wavelength that corresponds to a body with 3 Kelvin temperature – the 3 K cosmic background radiation. This temperature (actually, 2.8 K) was predicted by Erich Regener in 1933 (Regener 1933).

3 Optics and electromagnetic light theory

3.1 Rays, beams, and geometrical optics

We learnt in middle school how to understand and treat light as a bundle of rays that we can symbolize with straight lines. As light was implicitly considered as having a direction, we use arrows instead of just straight lines. The arrows point away from the light source. Often, the petals of a flower, like the yellow ones of a sunflower are implicated as the sun rays that point away radially from the plant, see Figure 3.1 below.

Fig. 3.1: Sunflowers resembling the sun with its rays.

Common daily life observations such as the occurrence of an image of an object, the mirror image in a mirror, the magnification of an object by a lens or the projection of an object on a screen, can all be mathematically described by what we call geometrical optics (Bethe 1938; Born 1933).

https://doi.org/10.1515/9783110629941-003

3.1.1 The rainbow

The human eye perceives the daylight as white light. The blue sky, the yellow sun, the green grass, and the various optical phenomena that produce colors, such as the rainbow, are consequences of the dispersion and demixing of white light into a spectrum of its components. Figure 3.3 shows a colorful spectrum that was obtained on a sunny day during a hiking trip in a suburb in Seoul, when the sunlight hit a glass on the table of the café. The glass worked like a prism, which split the white light into its components, which by coincidence fell on the left leg of my companion. You can see a purple–red stripe, followed by orange and yellow, a very thin green line, which is very hard to make out, then turquoise, light blue, and a broad stripe of purple and violet. Human eyes cannot sense the infrared range below the red stripe or the ultraviolet radiation above the purple stripe. However, with instrumentation we can measure them. The light diffraction effect of the prism is used for the monochromatization of light. Optical experiments typically require monochromatic light, which is light of only one wavelength λ, which should have a line width as small as possible. Light can be diffracted by very thin films (for example, an oil film on water) and small objects like aerosols (the tiny droplets, the water spray producing the rainbow), and they also produce colorful spectra.

In Figure 3.2, you can see a photo that I have taken in the winter of 2010 during my sabbatical at the University of Hawai'i at Manoa[1]. It shows a rainbow. The photo was taken when I was standing on the rooftop of the SOEST (School of Ocean Earth Science and Technology) building at the UH Manoa campus. The Manoa valley is 3 kilometers east of Honolulu and 2 kilometers north of Waikiki. Often, you can see foggy spray coming over the mountains in the north, which contains the water aerosols that produce the rainbow when the sun shines.

From the physical and mathematical points of view, the rainbow is a manifestation of Snell's law. A close inspection of the photo shows that there is a second rainbow, with a dimmer intensity. In the bottom panel of Figure 3.2, I show a geometrical construction that explains the formation of the rainbows.

Note the inset in the bottom panel, which is the rainbow photo from the top panel. On the right-hand side you can see my face and eyes, the detector so to speak. The blue disks should resemble the water droplets that form the mist in the Manoa valley. White light arrives from the sun on the droplets, but this is actually a composition from all wavelengths.

The fixed points for the geometric construction is the position of the sun as the light source S, the position of the water droplets as diffraction and dispersion object D,

1 Project "Oxide heterointerfaces in assemblies for photoelectrochemical applications" http://p3.snf.ch/Project-133944.

Fig. 3.2: Top: double rainbow over the Manoa valley in Honolulu, Hawai'i. Photo taken 21 January 2011 by Artur Braun. Bottom: geometrical construction of the double rainbow. The inset is the double rainbow from the top panel.

and the position of the observer's eye, *E*. The light beam along *SE* is diffracted when it enters the droplet because of Snell's law[2].

$$\frac{\sin \theta_o}{\sin \theta_i} = \frac{c_i}{c_o} = \frac{n_i}{n_o} \tag{3.1}$$

where θ_o is the angle under which the light beam hits the surface outside of the droplet, and θ_i the angle, under which it is diffracted "under" the surface, inside the water droplet. The "air" outside of the drop has a different diffraction index n_o than the water inside the droplet, $n_i = 1.333$ for water. The differences in the diffraction indices of the materials make that the speed of light in these materials is also different. This is the physical origin for the diffraction of light beam away from its original path.

The diffraction index is not a real constant but a dispersion quantity. It depends on the wavelength of the light and is characteristic for every material. Therefore, the diffraction angle is different for every wavelength of the light. Further, the angle under which the light exits a body, or an optical medium, is different for each wavelength. In the prism, this effect can be exploited for the separation of the white light from sun or other light sources into its spectral components. This is a simple monochromator, as shown in Figure 3.3.

Here is an exercise for the reader. Draw a circular droplet on paper and let a light beam fall onto it. At the curved surface, the beam will diffract away from the original beam direction towards the inside of the droplet. This diffracted beam will hit the inside wall of the droplet and be reflected back into the inside of the drop. When the reflected beam hits the drop's surface from the inside, it will be diffracted as it moves outside in the air where the diffraction index is different from water.

The incoming beam *SD* and the beam *DE*, therefore, form a virtual angle, which we call 2ϕ. You must extrapolate both beams behind the droplet; there is the intersection for 2β. The half-angle of this, ϕ, determines a horizontal line, which is useful for us as the geometrical axis.

The point inside the droplet, at the wall, at which the beam is reflected, is an intersection point where the diffracted beams meet. Here, the diffracted beam forms an angle β with the geometrical axis. With the incoming beam *SD* this geometrical axis

---

2 It is not always clear who was the first to make a discovery. What counts in the end is who is credited with it. Therefore, it is important to publish the discovery and also to disseminate the publication with your name – if you are the discoverer. "As soon as you have results, publish them". The diffraction law has Snell's name, but there are also other names credited, including Ibn Sahl, a mathematician from Persia in the ninth century (Rashed 1990; Kahl 1999). Debates about who is the founder or discoverer, therefore, sometimes occur, such as about the absorption law (Perrin 1948).

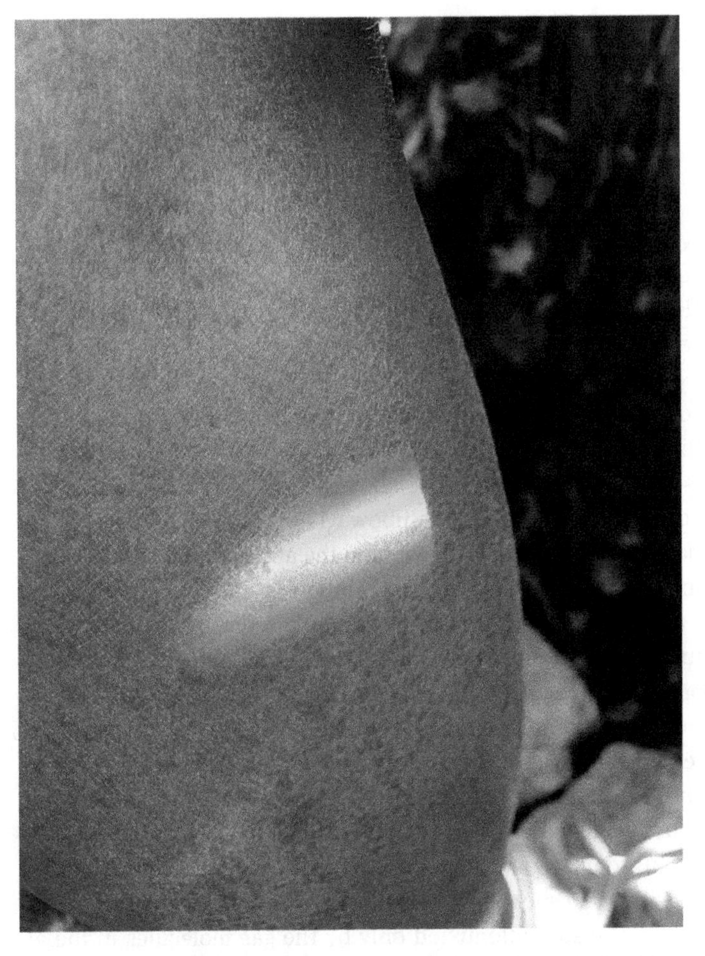

Fig. 3.3: An optical spectrum with visible colors of a rainbow, obtained from sunlight that through a beverage glass.

form an angle ϕ, as you can see when you draw a parallel to the geometrical axis through the point where the *SD* is being diffracted.

Now, consider the center of the droplet and from the center point draw a line through the point where the incoming beam *SD* is diffracted. With the geometrical axis this line forms the angle 2β. With the incoming beam *SD* it forms the angle $2\beta - \phi$. Using Snell's law, you can write

$$\sin(2\beta - \phi) = n_{\mathrm{i}} \sin\beta\,. \tag{3.2}$$

You can solve this equation for the angle ϕ and then try to find the maximum of this angle with respect to the angle β. This is the case when the first derivative disappears. This yields the condition

$$\beta_{\max} = \frac{1}{\cos\left(\frac{2\sqrt{n^2-1}}{n\sqrt{3}}\right)} \tag{3.3}$$

with a result for $\beta_{\max}$ of approximately 40.2°; this value can be inserted into Equation 3.2, which for the rainbow angle yields the value $2\phi_{\max} = 42°$. This is the angle formed by *SD* and *DE*, under which you see the principal rainbow, intense and with small radius, closer to Earth.

A few lines ago we made the assumption that the diffracted light beam in the water drop is reflected once and then diffracted again, as it leaves the drop. There is, however, the possibility that the light arrives under an angle, that the diffracted beam is reflected two times in the water droplet. The above geometrical construction will then look somewhat different, and a second rainbow will occur, with lower intensity because the optical path in the droplets is longer, and the absorption is, thus, bigger. The angle under which the same eye observes this rainbow is $2\phi^*_{\max} = 51°$.

I have taken the photo in Figure 3.4 in my bathroom after taking a shower. The bathroom was still full of moist water vapor, and I opened the window in order to ventilate the bathroom. It was the beginning of a sunny day, and the sunlight hit the walls of the house from a eastern direction, and the small holes in the window shades allowed the light to enter into the bathroom. Outside in bright daylight, you do not perceive the rays of the sun. You see only a diffuse brightness with the sharp objects around you and the blue sky above you.

Here, in the dark bathroom, light enters through small pinholes in the shades' blades. When you peek into one such single pinhole you will see the high-intensity light coming directly from the sun, attenuated only by the gas molecules in the atmosphere. When you move your eye away from the pinhole, you will not see the light unless it is being scattered by the moist water vapor, millions of tiny microdroplets per cubic centimeter volume. The droplets scatter the light in all directions; also right at your eyes as you stand there. The size of the droplets determines whether the light follows Mie scattering (particles larger than the wavelength) or Rayleigh scattering (particles much smaller than the wavelength of the light).

Figure 3.4 shows how a light beam comes in through every such hole. Where the density of the water vapor is very high, the beam looks more intense. Where there are regions in the bathroom with no water vapor clouds, the light beam looks interrupted. Clouds from dust or vapor are, therefore, very useful for the visualization of light beams. Similarly, this holds for virtually all objects in real life and the laboratory. Scattering is very important in physics and is technically used for the study of the structure of materials. We will come to this in Chapter 4 in this book. The Feynman

Fig. 3.4: Sunlight entering through pinholes in a foggy bathroom. The water droplets of the fog scatter the light diffusely and isotropically in all directions by Mie scattering.

diagrams basically represent the scattering processes in a simple schematic form. The form, however, is not arbitrary. Rather, there a particular rules that apply, the Feynman rules. We will learn about these in Section 3.9.2 on Feynman rules.

The light rays then, right after passing the pinhole, fell onto the window, the bathroom side of which was coated with a film of fine water droplets, condensed moisture from the shower, producing the image that you can see in Figure 3.5. You can see the color image from the house in the neighborhood, just upside down, and left turned right. This is the image of a pinhole camera. The entire dark bathroom is the camera (chamber), and the pinhole is one of the aforementioned pinholes. For comparison, a photo of the house through the open window taken with my Nikon camera on the same day is shown as well.

The law of reflection, for example, shows that the vector perpendicular (normal) to the reflecting surface is in the same plane as the incident light ray and the reflected light ray. Moreover, the angle between the incident light ray, θ_i and the normal vector is equal to the angle between the reflected light ray, θ_r and the normal vector: $\theta_i = \theta_r$ (angle of incidence equals angle of exit; in German: Einfallswinkel = Ausfallswinkel). A wide range of optics problems can be addressed and solved with the principles of geometrical optics.

Fig. 3.5: Right: pinhole camera image of a house in the neighborhood produced with the sunlight coming through the window shade pinhole and falling onto the window, the inside coated with water droplets after showering. Left: photo of the house taken with a Nikon camera.

The careful observer, however, will find that the behavior of light may deviate from the strict rules of geometrical optics, for example at boundaries, such as at the slit and double slit. The double slit interference experiment is attributed to Thomas Young and was supposedly carried out in 1801. It appears that the rays of light bend slightly at a corner of a wall, and this behavior cannot be described with the geometric approach.[3]

3.2 Determination of the Earth's radius by geometrical optics

The Greek philosopher Eratosthenes was aware that in the ancient Egypt city of Syene at the Nile River, today known as Aswan, there was a deep, dry well. At noon, the sun would shine vertically into the bottom of the well, and no shadow of the well walls would be cast onto the bottom. So, the sunlight would hit vertically into the well. Eratosthenes, by the way, was the director of the Great Library of Alexandria for 50 years (Canfora and Coleman 2016; Canfora 1999; Thiem 1979; Shcheglov 2018).

However, 800 km down the Nile up North in Alexandria, the sun would cast a shadow in a comparable well. The map in Figure 3.6 shows where Alexandria and Syene are located. Eratosthenes assumed that the sun was so far away from Earth and from himself, that its sun rays could be considered as parallel sun beams. With this assumption, he could sketch a geometrical situation that allowed him to determine the radius of Earth (Dutka 1993). You can see in the sketch in Figure 3.7 how the two wells in Alexandria and Syene are drilled vertically into Earth, as shown by the two straight lines pointing from the Earth's surface radially to the center of Earth.

3 Whenever a model fails to describe new empirical observations of reality, the model needs to be improved, adjusted, and when a better model is found, the old model maybe has to be discarded or rejected. Sometimes, incorrect models survive for didactic reasons. This, for example, holds for the Bohr atom orbital model, where the orbits are circular only.

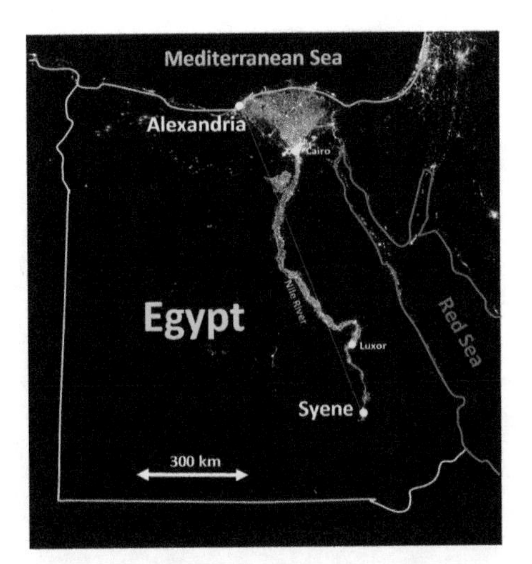

Fig. 3.6: A photo of a simple map of Egypt, taken from a satellite at night. The white intensity reflects the spots of light from civilization, basically the Nile river and the river delta. I have drawn the green line as terrestrial border and the red and blue lines as sea borders. The ancient city of Syene is today known as Aswan and lies 800 km south of Alexandria on the Mediterranean coast. Note that over such a long distance, the globe exerts a considerable curvature. The NASA Earth Observatory image by Jesse Allen and Robert Simmon, using VIIRS Day–Night Band data from the Suomi National Polar-orbiting Partnership. Suomi NPP is the result of a partnership between NASA, the National Oceanic and Atmospheric Administration, and the Department of Defense. Caption by Michon Scott.

Note that we and Erastothenes must assume here that the Earth is a globe, a sphere, instead of a disk. Then the two lines (with the as of yet unknown length R) representing the orientation of the two wells will include an angle θ, which spans a distance of 800 km (in old Egypt metrics this was called 5000 stadia). The light arriving from the sun, the sun rays, was considered a bundle of parallel rays; this, too, was an assumption that Eratosthenes made. While at noon the sun did not cast a shadow in the well in Syene, it did cast a shadow of known length in Alexandria. From this, he could determine the hypotenuse of the triangle built by the stick in Alexandria and its projected shadow on the ground. This would be the same angle θ. With the laws of trigonometry and proportionality he could determine the radius R of Earth.

Horst Janssen, a German painter (he called himself a *Grafiker*, graphic designer) puts it simply, like this (Janssen 1983): "Eratosthenes was able to determine the Earth radius by looking down a well"[4]. Goldstein believes and demonstrates that no accurate geometrical measurement was necessary for Eratosthenes to arrive at his results on the Earth's radius (Goldstein 1984). A repetition of the experiment is explained in Webb and Bustin (1988). You can reproduce the calculation as exercised in the paper by Brown and Kumar (2011).

4 Horst Janssen in his speech at the inauguration of the Alfred Mahlau exhibition at Kunsthaus Lübeck, in Lübeck 1980.

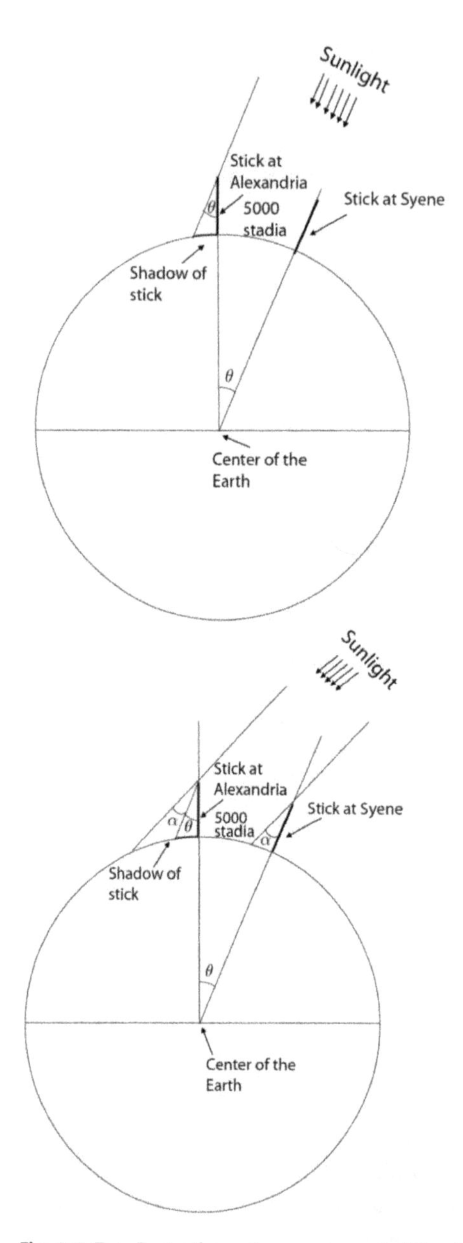

Fig. 3.7: Top: Eratosthenes' measurement of the circumference of the Earth as viewed when the shadow of the stick at Alexandria is at a minimum, at noon on the day of the summer solstice in Syene. Bottom: Eratosthenes' measurement of the circumference of the Earth as viewed when the shadow of the stick at Alexandria is off-minimum, either before or after noon on the day of the summer solstice in Syene (Brown and Kumar 2011). Reproduced from "A New Perspective on Eratosthenes' Measurement of the Earth" by Ronald A. Brown and Alok Kumar, The Physics Teacher 49, 445 (2011) https://doi.org/10.1119/1.3639158, with the permission of the American Association of Physics Teachers.

3.3 Wave phenomena and coherence

Geometrical optics can explain many observations in daily life. However, sometimes we notice that light deviates from the simple geometrical principles when we look at the corners of an object when it is hit by light. To explain the bending of the light rays at a corner, the concept of the scattering of the light rays is established. We assume that whenever a light ray hits a material particle, the particle will absorb the original light ray and emit a secondary one. This is a new model in comparison to the model of geometrical optics. Light is now considered a wave an no longer a ray. We can live with this model where the light is taken as an electromagnetic wave that propagates radially from the spot of its creation in all directions, and the normal vector to the wave azimuth is still the original ray direction. The superposition of two or more waves can cause the situations that we actually observe in nature.

Fig. 3.8: The agitation of a duck in the pond creates a curl with wave lines of roundish, azimuthal shape. Perpendicular to the circular wave lines are the radial rays.

Figure 3.8 shows a duck in a pond. The agitating movement of the duck excites a water wave that propagates in radial direction in the water plane. Even the corona of the sun is subject to such kinds of waves (Lee et al. 2015), both in radial and azimuthal directions. The wave is the profile of the height of the water level in radial distance from the center of the wave where the duck resides. The water level of the pond at rest is the reference height, and the wave is the evolution of water levels higher and lower than the reference level. This profile is typically approximated by a sinus wave form. This sinus is the solution of a differential equation that is based on a linear force field. The height of the higher (actually, the highest level, i.e., the maximum) water level versus the reference level is the amplitude A of the wave. The distance from one maximum to the next is the wavelength λ.

Fig. 3.9: Two ducks at two positions in a pond create waves with two wave centers. At a large distance, the two wavefronts add up by superposition to a new large wavefront.

Figure 3.9 shows two ducks at two positions in the pond, who both agitate and thus create waves with two centers of the waves being at either position of the ducks. At a large distance from the wave centers, near the boundary of the pond at the spectator's side, you can see large wave rings, which are composed of the smaller wave rings, each individually created by one duck. The two waves superimpose and, thus, form a new wave. We can imagine this effect now applied to a larger number of wave center, and then we have a general superposition of waves that form a large wavefront.

Now we briefly need to discuss the phenomenon of wave interference. Let us assume – for simplicity – that the two waves produced by the ducks have the same amplitude and same wavelength. The positions where two wave maxima coincide will superimpose to a higher maximum, and this is called constructive interference. At the positions where one maximum falls together (coincides) with one minimum, the amplitudes cancel each other out, and this is called destructive interference. On closer inspection, particularly when you draw a geometric construction on paper, you notice that the minima are placed on sets of lines that are a hyperbola. The focus of these hyperbola are in the centers of the two waves. We, therefore, have two point sources of a hyperbolic interference.

The hyperbola is made from all those points that have the travel time difference $t = \lambda/2$. The distance between the two apex $2a$ is the difference between the times T_1 and T_2 times the propagation velocity of the wave, v. Interference can occur in countless situations but only under some ordered conditions; the interference will amount to regular patterns. Think here also of the so-called Moiré patterns.

The human mind has always had a problem with understanding[5] wave phenomena. We certainly know waves in water, such as those made by the ducks in the pond. We also know the waves on rivers and in lakes from when we are on a boat or in the oceans, when we are on a ship or at the seashore. Waves can carry other objects away, in addition to the fact that a water wave certainly carries itself away, right?

Waves contain energy, and they also carry momentum (the kinetic energy E_{kin} runs over the momentum $\boldsymbol{p}$: $2mE_{kin} = |\boldsymbol{p}|^2$), which they can pass on to objects such as ships and swimmers. A very good example is surfing on waves, for example in Hawai'i.

3.3.1 How is a mechanical wave created?

How is a wave created? Let us stay with the water wave. From the physical perspective, we have to begin with an oscillating system – with an oscillator. We consider a physical body with a mass m (this can be a metal ball or any other object), which is fixed at a spring coil that hangs on the ceiling. The height of the mass m from the ground ($z = 0$) is $z = z_0$, for example 150 cm. Gravity pulls the mass of the body down and also the spring itself to some rest position. When you pull the mass down a little more out of the rest position, by some small increment in height δz, and then release it, the force of the spring will pull the mass back up. The body will reach the original equilibrium position and shoot up by a small increment to a maximum height and then fall back again and keep repeating this for a while[6]. The body with mass m is subject to the interplay and competition of forces from the spring, from gravity and also from inertia. The body is thus oscillating, and this is kept steady by the two opposing forces of gravity and spring.

If you make a stroboscopic series of photographs from the mass in such a way that you can determine the distance of the mass from its equilibrium height z_0, and plot $z(t)$ versus the time t, then you will notice that z varies like a sine or cosine function. The behavior of the spring coil to a large extent follows Hooke's law

$$F = k \cdot z \tag{3.4}$$

where k is the proportionality spring constant and z the spatial coordinate. The mass is also subject to Newton's law

$$F = m \cdot \ddot{z} \tag{3.5}$$

5 Here, we shall not confuse *understanding* with *being familiar*. Certainly many people have seen waves and know their beauty and danger.

6 The internal friction in the wire of the spring coil and the external friction between all moving parts and the surrounding air will use up kinetic energy and, thus, damp the oscillation and finally bring it to rest.

where inertia acts back on the forced motions of the mass by the gravity and by the spring force. The mechanics of the system can be approximated by equating the spring force with inertia deceleration as follows,

$$k \cdot z = m \cdot \ddot{z} \tag{3.6}$$

which is a second-order differential equation with the time t, see Equation 3.9.

With the *Ansatz*

$$z(t) = \hat{z} \sin(\omega t) \tag{3.7}$$

for solving this differential equation, we will find the solution as above in Equation 3.9.

This oscillation is also characterized by an amplitude and a frequency ω, which is given by the square root of the ratio between spring constant k and mass m:

$$\omega = \sqrt{\frac{k}{m}} \tag{3.8}$$

The frequency ω is here the Eigenwert, the eigenvalue. The mathematical expression for the solution of this oscillation would be more complicated if we had considered the damping contributions from the friction phenomena that occur in any real system.

Let us now allow for the interaction of the oscillating mass with the surrounding air; which it is doing anyway. The mass will kick the air molecules (80% N_2, 20% O_2) away in its rhythmic oscillation, and the air molecules will propagate with this rhythm in all directions. The air molecules pick up the momentum from the mass in the form of kinetic energy. This mechanical system is not much different from the skin of a drum, which kicks away air molecules, or the string of a guitar, violin, or piano; except that the music instrument has a resonance body from metal or wood or skin that amplifies the wave and, thus, gives a loud sound.

The air molecules leave the mass, skin, and strings in waves, and this constitutes a sound wave. It carries the oscillatory characteristics of the swinging body mass away as a sound wave with a particular frequency ω.

The wave is characterized by a wavelength and an amplitude, both of which depend on some boundary conditions. In general, waves are mathematically characterized by a second-order partial differential equation with respect to the time and spatial coordinate of the following fundamental type:

$$\frac{1}{c^2}\frac{\partial^2 u}{\partial t^2} - \sum_{i=1}^{n} \frac{\partial^2 u}{\partial x_i^2} = 0\,. \tag{3.9}$$

The Schrödinger equation and also the equation that describes Liesegang patterns, have a first-order time derivative.

3.3.2 Double-slit experiment

A well-known way to demonstrate the interference of waves is by placing a wall that has two slits in front of a wave, which is known as a double slit. Experience shows that the wave front of one wave will pass through the two slits. The wavefront of a second wave will also pass through both slits. Behind the slit, both wave fronts will meet and "interfere" and superpose each other and form a new wave. The original waves are diffracted at the boundaries of the slit, and from there secondary waves will evolve and propagate in all possible directions. It turns out that the intensity pattern behind the slits will have new positions for intensity maxima and minima. This experiment was first performed visible light by Thomas Young in 1802. On YouTube you can find a lecture by physicist Josef Gassner, where he explains the path integral formalism based on the double-slit experiment (Gassner 2018).

The experimental geometry is shown in Figure 3.10. The thick vertical black line on the left-hand side is a wall with two slits that have a slit width b each.

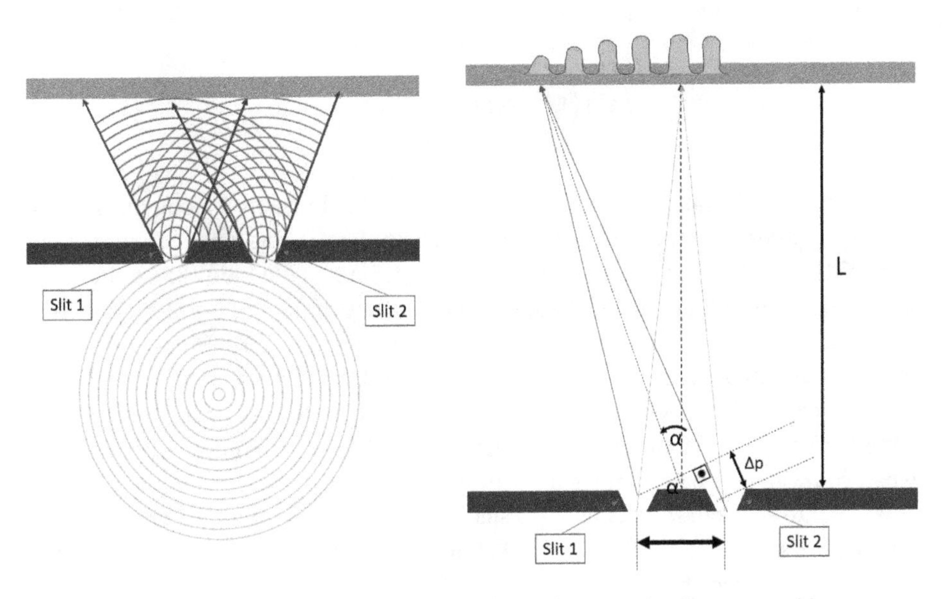

Fig. 3.10: Schematic of a double-slit with slit widths b and slit separation distance a, with a screen as the beam target for the interference pattern, along with the geometrical construction for derivation of the mathematical interference equation. Two rays on the left, indicated by horizontal arrows pointing to the right, interfere in the slits and merge with two new long arrows in position *alpha* from the optical axis.

The interesting finding is that this double-slit experiment will produce the same results, i.e., the same diffraction pattern behind the slits when it is done with lower energy microwaves and with higher energy x-rays, and even with particles such as electrons, neutrons, protons, heavier ions, and even larger particles, including golf balls. This means that particles in the ensemble, at a concert behave like waves, as De Broglie (1929) found. We know this physical phenomenon as particle-wave dualism. You do not have to decide whether what you are looking at is a wave, or a particle. Waves and particles do not fundamentally rule out each other. They may be two different representations of one physical process.

Figure 3.11 is a set of photographs that were taken at the Pacific Ocean in Bodega Bay, California. Two pairs of vertical yellow lines indicate the two slits between three rocks through which maritime waves, as indicated by the two horizontal red lines, propagate and get diffracted. The sequence of nine photos shows how the waves "break" at the rocks and recombine behind the rocks. A superposition pattern cannot be observed, though.

Close inspection of the three lower photos shows that the waves are backscattered into the ocean – though a few meters only. Such diffraction behavior is universal over all length scales.

$$I(\alpha) = I_0 \cdot \left(\frac{\sin\left(\frac{k}{2} \cdot b \cdot \sin\alpha\right)}{\frac{k}{2} b \sin\alpha} \right)^2 \cdot \cos^2\left(\frac{k}{2} \cdot a \cdot \sin\alpha\right) \tag{3.10}$$

Equation 3.10 calculates the angular distribution of the intensity I along the angle α, which is taken with respect to the optical axis, with the two slits perpendicular to this axis. This is indicated in the right panel in Figure 3.10 as the blue shaded intensity on the top. The width of the slit is b, and the distance of the centers of the slits is a. The wave vector $k = 2\pi/\lambda$ is the reciprocal of the wavelength λ. The sines and cosines in Equation 3.10 are raised to the second power in the brackets, and, therefore, the intensity can be 0 but never negative. It is an exercise for the reader to sketch $I(\alpha)$ on paper and discuss a number of questions: How will the interference pattern change when you widen or narrow the slit width b? What happens when you place the two slits closer to each other, i.e., making a smaller? You are working with monochromatic light with wavelength λ. When you switch to a higher or lower wavelength, will this alter the interference pattern? How?

In this section, we discussed the double-slit experiment in its classical theory based on geometrical optics. We also used Huygen's principle, which says that every material object or particle is the origin of a new wave, when hit by a wave. This principle is used in the geometric construction in the left part in Figure 3.10. An alternative treatment is to use Feynman's path integral formalism, which will be shown in Section 3.10.

Fig. 3.11: Rocks at the seashore at Bodega Bay Trailhead, acting like a double slit.

3.4 Ether theory

When experimental studies (Romer 1676) showed that the speed of light was an enormous 300,000 km per seconds, experts were wondering which nature was the medium that would allow for such a high velocity of light. This was a very well justified question. When I did my compulsory lab courses during my physics studies at RWTH Aachen, one experiment was the determination of the speed of sound. We would measure the speed of sound when the sound signal was transported, actually transmitted through some different gases kept in some tubes, and through some metal rods that had a length of over 1 meter and were of different metal elements (like copper, vanadium, or the like). We found that the speed of sound was incredibly fast through very heavy, very dense metal rods. The speed of sound depends on the medium transporting it. Very dense media have a high speed of sound. Media with a very low density have a very slow speed of sound. The vacuum does not allow for sound wave propagation, thus the speed of sound there is 0.

As the speed of light is so high, researchers naturally concluded that the medium of the light transportation, the ether must have an infinitely dense compactness. It was Michelson and Morley (1887) who wanted to test with their interferometer, that this was really the case.

3.5 A remark about time

In the very early 1970s, I saw the science fiction movie *Die Zeitmaschine* (Pal 1960) on German TV. The screenplay was based on the 1895 book *The Time Machine – An Invention*, authored by H.G. Wells (Wells 1895). The story plays in the late nineteenth century. An inventor builds a time machine and can travel to the past over tens of thousands of years away, and also to the future tens of thousands of years away. Wars have destroyed the world, but new human life has evolved into two classes, the young, white, blonde, dumb, and lazy, uncompassionate, and selfish humans, the Eloi, who receive food and care from the dark and dirty, but industrious human beasts, the Morlocks, who frequently capture, kill, and eat the Eloi. The inventor, now a time traveller, manages to show the Eloi compassion and can free them from the Morlocks.

Is time travel possible, in reality? Experience shows that we travel with time in a forward direction. We cannot speed it up, we cannot slow it down, and we cannot go backward. Time moves forward like an arrow. In 1956, philosopher Karl R. Popper published a short note in the magazine *Nature* with the title *The Arrow of Time*. In that short note he refers to the lecture "Natural Philosophy of Cause and Chance" given by Max Born in 1948 (The Wayneflete Lectures in Oxford, (Born 1949)). There, Born supposedly implied that classical mechanical processes were reversible in time. As a practical example, Popper suggests (Popper 1956) the experiment of a stone dropped in a water basin, where waves are created. If this experiment is recorded as movie, a reversible experiment should show that the waves would condense back into their origin, where they were created when the stone dropped. Experience shows this is not possible, or has never been observed. The process of wave formation and propagation is purely classic and yet irreversible, irreversible in time.

The mathematical equations that describe physical processes as a function of time t are still valid, in the mathematical sense, when t is replaced by $-t$, which means a reversal of the time arrow. This holds, for example, for the Maxwell equations with advanced and retarded potentials (Popper 1956). We say that the equations, or even the entire theory, is symmetric with respect to the time t. Also the Schrödinger equation is invariant to time reversal. E. Wigner introduced the time reversal operator, which is antilinear and unitary (Geru 2018). Based on the works discussed below by H.A. Kramers on magneto-optics and electro-optics, Wigner showed that Kramers' derivations implied the utility of a time reversal operator (Wigner 1993), the **T**-operator.

Kramers showed (Kramers 1930) that the energy levels of a quantum system with time-reversal symmetry are at least double degenerate, i.e., contain at least two identical energy levels, if the external forces acting on the system are of purely electric origin – provided that the number of particles is odd, not even, and the system has half-integer spin. Let us consider the hydrogen atom, which has one electron and one proton. So the spin is not half-integer, but integer. Thus, the lowest hyperfine energy levels of hydrogen are not degenerate. When, instead, we consider deuterium, as it has one neutron in the nucleus, the number of fermions and total spin is 3, and 3/2, respectively. The ground state of deuterium, therefore, has two ground-state components, and these are twofold and fourfold degenerate.

When the time-reversal symmetry decreases in systems with Kramers degeneracy, Kramers' theorem is violated, except for one case when the extent of lowering the time-reversal symmetry is not strong enough to remove Kramers degeneracy (Geru 2018). In his book Ion I. Geru writes (Geru 2018)

> Generally, the time-reversal symmetry violation remains the most mysterious symmetry violation, of which origin is not yet fully understood.

The invariance of the laws of physical kinetics and chemical kinetics towards time reversibility implies the so-called Onsager reciprocal relations (Onsager 1931).

In his lecture, Born talks about the two ways of application of causality, one being the timeless relation to dependency and the other depending on a fixed point in space and time. Born says it is always assumed that the cause precedes the action, the effect; and here the time comes in. However, causality can also be regarded as the mere logic in his understanding. Yet when you read his lecture, you may notice that Born allows for the relaxation of the necessity for a specific time, but nevertheless, even if no specific time is given, the logic implies that there is a relation of "before" and "after" that cannot be lifted when we talk about causality.

The physical equations of motion are invariant to the operator **T**, but introduction of the concept of probability and chance into the equations removes the inherent reversibility. For the derivation of the loss of reversibility, Born introduces the entropy S. Consider the distribution function f_1 of one single molecule or the distribution function f_N for a statistical ensemble of N molecules or particles, and the entropy defined as in the phase space span by p and q

$$S = -k \frac{\int f \cdot \log f \, \mathrm{d}p \, \mathrm{d}q}{\int f \, \mathrm{d}p \, \mathrm{d}q} . \tag{3.11}$$

The action of collisions among N particles can be analyzed with the Boltzmann collision equation and allows for determination of the time derivative of S and yields the result that it is never negative:

$$\frac{\mathrm{d}S}{\mathrm{d}t} \geq 0 . \tag{3.12}$$

The time derivative of the distribution function f for particles that do not interact can be written as a Poisson bracket and follows the Liouville equation (von Neumann equation)

$$\frac{\partial f}{\partial t} = [H, f] \tag{3.13}$$

As soon as a collision process occurs, a collision term has to be included in the mathematics of the entropy and distribution function, so that

$$\frac{\partial f}{\partial t} = [H, f] + C . \tag{3.14}$$

As Born expresses it after this derivation, the inclusion of *ignorance* is the reason why the irreversibility follows (Born 1949). Note also that in his paper on the quantum mechanics of collisions (Born 1926). Born made statements about the determinism of the collision process and the relevance of the scattering phases. He believes the indeterminism of the collision processes will remain, but acknowledges that those who criticize his opinion are welcome to include hitherto unknown parameters (Section 8.8.7) into the equations of motion and distribution, so that the determinism, like in classical physics of single particles, is maintained.

3.6 Electromagnetic light theory by Maxwell

When the electrical phenomena were explored in the early nineteenth century at the quantitative level, James Clark Maxwell figured that a handful of fundamental relations would describe what was known back then as ether. The empty space was called ether, and it was believed that it would constitute the medium that would allow the propagation of light, similar to water being the medium for the waves we just saw in the water or any sound wave that propagates through some fluid like gas and liquid, or solid matter. The state of this ether was defined by the electric field strength $\boldsymbol{E}$ and the magnetic field strength $\boldsymbol{H}$. When you bring in any matter in this ether, you have to take into account the phenomena of magnetic induction $\boldsymbol{B}$, the electric current density i, and the dielectric displacement current $\boldsymbol{D}$. These are all vector quantities, and the two magnetic ones $\boldsymbol{H}$, $\boldsymbol{B}$ are axial vectors, while and the other three are polar.

Carl Friedrich Gauss found that the electric charges were the sources of an electric field. Gauss' law for the electric field $\boldsymbol{E}$ reads (note, here the operator ∇ ("nabla") stands for the divergence, *div*)

$$\nabla \cdot \boldsymbol{E} = \frac{\rho}{\epsilon_0} \tag{3.15}$$

and literally means that the electric charge density ρ is the divergence of the electric field $\boldsymbol{E}$. I remember that one my theoretical physics teachers, Prof. Dieter Vollhardt, explained this by spreading out his arms, starting from the body's center and reaching far out, resembling rays coming from a center point and going into infinity[7].

Gauss' law is also known in integral form, particulary in mathematics, and should be understood as the volume integral over the enclosed electric charge being equal to the area integral of the sphere that encloses that charge and is truncated by the electric field vectors perpendicular to the surface of the area integral. The latter is the electric flux. Important here is that the surface encloses an arbitrary distribution of electric charges.

$$\oint\!\!\oint_{\partial\Omega} \boldsymbol{E} \cdot \mathrm{d}\boldsymbol{S} = \frac{1}{\epsilon_0} \iiint_{\Omega} \rho \, \mathrm{d}V \tag{3.16}$$

The magnetic induction $\boldsymbol{B}$, however, has no sources.

$$\nabla \cdot \boldsymbol{B} = 0 \tag{3.17}$$

and

$$\oint\!\!\oint_{\partial\Omega} \boldsymbol{B} \cdot \mathrm{d}\boldsymbol{S} = 0 \tag{3.18}$$

These two Gaussian laws are scalar relations.

The magnetic and electric vectors, E and D, and H and B are coupled in space and time and show the following relations. Faraday's law says the time change of the magnetic induction B will cause a rotation of the electric field. Practically, this means when you run a DC current through a wire[8], at the very moment when the magnetic induction begins to rise, an electric field with axial direction will rise.

7 From your early courses in geometry you remember that a line segment has a starting point and an end point and, thus, a length. The ray has a starting point but no finite end; it ends in infinity and, thus, has no finite length. The line has no finite starting point and no finite end point. Rather, the line begins at infinity and ends at infinity.

8 First we think naively about a metal wire from copper, silver, gold, or aluminum through which the electrons move. The electrons can also move through a metal oxide and also cause such magnetic induction. Even electron holes as positive charge carriers cause the same effect, although with a different sign or different orientation of the magnetic induction. We can even include massive ions with positive or negative charge that constitute an ionic current through a salt bridge or an "ionic wire" built by some electrolyte, be it solid or fluid. We also can go the extreme case and look at an electron beam in a vacuum, such as in a storage ring or a linear accelerator, like SLAC, CERN, or any synchrotron. The beam of charges (electrons, holes, protons, ions) constitutes a current.

When you switch on or switch off a magnetic induction $\boldsymbol{B}$ in a coil, then you have an induction changing with the time t, and this will induce the rotation *rot*, or *curl*, or $\nabla\times$ of an electric field $\boldsymbol{E}$:

$$\nabla \times \boldsymbol{E} = -\frac{\partial \boldsymbol{B}}{\partial t} \tag{3.19}$$

$$\oint_{\partial\Sigma} \boldsymbol{E} \cdot \boldsymbol{dl} = -\frac{\mathrm{d}}{\mathrm{d}t} \iint_{\Sigma} \boldsymbol{B} \cdot \boldsymbol{dS} \tag{3.20}$$

Ampere's law states that the rotation of the magnetic induction, B, is caused by the timely change of the electric field, plus the electric current density J:

$$\nabla \times B = \mu_0 \left(J + \epsilon_0 \frac{\partial E}{\partial t} \right) \tag{3.21}$$

In integral notation, we obtain

$$\oint_{\partial\Sigma} B \cdot \mathrm{d}l = \mu_0 \left(\iint_{\Sigma} J \cdot \mathrm{d}S + \epsilon_0 \frac{\mathrm{d}}{\mathrm{d}t} \iint_{\Sigma} E \cdot \mathrm{d}S \right) \tag{3.22}$$

Maxwell basically compiled the findings by Gauss, Faraday, and Ampere, added the displacement current, and there were the Maxwell equations. Today, we know the aforementioned as Maxwell equations. They are overall quite consistent. For example, the continuity equation follows immediately when we combine Ampere's law with Coulomb's law.

With some differential operator manipulations of the Maxwell equations we can arrive at a second-order differential equation that reads like the wave equation we know from classical mechanics. For this, we take the condition that the space has no electric charges, $\rho = 0$, and also there is no electric current flowing, $j = 0$. The rotation of the rotation of the electric field sounds like a complicated condition, but vector calculus reminds us of the Grassmann relation[9]:

$$\boldsymbol{a} \times (\boldsymbol{b} \times \boldsymbol{c}) = \boldsymbol{b}(\boldsymbol{a} \cdot \boldsymbol{c}) - \boldsymbol{c}(\boldsymbol{a} \cdot \boldsymbol{b}) \tag{3.23}$$

It is an exercise for the reader to apply this identity on the rotation of the electric field E and on the magnetic field B.

9 Hermann Grassmann was a German mathematician, who never attended a mathematics class at university. His lack of conformity with the mathematics community posed a large barrier for the acceptance of his theories. Compare also Bachelier, who, although a student of Henry Poincaré, had difficulties getting his PhD thesis on the mathematics of economy accepted. Grassmann is the founder of vector and tensor calculus (Lampert 2000). Grassmann used his new vector theory for the development of additive color compounding (Schubring 1996; Maxwell and Zaidi 1993; Grassmann 1854), which is still valid today. In your undergraduate studies on tensor calculus you may have come across the Grassmann identity, also known as so-called bac-cab rule.

In the end, you find the wave equation (see Equation 3.9)

$$\frac{\partial^2}{\partial^2 t}\boldsymbol{E} = \frac{1}{c^2}\frac{\partial^2}{\partial^2 \boldsymbol{r}}\boldsymbol{E} \tag{3.24}$$

The constant c in the wave equation is the speed of light, given to $c^2 = \mu\mu_0\epsilon\epsilon_0$. Equation 3.24 is a typical second-order linear differential equation with second derivatives for the spatial coordinate $\boldsymbol{r}$ and the time t.

We recall that we have neglected the presence of electric charges and currents. We physicists and engineers are interested in the solutions of this wave equation, whereas mathematicians are only interested in whether solutions exist at all. However, the solutions for this kind of wave equation are well known. The equation for sound waves has the same structure, but there exist some fundamental differences between a sound wave, which is a longitudinal wave, and a electromagnetic wave, which we will see now and which is a transversal wave.

The plane wavefunction of the type

$$\boldsymbol{E}(\boldsymbol{r}, t) = E_0 \cdot \frac{\boldsymbol{r}}{r} \cdot e^{i(kr-\omega t)} \tag{3.25}$$

is the standard solution for the wave equation. Note that we can write the exponential function with complex argument as the sum of a sine and cosine $e^{i\phi} = \cos\phi + i\sin\phi$.

The electromagnetic wave propagates with the speed of light c in forward direction $\boldsymbol{v}$. Perpendicular to the direction of the propagation are the electric vector $\boldsymbol{E}$ and the magnetic vector $\boldsymbol{B}$. This orthogonality condition makes the electromagnetic wave a transversal wave. We can abbreviate the unit vector or normal vector[10] to

$$\hat{\boldsymbol{r}}_{\mathbf{e}} = \frac{\boldsymbol{r}}{r} \tag{3.26}$$

The power P (in short, energy divided by time, $\Delta W/\Delta t$) of the electromagnetic wave points also in the direction of $\boldsymbol{v}$. The magnitude of this vector $\boldsymbol{P}$, and the Poynting vector, its modulus, is the vector product of the magnetic and electric components:

$$P = |\boldsymbol{E} \times \boldsymbol{B}| \tag{3.27}$$

3.7 Excursion: spectral analysis

3.7.1 Expansion of discrete transitions into Fourier series

When you look into the examples shown for the application of the Fourier series, you notice that we are dealing mostly with some sort of “sharp” transition that is analytically difficult to handle. We can basically consider these transitions as symmetry breaking.

10 The normal vector is a vector that is orthogonal to a plane area and has the size 1, unity.

In physical nature there is no spontaneous symmetry breaking in the sense that it could happen on the spot spatially or on the spot timely. There is always a spatial range over which the transition from one state or one phase to a different state or phase takes place, even if it extends only to sub-atomic scale. This takes some minimum amount of time, even if it is extremely short. Hence, all such transitions occur over a diffuse domain that spans the two or more of the aforementioned states and phases. One example is the surface of a perfect crystal. Below the surface is the bulk material of the crystal, and above the surface is the empty vacuum or some other phase, such as a liquid, water, some electrolyte, or a gas. Experiments show that the electronic structure of the subsurface in the crystal and the molecular structure of the double layer above the surface do not change discretely and continuously but do so over a finite transition range discontinuously.

In mathematics, these states and phases correspond to partially steady but different functions, which have to be matched at the very spot where they meet. The usual way to do this is the application of a Fourier series, which is a mathematical expansion of the two or more aforementioned functions into a periodical series of sine and cosine functions, many of which were found by Leonhard Euler (see, for example, Hewitt and Hewitt 1979).

Such a discrete situation is demonstrated in Figure 3.12, where we basically have a step function where the step height goes from $+h$ to $-h$ at the position $t = T/2$, as indicated by the two thick, blue horizontal lines. For t smaller than $T/2$, the step has the value $+h$. For t values larger than $T/2$, the height is at the base value $-h$. Mathematically, we express this situation as follows:

$$f(t) = \begin{cases} +h\,, & \text{if } 0 \leq t \leq T/2 \\ -h\,, & \text{if } T/2 \leq t \leq T \end{cases} \tag{3.28}$$

This function is known as the Heaviside function or Theta function (Θ function). A similar function is the Delta function (δ function), which does not define a step, but a peak. The Heaviside function and the Delta function are useful, for example, in x-ray and optical spectroscopy, where peaks and steps show up in the spectra as core level excitations and ionization thresholds. These can be mathematically modeled with density functional theory, for example, and the solutions of algebraic equations then yield the peak and step positions. You will see these solutions, for example, in Chapter 4. Actual spectra, however, have no sharp peaks, but peaks with a finite width, which originates from different physical effects. To account for broadening, one can convolute the Delta function and Heaviside function with a distribution function, such as a Gaussian or a Lorentzian distribution. This allows you, then, to fit the calculated spectrum to the experimental spectrum. A close inspection of Figure 3.12 reveals that the step function pattern is continued at the positions $t = 0$ and $t = T$, with the periodicity T, so that $f(t + T) = f(t)$.

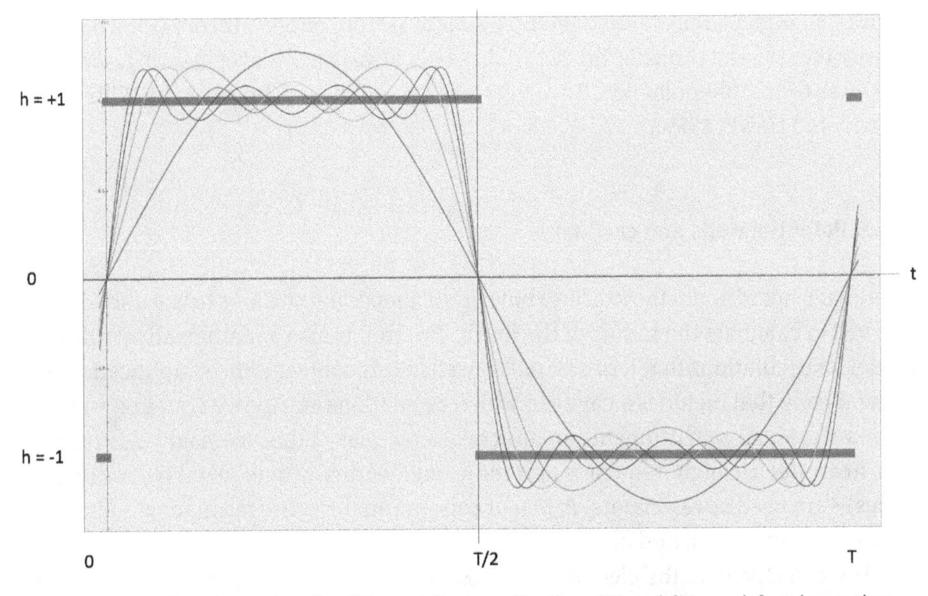

Fig. 3.12: A step function defined by $f(t) = +h$ for t smaller than $T/2$ and $f(t) = -h$ for t larger than $T/2$, as shown by the thick, blue horizontal lines. The red curve is the Fourier series with only one sine term. The orange, green, blue, and brown curves are the Fourier series with $k = 1, 2, 3$, and 4.

It has been known for over 200 years that some discontinuous functions can be approximated by sums of infinite series of sine and cosine functions. The expansion of this step function shown in Figure 3.12 into a Fourier series reads

$$\begin{aligned} F(t) &= \frac{4h}{\pi}\left[\sin \omega t + \frac{1}{3}\sin 3\omega t + \frac{1}{5}\sin 5\omega t + \frac{1}{7}\sin 7\omega t + \cdots\right] \\ &= \frac{4h}{\pi}\sum_{k=0}^{\infty}\frac{\sin((2k-1)\omega t)}{2k-1} \end{aligned} \tag{3.29}$$

I have plotted these Fourier series into Figure 3.12 for a suite of values k. For $k = 0$, the series in Equation 3.29 only has one single sine term, and the corresponding plot in Figure 3.12 shows one full sine oscillation (red color) with wavelength T. The orange curve is built from two sine terms in the Fourier series: $k = 0$, and 1. Note that the argument in the sine term also contains the $(2k - 1)$ factor, which makes the number of oscillations increase within the step function range T. You can clearly see this trend with increasing index k, when the number of oscillations increases, as you are adding up more terms in the series with increasing k. This trend enables you to increasingly better approximate the plateau of the step and the base line, and also the jump region becomes sharper, or at least narrower.

However, irrespective of how many terms you add to the Fourier series by increasing k, there will be an overshooting of the series, which makes it impossible to perfectly match the intensity at the step. The overshooting of the Fourier series over the actual step is considered a fundamental weakness[11] of Fourier's theory (for a book on Fourier theory read (Constantin 2016)). There is always a 9% overshooting, which is known as Gibbs' phenomenon. The effect was first published, however, by Wilbraham (Hewitt and Hewitt 1979).

3.7.2 Potential wells and cavities

We already heard about the well and how Eratosthenes found a way to use a shadow in the well to calculate the radius of the Earth. The well is also a mathematical physical model for oscillations that it can trap. The well in this context is the simplification of a general body that includes a cavity in which oscillations can form a standing wave by resonance. Musical instruments are practical examples of this. More advanced examples lie in electric high frequency and radio engineering, where metal bodies (tubes, cubes)[12] are used as resonators, *in lieu* of conventional electric resonant circuits built from capacitors, coils, and ohmic resistors.

You can calculate the eigenfrequencies f of such oscillators, or resonators, from the physical and geometrical dimensions of the resonator body. For a rectangular prism with the lengths a, b, and c, we can write:

$$f = \frac{c}{2\pi\sqrt{\epsilon_r \mu_r}} \sqrt{\left(\frac{n_a \pi}{a}\right)^2 + \left(\frac{n_b \pi}{b}\right)^2 + \left(\frac{n_c \pi}{c}\right)^2} \quad (3.30)$$

The constants ϵ_r and μ_r are the relative dielectric constant and magnetic permeability; n_a, n_b, and n_c are integers that run from 0 to ∞. As an exercise, you can calculate the frequencies for the above rectangular design for different long edges. You realize that the number of frequencies is infinitely high because of $n \to \infty$. With $n = 1$, you obtain the first-order frequencies. With higher n, the resonator produces higher-order frequencies.

This equation was originally derived by Lord Rayleigh (Rayleigh 2009b, c, a), and you can read up in the book by Pozar on how such microwave resonators can be designed (Pozar 2011). We will learn in Chapter 6 how cavities can be used also at the mesoscopic level and how quantum electrodynamics behaves at this level (Cottet et al. 2015). What is important here to learn that the sizes of the well, or the cavity, determine the frequencies.

---

11 I would not consider it a weakness, but rather a characteristic.

12 One of my neighbors in my childhood was a lumberjack. One day, he gave me a box made from white plastic. He had found it hanging in a tree in the woods. It was a radiosonde from a weather balloon of Bendix–Friez design, see, for example, Funk et al. (1966a, b). This is an application in so-called radio meteorology (Bean and Dutton 1966). I remember that there was a copper cylinder inside, in addition to the battery, electronics, a barometer, and so on. These radiosondes transmit their information in the 1000 MHz frequency range, for which cavity resonators are used.

For hydrogen atoms to fuse together as nuclear fusion and then release the huge amount of energy that we want to harness one day, the protons have to overcome the electrostatic Coulomb barrier, which is very high; once it is overcome, the nuclear forces will attract both protons. The potential between two nuclei can be approximated by the Morse potential, see, for example, Van Siclen and Jones (1986).

$$V_{\text{Morse}}(r) = V_{\text{D}} \left(1 - \mathrm{e}^{-a(r-R_{\text{e}}}\right)^2 \tag{3.31}$$

The Morse potential has the advantage that it can provide an exact analytical solution in the Schrödinger equation. In Figure 3.13, I have sketched a similar potential that combines a binding state and a nonbinding state between protons.

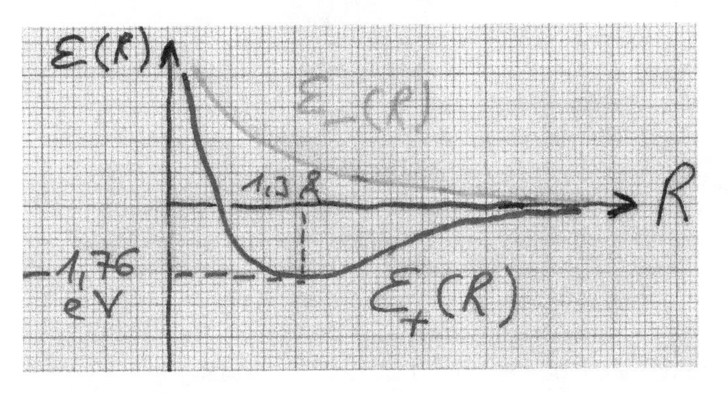

Fig. 3.13: The energies for a bound $\epsilon_+(R)$ orbital and the unbound $\epsilon_-(R)$ orbital in a pair of protons depending on the distance R of both protons.

The Hamiltonian for the potential shown in Figure 3.13 is built from two hydrogen atoms, which have a distance $\boldsymbol{R} = \boldsymbol{R}_2 - \boldsymbol{R}_1$, and are at positions given by the vectors $\boldsymbol{R}_1$ and $\boldsymbol{R}_2$. Their electrons shall have the positions $\boldsymbol{r}_i$. We can then write

$$\hat{H} = \frac{h^2 \Delta}{2m} - \frac{\mathrm{e}^2}{\boldsymbol{r} - \boldsymbol{R}_1} - \frac{\mathrm{e}^2}{\boldsymbol{r} - \boldsymbol{R}_2} + \frac{\mathrm{e}^2}{\boldsymbol{R}_1 - \boldsymbol{R}_2} \tag{3.32}$$

and know that this many-body-problem will not have an analytical and exact solution that we can determine. The approach for solving this unsolvable problem is that we start with some test wavefunctions from which we construct solutions by minimizing the resulting total energy. We can make a linear combination of the two wavefunctions ψ_i with $i = 1, 2$

$$\psi_i = \frac{1}{\sqrt{\pi a_0^3}} \mathrm{e}^{\frac{|\boldsymbol{r} - \boldsymbol{R}_i|}{a_0}} \tag{3.33}$$

The constant a_0 is Bohr's radius with $a_0 = \hbar^2/2\,\mathrm{m} = 0.529$ Å. This is half of an Ångstrom.

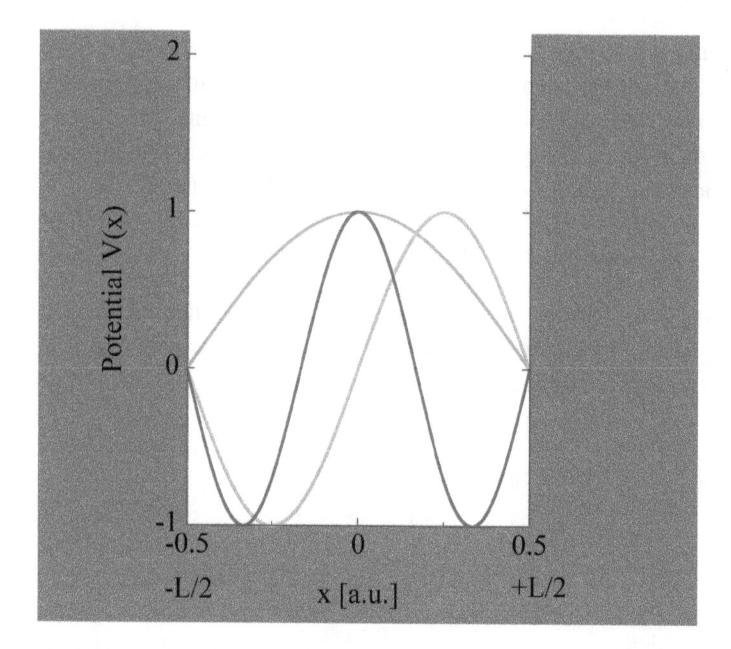

Fig. 3.14: A potential well of width L with three inscribed sine functions with arguments $V(x) = \sin(2\pi x + n\pi/2)$.

The expectation value for the Hamiltonian $\hat{H}$ is given by the ratio

$$\hat{H} = \frac{\langle\psi_1|\,\hat{H}\,|\psi_1\rangle + \langle\psi_2|\,\hat{H}\,|\psi_2\rangle \pm 2\,\langle\psi_1|\,\hat{H}\,|\psi_2\rangle}{2\left(1 + \int \mathrm{d}^3 r\,\psi_1^*(\mathbf{r})\psi_2(\mathbf{r})\right)} \tag{3.34}$$

The integral in the denominator in Equation 3.34 measures the overlap of the wavefunctions and is called the overlap integral. Here, it denotes the overlap of the 1s-orbitals. In the numerator, you can see three expectation values of the kind $\langle\psi_1|\,\hat{H}\,|\psi_1\rangle$. Maybe it is of interest for some readers to know how to code this in LaTeX; you write it like this

```
$\expval{\hat H}{\psi_1}$
```

For the mixed expectation value $\langle\psi_1|\,\hat{H}\,|\psi_2\rangle$, you type

```
\bra{\psi_1} \hat H \ket{\psi_2}
```

Schneider suggests that a rectangular potential barrier of finite height and width – you can see this in Figure 3.14 – can not prevent a significant fraction of particles from penetrating this barrier (Schneider 1989). For the derivation in Schneider's work, see Section 5.9. In other words, nuclei like protons or nuclei in hydrogen or helium would have the realistic chance of overcoming the barrier and, thus, fuse together. This would be nuclear fusion, but not thermonuclear fusion. This is an interesting aspect for cold nuclear fusion, the observation of which was claimed by Pons and Fleischmann in 1989 (Fleischmann and Pons 1989), but which has not been reproduced since by other researchers (Nagel 1998). After the 1989 workshop on the nuclear equation of state, Harley, Gajda, and Rafelski (Harley et al. 1989) wrote:

> The most significant problem facing cold fusion is the inability of some competent experimentalists to observe it.

3.8 Fourier transformation

One mathematical tool for solving a particular kind of differential equations is the Fourier transformation. For example, when you have differential equations in the Euclidean space (space domain, r, $\mathbf{x}$), and you can transform them into the momentum space (k, q, Q), and those then transform into algebraic equations. When we have a function $F(x)$, then the Fourier transform of F is obtained by integration of $F(x)$ along $\mathrm{d}x$ with an exponential, which has in its argument the product of momentum p and spatial coordinate x:

$$\tilde{F}(p) = \int_{-\infty}^{+\infty} \mathrm{e}^{\mathrm{i}px} F(x)\,\mathrm{d}x \tag{3.35}$$

Because of the general identity $\mathrm{e}^{\mathrm{i}\phi} = \cos(\phi) - \mathrm{i}\sin(\phi)$, we can write the right-hand side of Equation 3.35 as the sum of a Fourier sine transform and Fourier cosine transform. We can also back-transform the Fourier transform and obtain

$$F(x) = \frac{1}{2\pi} \int_{-\infty}^{+\infty} \mathrm{e}^{\mathrm{i}px} \tilde{F}(p)\,\mathrm{d}p \tag{3.36}$$

Generally, we say that $\mathcal{F}$ is the Fourier transform of F, and thus, $F(p) = \mathcal{F}\{F(x)\}$. Inversely, $F(p) = \mathcal{F}^{-1}\{F(x)\}$ is the inverse Fourier transform. The Fourier transformation is particularly helpful for solving problems in optics. As a matter of fact, *Spektralapparate*, optical spectrometers, carry out a Fourier transform when they are hit with light. We know the lens and the prism as optical apparatus. Another optical apparatus is the optical grating (*optisches Strichgitter*), the physics of which is based on the same phenomenon that we know from the double-slit experiment. The grating can be used as a monochromator for light composed of multiple wavelengths, certainly including "white light".

3.8.1 Fourier transform of the double-slit experiment

You can mathematically code the geometry of the double slit with slit positions x_1 and x_2 by convoluting it with the rectangular potential and obtain the relation

$$\delta(x \pm d) * \mathrm{rect}_b(x) = \left(\delta\left(x + \frac{a}{2}\right) + \delta\left(x - \frac{a}{2}\right)\right) * \mathrm{rect}_b(x)\,. \tag{3.37}$$

When a coherent light source shoots the photons in the slit direction, two photons become localized at the slits with states $\langle x_1|$, $\langle x_2|$ and go into a superposition of a new state, which can be calculated to

$$\langle \Psi| = \frac{1}{\sqrt{2}}(\langle x_1| + \langle x_2|) \tag{3.38}$$

We can employ plane waves as the solutions for the superposition problem.

Or you can define the position of the two slits on the x-axis with two slits with distance D and width W, where $f(x) = 1$ for the relation below

$$\frac{-(D+W)}{2} < x < \frac{-(W-D)}{2},\ \frac{(W-D)}{2} < x < \frac{(W+D)}{2} \tag{3.39}$$

and $f(x) = 0$ for the other x-values.

For this we can use the Θ function (as an exercise sketch the geometry of the double slit with coordinate x and the slit width and slit distance)

$$f(x) = -\Theta\left(x + \frac{D+W}{2}\right) - \Theta\left(x + \frac{W-D}{2}\right) + \Theta\left(x - \frac{W-D}{2}\right) - \Theta\left(x - \frac{D+W}{2}\right) \tag{3.40}$$

The reason for expressing the geometry in an analytical expression, at least as analytically as possible, has a simple practical purpose. As soon as you have identified that a system has mathematical, for example arithmetic, algebraic, or geometrical properties, then mathematical tools can be employed. The Fourier transform for the double-slit geometry then reads

$$\mathfrak{F}_x[f(x)](\omega) = \int_{-\infty}^{\infty} f(x) \cdot \mathrm{e}^{-2\pi \cdot x \cdot \omega}\,\mathrm{d}x = 2D \cdot \cos(\pi \cdot W \cdot \omega) \cdot \frac{\sin(\pi \cdot D \cdot \omega)}{\pi \cdot D \cdot \omega} \tag{3.41}$$

The expression of the type $\sin x/x$ in Equation 3.41 can be abbreviated as $\mathrm{sinc}(x)$ and is known as *sinus cardinalis*. In German, this function is also known as the *Spaltfunktion*, which translates to “slit function”. When we square the Fourier transform in Equation 3.41, we obtain the light intensity that we can probe with our eye or some other light detector, and thus, basically, obtain the same expression as shown in Equation 3.10.

3.8.2 From time domain to frequency domain: Fourier transformation

Impedance spectroscopy is an electroanalytical method for the determination of the electric charge carrier dynamics in electrodes, electrolytes, and entire electric systems, i.e., systems that have electric properties. You impose an alternating current with small amplitude on the system and record the voltage variation as a function of the frequency with which you tune the alternating current. Or the other way around.

The experimental data is the complex resistivity Z, which, because it is a complex number, has a real part $\mathrm{Re}(Z)$ and an imaginary part $\mathrm{Im}(Z)$ for every data point probed at a specific frequency f. As you scan the process with varying AC frequencies $\omega = 2\pi f$, the impedance is a dispersion quantity: $Z(\omega)$.

It is very useful when one can model the electric system with an electric circuit built from various resistivities R, capacities C, and inductivities L. One may then express the circuit by Kirchhoff's rules and determine the impedance Z and fit the mathematical model to the experimentally obtained data, the impedance spectrum. If this works out, then you can express your electrical system by the aforementioned circuit.

Sometimes it is not possible to carry out an impedance measurement – for various reasons. One reason can be that the electric system, the electric apparatus that you want to probe, delivers such high currents or voltages that the impedance analyzer (FRA, frequency response analyzer) cannot handle them. One technical way out of this dilemma is to carry out a switch-on or switch-off experiment with a strong but low-cost power supply and then measure the voltages and currents with simpler analytical tools like multimeters or cheaper meters. This provides us with a discharge curve of current versus time $I(t)$ or voltage over time $V(t)$. Now we can apply a Fourier transformation on this dataset and obtain the current and voltage as a function of frequency f. This is one practical application of the Fourier transformation, which I will lay out below.

3.8.3 Excursion to impedance spectroscopy

When I was a postdoctoral researcher in Berkeley, one of my colleagues in the Physical Biosciences Division cofounded the Berkeley Spectroscopy Club. Much of our work was about x-ray spectroscopy and being so close to the Advanced Light Source, we got to know many researchers, who used various kinds of such methods. So, he and a colleague suggested organizing regular seminars on spectroscopy, not only x-ray spectroscopy, but all kinds of spectroscopy. When you spend your scientific life in one field only, you easily narrow the world to your narrow field only. So, it was a great idea to run such club and invite all kinds of people and give seminars on their particular kind of spectroscopy. I had a Russian colleague who worked in air quality, and he did mostly mass spectroscopy. A PhD student in our battery group, where I worked in, did mostly nuclear magnetic resonance (NMR) spectroscopy. I had an Argentinean colleague in the aerosol group, and he did mostly infrared spectroscopy. Therefore, at one point, I thought I should give a talk at the club – on impedance spectroscopy.

I first learnt about impedance spectroscopy when I was a research intern at Philips Research Laboratories in Aachen, in their electroceramics section. There I used the method for studying piezoceramics for inkjet printers and for DRAM heterostructures, in a very high-frequency range of several hundreds of megahertz (MHz). In impedance spectroscopy, you apply an alternating current to a specimen and measure the resulting voltage. Or you apply an alternating voltage and measure the electric current response. Since this method deals with AC signals, you vary the frequency of the electric signal with the meter that you use. In doing so, you perturbate the specimen, and depending on the nature of the specimen, its characteristics may show electric resonances, which you can then identify in the impedance spectrum.

Practically, you record the current, the voltage, and the frequency that you employ. The result is three columns of data. The ratio of the voltage U versus the current I gives the resistance R. In Ohm's law, this R is a constant. When we deal with AC signals, we find that also capacitors and coils add to the resistance of a system, and then the resistance of such system depends on the frequency, and we thus call it impedance, rather than resistance. The impedance is thus a dispersion quantity. The impedance is thus a complex number with a real part and an imaginary part. Its analysis follows the theory of complex numbers.

Table 3.1[13] displays the three columns that you obtain from an impedance analyzer instrument. The data were recorded from a biofilm that was grown on an iron oxide photoelectrode. The electrolyte was phosphate buffer saline (PBS).

How do you treat the three-column data? You can, for example, plot the real part versus the frequency, and the imaginary part versus the frequency. As the frequency range extends over several orders of magnitude, it is necessary to plot the data on the logarithmic scale. You can also calculate the phase angle ϕ from the two components and plot this one versus the frequency. An often used representation is the Cole–Cole plot (Cole and Cole 1941) (also called the Nyquist plot), which is the imaginary part of Z versus the real part of Z.

The real part is a measure for the ohmic losses, and the imaginary part is a measure for the energy, which is in general shuffled between the capacitor and the inductor coil without such losses. One can compare this analysis to some extent to optical theory, where you have an absorption term and a dispersion term in the optical constant. This is why the angle of reflection and the diffraction angle depend on the wavelength of the light, as we can see from the rainbow in the sky and in the colorful spectrum in Figure 3.3.

---

13 If you are wondering why I did all the work to type this long table in LaTeX: I used an online tool that converted my table from .txt format to a .tex format table, which you find at www.tablesgenerator.com.

Tab. 3.1: The impedance data of a biofilm on an iron oxide photoelectrode in PBS electrolyte. The left-hand column is the frequency starting at 100 kHz and ending at 100 mHz. The middle column is the imaginary part of the impedance, $-\mathrm{Im}\,Z$. The right-hand column is the real part of the impedance, $\mathrm{Re}\,Z$.

Impedance data		
Frequency [Hz]	$-\mathrm{Im}\,Z$ [Ω]	$\mathrm{Re}\,Z$ [Ω]
100000	19.368	2.9776
63291	20.238	1.4436
40000	20.598	1.6030
25000	20.926	1.9634
15823	21.313	2.3305
10000	21.784	2.7790
6329.1	22.705	3.6571
4000.0	23.451	4.4651
2500.0	24.174	5.8349
1582.3	25.136	7.6561
1000.0	26.351	10.470
632.91	27.916	14.637
400.00	30.068	20.758
250.00	33.051	29.916
158.23	37.453	43.176
100.00	44.590	62.907
63.291	55.662	90.513
40.000	74.575	128.18
25.000	103.94	180.90
15.823	147.89	247.60
10.000	212.42	328.61
6.3291	296.74	431.92
4.0000	410.91	564.74
2.5000	557.72	732.37
1.5823	756.34	944.41
1.0000	1058.2	1171.9
0.63291	1473.4	1436.3
0.40000	2024.9	1650.8
0.25000	2688.8	1720.8
0.15823	3327.9	1621.7
0.10000	3877.3	1412.0

Wherever you have an oscillator, or an oscillating system, you can study it with impedance spectroscopy. This also holds for mechanical systems. I work at Empa in Dübendorf, and Empa is well-known for its research on building structures and bridges. I remember once seeing a model bridge construction in one of the large experimental halls, large enough to occupy much of the hall, and it was being measured with impedance spectroscopy. I guess there was a mechanical actuator that would excite the bridge with mechanical force impulses at some frequencies, and then the amplitudes of movement would be measured.

Electrochemical impedance spectroscopy (EIS) is used in electrochemical systems like batteries, supercapacitors, fuel cells, and electrolyzers, and photoelectrochemical cells and sensors. You employ two or three electrodes to the system and excite the system with the AC signal and record the impedance. Chemical reactions are relatively slow, as compared to processes that are known in condensed matter physics (solid-state phenomena). Thus, for conventional liquid phase experiments, impedance analyzers (frequency response analyzers, FRA) provide a frequency range of up to 300 kHz. A few may even go as far as 32 MHz. The lower frequency range may be in the 1 mHz range. Observe that a frequency as low as 1 mHz is 1/1000 per second. Take the reciprocal of this, and this makes a 1000 seconds, which is 16 2/3 minutes. This means the data point at this low frequency will take over 1/4 hour to be measured. A full spectrum over the entire accessible frequency range could thus take several hours, or one whole day, depending how many data points you want to measure.

EIS is sensitive to charge transfer processes between electrodes and fluids, between grains across grain boundaries, capacitances, and so on. The physical and chemical situation at the working electrode can be modeled as an electric circuit and mathematically computed according to Kirchhoff's rules, as a set of series and parallel resistors, capacitors, and maybe also inductor coils. By doing so, one can determine the charge transfer resistance, the double-layer capacity, the grain boundary resistance, and so on. You obtain these data by the least squares fit of the computed spectrum of the electric circuit model to the actual experimental spectrum.

A well-known software for the quantitative analysis of impedance spectra is ZView. However, since my favourite data analysis software is Igor Pro from Wavemetrics, Inc. (Busbey 1999), I was on the search for a macro for impedance analysis by using the Igor Pro platform. Eventually I found a paper by a physician at a hospital at a university on the US East Coast. He was doing studies on the pulmonary activity of the human lung; imagine the voluminous lung and the narrow throat through which the air must go, like a bottle and a bottleneck. A nice example with sketches and an electric circuit is shown in a different paper from a group in Hungary, see Hantos et al. (1997).

You can employ this method also with bio-organic systems, such as bio-electric interfaces. I did so for the iron oxide photoelectrodes in our Empa lab, which were coated with phycocyanin and biofilms (Braun et al. 2015a). The data in Table 3.1 are from such biofilm. A professional Igor Pro macro for impedance analysis is now available, which was developed by NIMS in Japan (Kobayashi et al. 2016).

Anyway, I volunteered at the Berkeley Spectroscopy Club with a presentation on impedance spectroscopy. About 15 to 20 interested people came, and I showed the basic principles of the theory and also a number of applications. At the end, when I was done, one of the seniors raised the questions whether what I had presented was legit and reasonable, or whether I was just making this up. I must admit that I was stunned over this kind of question from a senior. This gentleman was working at one of the best equipped laboratories in the world, both instrumentally and intellectually, and he was a senior, I guess close to retirement age. So he must have had quite some experience as researcher and scientist. My conclusion was then that I must have given a very dumb presentation to provoke such kinds of question.

For those interested in learning more about impedance spectroscopy, I recommend the book *Impedance Spectroscopy: Theory, Experiment, and Applications* by Barsoukov and Macdonald (2005). I received a free copy of the book and a cheque from the publishing house because I was one of the reviewers for a draft of this book when the 2nd edition was planned.

3.8.4 Electrochemical impedance spectroscopy

The measurement of complex resistances (impedance Z) with AC voltage techniques (electrochemical impedance spectroscopy, EIS) is an important tool for investigating the interface and volume properties of materials. When interfaces are studied, such as electrode–electrolyte interfaces, adsorption rates and reaction rates can be studied, as well as double-layer capacitance[14].

If a system is excited with AC voltage, say

$$U(\omega) = U_0 \cdot \exp(\mathrm{i}\omega t)\,, \tag{3.42}$$

with U_0 the amplitude, f the frequency, and

$$\omega = 2\pi f\,, \tag{3.43}$$

then the current $I(\omega)$ flowing through the system will, in general, be shifted against the voltage by a phase shift ϕ:

$$I(\omega) = I_0 \cdot \exp(\mathrm{i}\omega t + \phi)\,, \tag{3.44}$$

with I_0 the amplitude of the AC current. The impedance follows from Ohm's law:

$$Z(\omega) = \frac{U(\omega)}{I(\omega)} = Z_0 \cdot \exp(\mathrm{i}\phi) = Z_0 \cos\phi - \mathrm{i}Z_0 \sin\phi\,. \tag{3.45}$$

Such a phase shift between current and potential is illustrated in Figure 3.15.

The real part (imaginary part) of the complex impedance, $Z_0 \cos\phi$ ($Z_0 \sin\phi$), will be abbreviated with Z' (Z''). In many cases, systems can be modeled by an array of resistances R, capacitances C, and inductances L. Table 3.2 displays the frequency response of these simple elements.

14 You can also think of a thylakoid membrane in contact with the lumen or stroma, of which you want to measure the electric impedance.

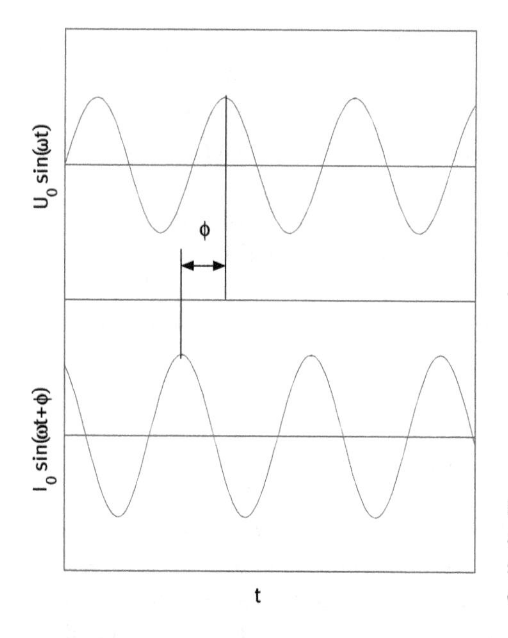

Fig. 3.15: Schematic of phase shift in impedance. The phases of current and voltage are shifted by the amount of $\Delta\phi$ against each other: $\Delta\phi = \omega\Delta t$.

Tab. 3.2: Resistivities of various electrotechnical elements. Coil and capacitor depend on the frequency ω of the AC voltage applied.

The complex resistances	
Ohmic resistance of a resistor	$Z_R = R$
Capacitance of a capacitor	$Z_C = \frac{1}{\mathrm{i}\omega C}$
Inductance of a coil	$Z_L = \mathrm{i}\omega L$

The impedance of a combination of R, L, and C can be calculated using Kirchhoff's rules. For a series circuit of two impedances Z_1 and Z_2, the impedance is

$$Z_{1,2} = Z_1 + Z_2 \,. \tag{3.46}$$

For a parallel circuit of two impedances Z_1 and Z_2, the impedance is calculated as follows:

$$\frac{1}{Z_{1,2}} = \frac{1}{Z_1} + \frac{1}{Z_2} \,. \tag{3.47}$$

3.8.4.1 Representation of electrochemical systems by electric circuits

Figure 3.16 displays a schematic of an electrochemical interface, which can be represented by the following electric circuit. In the schematic, R_T denotes the charge transfer resistance, which is attributed to chemical reactions occurring on the electrode surface.

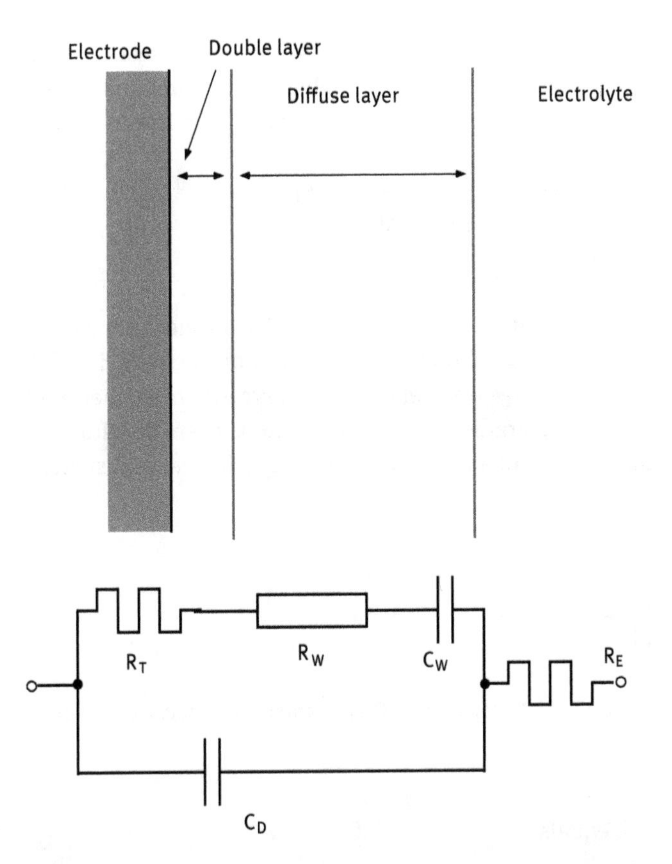

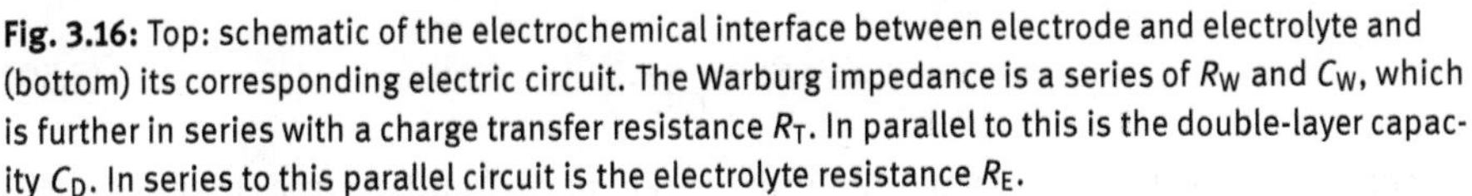

Fig. 3.16: Top: schematic of the electrochemical interface between electrode and electrolyte and (bottom) its corresponding electric circuit. The Warburg impedance is a series of R_W and C_W, which is further in series with a charge transfer resistance R_T. In parallel to this is the double-layer capacity C_D. In series to this parallel circuit is the electrolyte resistance R_E.

This resistance is infinite when no redox processes occur on the electrode. The electrochemical double layer has a capacitance C_D parallel to R_W. If chemical reactions occur, the concentration of ions may decrease in front of the electrode with the result that an additional resistance R_W in series with a capacitance C_W also occurs in the system, in series to the charge transfer resistance and parallel to the double-layer capacitance[15].

This element is known as the *Warburg impedance* and is characterized by a straight line with an angle of 45° against the real axis in the complex plane:

$$\mathcal{R}_W = R_W + \frac{1}{i\omega C_W} . \tag{3.48}$$

15 Although termed *occurrence of a resistance*, the correct terminology means the occurrence of a conductivity.

Finally, the resistance of the electrolyte, R_E, contributes to the overall resistance of the system in series. Using the complex resistances in Table 3.2 and Kirchhoff's rules (Equations 3.46 and 3.47), one finds for the impedance of the circuit in Figure 3.16:

$$Z = \frac{1}{\frac{1}{R_T + R_W + \frac{1}{i\omega C_W}} + i\omega C_D} + R_E . \tag{3.49}$$

Note that for high frequencies ω, the limit of the impedance Z is the electrolyte resistance R_E. In the case of an electrochemical double-layer capacitor electrode, if redox reactions do not take place, and if charge transfer resistances are very high, there will also be no remarkable concentration gradients of electrolyte ions. Then, the situation can be represented by a simple series of a resistance and a capacitance, as displayed in Figure 3.17.

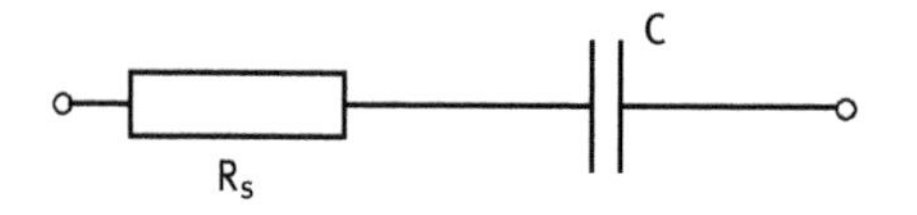

Fig. 3.17: Series circuit of a resistance R_S and a capacitance C representing the simplest capacitor electrode (single electrode).

The impedance of this circuit equals

$$Z = R_S - \frac{i}{\omega C} . \tag{3.50}$$

The complex impedance Z can be split into the real part Z' and the imaginary part Z'':

$$Z = Z' + iZ'' . \tag{3.51}$$

For the imaginary part we find

$$Z'' = -\frac{1}{\omega C} . \tag{3.52}$$

Thus, the double-layer capacitance of a system that is represented by the circuit in Figure 3.17 can be determined from the imaginary part of the complex impedance measured:

$$C = -\frac{1}{\omega Z''} \tag{3.53}$$

In real capacitors with two electrodes, a leakage current at an open circuit may occur with the result of a self-discharge of the capacitor.

The leakage current is taken into account by an additional parallel resistance R_P, as displayed in Figure 3.18. The real part and the imaginary part of the complex im-

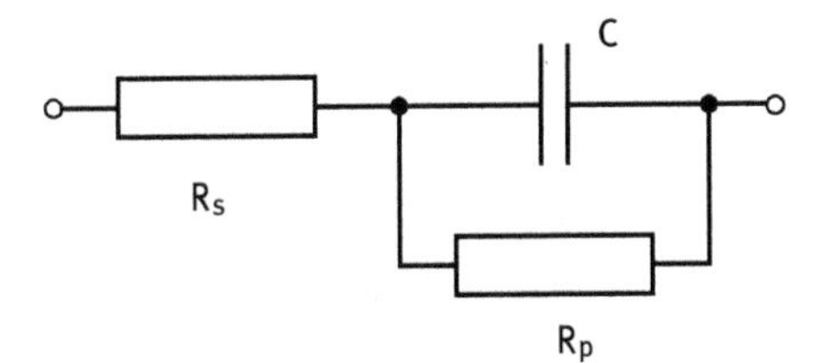
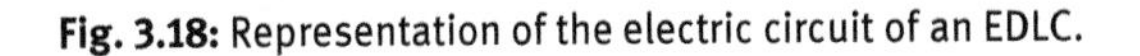

Fig. 3.18: Representation of the electric circuit of an EDLC.

pedance of this circuit are

$$Z' = R_s + \frac{1/R_p}{\omega^2 C^2 + 1/R_p^2}, \tag{3.54}$$

$$Z'' = \frac{-\omega C}{\omega^2 C^2 + 1/R_p^2}. \tag{3.55}$$

All information on R_s, R_p, and C can be extracted from experimental impedance data. From the high-frequency intercept of the impedance plot, the series resistance R_s can be extracted:

$$\lim_{\omega \to \infty} Z' = R_s. \tag{3.56}$$

The information on R_s is necessary to solve equations for the values for R_p and C:

$$C = -\frac{Z''}{\omega\,(-2\,Z'\,R_s + {R_s}^2 + {Z''}^2 + {Z'}^2)}, \tag{3.57}$$

$$R_p = -\frac{-2\,Z'\,R_s + {R_s}^2 + {Z''}^2 + {Z'}^2}{-Z' + R_s}. \tag{3.58}$$

Often experimental data deviate from the theoretical relations in so far as a pure capacitive contribution of C_D is not sufficient to describe the true behavior of the circuit in Figure 3.18. Then, phenomenologically a so-called *constant phase element p* is introduced:

$$Z_p = \frac{1}{(i\omega)^p C} \tag{3.59}$$

with the real number $0 \le p \le 1$. Often, p is close to 1. The expressions for R_p and C can be calculated in the same way as in the case without p. However, these expressions are more complex:

$$R_p = \frac{({Z'}^2 - 2Z'R_s + R_s^2 + {Z''}^2)\sin(\chi)}{R_s \sin(\chi) - Z' \sin(\chi) + Z'' \cos(\chi)}, \tag{3.60}$$

$$C = -\frac{Z''}{({Z'}^2 - 2Z'R_s + R_s^2 + {Z'}^2)\exp(1/2p\ln(w^2))\sin(\chi)}. \tag{3.61}$$

with

$$\chi = \frac{1}{2} p \,\mathrm{signum}(w)\pi. \tag{3.62}$$

3.9 Quantization of the electromagnetic field

In order to proceed with the mathematical description of the interaction of light with matter, we have to pursue the quantization of the electromagnetic field. Before doing so, we must move back one step. In the previous section, I introduced the electromagnetic field in the classical way, with the electric field vector $\boldsymbol{E}$ and the magnetic field vector $\boldsymbol{B}$. From classical mechanics we know that the fields are gradients of potentials. Recall that the gradient is the first derivative of a function with respect to the spatial coordinate $\boldsymbol{x}$, or $\boldsymbol{r}$.

We remember from electrostatics that the electric field $\boldsymbol{E}$ is the gradient of the electric potential, Coulomb potential ϕ:

$$\boldsymbol{E} = -\frac{\mathrm{d}\phi}{\mathrm{d}r} = -\operatorname{grad}\phi \tag{3.63}$$

It is a natural question to ask "what then is the potential of the magnetic field $\boldsymbol{B}$?" This was a difficult question because the magnetic field, unlike the electric Coulomb field with positive or negative electric charges, has no sources; $\boldsymbol{B}$ has no divergence! We then introduce as help the so-called vector potential $\boldsymbol{A}$, which is implicitly defined as follows

$$\boldsymbol{B}(\boldsymbol{r}) = \operatorname{rot}\boldsymbol{A}(\boldsymbol{r}) = \nabla \times \boldsymbol{A}(\boldsymbol{r})\,. \tag{3.64}$$

A vector potential can, for example, look like this:

$$\boldsymbol{A}(r, t) = \frac{1}{\omega}\boldsymbol{E}_0 \cos(\boldsymbol{k}(\omega) \cdot \boldsymbol{r} - \omega t) \tag{3.65}$$

It is specific to potentials that they are not considered in absolute terms, absolute values, unless one introduces a calibration reference potential, with respect to which a potential difference can be established. When we calibrate the system by the conditions $\phi = 0$ and $\nabla \cdot \boldsymbol{A} = 0$, which is called Coulomb calibration, then we obtain the wave equation for the vector potential. The wave equation holds also for the vector potential $\boldsymbol{A}$ as the wave variable and looks as follows, provided that we have

$$\nabla^2 \boldsymbol{A}(\boldsymbol{r}, t) - \frac{1}{c^2}\frac{\partial^2}{\partial t^2}\boldsymbol{A}(\boldsymbol{r}, t) = 0 \tag{3.66}$$

Note that this equation describes the electromagnetic field in the absence of charges and currents. For the solution of this equation, it will be convenient to consider the field in a finite volume V, which one can chose in Cartesian coordinates or spherical coordinates. Let us consider a cube with length L for the x, y, and z coordinates, and, thus, $V = L^3$ (Greiner 1989).

We further need to consider boundary conditions for this volume and find it may be convenient when the vector potential $\boldsymbol{A}$ is periodic, with one or two periods equaling the length L of the cube. We then solve the wave equation with the Ansatz of a plane wave of the form $\boldsymbol{A}(x, y, z)\mathrm{e}^{\mathrm{i}\omega t}$.

The "nabla-operator" ∇ (differential operator) is known as the second derivative with respect to the spatial coordinates x, x_i, r, and r_i, and similar. In the wave equation, it also comes with the second derivative of the time t. Some people summarize this and abbreviate it with the new symbol □, which is then called the "delta operator". This step does not involve new mathematics. It is only used to save time and space during calculation.

The wave equation can thus be simplified to

$$\left(\Delta + \frac{\omega^2}{c^2}\right) \boldsymbol{A}(x, y, z) = 0 \tag{3.67}$$

The angular frequency ω and light speed c are related to the wave vector k as follows:

$$\omega_k^2 = c^2 k^2 \tag{3.68}$$

The wave vector $\boldsymbol{k}$ should be fitted in a way that it conforms with the size of the cube with length L; to form the scalar product below correctly we need to also declare L as vector $\boldsymbol{L}$. In the reciprocal space picture, this requires the condition

$$\boldsymbol{k} \cdot \boldsymbol{L} = 2\pi \cdot \{n_1, n_2, n_3\}, n_1, n_2, n_3 \in \mathbf{Z} \tag{3.69}$$

The plane wave that describes the electromagnetic field then has the following structure:

$$\boldsymbol{A}_{\boldsymbol{k}\sigma} = N_k \boldsymbol{\epsilon}_{\boldsymbol{k}\sigma} \mathrm{e}^{\mathrm{i}\boldsymbol{k}\cdot\boldsymbol{x}} \tag{3.70}$$

The index σ denotes 1 and 2 for the two polarization vectors $\boldsymbol{\epsilon}_{\boldsymbol{k}1}$ and $\boldsymbol{\epsilon}_{\boldsymbol{k}2}$, which we set perpendicular to each other much like the electric vector (called the light vector) and the magnetic vector in classical electrodynamics. The wave vector $\boldsymbol{k}$ stands perpendicular to the plane that is spanned by the two polarization vectors. We can call this a condition of transversality, which reads

$$\boldsymbol{\epsilon}_{\boldsymbol{k}\sigma} \cdot \boldsymbol{k} = 0 \tag{3.71}$$

and with the Kronecker delta δ, which is 0 when its arguments are equal:

$$\boldsymbol{\epsilon}_{\boldsymbol{k}\sigma} \cdot \boldsymbol{\epsilon}_{\boldsymbol{k}\sigma'} = \delta_{\sigma\sigma'} \tag{3.72}$$

Next, we write the general field $\boldsymbol{A}$, which can be the vector potential or the electric field and the magnetic field as a Fourier series with following normal modes

$$\boldsymbol{A}(\boldsymbol{x}, t) = \sum_{\boldsymbol{k}, k_z > 0} \sum_{\sigma=1,2} N_k \boldsymbol{\epsilon}_{\boldsymbol{k}\sigma} \left(a_{\boldsymbol{k}\sigma}(t) \mathrm{e}^{\mathrm{i}\boldsymbol{k}\cdot\boldsymbol{x}} + a^*_{\boldsymbol{k}\sigma}(t) \mathrm{e}^{-\mathrm{i}\boldsymbol{k}\cdot\boldsymbol{x}}\right) \tag{3.73}$$

A close inspection and comparison with the previous equations shows that the time-dependent factors $\mathrm{e}^{\mathrm{i}\omega t}$ are no longer obvious; they are now included in the Fourier coefficients $a_{\boldsymbol{k}\sigma}$. We must keep these Fourier coefficients in mind because they will later be the generation operators $a^*_{\boldsymbol{k}\sigma}$ and annihilation operators $a_{\boldsymbol{k}\sigma}$ for the photons.

The Fourier coefficients a of the normal modes A can be considered as the coordinates of the electromagnetic field, and the equations

$$\frac{\mathrm{d}a_{\boldsymbol{k}\sigma}(t)}{\mathrm{d}t} = -\mathrm{i}\omega_k a_{\boldsymbol{k}\sigma}(t) \tag{3.74}$$

can be considered as the equations of motion of the electromagnetic field. They can be derived from a Hamilton function that is identical with the total energy of the electromagnetic field. You may recollect that the energy stored in the electromagnetic field is given by

$$H = \frac{1}{8\pi}\int_{L^3} \mathrm{d}^3x(\boldsymbol{E}^2 + \boldsymbol{B}^2) \tag{3.75}$$

when L is the length of the cube that contains the electromagnetic field.

When you substitute the vectors of the electric and magnetic fields with the corresponding Fourier coefficients a, you can write the total energy of the electromagnetic field as

$$H = \sum_{\boldsymbol{k},\sigma} \frac{\omega_k^2}{4\pi c^2} N_k^2 L^3 \left(a_{\boldsymbol{k}\sigma}a_{\boldsymbol{k}\sigma}^* + a_{\boldsymbol{k}\sigma}^* a_{\boldsymbol{k}\sigma}\right) \tag{3.76}$$

We can make this equation somewhat easier to read when we summarize the prefactor before the parentheses in the sum by

$$N_k = \sqrt{\frac{2\pi\hbar c^2}{L^3\omega_k}} . \tag{3.77}$$

This can be used as normation, normalization for N_k, so that this prefactor becomes unity, and then the equation simplifies to

$$H = \frac{1}{2}\sum_{\boldsymbol{k},\sigma} \hbar\omega_k \left(a_{\boldsymbol{k}\sigma}a_{\boldsymbol{k}\sigma}^* + a_{\boldsymbol{k}\sigma}^* a_{\boldsymbol{k}\sigma}\right) . \tag{3.78}$$

With a keen eye you realize that this equation can be interpreted as the energy of an ensemble of harmonic oscillators. This ensemble of oscillators represents the electromagnetic field – in the vacuum; in the empty space. The frequency of the oscillators is ω_k, and their energy is $\hbar\omega_k$. Be reminded of the relation $E = h\nu$ (ν is the frequency of the light of the photons), and with $\omega = 2\pi\nu$, you then realize the identity of the expressions, because $\hbar$ is defined as $h/2\pi$.

We know already that we can interpret the Fourier coefficients a as creation and annihilation operators. Let us write from now on $a_{\boldsymbol{k}\sigma}^*$ as $\hat{a}_{\boldsymbol{k}\sigma}^\dagger$. Correspondingly, we define that the operator $\hat{a}_{\boldsymbol{k}\sigma}^\dagger$ and its adjunction $\hat{a}_{\boldsymbol{k}\sigma}^\dagger$ satisfy the commutator rule (for adjunction, see Karakostas and Zafiris 2015)

$$[\hat{a}_{\boldsymbol{k}\sigma}, \hat{a}_{\boldsymbol{k}\sigma}^\dagger]_- = \delta_{\boldsymbol{k}\boldsymbol{k}'}, \delta_{\sigma\sigma'} \tag{3.79}$$

As a further requirement we ask that

$$[\hat{a}_{\boldsymbol{k}\sigma}, \hat{a}_{\boldsymbol{k}'\sigma'}]_- = [\hat{a}_{\boldsymbol{k}\sigma'}^\dagger, \hat{a}_{\boldsymbol{k}'\sigma'}^\dagger]_- = 0 . \tag{3.80}$$

As an exercise, try to interpret this relation physically!

The aforementioned definitions now allow that the Hamilton function for the total energy of the electromagnetic field can be written as a Hamilton operator for the radiation field

$$\hat{H} = \sum_{k,\sigma} \hbar\omega_k \left(\hat{a}^\dagger_{k\sigma} \hat{a}_{k\sigma} + \frac{1}{2} \right) \tag{3.81}$$

We can now apply the Schrödinger operator to the creation and annihilation operators a, and this yields

$$-\frac{\hbar}{\mathrm{i}} \frac{\partial \hat{a}_{k\sigma}}{\partial t} = [\hat{a}_{k\sigma}, \hat{H}] = -\mathrm{i}\omega_k \hat{a}_{k\sigma} \tag{3.82}$$

The Schrödinger equation goes with the first derivative to time. This is a deviation from the common understanding that when we look at a wave equation we have the second derivative to time. This discrepancy has been the basis for some researchers to consider the Liesegang rings, spectacular observations made in gels where chemical reactions and diffusion compete and precipitate in a patterned structure, a manifestation of a quantum mechanical nature of effects measurable at the nonmicroscopic scale. Reinhold Fürth discusses the analogy between the formulation for classical diffusion and the Schrödinger equation in Fuerth (1933). See, also, Migulin and Menger (2001); Kuzmenko (1969).

$$-\frac{\hbar}{\mathrm{i}} \frac{\partial \psi}{\partial t} = H\psi \tag{3.83}$$

As the electromagnetic field is composed of the normal oscillations of each oscillator, and we have many of those, the Hamilton operator will certainly be composed by the sum of all individual Hamilton operators of the oscillators. The Zustandsvector must, therefore, be separable with respect to the individual normal modes. The *Zustandsvektor* of the radiation field is, thus, a product of the Zustandsvectors of the electromagnetic field oscillators.

The state of the electromagnetic field oscillator writes as $|n_{k\sigma}\rangle$, and all the many oscillators' states "add up" in the *product* as follows:

$$|\ldots n_{k\sigma} \ldots n_{k'\sigma'} \ldots\rangle = |\ldots\rangle \cdot \ldots \cdot |n_{k\sigma}\rangle \cdot |n_{k'\sigma'}\rangle \cdot \ldots \tag{3.84}$$

We can apply the particle number operator $\hat{N}_{k\sigma}$ on the state $|n_{k\sigma}\rangle$ of the oscillator, which is identical with the annihilation and creation operators a and, thus, reads

$$\hat{a}^\dagger_{k\sigma} \hat{a}_{k\sigma} |n_{k\sigma}\rangle \tag{3.85}$$

With $n_{k\sigma} = 0, 1, 2, \ldots$, we can, thus, formulate a simple eigenwert equation for the particle number operator, $\hat{N}_{k\sigma}$, which reads

$$\hat{N}_{k\sigma} |n_{k\sigma}\rangle = n_{k\sigma} |n_{k\sigma}\rangle \tag{3.86}$$

The state vectors $|\ldots n_{k\sigma} \ldots n_{k'\sigma'} \ldots\rangle$ are also eigenvectors for the Hamilton function H, and we can, thus, write

$$\hat{H} |\ldots n_{k\sigma} \ldots n_{k'\sigma'} \ldots\rangle = E |\ldots n_{k\sigma} \ldots n_{k'\sigma'} \ldots\rangle \tag{3.87}$$

and the energy of the radiation field is given by the sum of all oscillators that emit particles, i.e., photons, given by

$$E = \sum_{\boldsymbol{k},\sigma} \hbar\omega_k \left(n_{\boldsymbol{k}\sigma} + \frac{1}{2} \right) = \sum_{\boldsymbol{k},\sigma} \hbar\omega_k \cdot n_{\boldsymbol{k}\sigma} + \frac{1}{2} \sum_{\boldsymbol{k},\sigma} \hbar\omega_k \tag{3.88}$$

We have to remark here that the energy of the electromagnetic field written in this scheme allows the conclusion that when $n_{\boldsymbol{k}\sigma} = 0$, the energy at zero point (zero-point energy, see Milonni 2009) is $\sum_{\boldsymbol{k},\sigma} (\hbar\omega_k)/2$. Since the index k runs from 0 to ∞, the energy will diverge to infinity, too. The index σ is written in the sum formula, but it does not apply at all, as you can see. This appears to us as an unphysical result. The origin of this strange result is rather the mathematical formalism, and in most physical problems, this zero-point energy will simply not show up.

In analogy to the classical electrodynamics, we can define a Poynting vector for the power, the momentum of the electromagnetic field, which is

$$\boldsymbol{p} = \sum_{\boldsymbol{k}\sigma} \hbar\boldsymbol{k}\hat{a}^{*}_{\boldsymbol{k}\sigma}\hat{a}_{\boldsymbol{k}\sigma} \tag{3.89}$$

for which we can define the momentum operator $\hat{\boldsymbol{p}}$, which, provided that we make the proper corresponding transformations in previous equations,

$$\hat{\boldsymbol{p}} = \sum_{\boldsymbol{k}\sigma} \hbar\boldsymbol{k}\hat{n}_{\boldsymbol{k}\sigma} \tag{3.90}$$

Careful inspection of the previous equations and relations shows the analogy that holds for the momentum operator and yields eigenvalues of $\sum_{\boldsymbol{k}\sigma} \hbar\boldsymbol{k}n_{\boldsymbol{k}\sigma}$.

When we now review all mathematical transformations exercised in this section, we gain a picture about the physical meaning and interpretation of the relevant quantities. The electromagnetic field can be interpreted as a state comprised of $n_{\boldsymbol{k}\sigma}$ photons, which are the electromagnetic field quanta. These photons have a momentum vector $\boldsymbol{p} = \hbar\boldsymbol{k}$ over which an energy of $E = \hbar\omega_k = h\nu$ persists. The number of photons per volume is very large. The operator $\hat{a}^{\dagger}_{\boldsymbol{k}\sigma}$ increases the number of photons of type $\boldsymbol{k}\sigma$ by 1; correspondingly the adjunction decreases the number of photons of that state. This is why they are called the creation operator and the annihilation operator. With every application of these operators, the number of photons increases or decreases accordingly for the particular photon sort. This is why $\hat{n}_{\boldsymbol{k}\sigma}$ is called the number operator. We have thus accomplished the quantization of the electromagnetic field. We can interpret this as the radiation field. Note that this radiation field is described in the empty space, the vacuum, so to speak.

3.9.1 Interaction of the electromagnetic field with matter

Light is scattered by matter. We know this from the experiments where light rays and beams pass slits and double slits. Light does not propagate straightforwardly when it hits on atoms. The interaction of light occurs with the electrons, with the Coulomb field of the electrons of the atoms and ions and molecules. We must, therefore, investigate the interaction of an electromagnetic field with a Coulomb field.

3.9.2 Feynman rules

For the transition matrix element, the following Feynman rules apply (Schilcher 2019). First, sketch all connected and topologically nonequivalent diagrams. Second, for every internal electron, you add one factor

$$\frac{\mathrm{i}}{\not{p} - m + \mathrm{i}\epsilon} = \frac{\mathrm{i}(\not{p} + m)}{p^2 - m^2 + \mathrm{i}\epsilon} . \tag{3.91}$$

Third, for every line representing a photon, add the factor

$$\frac{\mathrm{i}}{k^2 + \mathrm{i}\epsilon}\left[-g_{\mu\nu} + (1-a)\frac{k_\mu k_\nu}{k^2 + \mathrm{i}\epsilon}\right] . \tag{3.92}$$

Fourth, add a factor $-ie\gamma_\mu$ for any vertex. Fifth, for the outer lines the following factors are added: for an incoming electron e^-, the factor $u^{(r)}(p)$, and for the outgoing electron, the factor $\bar{u}^{(r)}(p)$. Accordingly, for positive charges, such as positrons e^+, we write $v^{(r)}(p)$ and $\bar{v}^{(r)}(p)$. The same would hold for electron holes h^+. An incoming photon will need the factor $\epsilon_\mu^{(s)}(k)$, and an outgoing photon the factor $\epsilon_\mu^{(s)*}(k)$.

For a closed Fermion loop, this is the sixth rule, a factor of (−1) is applied. It is an exercise for the reader to find out why this is required. The Wick contraction might give you a hint. The seventh rule brings in another factor (−1) among two diagrams, when they differ only by the exchange of electrons as a result of the Fermi statistics.

Finally, if there are loop momenta, we need to carry out an integration for every such loop as $\int (\mathrm{d}^4k)/(2\pi^4)$. As long as the momentum conservation is warranted in the equations, the direction of the momentum does not matter.

3.9.3 Examples of Feynman graphs

The Feynman diagrams (see *Space–Time Approach to Quantum Electrodynamics*, Feynman 1949) are typically linked as graphical representations of reactions between fermions and electromagnetic waves in quantum electrodynamics. On the cover page of this book, I have simply put some graphs on the photo with the sunflowers in order to make it appear some more mathematical. The graphs may look arbitrary and schematic only, but in fact, the Feynman diagram is a genuine x–t-diagram, which records the location of an event versus the time of an event. We would naturally think that the distance on the spatial coordinate x is somehow in the microscopic range.

As the Feynman graphs are an application of graph theory, a branch of mathematics, there should be no fundamental reason that the application of the theory would be limited to the microscopic scale (scale invariance). The mathematical formalism should be the same. An application of the Feynman graph theory can, for example, be found in a paper on the interaction of seismic waves and ocean waves and currents by Hasselmann, which he published in *Reviews of Geophysics* (Hasselmann 1966), a journal that we would not immediately associate with the submicron scale. The scale, however, is hardly ever explicitly or quantitatively mentioned in the Feynman diagrams. The distance of events is also not given in Hasselmann's graphs. We do not know if it is 10^{-10}, 10^{-3}, or 10^3 m.

3.9.4 Diagrams for changes of electronic state upon absorption and emission

Before we come to the Feynman diagrams, which explicitly mentions space and time, we discuss another representation, specifically, the absorption and emission of light and the corresponding change of the electronic states of electrons.

Figure 3.19 illustrates the absorption process of a photon by an atom, where the energy levels of an electron are considered, but where the momentum vector $\boldsymbol{k}$ (which relates to the energy $E = h\nu = \hbar\omega_k$) is completely ignored. The beam of light, the

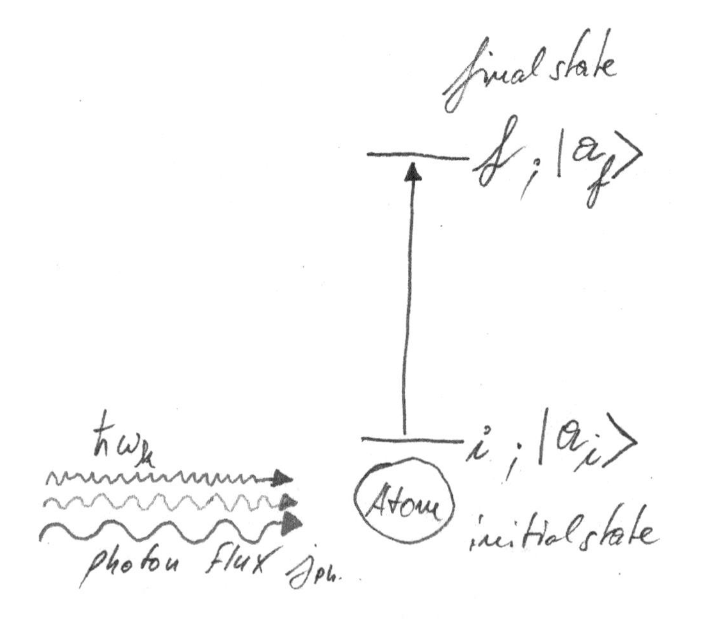

Fig. 3.19: Matter is irradiated with light. An atom will absorb the radiation by virtue of its electrons, which become excited from an initial state $|a_i\rangle$ with a particular energy to a final state $|a_f\rangle$ with higher energy. However, only particular photon energies and wavelengths can be absorbed by electrons with specific states.

photon flux j_{ph}, is represented by wiggly arrows on the left-hand side, the photons that come in as a light wave, as an electromagnetic wave. The blue arrow has denser wiggles and stands for a high-energy photon, corresponding to a short wavelength, in contrast to the red arrow with fewer wiggles per arrow length and a longer wavelength, which means a lower energy photon. The green arrow in between has an intermediate photon energy and corresponding wavelength. The atom absorbs the radiation by virtue of its electrons, which become excited from an initial state $|a_i\rangle$ with a particular energy to a final state $|a_f\rangle$ with higher energy. However, only particular photon energies and wavelengths can be absorbed by electrons with specific states.

Provided that the photon has the proper energy, it will be absorbed by an appropriate electron of the atom, which has an initial state $|a_i\rangle$. With the energy of the photon, the electron will rise to the initial state $|a_f\rangle$. In common terminology, we say that the atom is excited, or the molecule, and not the electron. Excitation requires an excitation energy.

Figure 3.20 is an illustration of the absorption and the subsequent[16] re-emission of a photon by an electron with initial state level i and final state level f. The left panel

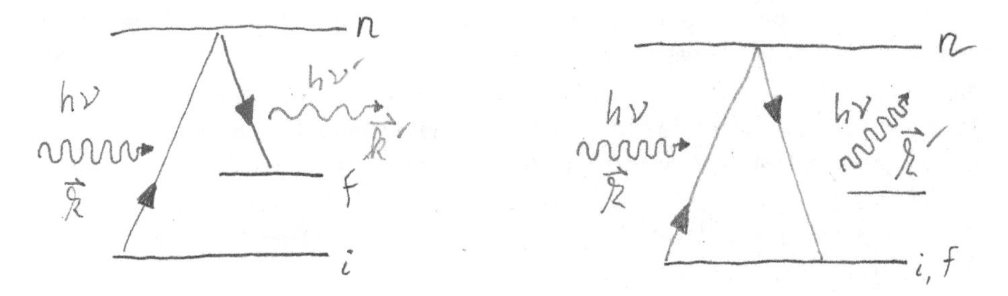

Fig. 3.20: Left: a photon with energy $h\nu$ (blue wiggly arrow) and momentum k-vector $\boldsymbol{k}$ excites an electron from the initial state i to excited state n and then relaxes to a final state f with emission of a photon with lower energy $h\nu'$ (green wiggly arrow) and momentum $\boldsymbol{k}'$. This is the process of inelastic scattering. Right: a photon with energy $h\nu$ (blue wiggly arrow) and momentum k-vector $\boldsymbol{k}$ excites an electron from the initial state i to excited state n and then relaxes to a final state f, which is identical with the initial state i with emission of a photon with the same energy $h\nu'$ (blue wiggly arrow) and momentum $\boldsymbol{k}'$. This is elastic scattering, where the momentum transfer is k-vector $\boldsymbol{k} - \boldsymbol{k}' \neq 0$, but the energy transfer is $h\nu - h\nu' = 0$.

16 The term "subsequent" implies that we first have the absorption before the re-emission can take place. I am not so sure whether this classical causalism is required or valid. Think of whether there is a tunneling-like process possible, where the atom senses the photon coming and emits already a photon before the primary photon arrives. Compare it to a bank, which issues a loan to some person B before some person a comes with a savings account deposit. A somewhat related issue was brought up by Slater in a Comment published in Slater (1925b); this issue concerned the potential emission of a virtual field prior to emission of a corpuscle. Slater begins his Comment, titled *The Nature of Radiation*, with the remark that Bothe and Geiger measured that the recoil electron and the photoelectron appeared simultaneously.

shows the absorption of a photon with wave vector k and energy $h\nu$, and the absorber electron is excited to state n. From state n, the electron will emit a photon with wave vector k' and energy $h\nu'$. This is inelastic photon scattering, such as Raman scattering. Note that, here, the energy and the momentum vector are changed during the processes. The right panel shows an example of elastic photon scattering, where the electron is excited from level i to level n by absorption of a photon, and the excited state decays to the final state f, which has the same level as the initial state i. The momentum vector of the photon changes but the energy remains the same.

3.9.5 Jablonski diagrams

Experiments show that absorption and fluorescence can occur between various ground states and excited states. When you do spectroscopy experiments, you plot the spectral intensity against the energy (or wavelength, or frequency, or wave numbers). Transitions show up as intensity peaks, and those peaks have a spectroscopic origin that can be plotted and rationalized in the Jablonski diagram (Frackowiak 1988).

An excited atomic or molecular state is not an equilibrium state, and the excited state can relax to a state with lower energy. There may be metastable states that discharge fast, others have a longer lifetime. There are many ways by which the energy state of an atom or molecule can be changed. These can be shown in a Jablonski diagram (Jablonski 2013). The Jablonski diagram has one axis, and this is the vertical energy axis. Figure 3.21 shows three energy levels. The lowest level is ${}^{1}\mathrm{A}$, from which the system is excited to level ${}^{1}\mathrm{A}^{*}$, as shown by the upward arrow. You may have a fluorescence process from energy level ${}^{1}\mathrm{A}^{*}$ back to level ${}^{1}\mathrm{A}$.

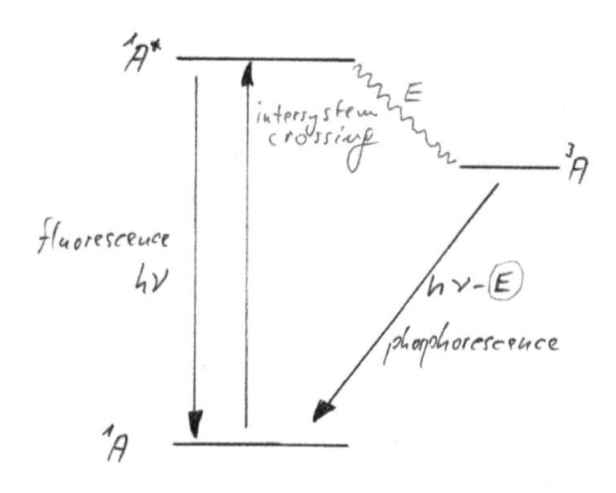

Fig. 3.21: A Jablonski diagram showing the excitation of molecule A to its single excited state (1A*) followed by intersystem crossing to the triplet state (3A) that relaxes to the ground state by phosphorescence.

It may be possible that part of the excitation energy is passed on to donor levels in the system (summarized as S_1), not by photons but by vibration energy or electromagnetic field energy (Figure 3.22). Such nonradiative transitions from the excited energy

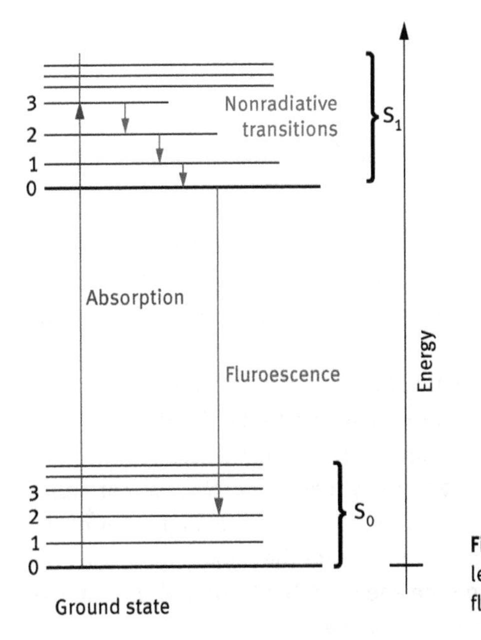

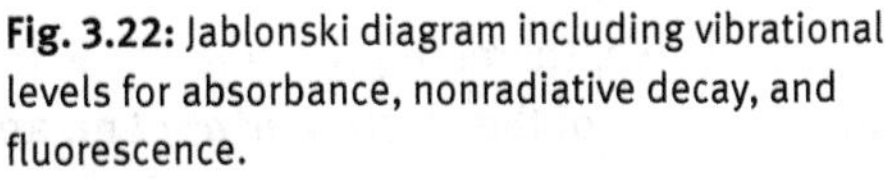
Fig. 3.22: Jablonski diagram including vibrational levels for absorbance, nonradiative decay, and fluorescence.

level may be the reason for a number fluorescence processes to energy levels near the ground state, summarized as S_0 with release of photons with different energy.

However, there may be also situations where the radiation system, the acceptor and donor, and the molecule may have a state that is available for a transition, which is known as *intersystem crossing*. By the release of an amount of energy E from the excited level ${}^1A^*$, the lower level ${}^3A^*$ becomes excited and can then release a photon with energy $h\nu - E$, and this process is known as phosphorescence.

Today, the Jablonski diagram is an established graphical tool for the illustration of energy transitions in molecules and condensed matter. Modern laser-optical methods and also synchrotron and free electron laser-based experiments provide optical spectra and x-ray spectra with a very high time resolution in the femtosecond (fs) time domain. It is, thus, possible to experimentally observe and then sketch a wide variety of hitherto unknown metastable states relevant for molecular photophysics (Kasha 1999).

3.9.6 Photoelectric effect for the hydrogen atom

Let us look at an exercise for the emission of an electron by the hydrogen atom upon photoelectric stimulation, as demonstrated in the book by Walter Greiner (Greiner 1989). This is only one of the many situations where an electric current can be produced by irradiation with light, provided that the photon energy is large enough. For this to calculate, we need to begin with the transition probability, which per time is given by

$$\frac{2\pi}{\hbar}\left(\frac{e}{mc}\right)^2\left(\frac{2\pi\hbar c}{L^3\omega_k}\right)n_{k\sigma}\left|\langle a|\,\hat{\boldsymbol{p}}\cdot\boldsymbol{\epsilon}_{k\sigma}\mathrm{e}^{\mathrm{i}\boldsymbol{k}\boldsymbol{r}}\,|b\rangle\right|^2\delta(E_b+\hbar\omega_k-E_a) \tag{3.93}$$

This is a product of three terms, which includes as the main factor the square of the matrix element, $|M|^2$ for the transition from an initial state $|a\rangle$ to a final state $|b\rangle$. On the left-hand side, there is a number of constants, and on the right-hand side a δ function, the argument of which contains the energies.

The quantitative goal in this exercise is to find the differential scattering cross section $\mathrm{d}\sigma$ for the emission of the electron into the angular segment $\mathrm{d}\Omega$: $\mathrm{d}\sigma/\mathrm{d}\Omega$.

For the $|b\rangle$ state, we take the 1s-orbital of the hydrogen atom, which is $|1\mathrm{s}\rangle$ with the wavefunction representation $\psi_{1\mathrm{s}}(\boldsymbol{r}) = \mathrm{e}^{-r/a}$ and the energy $E_b = E_{1\mathrm{s}}$. We normalize this wavefunction by dividing it by the square root of $\sqrt{a^3}$. In the argument of the δ function, we, therefore, the sum of the three energies for the photon $h\nu = \hbar\omega_k$, the kinetic energy of the electron $\hbar^2 q^2/2\,\mathrm{m}$, and the energy of the 1s-orbital, $E_{1\mathrm{s}}$. Where these become 0, the δ-function produces the scattering event.

For $|a\rangle$, we have a plane wave of the form $\psi_q = \mathrm{e}^{\mathrm{i}kr}$, which we normalize by dividing it by the square root of the volume element (or voxel) L^3. The energy of the plane wave is $E_a = \hbar^2 q^2/2\,\mathrm{m}$.

The process is illustrated in Figure 3.23, where the photon comes in as wave with momentum vector $\boldsymbol{k}$, hits the atom, and kicks out an electron from the 1s-orbital. This is an ionization process because the neutral hydrogen atom will become a proton after the electron emission. The trick used here in this exercise is that we assume the incident wave as a high energetic so that it can be considered for the electron wave as a plane wave.

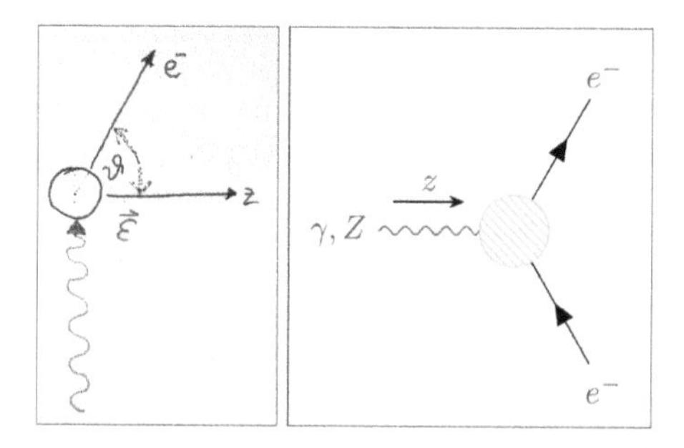

Fig. 3.23: Demonstration of the geometry for the photoelectric emission of an electron. The photon with wave vector $\boldsymbol{k}$ hits the atom, which releases, actually emits an electron under the angle θ versus the spatial coordinate z, where the polarization vector $\boldsymbol{\epsilon}$ points; this shall be our choice in the coordinate system for convenience, in the z-direction. On the right you can see the corresponding Feynman diagram, which was built with the TikZ-Feynman package (Ellis 2017).

We can nicely illustrate this also with the TikZ-Feynman package (Ellis 2017) in LaTeX, as is shown in Figure 3.23 and in the LaTeXsource code below:

```
\tikzfeynmanset{
every vertex={red, dot},
every particle={blue},
every blob={draw=green!40!yellow!, pattern color=green!40!yellow},
}

\feynmandiagram [horizontal=a to b] {
a [particle={\(\gamma, Z\)}] -- [boson, momentum=\(z\)] b [blob],
c [particle=\(e^-\)] -- [fermion] b -- [fermion] d [particle=\(e^-\)],
}.
```

3.9.7 Photons and the "free" electron

In a quantum mechanics course or an atom physics class you may frequently hear about "forbidden transitions". This is harsh language, similar to the terms "master" and "slave" in logic circuit boards for computing. Harsh language. But what about the "forbidden transitions"? This is the case for the free electron, for example. Free electrons cannot absorb and cannot emit photons. An electron must be *kept* in some orbital or some force field so that one can force it to accept or release a photon. Figure 3.24 shows the situation where a photon with wave vector $\boldsymbol{k}_i$ hits an electron that has an initial state q_{initial}. After the impact, the electron has the final state q_{final}, but the photon is not absorbed. This is an elastic scattering process.

In physics, we require momentum conservation, which for the problem in Figure 3.24 reads

$$\hbar \boldsymbol{q}_{\text{final}} = \hbar \boldsymbol{q}_{\text{initial}} + \hbar \boldsymbol{k}_i \tag{3.94}$$

and also the conservation of the energy, $E_{\text{final}} = E_{\text{initial}}$, which we derive as transition probability by Fermi's golden rule

$$\frac{(\hbar \boldsymbol{q}_{\text{final}})^2}{2m_e} = \frac{(\hbar \boldsymbol{q}_{\text{initial}})^2}{2m_e} + \hbar\omega_k \,. \tag{3.95}$$

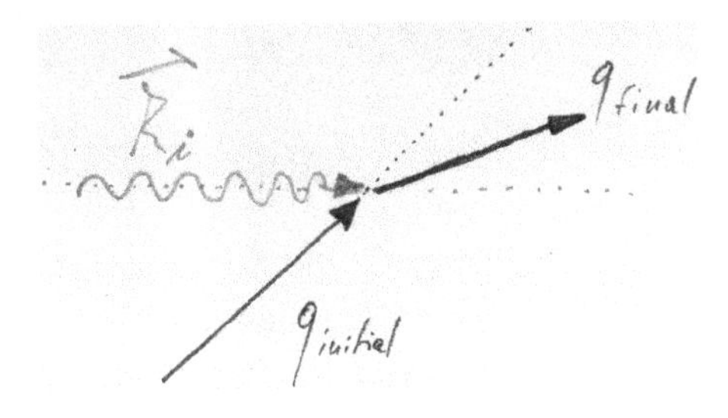

Fig. 3.24: The photon with wave vector $\boldsymbol{k}_i$ hits a free electron with state q_{initial}, which gets excited to a final state q_{final} without absorption of the photon.

It is an exercise for the reader to show that Equations 3.94 and 3.95 have no solution. As a hint, you can argue by using the "four-momentum" $p = (p^0, p^1, p^2, p^3) = (E/c, p_x, p_y, p_z)$, where you use the momentum for the electron, p_e and for the photon, p_γ. In the end, you arrive at the result $E_\gamma = 0$, the correct interpretation of which is that there is no photon absorption possible. The photon is scattered by the free electron but not absorbed. Hence, we are dealing here not with an inelastic scattering process, but with an elastic scattering process.

I need to add a remark about the condition of the free electron. Some readers may know about particle accelerators, such as the Stanford Linear Accelerator Center (SLAC), where electrons are accelerated on a linear racing track by the use of strong electric fields. When the high-speed electrons hit the target, they become decelerated, and this causes them to emit Bremsstrahlung. Here, the electrons are *kept* in a decelerating field, which makes them release their high kinetic energy by emitting high energy photons. This "white light" electromagnetic radiation is not to be mistaken for the photons that get released when a particular electron in the target is excited to a higher energy state, which is the element specific characteristic radiation. Similar holds for circular storage rings, in which high-energy electric fields accelerate electrons, and magnetic fields force them onto a circular track, at which synchrotron radiation is produced.

3.9.8 Inelastic scattering of a photon by an electron

There are situations where an object is illuminated with light of one color, and you notice it will shine back in a different color, with a larger wavelength and lower energy. This is observed in Raman scattering and in Brillouin scattering. Figure 3.25 shows the Feynman diagram of a process where an electron with state $|q_i\rangle$ is hit by a photon with the wave vector $\boldsymbol{k}_i$ and absorbs this very photon. After the collision, the electron has the state $|q_f\rangle$ and emits a photon with a wave vector $\boldsymbol{k}_f$. This is an inelastic scattering process.

The process can be conceptually described by Schilcher (2019)

$$\nu(k_i) + e^-(q_i) \rightarrow \nu(k_f) + e^-(q_f) \tag{3.96}$$

and further quantum mechanically formulated by

$$\langle \nu e^- | S | \nu e^- \rangle = \frac{1}{2}(-ie)^2 \int d^4x\, d^4x' \, \langle p_2, \eta_2 | T(: \bar{\Psi}\nu^\mu \Psi :: \bar{\Psi}'\nu^\nu \nu \Psi' :) | p_1, \eta_1 \rangle$$
$$\times \langle k_2, \sigma_2 | T(A_\mu A'_\nu) | k_1, \sigma_1 \rangle \tag{3.97}$$

There, $|p, \eta\rangle$ is a single electron state with four-momentum p and the spin eigenvalue η, and $|k, \sigma\rangle$ is a single-photon state with four-momentum k and polarization σ. The Raman process also falls in the category of the scattering of a photon by an electron and can be treated in a similar way with Feynman diagrams (Yariv 1977).

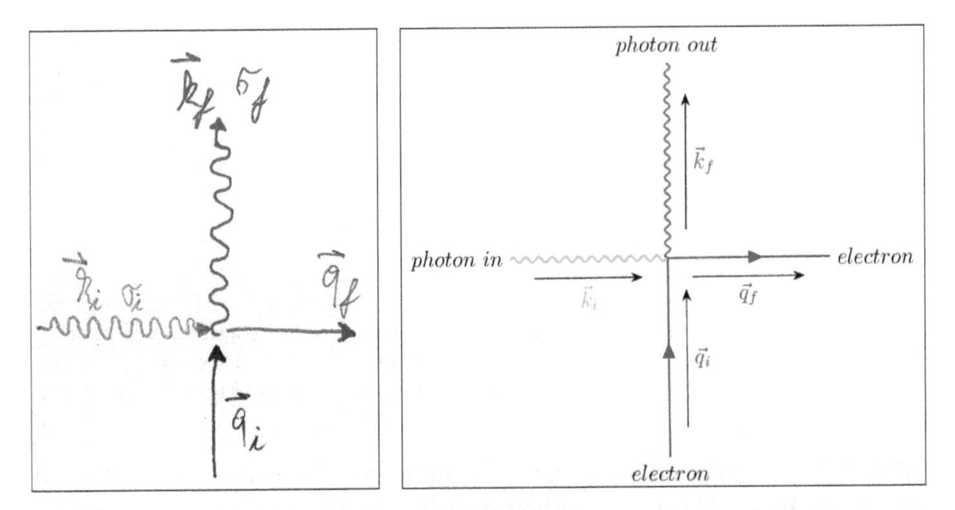

Fig. 3.25: Feynman diagram of a complex scattering problem where an electron with initial state $|q_i\rangle$ (black) absorbs the photon with wave vector $\boldsymbol{k}_i$ (green) and becomes scattered into the excited state $|q_f\rangle$ (blue), while it emits a photon with wave vector $\boldsymbol{k}_f$ (red). On the right-hand side you can see the Feynman diagram sketched with the TikZ-Feynman package (Ellis 2017).

3.9.9 Virtual processes and line width

In Section 3.11 on absorption spectroscopy, we learnt that the absorption peaks are practically not at one discrete energy or wavelength. Rather, the peaks have a finite width, which depends on the design of the spectrograph. The same holds for the emission peaks. However, even if the spectrographs would be so perfectly designed that they allowed for sharp peaks with a $\Delta\omega = 0$, we would find that the peaks still have some actual finite width. The very sharp lines are mathematically expressed with a delta function of the form $\delta(E_{a_i} - E_{a_f} - \hbar\omega_k)$ for the transition probability[17]. The reason for this is more of a fundamental physics nature, which is understood as the natural line width[18].

When a photon is emitted, the corresponding wave has a limited duration and, thus, a limited length. This is because the excited state only has a limited lifetime τ. According to Heisenberg's uncertainty principle, the limited lifetime τ thus amounts to an energy width of $\Delta E \approx \hbar/\tau$. This is the fundamental reason why emission lines are not infinitely sharp, but broad. In the Jablonski diagram, we, therefore, should be aware that the horizontal lines that signify the energy levels are not meant to be

17 We recall the paper of J.C. Slater from 1925, where he shows how to determine the transition probabilities from an absorption spectrum (Slater 1925a).

18 Here, I follow a treatment that you can look up in the textbook of Walter Greiner (Greiner 1989).

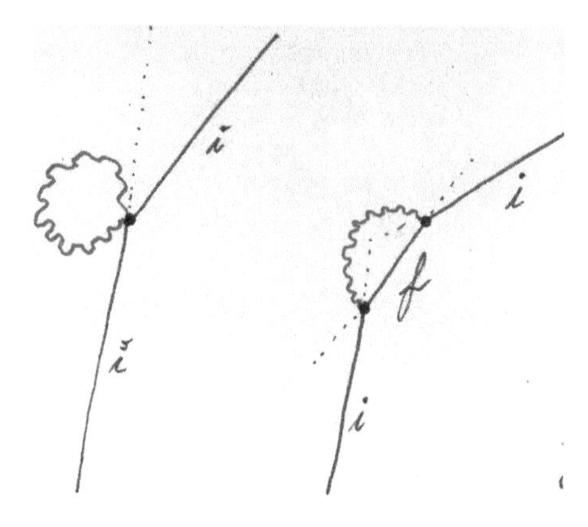

Fig. 3.26: Two paths, graphs, for the self-energy of a bound electron. The wave lines denote virtual photons, background fluctuation photon fields. A photon is emitted and then reabsorbed.

infinitely thin, but should have some finite thickness, some energy width along the energy axis. We would not see these horizontal lines in the sketch anyway, if they were infinitely sharp.

Figure 3.26 shows two Feynman diagrams for the self-energy of a bound electron. On the left-hand side, we see only two initial states $\langle i|$ (or should we write $|i\rangle$ as the final state?), and no final state $|f\rangle$. This is relevant for the emission and re-absorption of a photon by an electron. The energy conservation law need not be satisfied by this kind of process, which is then called (Greiner 1989) a *virtual process*[19]. We can try to imagine this as an intermediate spontaneous dissociation of an actual physical atom in an atom and a virtual photon. All potential momentum and polarizations of the atom that correlate with the corresponding situations of the atom account for an energy shift, a spread between E_f and E_i, which is visualized in Figure 3.27.

Eugene Wigner and Victor F. Weisskopf were the first to "work in" the line width into the quantum mechanical formalism (Weisskopf and Wigner 1930b). It is a result of the energy shift ΔE_i, which is a complex quantity with a real part and an imaginary part. Here, I will follow the formulation as exercised in (Greiner 1989)

$$\mathrm{Re}(\Delta E_i) = \sum_{\boldsymbol{k},\sigma} \sum_{f} \frac{|\langle f \boldsymbol{k} \sigma | \hat{H}'_{\mathrm{int}} | i_0 \rangle|^2}{E_i - E_f - \hbar\omega_k} \tag{3.98}$$

$$\mathrm{Im}(\Delta E_i) = -\pi \sum_{\boldsymbol{k},\sigma} \sum_{f} |\langle f \boldsymbol{k} \sigma | \hat{H}'_{\mathrm{int}} | i_0 \rangle|^2 \delta(E_i - E_f - \hbar\omega_k) \langle a' | | a \rangle \tag{3.99}$$

As the summation over the multiple final states in Equation 3.98 is not limited, the law of conservation of energy does not apply. We, therefore, cannot consider this a

19 The conservation law seems to be so intuitive in human perception that it has been considered in the development of science, although it has not necessarily been formulated, see, for example, Katzir (2019).

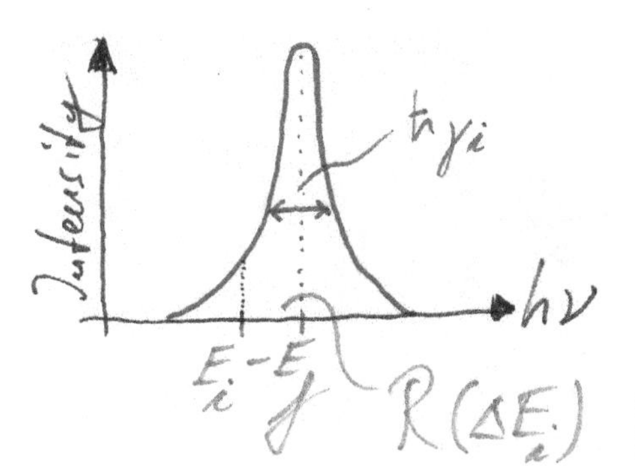

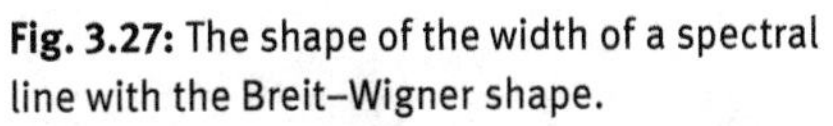

Fig. 3.27: The shape of the width of a spectral line with the Breit–Wigner shape.

real process, but a *virtual* process, which we can imagine as an atom that dissociates in an atom plus a virtual photon. These virtual photons are not specified in terms of momentum and polarization. Consequently, this can be considered a broadening ΔE around the central energy E_i.

The summation in Equation 3.99, however, is limited by the condition that the energy conservation $E_\mathrm{i} = E_\mathrm{f} + \hbar\omega$ be satisfied. These photons are real, in contrast to the aforementioned virtual photons. The shape of the line width was calculated exemplarily for the harmonic oscillator by Weisskopf and Wigner (1930a) and reads

$$w(E)\,\mathrm{d}E = \frac{1}{h}\frac{\gamma^A\,\mathrm{d}E}{\left(\frac{1}{2}\gamma^A\right)^2 + 4\pi^2(E - E_\mathrm{f})^2} \tag{3.100}$$

$$J(\nu)\,\mathrm{d}\nu = \frac{\gamma^A_{B'}\,\mathrm{d}\nu}{\left(\frac{1}{2}\gamma^A\right)^2 + 4\pi^2(\nu - \nu^A_{B'})^2} \tag{3.101}$$

Figure 3.27 shows a plot of this Breit–Wigner formula for the natural line width from photoemission. The photoemission intensity is plotted versus the energy $E = h\nu$, and the width is given as $\hbar\gamma_\mathrm{i}$.

3.9.10 Exercise: two-photon decay of the 2s state in H

The 2s state is one of the excited states of the hydrogen atom. Figure 3.28 displays a photo that I took on a dark night in the summer of 2018, on 13 August 2017 at 01:15. I used a so-called fish-eye objective, which captures the image over a very wide angle and also captures enough light at night. The lens focus was 10.5 mm with an opening aperture of 2.8 and 30 seconds of illumination time, and ISO 2000 sensitivity. At that time, the occurrence of shooting stars was announced. In the middle of the photo you can see a yellow thin circle, which encircles the Andromeda nebula, a diffuse oval-shaped spot. The nebula contains huge amounts of hydrogen, which emits several spectroscopic signatures, including the decay of excited states of hydrogen. A very

Fig. 3.28: Photo of the Milky Way. The shape enclosed by a wide thin yellow line denotes part of the Milky Way visible at night at 01:15 am. Captured with Nikon D7500, FishEye lense f/2.8, ISO 1600. Photo 13 August 2018.

thin line across the circle shows a shooting star. Another yellow circle between the trees shows another shooting star trace. The wide yellow encircled region denotes our Milky Way galaxy.

About 370,000 years after the Big Bang (Chapter 6), electrons and protons in the universe would chemically react to form neutral hydrogen atoms, which is called recombination (Tanabashi 2018). The hydrogen from this era was in an excited state and would discharge its energy by the release of a Lyman α series photon from the 2p state.

There exist, however, an alternative path with the release of two photons from the 2s level. This latter process is, however, very slow (Chluba and Sunyaev 2008; Tung et al. 1984).

This is the light that we as astronomers see from deep-space objects like the Andromeda galaxy or the Orion nebula, when we look up into the dark starry night, either with the naked eye or with optical instruments. With an astronomical telescope and a spectrograph, one can record the spectrum of the objects, which allows for some chemical analysis. The speed at which the source of the light moves, will give rise to Doppler effects and, thus, helps us to measure times and distances in the universe.

80 years ago, Breit and Teller pointed out that the electric field existing in the interstellar space, though extremely small, would have a significant effect on the lifetime of the 2s excited state (Breit and Teller 1940). For the lifetime of a state with two involved photons $h\nu$, we write (Greiner 1989)

$$\frac{1}{\tau} = \sum_{\text{final states}} \frac{2\pi}{\hbar} \left| \langle 1\mathrm{s} | \hat{H}_\mathrm{I} | 2\mathrm{s} \rangle \right|^2 \delta(E_{2\mathrm{s}} - E_{1\mathrm{s}} - h\nu_1 - h\nu_2) \tag{3.102}$$

where the Hamiltonian is combined from the first and second-order perturbation theory contributions, $\hat{H}_\mathrm{I} = \hat{H}' + \hat{H}''$. For example, for the two-photon generation process, we have

$$\hat{H}'' = \frac{e^2}{2mc^2} \sum_{\boldsymbol{k},\sigma,\boldsymbol{k}',\sigma'} \left(\frac{2\pi\hbar c^2}{L^3} \right) \frac{\boldsymbol{\epsilon}_{\boldsymbol{k}\sigma} \cdot \boldsymbol{\epsilon}_{\boldsymbol{k}'\sigma'}}{\sqrt{\omega_k \omega_k'}} \hat{a}^\dagger_{\boldsymbol{k}\sigma} \hat{a}^\dagger_{\boldsymbol{k}'\sigma'} \mathrm{e}^{-\mathrm{i}(\boldsymbol{x}_1 + \boldsymbol{x}_2)\cdot \boldsymbol{x}} \tag{3.103}$$

for which we only need to consider the matrix element $\langle f | \hat{H}'' | i \rangle$ in first-order perturbation theory. However, the Hamiltonian $\hat{H}'$ produces only one photon. As a side note, $\omega = ck$ holds for the circular frequency, and the classical electron radius is $r_\mathrm{e} = e^2/(mc^2)$. It is, therefore, necessary to produce photon number 2 after photon number 1, which requires a second-order process in the perturbation theory (Greiner 1989). The chronology for the production of photon 1 and photon 2 in this two-photon process is certainly important and is illustrated in the Feynman diagram shown in Figure 3.29, as the consequence of two contributions from second-order perturbation theory.

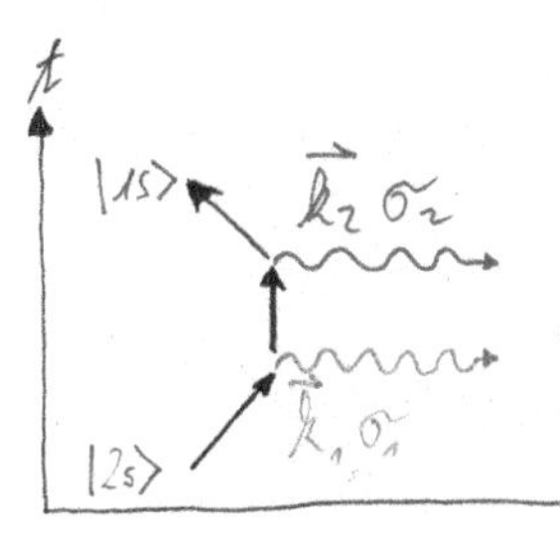

Fig. 3.29: Feynman diagram for the two-photon decay of the 2s-state in hydrogen. The excited state $|2\mathrm{s}\rangle$ decays to the ground state $|1\mathrm{s}\rangle$ by sequentially emitting two photons with wave vectors $\boldsymbol{k}_1$ and $\boldsymbol{k}_2$.

The time axis t goes in the vertical direction and begins with the excited state $\langle 2s|$, and releases one photon with momentum $\boldsymbol{k}_1$ and polarization σ_1, as indicated by the horizontal wiggly line. This is followed by release of the second photon with momentum vector $\boldsymbol{k}_2$ and polarization σ_2; now the hydrogen atom is in the state $\langle 1s|$.

After a number of simplifications and approximations, Equation 3.102 becomes

$$\frac{1}{\tau} = \frac{r_e^2}{2\pi} \int k_1 \, dk_1 \int k_2 \, dk_2 \delta(k_0 - k_1 - k_2) = \frac{r_e^2 c k_0^3}{12\pi} \tag{3.104}$$

The energy difference $E_{2s} - E_{1s}$ in Equation 3.102 is related to k_0 to $\hbar c k_0$. The lifetime for the 2s-state in excited hydrogen turns out to be $\tau_H = 0.114$ sec in good approximation, where the matrix element from $\hat{H}'$ was not fully calculated. With fully taking into account all contributions, the lifetime of the excited 2s-state in an element with atom number Z reads $\tau = 0.1216/Z$ sec (Shapiro and Breit 1959).

3.10 Path integral formalism

In this section, we come to an important formalism and theory of quantum electrodyamics: the path integral formalism. The central physical quantity for this formalism is the action S, and the action integral, or better, action functional, which is defined as

$$S = \int_{t_a}^{t_b} dt' L(x(t'), \dot{x}(t'), t') \, . \tag{3.105}$$

Because the integration borders are not finite, we call it a functional and not an integral. Although you integrate the function, the integral is then by itself again a function(al). Some may consider this just semantics, but it is important during an integration with a constant to realize whether we are dealing with a parameter or with a variable. We will see this further down below in this section.

L is here the Lagrange function, which we know from classical mechanics and which contains basically the spatial temporal details of the trajectory of a moving body. Look at Figure 3.30, which shows an x, t diagram. Actually, it is a t, x diagram[20], which shows the abscissa as the time axis and the ordinate of the spatial coordinate. We are, thus, looking at data points (t_i, x_i). Usually, particularly in experimental physics, the time axis is the ordinate, and the spatial coordinate is the abscissa. In the Feynman diagrams, the time axis is usually shown vertically upwards. However, I think we are not bound to any convention here. What we shall be looking at first

20 We learn in physics class in middle school how the distance is plotted on the abscissa versus the time in the ordinate axis, when we define the velocity of an object in kinetics or mechanics. This appears to be a typical convention. In quantum mechanics, we typically plot the time on the abscissa, and the spatial coordinate is the ordinate axis.

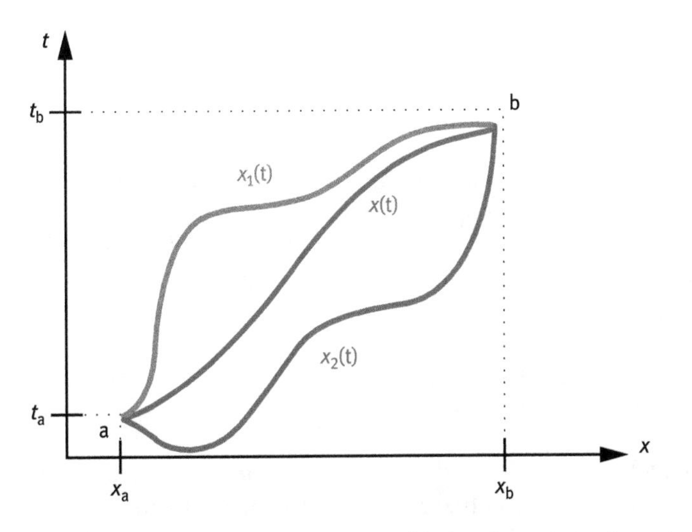

Fig. 3.30: The so-called classical path $x(t)$ (blue) and two alternative paths $x_1(t)$ (green) and $x_2(t)$ (red). There can be a manifold of alternative paths.

are the two locations a and b, which shall be the positions of an object, a particle. At time t_a, the particle is at position x_a, and at time t_b, it is at position t_b. The blue line $x(t)$ shows the trajectory of the particle from position a to b.

In real life, this may be the problem of finding a path from point a to point b. Let us say that we want to go from Zürich to Lugano, for which we have to cross the Alps. All ways may lead to Rome. According to Wilhelm Tell, there was only one way to the town of Küssnacht. Moreover, there are less than a handful of ways from Zürich to Lugano, one directly over or under the Gotthard Pass, one along the Rhein Valley over the San Bernardino Pass, and one along the Rhone Valley over the Nufenen Pass. Geography shows which path nature, animals and men, has taken over thousands of years. Some are likely to be taken, some are unlikely to be taken.

Figure 3.30 thus shows an alternative red path $x_1(t)$ and an alternative green path $x_2(t)$. Let us call the blue path the classical path, which would not be, if you do not mind, the straight way over the highway through the 17 kilometer long Gotthard tunnel, but the curvy road over the Gotthard Pass.

An electric current from lightning will also be "looking" for the best way from the sky to the ground, and as an experienced lightning watcher you know that the lightning currents can be very curvy. There may be countless alternatives and ways for a trajectory, but the classical trajectory is the one for which the action integral is minimum, the condition for which is given as

$$\frac{\partial S}{\partial x} = 0 \,. \tag{3.106}$$

In classical mechanics, we consider the Euler–Lagrange equation

$$\frac{\partial L}{\partial x} - \frac{\mathrm{d}}{\mathrm{d}t}\frac{\partial L}{\partial \dot{x}} = 0\,. \tag{3.107}$$

The path integral formalism has historically been important for the development of quantum electrodynamics. In one of the popular books about Feynman's life (It may have been "Surely you're joking, Mr. Feynman", or "The pleasure of finding things out"), it is quoted that he found a book in the library, which was about calculus, which made him excited (Baldwin 2017):

> When he found a subject that interested him, he was not about to wait for the right teacher to come along; he was determined to master it himself.

That book was authored by Thompson and had the title *Calculus for the Practical Man*, and you can find it at www.archive.org for free download (Thompson 1946).

The application of the path integral can be demonstrated with the double-slit experiment, where waves can interfere and form an interference pattern. The path integral is not specific to quantum mechanics, however; it is the summing up of all path elements in a trajectory. It measures the length of a curve (in German: *Linienintegral*), whereas we usually know the integral for measuring the area under a curve (in German: *Flächenintegral*). It may not be relevant for the topic of this book on photosynthesis, but I want to mention here a paper of mine (Braun 2003), where I wanted to calculate a path integral from the experimental data of ultrathin epitaxial films that I had gathered in my diploma thesis research in 1995. This exercise was the conversion of a distance that was given in atomic monolayers to a distance required in Ångstrom, which I did during a boring synchrotron beamtime campaign at SLAC, actually the Stanford Synchrotron Radiation Laboratory (SSRL) in Menlo Park. I wrote the entire paper over a few nights[21], but it did not include any quantum electrodynamics.

Figure 3.31 shows an arrangement for a double-slit experiment not much different from the other examples for double slits such as those shown in Figures 3.10 and 3.11. There is a radiation source at location a, to the right of it the wall with the two slits, and on the right the target screen, with an arbitrary position b, where the radiation, where waves interfere to an intensity that contributes to the diffraction (interference)

21 In 2005, I had to spend many hours at the airport in Madrid between flights, so I wrote a short Letter to the Editor (Braun 2006). On another instance, we visited Paris, and as dogs are not permitted in the Louvre museum, I chose to guard our dog while sitting in a cafe, while my wife and children explored the Louvre for several hours. I used the time to write a Short Communication (Braun et al. 2007). By the way, in the basement of the Louvre, there is a research center, the Center for Research and Restoration of the Museums of France (C2RMF), Centre de recherche et de restauration des musées de France, which provides service for the 1200 museums in France. They have a MeV particle accelerator in there for chemical analyses, for example, see Radepont et al. (2018).

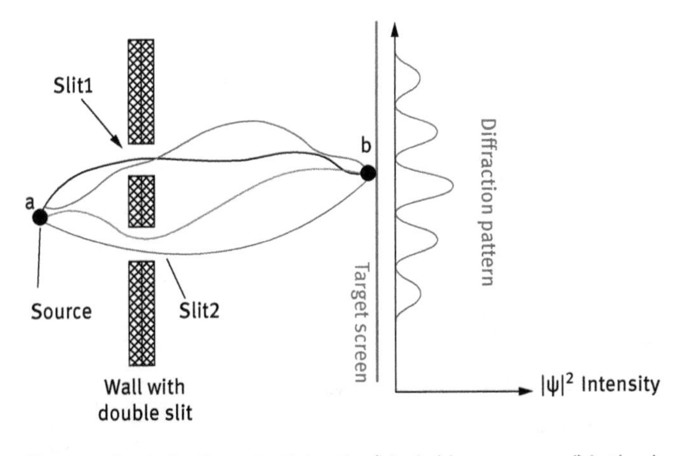

Fig. 3.31: Examples for potential paths (black, blue, green, red) in the double-slit experiment. Note the difference compared to the geometrical construction of the double-slit experiment in Figure 3.10, and refer to the YouTube lecture by Dr Josef M. Gassner (Gassner 2018).

pattern, given by the square of the probability amplitude, $|\psi|^2$. However, in this case, the radiation is not shown to propagate as straight lines, in contrast to the geometrical optics in Figure 3.10.

The distance between the radiation source and the target screen is overcome by four different paths, displayed by the black, blue, red, and green curvy curves, which are not straight at all. Maybe they remind you of curvy paths somewhere through nature, in the Swiss Alps or something similar, where a straight walking path is impossible. There may be countless path alternatives from a to b with a probability $P(a, b)$, so that you get from point a with coordinate x_a at time t_a to point b with coordinate x_b at time t_b. The contribution of one single path to the overall action S has the phase (this is also a functional)

$$\phi(x(t)) = c \cdot \exp\left(\frac{i}{\hbar} S(x(t)\right) . \tag{3.108}$$

Note that the argument of the exponential in Equation 3.108 contains the action integral S as defined above in Equation 3.105. The amplitude is given as

$$Q(b, a,) = \sum_{\text{all paths from } a}^{\text{to } b} \phi(x(t)) \tag{3.109}$$

and the probability P to go from point a to b is given by the square of Q, so that $P(b, a,) = |Q(b, a)|^2$.

The next step in treating the paths is the consideration of the discrete segments that add up to the trajectory. Figure 3.32 shows one such curvy path, which is approximated by a number of line segments formed by the positions x_{i-1}, x_i, x_{i+1} at times t_{i-1}, t_i, and t_{i+1}. To obtain the length of this path, which is made by a sum of $\mathrm{d}x_{i+1}$, we can

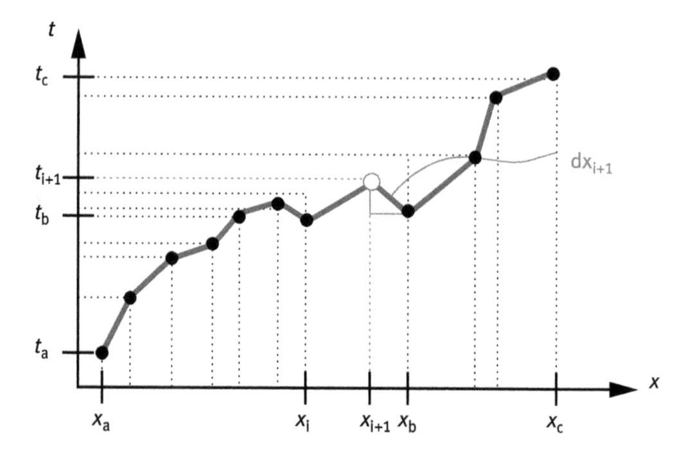

Fig. 3.32: A discretized path from locations (x_a, t_a) to (x_b, t_b). Note the time axis being the abscissa and the position axis being the ordinate.

formulate an integration that basically reads

$$P(b, a) \sim \int \prod_{i=1}^{N-1} \mathrm{d}x_i \phi(x(t)) \, . \tag{3.110}$$

When you insert the ϕ functional in the integrand and apply a normalization factor, which forces the convergence of the summation over the i, and carry out the infinitesimal steps $N \to \infty$, the integral reads

$$P(b, a) = \int_b^a \prod_{i=1}^{N-1} c \cdot \exp\left(\frac{i}{\hbar} S(x(t))\right) \mathrm{d}x_i \, . \tag{3.111}$$

Be reminded that we are treating here the transition amplitude Q, the absolute square of which stands for the probability P, that the distance from point x_a to x_b can be made in the time interval $t_b - t_a$. The action functional S includes the Lagrange function L, which can be given for any specific physical system with a kinetic energy T and some potential V. The x-dependency of the amplitude Q is given by the x-dependency of the Lagrange function: $L = T - V$.

The trajectories that associate with this Lagrange function translate into the action functional because of its "transitivity". Imagine now an additional position x_c on a trajectory that is reached at time stamp t_c. We, therefore, have $t_a < t_b < t_c$, and transitivity means that for the action functional $S(c, a) = S(c, b) - S(b, a)$ follows. Look, therefore, at the Figure 3.33, which shows a t, x-diagram for three consecutive events at times t_a, t_b, and t_c.

The integration for obtaining the transition amplitude takes place in two steps. One step is the integration over all $\mathrm{d}x_i$

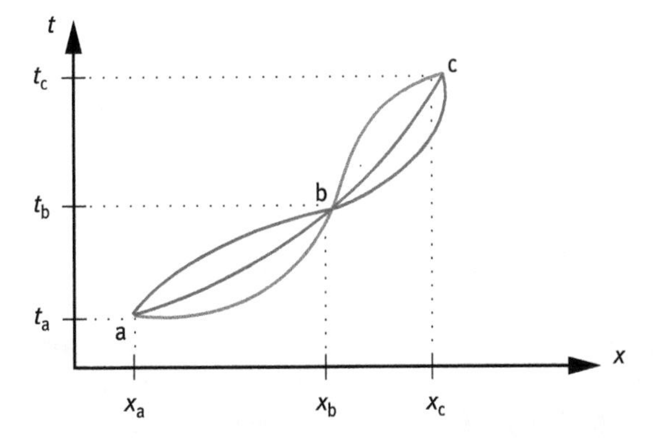

Fig. 3.33: Two consecutive events (x_c, t_c) and (x_b, t_b) and the corresponding paths.

3.11 Absorption spectroscopy

Everything you see in Figure 2.1 in the Chapter 2, the plants, the road, and the air in the blue sky, and the water in the clouds attenuate light in particular and radiation in general. It is a natural observation that the light is attenuated when it passes through matter. What in the light exactly, is attenuated? Light is defined by the (i) wavelength λ or the corresponding energy $E = hc/\lambda$ of the photons, (ii) the number of photons for a particular wavelength (spectral distribution, for example the solar spectrum in Figure 2.15), and the (iii) polarization of the photons. It is the intensity of the light that is being attenuated, the number of photons[22].

When an atom absorbs a photon with the energy $h\nu$, the following relation holds according to Bohr's atom model: $E_f - E_i = h\nu$; E_i is the energy of the initial state of the atom, and E_f is the energy of the final state of the atom, after absorption of the photon.

22 Just as a side note, but important for photosynthesis research: you can count the number of photons in multiples of Avogadro's number (in German: *Loschmidtsche Zahl*). In 1811, Avogadro postulated that equal volumes of ideal gases, irrespective of their chemical nature, should contain the same number of molecules. In 1865, Josef Loschmidt was able to determine this number as the number of molecules in a volume of gas under standard conditions of temperature and pressure to $n_0 = 2.686780111 \cdot 10^{19}$ per cm^3. This *constant* is related with Avogadro's constant N_A via the universal gas constant R as $n_0 = N_A p_0/(R \cdot T_0)$ and equals $6.02214076 \cdot 10^{23}$. This is 1 mol. Since the mol is used as a number of atoms, molecules, or objects in general, it can be helpful to count photons and electrons also in terms of mol, or multiples of mol.

When light is absorbed by molecules, the absorption spectrum is not a sharp line but a somewhat broadened absorption band, mostly because some of the photon energy is being converted into molecule vibrations (this extra broadening is accounted for mathematically by the Debye–Waller factor). Also, the lifetime of an excited state is not infinitely long, rather the excited state decays after some time, and in photoemission spectra, this limitation of lifetime causes a broadening of the peak position.

The absorption process is illustrated in Figure 3.19 by a simplified Feynman diagram (a very general one), which signifies two energy levels of the electron. We have a photon flux j_{ph} of photons with various energies $\hbar\omega_k$ impinging on atoms, only one of which shown. Before exposure of photons, the atom is in its initial, not excited state i, with state $|a_i\rangle$. The horizontal line at level i is level i on the energy axis. One particular energy of photon $\hbar\omega_k$ is absorbed by the atom, and the atom will be excited with an electron excited to the final state $|a_i\rangle$, as shown by the horizontal line with level f on the energy axis. Note that in this sketch, the energy axis is not explicitly shown. The arrow in the up direction implies the energy axis. This sort of graphical representation is called a Jablonski diagram (Jablonski 1933) and is comparable to the Grotrian diagram (Grotrian 1921; Bashkin and Stoner 1975), both of which were invented in the 1920s and 1930s. The process described here produces an absorption peak in an absorption spectrum. Generally, all quantum numbers, i.e., the principal quantum number, the rotational quantum number, and the spin quantum number of the atom can be affected by this process. Spectroscopy can help to experimentally determine these quantum numbers.

How do we count photons? We can use a photodiode, which works based on the photoelectric effect.[23] When the photodiode is hit by light, it will generate a photocurrent j, which scales with the number of photons. Few photons will generate a small photocurrent. Many photons will generate a large photocurrent. The photodiode is a light sensor and, thus, should have a linear sensor characteristic in a defined photon intensity range.

Figure 3.34 shows the photoresponsivity of a commercial photodiode. This is the photocurrent as a function of the wavelength, normalized to Ampere photocurrent versus Watt of photon power. For example, when you shine green light with $\lambda = 530$ nm on the photodiode at a power of 1 Watt, you will measure a photocurrent of 0.58 Ampere. When you measure a lower current, then the power of the light is not 1 Watt but

---

23 When there is ionizing radiation such as high-energy UV, x-rays, or γ-rays, the use of a gas ionization chamber becomes practical. Every photon impinging in a gas volume will ionize the gas molecules and, thus, charge carriers are created, which can be detected as an electric current. Historically, the power of the white light from all wavelengths is measured by absorption of the photons on a black material, which becomes hot. The instrument is the bolometer (Hale 1900). Maybe it is easier to memorize the name bolometer when you call it "boil-ometer", because it becomes warm or hot.

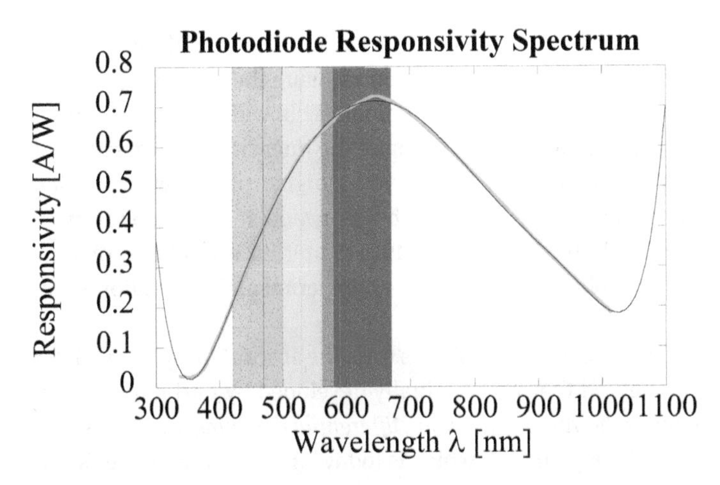

Fig. 3.34: The photoresponsivity A/W (current in Ampere per Watt power of incident light of a particular wavelength) of a THORLABS photodiode from 350 to 1000 nm, as shown by the green solid line. The thin black solid line is a ninth-order polynom least squares fit. The colored vertical bars indicate the color of the corresponding wavelength. The maximum responsivity is in the red range at 650 nm.

less than that. It is important that the diode is properly calibrated, so that one can determine the light power from the measured electric photocurrent[24].

A simple absorption spectroscopy experiment then looks as follows: you will need a monochromatic light source that gives you photons of a specific wavelength. You will need a photodetector like a photodiode. You can also use a so-called charge-coupled device (also known as a CCD chip). Moreover, you will need the sample from which you want to obtain the absorption spectrum.

The monochromatic light source can be a freely tunable laser, which allows changing the wavelength from, say, 350 to 700 nm. This, however, can be very expensive. Another alternative could be sunlight; sunlight is “white light” composed of an extremely wide range of wavelength or photon energies, extending from the very long radio waves to the highly energetic γ-rays. You guide the sunlight through an optical monochromator, such as a prism, and thus split the white sunlight into a colorful rainbow (Figure 3.3), and you select each wavelength to the best of your capacity

24 Some suppliers provide the spectrum of their light sources as enclosed paper with their light source, or they post the data online. When you buy instruments, you should not rely on the data which are disclosed, but you should check and measure them by yourself. I purchased a dozen light emitting diodes (LED), where the spectrum of the light source was off by 5–10 nm for two of them, as a PhD student in my group found. The supplier later corrected the data sheet after the PhD student notified them of the error.

and guide it on the sample. Then, "one color" (for example, green) of light passes through the sample, it becomes attenuated, and you measure the intensity of the attenuated light *behind* the sample with the photodiode. This will give you the photocurrent after attenuation j_a. Then, you will measure the intensity of the green light *before* it passes the sample. This is likely to give you a higher photocurrent, which we call j_0. An early quantitative treatment of such absorption spectroscopy is mentioned in a paper by theorist[25] John C. Slater from 1925 (Slater 1925a), actually on how to determine the transition probabilities A_{ij} from an experimental absorption spectrum.

In his derivation of the transition probabilities from the absorption spectra, Slater refers to a mathematical method that was originally developed by Carl Runge (Potsdam 1894, *Über die numerische Aufloesung von Differentialgleichungen*; On the numerical solution of differential equations), which is today known as the Runge–Kutta approximation (Martin Wilhelm Kutta extended the four-step solution to a generalized s-step solution). This is a numerical approach for the solution of ordinary differential equations where the differentials are approximated with differential quotients (Paschen 1897; Runge 1895).

An example for optical absorption spectra is shown in Figure 3.35. The two spectra are from biological samples (Kansy et al. 2017), from the *Cyclotella meneghiniana* plankton diatoms (Willen 1991), which have a flat disk-like shape. Spectra were collected from the *intact* cells (the black spectrum) and from their extracted thylakoids (the red spectrum). For both samples, we notice a relatively sharp absorption peak at 675 nm with a width of around 50 nm. There is a very broad absorption region from 350 to 550 nm, but the thylakoid spectrum appears sharper and better resolved than the spectrum from the intact cells. This is maybe not so surprising, because the thylakoids are extracted from the cells and, thus, constitute a spectroscopically more homogeneous material than the cells, which also include other material, thus causing more diffuse intensity in the spectrum.

The work of the spectroscopist is, then, to carry out a spectroscopic assignment of the peaks. Practically, what you can do is to separate the material into its components as much as possible, and record spectra of every component and, thus, a data base, based on which you may later determine the composition of the sample – to some extent. This is necessary, because real-life samples, particularly biological specimens, are large and very heterogeneous.

As an experimenter you will notice that the constitution of the sample has a large influence on the measurement. When the sample is too thick, no light comes through, and, thus, j_a is too small or equals 0. When the sample is too thin, the absorption is too low, and the difference between j_0 and j_a does not allow satisfactory results.

25 See the section therein on experimentalist versus theorist.

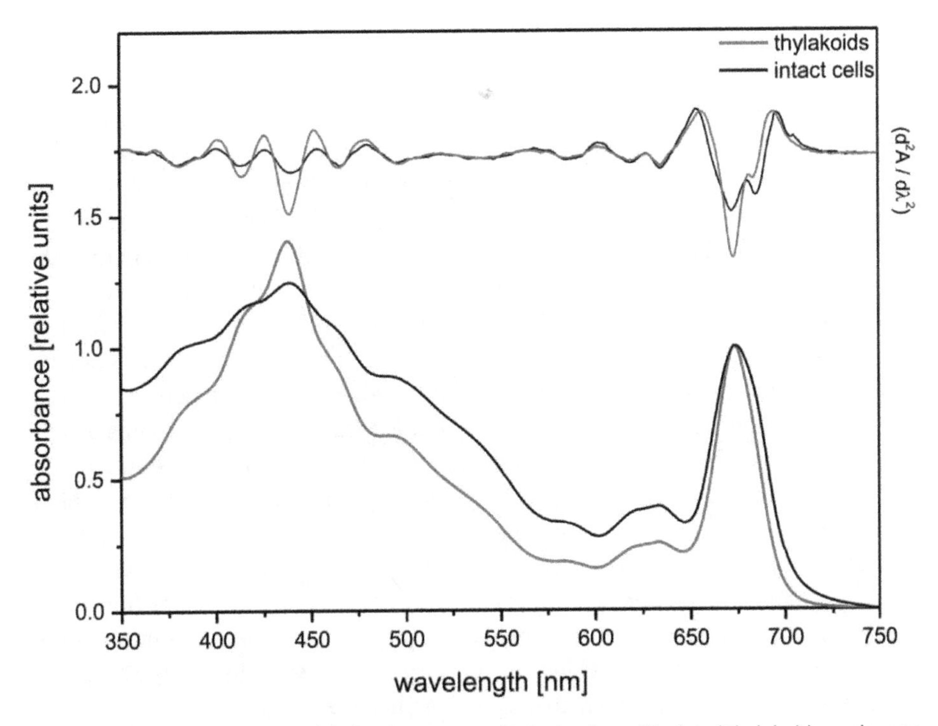

Fig. 3.35: Absorption and second derivative spectra of intact cells and isolated thylakoid membranes of *C. meneghiniana*. The Chl concentration for the absorption measurements was 2 µg/mL. For further measurement details, see the Methods section. The figure shows representative spectra reprinted from the article "An optimized protocol for the preparation of oxygen-evolving thylakoid membranes from Cyclotella meneghiniana provides a tool for the investigation of diatom plastidic electron transport" by Marcel Kansy, Alexandra Gurowietz, Christian Wilhelm, Reimund Goss, published in BMC Plant Biology 17/221, (2017) and distributed as an open-access article under the terms of the Creative Commons Attribution 4.0 International License (http://creativecommons.org/licenses/by/4.0/), (Kansy et al. 2017).

Thus, you have to adjust the thickness of the sample[26]. You will also notice that the measurements look particularly nice when the sample has a homogeneous thickness and density distribution.

It is important to realize and accept that the selected color of the light, the wavelength, and the photon energy is not an exact sharp, finite number. It is typical for physics to declare a physical quantity such as the wavelength λ, frequency ω or $2\pi\nu$, or energy E[27] comes with an experimental, technical uncertainty $\Delta\lambda$ and ΔE. It is tech-

26 We will see later that the sample has an optimum thickness for absorption measurements when it equals one absorption length. This is the length, or thickness, where the absorbed intensity is $1/e$ of the original intensity j_0; e is Euler's number 2.71828 $\sum_{k=0}^{\infty} 1/k!$.

27 We recall the relations $E = h\nu = hc/\lambda$, and $\omega = 2\pi\nu$.

nically not so easy to manufacture instruments with an absolute error of 0. It is impossible. The first question a technician in the machine shop who manufactures a tool for you, or say a thin metal sheet of 0.5 mm thickness D for some scientific instrument will ask you is "what should the tolerance ΔD for the thickness of the sheet be?" So the thickness is then given as $D \pm \Delta D$, for example 0.5 ± 0.01 mm. In a complete instrument and in a final experiment under given variables, all errors will propagate into the final physical quantity that we want to determine, such as the wavelength of the light that comes out of the monochromator: 530 ± 7.5 nm, for example. Sometimes the wavelength is approximated with a Gaussian curve, and this may have a different physical origin, for example when the light originates form a molecule that emits photons. Emission spectra from gas atoms and molecules are typically characterized as sharp lines, but the lines also have a finite width and a particular shape. The line shapes can be measured quite well, when the spectrometer has the sufficient resolution, and be reproduced by mathematical physical models. The molecular environment of an emitting molecule can influence the line shape of the emission light (Spano 2009).

What sample are we looking at anyway? We can take colored plastic foil, for example. As I am working in a ceramics laboratory, we often use thin films from metal oxides that are coated on glass. Iron oxide Fe_2O_3 has a red color, for example. Tungsten oxide WO_3 is yellow, unless it is oxygen substoichiometric, like $WO_{3-\delta}$, then it is blue. Later, we will look at green chlorophyll in Chapters 4 and 6.

When you have found a good sample to work with, you can pile up the samples (like making a stack of papers) and, thus, increase the overall thickness. When you then measure the absorption, you will notice that the absorbed intensity scales exponentially with the sample thickness d.

Figure 3.36 is redrawn from a historical document from Bouguer's work on light absorption in 1729. He recorded the intensities of light beams after they had passed absorbers with various thicknesses, actually liquids filled with dyes. Figure 3.36 shows an absorber of the size ABCD. The horizontal lines indicate layers and stand for the thickness levels. The length QB indicates the intensity of the incident light beam.

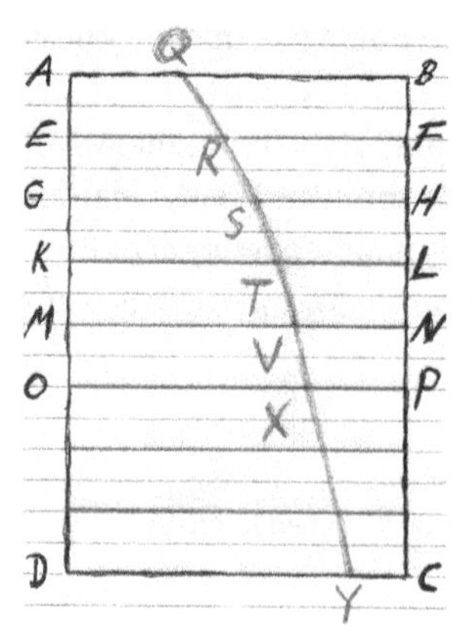

Fig. 3.36: Adopted from the original drawing of Bouguer (1729).

When the light has passed the first layer, the remaining light intensity is RF. After the second layer is passed, the remaining intensity is SH. You can consider the bottom line BC as the x-axis, the thickness axis.

Bouguer discovered that the relation between the thickness of the absorber and the intensity did not follow an arithmetic series but a geometric progression series. This is why the thickness and intensity are in a logarithmic relation and not a linear relation. For statistical analysis and outcome, it is important to realize whether a process follows an arithmetic progression or a geometric progression. The mathematical description for the absorption is that the change of the intensity I over an infinitesimal change of the thickness of a material characterized by a materials constant μ is $\mathrm{d}I$, which is described as following linear differential equation:

$$-\,\mathrm{d}I = \mu I \,\mathrm{d}x \tag{3.112}$$

Divide by I and you can integrate this differential equation easily and come to an exponential relation

$$\ln\left(\frac{I_0}{I}\right) = \int_0^d \mu \,\mathrm{d}x \tag{3.113}$$

which, in the end, gives you the Lambert–Beer–Bouguet absorption law, which is a function of the thickness and the wavelength:

$$I(d,\lambda) = I_0 \exp(-\mu(\lambda)\cdot d) \tag{3.114}$$

The “constant” μ is the linear absorption coefficient. It is not a real constant because it depends on the wavelength λ of the light. This is the reason that we can use the method of absorption spectroscopy for the analysis of materials. What we observe here is that out of the 100% of photons that arrive at the sample surface, a number will be absorbed by the atoms or molecules in the material. When you plot the intensity versus the thickness, you will find an exponential decay curve. When you plot it on a semilogarithmic scale, the slope of the curve equals the linear absorption coefficient μ. The technical setup of an optical absorption experiment is illustrated in Figure 3.37.

In a paper from 1925 (Slater 1925a), Slater exercised how to determine the transition probabilities A_{ij} from an experimental absorption spectrum. When the radiation is absorbed by the matter to the extent that the radiation energy is so high that further photon emission is created in the matter, absorption corrections may become necessary (Abitan et al. 2008).

Now we move to the next experimental step. We use the monochromator and tune the wavelength to a different color by some small increment of $\Delta\lambda$ and then measure the intensity again, as it was transmitted by the sample. We should then repeat the measurement at the same wavelength without the sample – in order to obtain the photocurrent for background subtraction. For practical purposes, we will not always switch the sample position; rather, we will again select the next wavelength and measure the new photocurrent, and so on. In the end, we have a spectrum of photocurrent

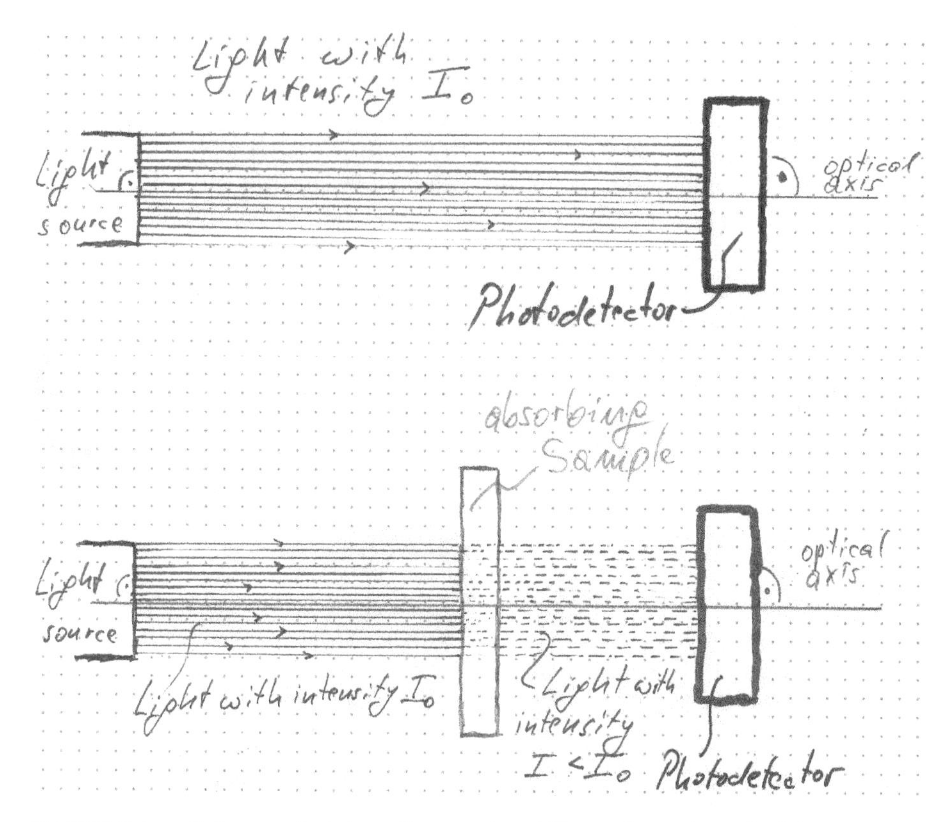

Fig. 3.37: Schematic for an optical absorption experiment. The top figure shows the light source on the left, emitting a bundle of parallel rays of light (light with incident intensity = photons with number I_0) to a sample (thin green rectangle), which absorbs part of the photons, whereas the rest of the photons (light with intensity I_{sample}) is transmitted to and absorbed by the photodetector (red thick rectangle). The long horizontal line through the middle of the apparatus is the optical axis, along which the planes of light source, sample, and detector should be aligned, preferably and for simplicity of later data analysis with 90°. The intermitted lines between sample and detector indicate the light intensity < 100% after absorption by the sample. The bottom figure shows the situation where the photodetector is moved from behind the sample right before the sample, so that the detector receives 100% light intensity. This is a reference measurement for the determination of I_0.

versus wavelengths when the sample is in the beam. Now, we repeat the entire wavelength scan without the sample in the beam. The difference of both spectra gives the absorption spectrum of the sample.

I will point you to another issue. When you make the experiment completely dark so that no light, no photons will hit the photodetector, you may notice that the detector will still send out some very small current. We cannot call it photocurrent because there are no photons that obviously hit the detector. Yet there may be cosmic rays and particles that trigger very rare events in the detector, which produce a small current.

This is a dark current that also needs to be subtracted from the spectrum. This is a constant number, which must be subtracted from the spectrum for every wavelength. We recall the responsivity of the photodiode, which is also a function of the wavelength. In the red range, the diode shows a higher response than in the blue range. This means that the effect of the photons in the blue range that pass the sample and then hit the detector is underestimated, whereas the effect of the red portion of the spectrum is overestimated. The effect of this correction might be that the intensities shift up and down in various regions of the spectrum. It is not always necessary to carry out such corrections. Often, researchers are only interested in the positions of the absorption peaks or the emission peaks, and not so much in their absolute or relative intensities. Further, often, the spectrometers are equipped with very good calibration and correction routines, which make it easy to account for these corrections with ease.

3.12 Experimentalist versus theorist

It is an unfortunate development in the physical sciences that a distinction is made between experimentalists and theorists. I am saying this because when writing this book, I had a conversation with some theorists who have been working with me on a paper on the electronic structure of metal oxides, and their experimental assessment with soft x-ray absorption spectroscopy (NEXAFS; near-edge x-ray absorption fine structure spectra), and their *ab-initio* calculation with a novel method of density functional theory (DFT). The NEXAFS spectra were recorded by my PhD student at the Advanced Light Source (ALS), a synchrotron facility in Berkeley, California. These experimental spectra show absorption peaks but also one or more step functions, latter of which arise from passing the ionization threshold of the probed atom in the sample. You can simulate these NEXAFS spectra with DFT calculations provided that you have a good enough crystallographic model of the sample. The result of the calculations is the electronic band structure from which you can derive the density of states (DOS) and the absorption spectra, the absorption peaks. However, you never see the step functions in any of these calculated spectra. This is likely to be because they are not relevant for the modelers. You can work them into the calculations, but nobody does this, to the best of my knowledge. In order to better compare the experimental and the calculated spectra, I, therefore, routinely extract the step functions by applying a quite laborious least-squares fitting routine, which includes a very good fit of the step, and then I later simply plot only the peaks, but not the step. For theorists, it was difficult to handle this issue because they were not there when the spectra were recorded, and they simply had no idea how such NEXAFS experiments are performed in reality. Therefore, it is a pity when theorists are free from any experimental work, and experimentalists focus on the experiments and let other people do the calculations. A few years ago I had an exchange with another theorist, who was a senior expert on the DFT calculations of metal oxides. We were partners in a bilateral project on lithium ion battery research. When I shared the resonant photoemission spectra and the NEXAFS spectra that two of my PhD students had measured at SSRL (Stanford Synchrotron Radiation Laboratory) in Menlo Park in California, the senior theorist had no use for these valuable and very rare experimental spectra because he was not aware that a DOS of a battery electrode material could actually be experimentally measured with high accuracy and precision. He did not understand the concept of these spectra. I had a way better experience with a theorist about 25 years ago during my time as a physics diploma student at KFA Jülich in Germany. There, I had carried experiments on ultrathin magnetic metal films, and one of my analytical methods was low-energy electron diffraction (LEED), a method which allows for very precise determination of the crystallographic structure of a surface and subsurface. My supervisor gave me the opportunity to spend 2 weeks in Grenoble in France at CNRS, with a computational scientist. He taught me how to perform full dynamic LEED I-V calculations and he had asked me to bring my experimental data as raw as possible. He wanted to know all experimental details on how and under what conditions the data were recorded with which instruments. Because he knew of all the effects that experimental conditions can have on the data and also which instruments have what specific weaknesses that inflect on the data quality. You learn these things only when you spend time with the other people on their own playing field. Going this way certainly extends your time of education and training, but it is a very good investment for a scientist who wants to do an above-average job.

4 The interaction of light with matter

4.1 Gedankenexperiment on how to observe the invisible

The image that you saw when you were the hiker in the scene in Figure 2.1 is the light that originally came from the sun, but has been scattered by the blue sky, green plants, and dark asphalt road. Before the light arrives to your eyes, it has interacted with the atoms and molecules in the ionosphere, troposphere, atmosphere, and biosphere (Figure 2.4). On arriving at your eyes, it certainly interacts with the retinals in your eyes, the rhodopsin molecules that are important for light sensing in our vision apparatus. Part of the light is absorbed, and part of the light is scattered to our eye.

When I was in elementary school, maybe 8 years of age, our teacher told us a way to spot the man who was wearing an invisibility cloak. The invisibility cloak is what you want to have if you want to go in stealth and not have anybody see you. At that age, during breaks or sports at school, we would play *Völkerball*, dodgeball. The playing field is divided into two halves, each of the two teams is restricted to their own half, one large ball is used, and you want to hit a player of the opponent team, while all players of the opponent team want to avoid being hit. If you hit an opponent, he is out, and thus you weaken the opponent team. If no one is hit, the other team can pick up the ball and now try to hit you in your field, and you want to avoid being hit and run away from the ball. Now, if the other team were wearing invisibility cloaks, you would not see them. You would not even know they were there, if they were silent. You would not be able to count how many there are. The teacher said, if you throw the ball often enough, chances would be that you would hit the invisible person once, and then the ball would be scattered in an unpredicted direction. Therefore, you can probe invisible things with scattering experiments. Of course, seeing the opponent player with your eyes is also a scattering experiment, because the light is, here, the photons are the probes that scatter the information to the observer. However, in the aforementioned Gedankenexperiment, the ball is the probe to detect the man under the invisibility cloak.

4.2 Excursion into scattering processes

Scattering is here one import key word to consider. Let us begin simply and refer to billiard balls on a table[1]. When we shoot a billiard ball onto other billiard balls, they will "scatter". Billiard ball 1 has a position vector r_1 and a momentum p_1 and will hit ball 2, which may be at rest with $p_2 = 0$ at position vector r_2 at time $t = 0$. Several

1 Billiard problems are canonic in physics, see Fernandez-Alonso and Zare (2002).

https://doi.org/10.1515/9783110629941-004

scenarios are possible after the collision. Ball 1 stops ($p_1 = 0$ at r_2) and passes on the momentum to ball 2, $p_2 \rightarrow p_1$. So, ball 2 will have the same direction as ball 1. The other scenario is that ball 1 changes its direction, and ball 2 picks up momentum from ball 1 and goes in some other direction. This is the scenario that typically occurs in billiards.

Yet another scenario would be that the ball 1 bounces back to exactly where it came from, and ball 2 remains at rest. Is this case possible? Not with balls of the same mass. It could be possible if ball 2 has a very large mass, and ball 1 has a comparably very small mass. These are the scattering events that we can think of in billiard-like collision experiments. The physical laws of conversation of momentum p and energy E govern these processes, as we know from classical mechanics. We can also think of shooting a bullet with a rifle onto the ball; the ball will be destroyed by the heavy impact of the high-speed bullet, and the fragments of the ball and the bullet will scatter in various directions. Instead of a billiard ball or a pumpkin, think of a very large atom such as uranium U, for example. You build a particle collider in which you accelerate ions at high speed and let them collide with the U atom. Depending on the momentum of the ion, the U atom may split or decay into several fractions.

When a spaceship is to return from Moon or from some orbit back to Earth, it will do so with a very high speed. The spaceship must approach Earth in a particular angle of inclination. Chances are that when the angle is too low, the spaceship will bounce off the Earth's atmosphere and get lost in space. Another scenario is that the angle will be too small, and the spaceship will crash steeply onto Earth. The proper angle will help the spaceship experience friction by the atmosphere during landing, heat up very much without catching fire, and get decelerated to a speed where a controlled touch-down is possible. The atmosphere, the layers around the Earth that are depicted in Figure 2.4, is what the spaceship will "sense" and to which it will react.

Neutrons will do something similar when they are used in neutron scattering experiments and shot onto a sample of matter. The neutrons sense the atoms that build up the sample and interact with them and get scattered in particular directions and form a scattering pattern that can be recorded with a detector. The same holds for x-rays, which are scattered in diffraction experiments, such as in Bragg diffraction. X-ray scattering and neutron scattering are important methods for the determination of the microstructure and crystallographic structure of materials. Similar holds for scattering with visible light and ultraviolet light.

Let us look at Figure 4.1, which illustrates the Compton effect. An electron with momentum $p = 0$ and rest mass energy $E = m_0c^2$ shall be our target electron. Then we hit it with an incident photon, which has the energy $E = h\nu$ and momentum $p = h\nu/c$. After the "collision", the target electron is scattered (kicked) into some direction, which includes an angle ϕ with the direction of the incident photon, which amounts to a new momentum p and an energy $E^2 = m_0^2c^4 + p^2c^2$. The momentum conservation law would not be preserved if there was no additional object involved in the process. In Compton scattering, this is the scattered photon that has a lower photon energy

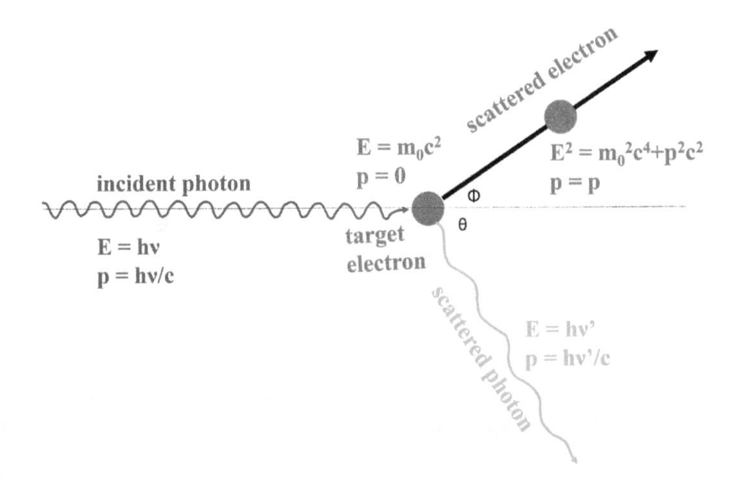

Fig. 4.1: Schematic for the scattering of a photon on an electron. After the kick, the photon has a different energy and different momentum, and the electron has different energy and different momentum, too.

now, $E = h\nu'$, and which carries away a momentum $p = h\nu'/c$ under the angle θ with the direction of the previous incident photon. This is only one example of a relevant scattering process in molecular physics.

In the field of human interaction and social behavior, also scattering events are possible. Some people are simply "attractive" and we like to approach them with ease and socialize with them, provided that we feel attracted. Interaction is "inter action" and always requires two. There is the case of repulsive interaction as well. You may try to avoid some people, and when by chance you cross their ways, you try to avoid them. Similar would hold for an aggressive animal, like a big scary dog that you see coming towards you, on the leash of some man. Because you are scared, you may chose to cross the road and walk on the pedestrian path on the other side of the road – and increase the distance between you and the dog.

It is the interaction between two and more bodies that modulates the scattering. The interactions are the forces that we know, such as gravitation, electric forces, electroweak forces, and strong forces. One central metric in the scattering process is the scattering cross section. Practically, the scattering cross section is taken as differential scattering cross section (in German: *der differentielle Wirkungsquerschnitt*) and measured as an area, for example, in cm^2.

We remember the dodgeball story , where the ball is the probe for detecting invisible people, if they wear the invisibility cloak. This the same as in a billiard game, we are playing with probes and objects to be probed, around which a momentum evolves and changes. You kick the ball and give it momentum, and when it hits an object, the momentum will change, and from the momentum change you derive information. With the change of momentum, we typically associate forces, because $F = ma = m\,\mathrm{d}v/\,\mathrm{d}t$, because $p = mv$.

We remember that the forces result from the derivatives of the momentum versus the time, $F = \mathrm{d}p/\mathrm{d}t$. In the scattering experiment, the scattering is the result of the force: the action. We cannot see forces, but we can see the action of the force. This is one reason why the action, or the action integral, has turned out to be a fundamental quantity in physics. The action is an integral quantity and provides a global description of the physical system with the Hamilton function. The action is the product of canonically conjugated variables, such as momentum and spatial variable $p \cdot q$, or work (energy) and time $E \cdot t$, or angle and angular momentum $\phi \cdot L$. The canonically conjugated variables span the phase space (in German: *Phasenraum*), in which the dynamic system is defined, and the action is the integral in the phase space.

Related with the action is the action cross section (Σ, or σ, with the dimension of a geometric area like cm^2), which is the measure of the probability that an event takes place. While the Hamilton formalism only shows the global properties of a physical system, it is the Lagrange equations that describe the actual path of an object on its trajectory. The minimization of the action integral, along with the initial conditions and boundary conditions, is the way to determine this trajectory. In this respect, forces do not provide us with information. This amounts, then, to the more philosophical question of whether that what we do not see really exists, and whether what we cannot reach can actually be real[2].

We can rank the interaction of light, of photons with matter, depending on the energy that is involved in this interaction. The kicking out of an electron (electron emission) from an atom, molecule, or condensed matter requires rather low energy. The photoelectric effect falls in this low-energy category. Important to note is also the following. In the billiard example, we are in classical physics. When light comes to the scattering problem, we are dealing with a relativistic physics problem and need to take into account the 4-vector properties of the system. Similar would hold when the particles approach the speed of light. Then we are in high-energy physics. So the range of energy in which processes may take place is a matter of the classification of phenomena.

With increasing energy, and this can be the energy of the electron or photon or both, we come to mid-energy phenomena. Thomson scattering falls in this category. An incoming photon bounces off an electron or some charged particle in some direction, and the electron is kicked in some other direction, so there is an exchange of momentum $\boldsymbol{p}$ but no exchange of energy; $\delta E = 0$. Hence, the wavelength of the photon does not change. Thomson scattering plays a role in plasma physics. The Earth's ionosphere is a plasma sphere, and here the sunlight may scatter with it. About 10 years ago, it was shown that the Thomson scattering of sunlight with the ionosphere can produce signals that can be used for so-called geospace imaging (Meier et al. 2009). Compton scattering is also a mid-energy effect.

Pair production is considered the consequence of a high-energy photon–particle interaction, but I do not fully agree with it for the following reason. When a semicon-

---

2 In German: *Über die Existenz des Unsichtbaren und die Wirklichkeit des Unerreichbaren.*

ductor is struck by photons with an energy higher than the semiconductor bandgap energy, there will be a pair formation, and this is the pair of an electron in the valence band and its corresponding hole. The same holds in molecules, where there is have no valence band or conduction band, but the highest occupied molecular orbital (HOMO) and the lowest unoccupied molecular orbital (LUMO). We have learnt about the recombination of protons and electrons in stellar gas clouds, which would imply a previous pair production, but this has not taken place. The term recombination in the 2s decay in hydrogen is a misnomer and nomenclature in the history of early astrophysics.

4.3 Born's approximate treatment of collision processes

In an early paper from 1926 (Born 1926), Born acknowledges that the collision processes give convincing experimental proof for the basic assumptions of quantum theory and are also suitable for the elucidation of the formal laws of quantum mechanics. He criticizes, however, that Heisenberg's theory is based on the assumption that an absolute presentation of the processes in space and time is impossible, and the physical observables are only interpretable as properties of physical movements in the classical limit.

Moreover, Schrödinger's theory, he argues, is based on de Broglie's wave theory, which suggests that packets of waves represent the physical corpuscles. Neither representations are convincing for Born, and he suggests a third interpretation, which is also useful for collision processes. Then he refers to Albert Einstein, who suggested that waves guide the corpuscular light quanta, which we know as photons. For this, Einstein used the word *Gespensterfeld*: ghost field, the guiding field, so to speak.

Born proposes to continue the development of quantum mechanics with investigations of the aperiodic stationary processes, and the definition of the proper terminology will only be possible once the mathematical formulation of the problem has been done with success. Born acknowledges the assistance of Norbert Wiener in this matter and begins the formulation, before he comes to the actual physical hypothesis in the end of his work (Born 1926).

The formulation of a scattering problem with wavefunctions $\psi(r)$ and $\psi(r)'$ in a potential $V(r)$, which is sketched in a space with spatial variables $\boldsymbol{r}$ and $\boldsymbol{r}'$, can be made as an integral equation with Green's function (Hecht 2000):

$$\psi_{\boldsymbol{k}}(\boldsymbol{r}) = \phi_{\boldsymbol{k}}(\boldsymbol{r}) + \int \mathrm{d}\boldsymbol{r}'G^{+}(\boldsymbol{r},\boldsymbol{r}')\frac{2\mu}{\hbar^2}V(\boldsymbol{r}')\psi_{\boldsymbol{k}}(\boldsymbol{r}') \tag{4.1}$$

The problem here is that $\psi_{\boldsymbol{k}}(\boldsymbol{r}'$ is not known (this is a known unknown). When the potential is weak, or when it is weak compared to a large kinetic energy $p^2/(2m)$, it can be treated as a perturbation and made subject to perturbation theory. So the condition for doing so is

$$|V(\boldsymbol{r}')| \ll \frac{\hbar^2 k^2}{2\mu} \tag{4.2}$$

The trick is now to approximate the unknown function $\psi_{\boldsymbol{k}}(\boldsymbol{r}')$ by a plane wave to yield the approximate solution, which is known as the first Born approximation:

$$\psi_{\boldsymbol{k}}(\boldsymbol{r}) \sim \frac{1}{(2\pi)^{\frac{2}{3}}}\left(\mathrm{e}^{\mathrm{i}\boldsymbol{kr}} - \frac{1}{4\pi}\,\mathrm{d}\boldsymbol{r}'\frac{\mathrm{e}^{\mathrm{i}k|\boldsymbol{r}-\boldsymbol{r}'|}}{|\boldsymbol{r}-\boldsymbol{r}'|}\frac{2\mu}{\hbar^2}V(\boldsymbol{r}')\mathrm{e}^{\mathrm{i}\boldsymbol{kr}}\right). \tag{4.3}$$

When the potential has the shape

$$V(r) = -g^2\frac{\mathrm{e}^{-\frac{mc}{\hbar}|\boldsymbol{r}|}}{|\boldsymbol{r}|}, \tag{4.4}$$

this is the Yukawa potential with the coupling parameter g. Then the solution for the so-called differential scattering cross-section yields

$$\frac{\mathrm{d}\sigma}{\mathrm{d}\Sigma} = |f(\theta)|^2 = \frac{4\mu^2 g^4}{\hbar^4}\frac{1}{\left((\frac{mc}{\hbar})^2 + 4k^2\sin^2\frac{\theta}{2}\right)^2}. \tag{4.5}$$

4.4 Born–Oppenheimer approximation

When you want to determine the electronic energy levels of a molecule, you may assume that the nuclei of the atoms that constitute the molecule are at rest. In reality, the nuclei may be vibrating, of course. However, we ignore this (for the time being; later we will also consider vibrations) given that the large mass of the nuclei, we call them M, is considerably larger than the electron mass m. The error that originates and propagates from the vibration of the nuclei is of the order of the root of the ratio of the two different masses, $(m/M)^{1/2}$. This is how Max Born and Robert Oppenheimer[3] investigated the problem (Mott 2008; Born and Oppenheimer 1927).

Consider the geometric construction of an atom, with nucleus and two electrons as shown in Figure 4.2. The heavy nuclei and the light electrons, thus, move on different timescales, which allows the equations of motion of the light electrons to be

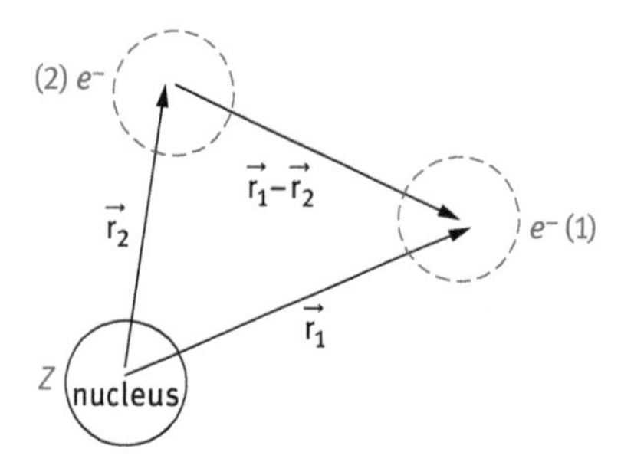

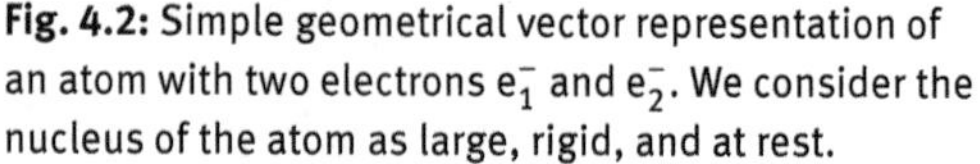
Fig. 4.2: Simple geometrical vector representation of an atom with two electrons e_1^- and e_2^-. We consider the nucleus of the atom as large, rigid, and at rest.

3 Oppenheimer was nominated three times for the Nobel Prize in Physics. He also did some fundamental work on photosynthesis, (Arnold and Oppenheimer 1950; Oppenheimer 1941).

solved without including the slow motion of the heavy nuclei. It is, thus, possible to separate the Schrödinger equation into *fast* and *slow* components.

The Born–Oppenheimer approximation (Born and Oppenheimer 1927) is typically employed for problems with molecules and condensed matter, as these contain at least two atoms (which have a large mass) and a large number of electrons (which have negligible mass). In physical chemistry, the approximation is also useful because exact analytical solutions for the Schrödinger equation are only known for very simple systems, like the hydrogen atom. Note that the approximation provides good results for molecules in the ground state only, particularly when the nuclei have a large mass. For the excited state, the results are not as good.

4.5 Fermi's golden rule

The absorption and emission of photons was treated in 1927 by Paul Dirac as problems of quantum mechanics (Dirac 1927). The molecule or an atom is considered as a quantum system, and it is excited by an electromagnetic wave, dipole radiation. This is considered a perturbation, which is represented by a perturbation operator. The quantity of interest is the transition dipole moment.

According to Dirac, the transition probability is given by

$$w = \frac{2\pi}{\hbar}|\psi|^2\frac{\mathrm{d}n}{\mathrm{d}E} \tag{4.6}$$

but it was Enrico Fermi in his textbook (Fermi 1949) on nuclear physics who coined the term "golden rule". Hence it is actually Dirac's golden rule, but it is attributed to Fermi because he was supposedly the first to call it that, in his book or in the lecture series that he gave at University of Chicago.

You begin with a quantum system that has a Hamiltonian H_0 and eigenstates $|a^0\rangle$, so that the Hamilton equation reads[4]

$$H_0\,|a^0\rangle - E_a^0\,|a^0\rangle = 0\,, \tag{4.7}$$

and then wonder how the system responds to a small perturbation $H_1(t)$ (or $V(t)$), which is not constant but time dependent. The physical nature of such a perturbation can be manifold. It can be an electric field $\boldsymbol{E}$ or an electromagnetic wave with frequency ω, actually photon energy $h\nu$, and the perturbation varies with the frequency ω as $H_1(t) \sim \cos(\omega t)$.

In order to know, for example, the probability $W(t)$ that the quantum system is in a different eigenstate $|b^0\rangle$ at some arbitrary time t, one must apply the time-dependent perturbation theory with a $|\psi(t)\rangle$, for which at time $t = 0$ it hold that: $|\psi(0)\rangle = |a^0\rangle$.

4 I refer here to a script from Andreas Wacker at Lund University (Wacker 2016).

This probability is calculated as

$$W(t) = |\langle b^0|\psi(t)\rangle|^2 \,. \tag{4.8}$$

It is an exercise for the reader to write down the time dependence for $\psi(t)$ when there is no perturbation H_1, i.e., $H_1 = 0$. Here is a hint: there is an exponential term involved. Over a long time under weak perturbation, the transition probability w increases linearly with time t. It is then interesting to calculate the rate for the transition from state a to state b, the transition rate $w = \mathrm{d}W/\mathrm{d}t$:

$$w(t) = \frac{2\pi}{\hbar} |\langle b^0|\, H_1 \,|a^0\rangle|^2 \delta\left(E_b^0 - E_a^0\right) \tag{4.9}$$

The delta function δ in Equation 4.9 mandates that a transition from state a to state b is only possible when they have the same energy: $E_a^0 = E_b^0$.

The reciprocal of the transition rate w is the average lifetime τ of the initial state a, provided that there are no additional transitions possible. "Dirac's" golden rule turned out to be valid for many problems in physics where particles are created and annihilated, the particles being photons, and quasi particles like phonons and spin waves. Dirac finishes his paper (Dirac 1927) with the gratifying conclusion "The theory leads to the correct expressions for Einstein's A's and B's."

Absorption and emission of light, spontaneous nuclear decay and decay of elementary particles, and the determination of the interaction cross section of any reacting species are applications for Fermi's golden rule. A recent application of Fermi's (Dirac's) golden rule was shown by Manna and Dunietz, who calculated the charge transfer in arrays of porphyrin molecules (dyades) (Manna and Dunietz 2014). However, there are more methods to calculate the transition rates and probabilities, see Kananenka et al. (2018).

The golden rule is also useful for investigating processes in semiconductor photoelectrochemistry, such as charge transfer from the electrode to the electrolyte. However, when such charge transfer occurs in the ultrafast time domain of 100 femtoseconds and faster, the second-order perturbation theory, which is the basis for the golden rule, no longer holds, and fourth-order perturbation methods are necessary (Li et al. 2014).

Tokita et al. (2008) studied the mechanism of intramolecular electron transfer when a zinc substituted cytochrome c (Zn-cyt c) was excited with photons. The electron transfer (ET) rates were determined using Fermi's golden rule. Of general interest is the long-range ET in proteins, which are large molecules. One question is which of the molecular orbitals would be able to perform the charge transfer. The researchers found that the intramolecular charge transfer is carried out by electron holes. The researchers carried out experiments in which they coated gold electrodes with a self-assembled carboxyl layer and then coated the cytochrome c on it, which was then subject to photoelectrochemical studies.

Specifically, they recorded the photocurrent under illumination with 420 nm light (with a chopper for on/off dynamics) where the zinc substituted cytochrome c has an absorption maximum that coincides well with the photoaction spectrum of the molecule, “indicating that the photocurrent in the UV–vis region originated from its absorption and that Zn-cyt c fixed on the HOOC-SAM electrode retains its original coordination structure”, (Tokita et al. 2008). They modeled the molecule[5] and obtained k as the rate constant:

$$k_{i \to f} = \frac{2\pi}{\hbar} H_{fi} \delta(E_{fi}) \,. \tag{4.10}$$

To arrive at this simple result, a number of steps and considerations are required. When the molecular orbitals (MO) do not overlap, their overlap integral is zero, and, thus, the matrix elements remain zero. In accordance with the classical notation, such molecular orbitals are called *canonical*. Then, the time-dependent perturbator Hamiltonian H' can be approximated by a stationary, time-independent, H. Tokita et al. write F as the δ function of the energy difference between final states E_f and initial states E_i. For a complete derivation, see the article (Renger 2007).

Another assumption was that the so-called Franck–Condon factor (FC) is unity (1). The FC is an overlap integral from wavefunctions. The Franck–Condon effect means that in the course of electronic transitions, changes from one vibrational energy level to a different such level are more likely to occur when the corresponding wavefunctions for the vibration levels have a significant overlap.

Let us look at the hydrogen atom H with the principal quantum number n, rotation momentum l, magnetic quantum number m, and spin quantum number m_s. These are the eigenstates $\langle a^0|$, which can also be written as $\langle n, l, m, m_\mathrm{s}|$. The atom will be hit by light, an electromagnetic wave. The wave has a vector potential A, see, for example, Equation 3.65, and a Coulomb potential ϕ. This is an example that has been demonstrated by Andreas Wacker (Wacker 2016).

5 Scientists and engineers and some others typically consider mathematical modeling as modeling. This is not correct. The model is first a mind model, where you make up some structure in your mind. Today, as we know that molecules look like spheres connected by sticks, the spheres standing for the atoms and the sticks for the chemical bonds, when we think of a *new* molecule, then this will only be a variation of an already existing *old* molecule. The next step is, then, to make a physical model by sketching the atoms and the bonds, and the required bond angles, and the like. You can do this on paper or with tennis balls and sticks, or by sketching it with computer software. The mathematical model then comes relatively late, where you translate the physical model into mathematical equations. The same holds when the automotive industry hires engineers and technicians – as designers! – who model cars. You may use the physical model for test driving and even crash tests. You can do the same with computer programs, the crash test. Maybe as a last note: synthetic chemists sometimes make structural models of molecules, such as hydrogenases, and then at a later stage the functional model. You maybe understand the difference. The new molecule may in part look like a natural hydrogenase, but it cannot perform the chemical task. Only further work on the structure of the molecule may eventually make it a functional model. I explained such a case in my previous book in chapter 7.3.3. Protein spectroscopy, (Braun 2017).

The Hamiltonian for this system reads

$$\hat{H} = \frac{[\hat{p} + eA(r,t)]^2}{2m_e} - \frac{e^2}{4 \cdot \pi \cdot \epsilon_0 |r|} + g_e \frac{e}{2m_e} \hat{S} \cdot B(r,t) \tag{4.11}$$

The scalar product of the spin operator $\hat{\mathbf{S}}$ and the magnetic induction $\boldsymbol{B}(\boldsymbol{r}, t)$ from the electromagnetic wave of the light is an effect of second order only and can be neglected. From the wavelength λ of the visible light, one can determine the wave vector k because of the relation $k\lambda = 2\pi$. Bohr's atom radius is $r_B = 0.053$ nm. So the product $k \cdot r$ is very small in the arguments of the exponents and can be neglected. These are some approximations we can make. The quantum mechanical transition rate from energy term a to b is given by (Renger 2007)

$$\Gamma_{a \to b}(t) = \frac{2\pi}{\hbar} \left| \frac{e}{2m_e\omega} \langle b^0 | E_0(\omega) \cdot \hat{p} \, | a^0 \rangle \right|^2 \left[\delta_t(E_b^0 - E_a^0 - \hbar\omega) + \delta_t(E_b^0 - E_a^0 + \hbar\omega) \right] \tag{4.12}$$

Note the terms $-\hbar\omega$ and $+\hbar\omega$ in Equation 4.12; these are the portions of energy, the *quanta*, which the electromagnetic wave exchanges with the H atom by emission or absorption. Here, the dipole approximation is being applied, and the result in Equation 4.16 contains the dipole matrix element $\mathbf{M}_{ab}^{\text{dip}}$. This will become 0, unless the quantum numbers for the eigenstates $\langle a^0|$, see above, and $\langle b^0|$ satisfy particular algebraic conditions; $\langle b^0| = \langle n', l', m', m_s'|$:

$$l' = l + 1\,, \quad \text{or} \quad \Delta l = \pm 1 \tag{4.13}$$

$$m_s' = m_s\,, \quad \text{or} \quad \Delta m = 0, \pm 1 \tag{4.14}$$

$$m' =\in \{m-1, m, m+1\}\,, \quad \text{or} \quad \Delta m_s = 0 \tag{4.15}$$

The conditions above are the so-called *selection rules*, which cause the absorption and emission of photons of energy $\hbar\omega$ between the energy levels a and b with the rate

$$\Gamma_{a \to b}(t) = \frac{2\pi}{\hbar} \left| \frac{\mathbf{E}_0 \cdot \mathbf{M}_{ba}^{\text{dip}}}{2} \right|^2 \left[\delta_t(E_b^0 - E_a^0 - \hbar\omega) + \delta_t(E_b^0 - E_a^0 + \hbar\omega) \right] . \tag{4.16}$$

4.6 Scattering resonances in chemical reactions

I used the comparison of the billiard balls with scattering experiments. In this book on photosynthesis and quantum electrodynamics, the scattering concerns mainly the light scattering phenomena (photons, or bosons in general), including absorption and emission. The scattering of elementary particles like electrons, positrons, and protons follows the same principles. A not well-known field is the scattering theory in chemical reactions' reactive scattering resonance; examples can be found in the review paper (Fernandez-Alonso and Zare 2002), which deals with the chemical reactions of H and H_2 and D_2, and the like. Such reactions can occur in hydrogen deuterium exchange

(H–D exchange) experiments, which are very important for the hydro-desulfurization in petrochemistry. Practical examples are the removal of H_2S in mineral oil and natural gas (Katsapov and Braun 2010).

Scattering theory applies so well to the chemical reactions that, for example, Breit–Wigner line shapes can be observed or so-called Feshbach resonances, which are usually known in x-ray spectroscopy. Fernandez–Alonso and Zare say: "In this context, reaction thresholds associated with maxima of vibrationally adiabatic curves have been related to barrier resonances." (Fernandez-Alonso and Zare 2002)

4.7 Scattering matrix

The previous sections dealt with scattering problems where an initial state is promoted to a final state. The mediator that produces the objects from the initial state to the final state is mathematically summarized and expressed by an analytical function, which is called the scattering matrix, or for short, the S-matrix. It is considered that the time before the scattering event was an infinitely long time ago; this is when the object was sent to the target. The other consideration is that the scattered object is in a final state when an infinitely long time has passed after the scattering event.

$$S = \langle b_{t\to\infty} | \, | a_{t\to-\infty} \rangle = \langle b_{\text{out}} | \, | a_{\text{in}} \rangle \tag{4.17}$$

It is necessary here to mention that the initial states and the final states are considered free (this means: not mutually interacting) at asymptotic distance. This is a necessary condition for the scattering matrix. The S-matrix includes the physical properties of the scatterer and yields the scattering amplitudes. To some extent you can consider the scattering matrix as the response function.

The matrix formalism for treating scattering problems was originally introduced by John Archibald Wheeler, a colleague of Feynman, when he worked on scattering problems in nuclear physics. Later, Werner Heisenberg introduced the matrix formalism known as *Matrizenmechanik*, in the scattering of elementary particles (Heisenberg 1943).

Matrix formalism in itself is nothing new. It is well established in the mechanics of continua, the solid matter. Particularly in crystalline materials, the properties of the material depend on the crystallographic axes. The relation between mechanical displacement and stress and strain is given by the stress tensor and the strain tensor[6], which contain the elastic constants. In theory, this tensor is a 9×9 matrix with 81 entries. However, because of the simplest symmetry constraints, the matrix reduces to 27 entries for the practically most complicated case. In many cases, for crystalline materials, a simple 3×3 matrix will do, and in the simplest case, we are reminded of Hooke's law, where there is a nonvectorial $F = k \cdot x$.

6 https://www.sciencedirect.com/topics/engineering/strain-tensor.

Also, the optical, electric, magnetic, and dielectric properties of materials depend on the crystallographic axes and are expressed by tensors, which are matrices. It is the fine structure of the atoms, given by the atomic and molecular orbitals, their spin states and overlaps, which forwards the anisotropy of the atoms into the crystal structure. This is why a tensor is necessary for the description of the interaction of external forces, fields, and photons with the matter.

Also, the nuclei of the atoms have a fine structure, which means that the scattering can be an anisotropic process. This is why the scattering matrix is useful for the description of the scattering process. From the mathematical point of view, the S-matrix is a unitary matrix, which connects a manifold of particle states that are asymptotically free. These states are defined in the Hilbert space with infinite dimension. The S-matrix provides the scattering amplitudes. The square of the entries of the S-matrix determine the probability, that the initial state of a scattering is mapped into the final state.

The equation below shows the role of the S-matrix, which transforms the incoming wave vector to the outgoing wave vector

$$\begin{pmatrix} \Psi^{-}_{-L}(x=0) \\ \Psi^{+}_{+L}(x=L) \end{pmatrix} = \begin{bmatrix} S_{11} & S_{12} \\ S_{21} & S_{22} \end{bmatrix} \cdot \begin{pmatrix} \Psi^{+}_{-L}(x=0) \\ \Psi^{-}_{+L}(x=L) \end{pmatrix} \tag{4.18}$$

When there is no interaction, the S-matrix becomes unity, $S = 1$ or $S - 1 = 0$. When we consider the interaction, the relation would read $S - 1 = R$, where $|R|^2$ is the transition probability. The nontrivial part of the S-matrix is called the transfer matrix, T-matrix.

The quantum field theory as it was developed by its founders lacks in some essential issue, and this is the existence of divergences that require a renormalization, which turns out to be impractical. In 1955, an alternative protocol for the calculation of the entries of the S-matrix was developed by Lehmann et al. (1955).

4.8 Foundation and early development of quantum electrodynamics

The theory of quantum electrodynamics can describe all phenomena that arise from the interaction between photons (with energy $h\nu$, wavelength λ, and frequency ν, such as visible light, infrared and UV light, x-rays, and gamma rays) and charged particles, such as electrons, positrons, and holes. In general, these effects include photon absorption and annihilation, photon creation and emission, light scattering, and electron–hole pair formation.

Quantum electrodynamics also includes the classical electrodynamics in the limit of strong electromagnetic fields and high energies, where the experimental data can be considered as continuous numbers, in contrast to quantum discrete numbers for the other cases. As we will see later, formalisms of quantum electrodynamics can also be applied to other systems, such as describing problems on the stock market.

Werner Heisenberg founded the matrix formulation of quantum mechanics (Heisenberg 1925) and further developed it with Ernst Pascual Jordan and Max Born. His theory and their formalism assumes that it is, by principle, impossible (Heisenberg's uncertainty principle) to make an absolutely accurate determination of the mechanical processes in time and space. Experimental observables are only in the classical limit manifestations of physical movements.

Erwin Schrödinger, however, follows the rational of Louis de Broglie and considers the waves of particles, the particle waves as a reality, which are similar to the wave nature of the light. Wave packets with extremely small size in all dimensions constitute the moving particles.

In his paper *Quantenmechanik der Stossvorgänge* (Born 1926), Max Born expresses his discomfort with both theories and offers a third interpretation and shows its utility in the explanation of *Stossvorgänge*, collision experiments. Born refers to Albert Einstein, who once made a remark about *Gespensterfeld*, the ghost field, which would guide the particles but would not contain any energy or momentum.

Given that the electron as particle and the light quantum as particle are analogous, the movement of the electron should follow the same principle as the movement of the light quanta. In this case, Born considers the de Broglie wave of the electron as the guiding field for the movement of the electron.

It was Dirac who expanded (Dirac 1927) Maxwell's classical theory of electromagnetism, which is also the foundation for the classical light theory (see Max Born's book *Optik. Ein Lehrbuch der elektromagnetischen Lichttheorie*, (Born 1933)), with the modern ideas of light quanta and energy quanta, which were developed by Albert Einstein (photoelectric effect 1905) and Max Planck.

It is a noteworthy detail that quantum electrodynamics is able to predict the anomalous magnetic moment of the electron at an accuracy of 10^{-11}, in comparison with the experimentally determined value, the so-called Landé factor gL. Let us, therefore, look at this. The quantity g stands for gyromagnetic factor (or ratio) and constitutes the proportionality constant of the measured magnetic moment μ and the angular magnetic quantum number (here represented by the quantum number S of spin S) and a unit magnetic moment, such as the Bohr magneton:

$$\boldsymbol{\mu} = g\frac{\mathrm{e}}{2m}\boldsymbol{S} \tag{4.19}$$

The Landé factor must take into account the contribution from the angular momentum $\boldsymbol{L}$ of the electron and its spin $\boldsymbol{S}$.

$$\boldsymbol{\mu} = -g_J\frac{2\pi\mu_\mathrm{B}}{h}\boldsymbol{J} \tag{4.20}$$

We know from physics class that the angular momenta couple by the principles of vector algebra. For the Landé factor, the coupling of L and S yields

$$g_J = g_L\frac{J(J+1) - S(S+1) + L(L+1)}{2J(J+1)} + g_S\frac{J(J+1) + S(S+1) - L(L+1)}{2J(J+1)} \tag{4.21}$$

It is an exercise for the reader to calculate the g_J when we have the orbital g_L being 1 and approximate the g_S to 2.

When moving from classical mechanics to quantum mechanics we have to express our physical system in terms of the Schrödinger equation, which we well know is written as (Rebhan 2003)

$$\mathrm{i}\frac{h}{2\pi}\frac{\partial}{\partial t}|\psi\ (t) = H\,|\psi\ (t) \tag{4.22}$$

where the Schrödinger operator is applied to the Hamilton function H. As the photons that interact with the electrons come and arrive with the speed of light, we have to apply the relativistic Schrödinger equation. This will complicate the mathematical formalism considerably, and we must do this here step by step.

It is also characteristic for a relativistic theory that we have to account for particle creation and particle annihilation. The particles under consideration here are the photons, which can be absorbed and create electron hole pairs, or which can be created, as we know, from emission processes.

While quantum mechanics could explain some basic observations in physics and nature, it was unable to take into account the velocity of the electromagnetic interactions with matter (in German: *elektromagnetische Kraftwirkungen*). This was the motivation for Werner Heisenberg and Wolfgang Pauli to reformulate quantum mechanics towards quantum electrodynamics in two joint papers in 1929 (Heisenberg and Pauli 1929, 1930). Dirac had found already that the one-electron problem resists the relativistic invariant mathematical treatment.

We can consider the Lagrangian density (Milton 2009) as the starting point for the new theory

$$\mathcal{L} = -\frac{1}{4}\mathfrak{F}^{\mu\nu} - \bar{\psi}\left[m + \gamma^{\mu}\left(\frac{1}{\mathrm{i}}\partial_{\mu} - \mathrm{e}A_{\mu}\right)\right]\psi \tag{4.23}$$

To understand the concept of Lagrangian density, simply consider, when you integrate the Lagrange density $\mathcal{L}$ over some volume element of arbitrary dimension n, that the result of the integration will be the Lagrange function L itself:

$$L = \int \mathrm{d}^n r\,\mathrm{d}\mathcal{L} \tag{4.24}$$

The purpose of defining and using $\mathcal{L}$ is for the description of the equation of motion for fields. This is particularly important for relativistic effects, where the Lagrange function should be formulated in covariant representation. The action S is then written as

$$S = \int \mathrm{d}^4 x\,\mathrm{d}\mathcal{L} \tag{4.25}$$

and it is invariant against the Lorentz transformation.

Looking at Equation 4.25, we define the parameters: A_μ is the four-vector potential for the photon. The strength of the electromagnetic field is built A_μ as $F_{\mu\nu} = \partial_\mu A_\nu - \partial_\nu A_\mu$. The electron wavefunction is ψ.

4.9 Lippmann–Schwinger equation

Collisions of bodies such as the billiard balls, or cars in a traffic accident, are the matter of classical mechanics and can be described with the conservation laws of momentum and energy. For processes in the quantum domain, such as the scattering of elementary particles and atoms and photons, quantum mechanics is applied. Bernard Lippmann and Julian Schwinger have begun to treat such a molecular level using Lagrange's variation principles.

Their approach is that two bodies when spatially separated would have zero or close to zero interaction energy. The Hamiltonian of such a system is given by a part H_0, which is not perturbed by interaction, and the part H_1, which describes the interaction between the two bodies. The corresponding Schrödinger equation, thus, reads like Equation 4.26

$$i\hbar\frac{\partial\psi'}{\partial t} = (H_0 + H_1)\psi'(t) \tag{4.26}$$

As we are interested in the scattering problem, as this emerges from the Hamiltonian part H_1, which describes the interaction (we can also call this interaction Hamiltonian V), Lippmann and Schwinger used the trick of separating away the irrelevant part H_0 by using a unitary transformation of the type

$$\psi'(t) = \exp\left(\frac{-\mathrm{i}H_0 \cdot t}{\hbar}\right)\psi(t) \tag{4.27}$$

It is an exercise for the reader to carry out this transformation and show that the Schrödinger equation then reads

$$i\hbar\frac{\partial\psi(t)}{\partial t} = H_1(t)\psi(t) \tag{4.28}$$

$$H_1(t) = \exp\left(\frac{\mathrm{i}H_0 \cdot t}{\hbar}\right)H_1 \cdot \exp\left(\frac{-\mathrm{i}H_0 \cdot t}{\hbar}\right). \tag{4.29}$$

The so-called Lippmann–Schwinger equation in Equation 4.30 describes the interaction of two colliding objects

$$|\psi\rangle = |\phi\rangle + \frac{1}{E - H_0 \pm \mathrm{i}\epsilon}H_1\,|\psi\rangle \tag{4.30}$$

with the readily explained contributions from H_0 and H_1.

The Lippmann–Schwinger equation is used in treating a wide range of scattering problems in particle physics and optics, and also in mass transfer and diffusion, such as in physiology and the medical sciences[7].

---

7 During my graduate studies at RWTH Aachen I had a friend who did his diploma thesis, and later also his doctoral thesis (Hermanns 2004), in theoretical physics. In the beginning, he dealt with scattering problems, and from him I learnt about the Lippmann–Schwinger equation as one of the fundamental pillars of scattering theory. He, however, based his work on the Navier–Stokes equations.

When we consider a quantum mechanical scattering state $|\psi_p\rangle$,where p is the momentum, then we can write the Lippmann–Schwinger equation as

$$|\psi_p^{\pm}\rangle = |\psi_p^0\rangle + G^0(E_p \pm \mathrm{i}\epsilon)V\,|\psi_p^{\pm}\rangle \tag{4.31}$$

The ± sign above the scattering state $|\psi\rangle$ stands for the in-going wave and the out-going wave, or particle, as boundary conditions; G^0 is Green's function of the free particle and $G^{(+)}$ Green's function for the out-going free particle.

$$G^{(+)}(r,r') = -\frac{2m}{\hbar^2}\frac{\mathrm{e}^{\mathrm{i}k|r-r'|}}{4\pi|r-r'|} \tag{4.32}$$

The presence of the $|\psi_p^{\pm}\rangle$ term in Equation 4.31 makes solving the equation somewhat problematic. The solving of the problem is simplified by inferring the Born approximation whence the new Equation 4.31 reads

$$|\psi_p^{\pm}\rangle = |\psi_p^0\rangle + G^0(E_p \pm \mathrm{i}\epsilon)V\,|\psi_p^0\rangle \tag{4.33}$$

After a lengthy derivation, Lippmann and Schwinger arrive at the equation for the differential scattering cross section $\mathrm{d}\sigma/\,\mathrm{d}\Omega$ as a function of the scattering angle θ

$$\frac{\mathrm{d}\sigma(\theta)}{\mathrm{d}\Omega} = \frac{1}{k^2}|\sum_l (2l+1)\sin\delta_l \mathrm{e}^{\mathrm{i}\delta l}P_l(\cos\theta)|^2\,, \tag{4.34}$$

where $P_l(\cos\theta)$ stands for the Legendre polynomials, and l is the running index over which the summation is taken.

The elastic scattering of electromagnetic radiation by a free charged particle is what we call Thomson scattering. This is the low-energy limit of Compton scattering. The kinetic energy of the particle and the wavelength of the photon are not changing by the scattering process. The low energy limit is valid provided the photon energy is considerably smaller than the mass energy of the particle: $\nu \ll mc^2/h$. This corresponds to the condition that the wavelength of the light is much greater than the Compton wavelength of the particle.

4.10 Thomson scattering

We recall the discussion of light scattering in the atmosphere from Chapter 1. It is our empirical daily observation that the sky is blue in whatever direction we look, whereas yellow or red sunlight does not come from all directions but directly from the sun. The light is, the photons are scattered by the atoms and molecules in the air in various directions. I remember from my undergraduate physics studies at RWTH Aachen the lab course where we recorded the intensity of light from a light source, a lamp, which was passing some medium. A light detector (photodiode) was mounted on a rotor, which allowed us to record the photocurrent of the photodiode depending at which angle versus the optical axis the photodiode was positioned. Thus, we got an angle-dependent intensity profile of the scattered light. A similar experiment will be shown in Chapter 4.11.

The underlying effect is known as Thomson light scattering. The photons change their momentum $\boldsymbol{p} = \hbar\boldsymbol{k}$ but not their energy $h\nu$. When the target particle is a "free electron", the electron may change its momentum too. Like in any closed system, the momentum and energy are conserved. We learnt in the previous Chapter 3.9.7 that the free electron (and the very weakly bound electron) cannot be brought to an excited state by the absorption of a photon. However, when the photon strikes on an electron, it may give that electron a kick and change its momentum $\boldsymbol{p}$ by an $\Delta\boldsymbol{p} = \boldsymbol{p}' - \boldsymbol{p}$, while the energy $h\nu$ of that photon remains constant.

Let us consider the "free electron" that is being hit by a beam of light, by photons. Figure 4.1 illustrates how an electron that may have a momentum $\boldsymbol{p} = \boldsymbol{0}$ and rest energy $E = m_0c^2$ is struck by a photon. The photon may have the wavelength λ and, thus, the energy $E = h\nu$ and momentum $p = h\nu/c$. You notice in Figure 4.1 that there is a horizontal line, which we simply consider the optical axis. When you carry out angle-resolved optical and electron spectroscopy experiments on single-crystal surfaces, you are able to detect in which directions the electrons escape, and the photons escape after the collision process. The photon may be scattered in direction θ with the optical axis, and the photon then has the momentum $p = h\nu'/c$, and it may also change its energy from the previous E to $E' = h\nu'$. Such an experiment will certainly be done with a monochromatic light source and a detector, which can detect the wavelength of the scattered photon.

The experimental result is then that the light that is made by the photons changes its color from, say, high energy blue to lower energy green. We can argue over whether the original (primary) photon simply changes its energy and momentum, or whether the original photon is absorbed by the electron, and as a result of the collision (absorption), a new (secondary) photon with different energy and momentum is created. The electron will pick up momentum from the collision and be kicked in some direction with angle ϕ versus the optical axis. The momentum then changes to some $p \neq \boldsymbol{0}$.

The energy of the electron after collision changes accordingly to $E = \sqrt{m_0^2c^4 + p^2c^2}$. The zero-point energy (*Ruhenergie*) and the kinetic energy obviously arise from two vectors, which are orthogonal, and their modulus is the total energy E. This is an inelastic scattering process. In their paper from 1976, Pradhan and Khare show how atomic collisions can be calculated based on Feynman diagrams (Pradhan and Khare 1976). A review on the quantum theory of Thompson scattering can be found in Crowley and Gregori (2014).

4.11 An example of light scattering

If you want to measure such scattering patterns, then you need a spherical detector, or a detector that you can move in any angular position around the sample and then record the scattered intensity at the various angles with respect to the optical axis while not changing the radius from the scatterer (the sample). An example of such a light scattering experiment is shown in Figures 4.3 and 4.4.

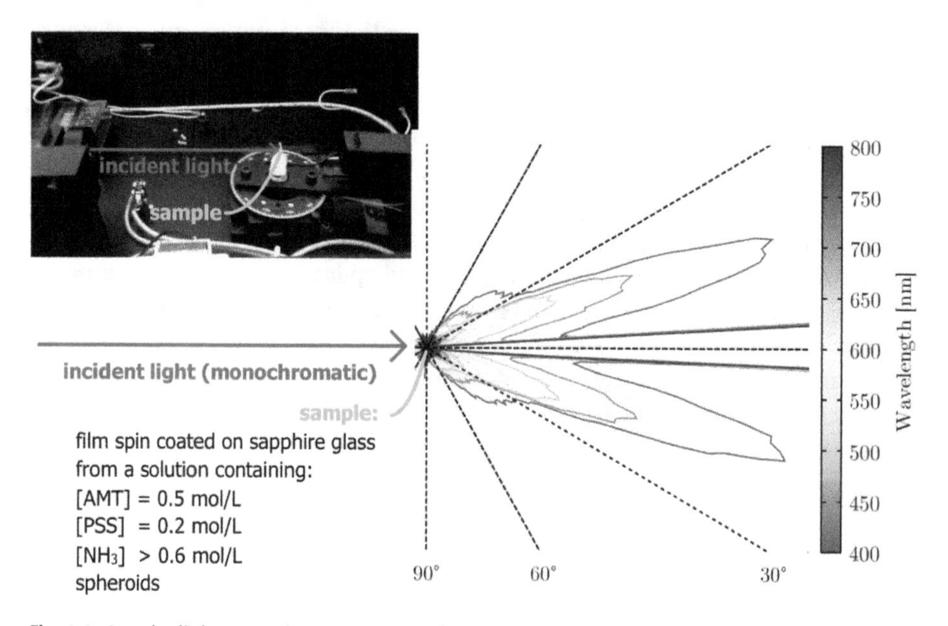

Fig. 4.3: Angular light scattering patterns as a function of the incident laser wavelength from a film composed of microspheroids.

Figures 4.3 shows on the top left a photo of the instrument, which was custom built by Empa scientists and engineers. Experiment and data analysis was made by PhD student Florent Boudoire (Empa and University of Basel, see Boudoire 2015). This light scattering instrument was designed and made at Empa Dübendorf. For my colleagues in the Functional Polymers Laboratory – they work on organic solar cells – it was important to have a light scattering instrument that can use light sources with a whole range of wavelengths. The commercial apparatus would typically have only a laser with one wavelength. In our case, the custom built instrument was designed and built, and as light source we had a strong xenon lamp (500 Watt), which shone into a computer controlled monochromator. With the computer you can select a wavelength and also scan the monochromator, so that the wavelength is scanned over time.

Then, one selected wavelength (incident monochromatic light) is guided into the light scattering instrument, shines on the sample, and the detector scans in azimuthal direction and records the scattered light. You can see the recorded intensity in Figure 4.3 with the wavelength color bar on the right. The red light is strongly scattered in the forward direction of 0°. The other wavelengths are scattered in other angles. This is further schematized in Figure 4.4, where the scattering with 550 and 800 nm light is compared for an FTO substrate and the nanostructured electrode from mWO_3 pillars (actually, flat plates rather than pillars) with a hematite layer. The scattering, thus, depends on the wavelength of the scattered light and on the structure of the electrode.

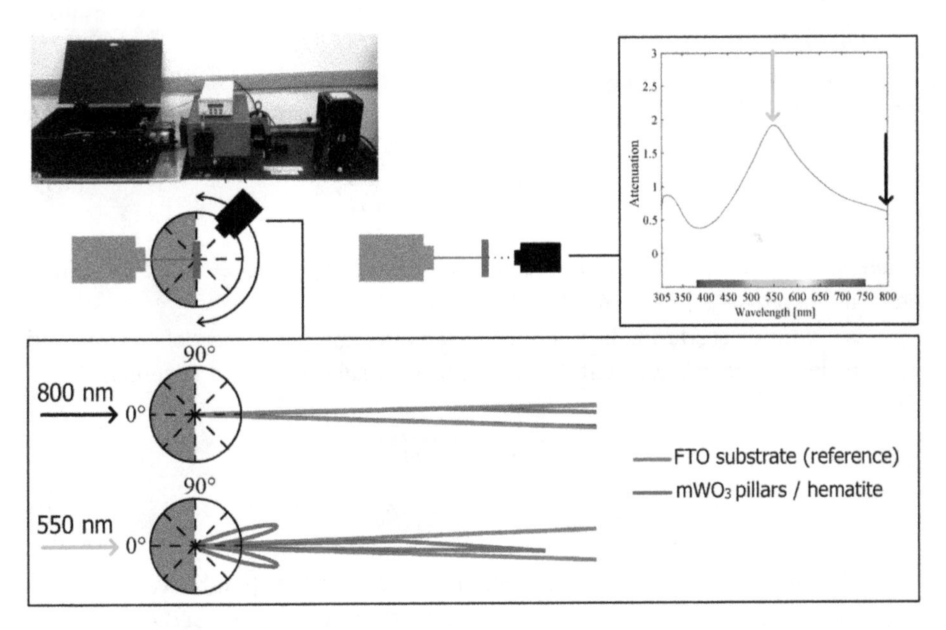

Fig. 4.4: Angular light scattering patterns as a function of the incident laser wavelength from a film composed of microspheroids. Experiment and data analysis was made by PhD student Florent Boudoire (Empa, and University of Basel), see Boudoire (2015).

The upper right panel in Figure 4.4 shows the light attenuation intensity as a function of wavelength. Here, we see that the green light with 550 nm wavelength causes a maximum attenuation. To a large extent, this light stays in the photoelectrode and can be used for energy conversion of photons into charge carriers. This is enhanced by the structuring of the electrode material.

4.12 Compton scattering

It turns out that when the energy of the photons is large enough (the wavelength is short enough), the light that is scattered by an electron has a larger wavelength and the photons a smaller energy than the primary incident light. The obvious lost energy is then found in the gain of the electron energy. This effect was discovered by Arthur Holly Compton (Roy 2018). The *direct* Compton effect is when a photon with momentum q_1 and polarization vector e_1 hits on an electron that has the initial momentum p_1. A new photon then results from the scattering process, with momentum q_2 and polarization vector e_2.

Technically, you see that the light is scattered by an angle θ, and that it also changes its wavelength from ν to ν'. This is mathematically formulated by the relation

$$h\nu' = \frac{h\nu}{1 + \frac{h\nu}{m_0 c^2}(1 - \cos\theta)}, \tag{4.35}$$

where $m_0 c^2$ is the rest mass of the recoiling electron (Higgins 2012). We saw a similar effect in the previous Section 4.11.

The Compton scattering is described by a matrix and parameters of electron and photon yield the matrix element $W = R + S$, which read (Brown and Feynman 1952)

$$R = \frac{e_2}{p_1 + q_1 - m} e_1 \tag{4.36}$$

$$S = \frac{e_1}{p_1 - q_2 - m} e_2 \tag{4.37}$$

The differential cross section for the Compton scattering scales parabolically with the expectation value of the matrix element W, weighted by the squares of the ratio of the energies of incident photon and scattered photon

$$\frac{\mathrm{d}\sigma}{\mathrm{d}\Omega} = \mathrm{e}^4 \left(\frac{\omega_2^2}{\omega_1^2}\right) |\langle W\rangle|^2 \tag{4.38}$$

When you do x-ray fluorescence scattering experiments, you may detect the signatures from characteristic element-specific processes but also an incoherent scattering background, which is due to the Compton process, see, for example, my book (Braun 2017). The PhD thesis by Higgins (2012) is a practical example where Compton scattering is measured and analyzed. Compton scattering is a spectral background signal that is not useful for chemical analysis by x-ray methods. Yet the signature may be there. When we know the mathematical description of the Compton process, we can include it in the scattering model and substract this signal from the spectrum and, thus, obtain a "corrected" spectrum that yields information with higher accuracy.

Compton scattering is not only an important step for the understanding of spectra; it is also considered an entry to quantum electrodynamics theory (Brown 2002).

4.12.1 The Klein–Nishina formula for Compton scattering

We learnt that Compton scattering is an important scattering process. Of interest is the intensity of this scattering, which certainly includes the intensity distribution. As Dirac had developed the relativistic quantum mechanics, Oskar Klein and Walter Gordon came to include a relativistic term in the Schrödinger equation, which then reads

$$\frac{1}{c^2}\frac{\partial^2}{\partial t^2}\psi - \nabla^2\psi + \frac{m^2c^2}{\hbar^2}\psi = 0 \tag{4.39}$$

and which is known as the Klein–Gordon equation.

Klein and Nishina included Dirac's theory in the derivation of the intensity of Compton scattering, i.e., the scattering of the radiation by free electrons (Klein and Nishina 1928). They found that their solution deviates from the scattering intensity when it is calculated by the formulation of Dirac and Gordon. Specifically, the new formulation brings about a second-order correction with respect to the ratio of the energy $h\nu$ of the photon and the energy from the rest mass of the electron $E = mc^2$.

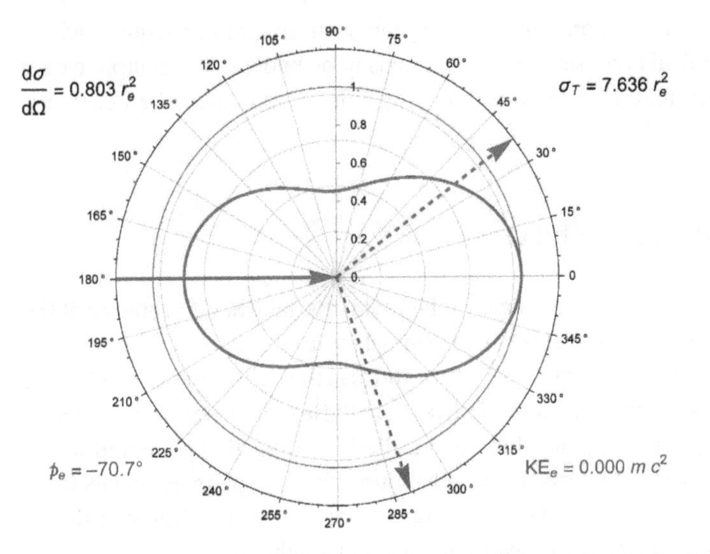

Fig. 4.5: The azimuthal distribution of the scattered photons calculated by the Klein–Nishina formula with the Mathematica software.

Figure 4.5 shows the azimuthal angular distribution of scattered radiation calculated according to (Klein and Nishina 1928) as

$$S = \frac{2\pi Ne^4}{m^2c^4}\left\{\frac{1+a\alpha}{\alpha^2}\left[\frac{2(1+\alpha)}{1+2\alpha} - \frac{1}{\alpha}\log(1+2\alpha)\right] + \frac{1}{2\alpha}\log(1+2\alpha) - \frac{1+3\alpha}{(1+2\alpha)^2}\right\}. \tag{4.40}$$

This is a correction by the order of α^2 in comparison to Dirac's pioneering work. Brown and Feynman later provided further corrections to Dirac's theory and the Klein–Nishina formulation by the order of α^6 (Brown and Feynman 1951). You may wonder why the Klein–Nishina formula has any relevance. During my time as a scientist at Berkeley, we frequently used a custom built x-ray spectrometer to detect fluorescence data for x-ray photoemission experiments on samples relevant for photosynthesis and for energy applications, including renewable electrochemical storage and conversion and also problems from the fossil fuel industry.

Monochromatic x-rays from the synchrotron are shot on the sample, which is considered a point source. The fluorescent x-rays from the sample are detected by a circular array of silicon crystals, which reflect the x-rays in their azimuthal focus, where the x-ray detector is located. Motors move the silicon crystals, which have a diameter of around 10 cm, so that the Bragg condition selects particular wavelengths of the x-rays scattered from the sample. I showed a photo of the spectrometer in one of my recent books (Braun 2017; Braun et al. 2015b). The motor movement means the energy scanning of the x-rays. This is how an x-ray emission spectrum of the sample is recorded with azimuthal resolution.

The x-ray spectrum also contains the Compton scattering of the sample, which is irrelevant for chemical analysis, but which needs to be known for a complete data analysis in the spectrum, and it contains the x-ray Raman signatures of all elements in the sample.

4.13 The photoelectric effect

When light hits matter, electrons are emitted by that material when the energy of the photons is high enough. This effect is called photoelectric emission. For this effect to happen, a minimum photon energy is necessary and also the right material needs to be selected. Pioneering studies by Wilhelm Hallwachs in the late nineteenth century were done on metal zinc plates; he sent ultraviolet radiation to the zinc plate and found that its electric properties changed. Hallwachs found that the plate emits electrons when irradiated with UV light. The kinetic energy E_{kin} of the electron follows a simple relation, actually a relation linear in the frequency of the light ν:

$$E_{\text{kin}} = h\nu - W \tag{4.41}$$

where h is Planck's constant and ν the frequency of the light (the product is the photon energy $E_{\text{photon}} = h\nu$), and W is known as the so-called work function; W is the minimum energy required to remove an electron from the metal surface. We can call ν_0 the threshold frequency of the metal and then write

$$E_{\text{kin}} = h(\nu - \nu_0) \tag{4.42}$$

Part of the photoelectric effect is the absorption of light by the matter, actually the absorption of the photon by the atom. We will follow here the treatment that was first published by Dirac in 1927 (Dirac 1927). Dirac considers an atom enclosed in a volume with a radiation field that is also enclosed, so that the degrees of freedom are limited. The Hamiltonian for this system is composed of the energy from the atom H_0 and the sum of all Fourier components of the radiation field, $\sum_r E_r$:

$$H = H_0 \sum_r E_r \tag{4.43}$$

This implies that there is no interaction term that would account for an interaction between the atom and the radiation field.

4.13.1 Dirac's quantum theory of the electron

The interaction of electrons[8] with the electromagnetic field was described classically by Maxwell's theory. When ordinary quantum mechanics is applied to a point charge electron, it would not reproduce the experimental result of what Dirac called the "duplexity phenomenon" (Dirac 1928), and which was the differentiation of the electron state with an angular momentum, a spin in one and the other direction, which we know today as spin up ↑ and spin down ↓. These are the spin quantum numbers $s = -1/2$ and $s = +1/2$.

Dirac wondered almost 100 years ago why "Nature had chosen" this duplex description of the electron and not the point charge electron description. It turns out that the correct quantum theoretical description of the electron requires that it is compatible with the relativity theory and compatible with the general transformation theory of quantum mechanics[9].

---

8 In the literature, we always read about electrons. However, we can generalize to any electric charges, which are electrons e^-, ions like O^{2-}, Na^+, the proton H^+, and the lightest isotope of hydrogen with almost zero mass, the positron p^+, and so on. The positron is actually an invention or discovery made by Dirac in his paper from 1928.

9 I do not know whether Dirac ever became explicit about his transformation theory. What he means is that the quantum mechanical states, which are expressed and defined by the various quantum numbers of atoms such as the principal quantum number n, the angular momentum L, and the spin S, are defined in a Hilbert space and subject to particular mathematical, algebraic rules such as linear combinations of the L and S, which also constitute valid quantum states. The interaction of quantum mechanical objects is, thus, subject to the mathematical rules that we know as vector algebra. If the electron, for example, did not have a quantum number S, we would not be able to measure the Russell–Saunders coupling – as an example that there is a coupling of angular momentum. (Russell and Saunders 1925) "5. Remarks on spectroscopic notation.—The present state of notation for spectroscopic terms is chaotic." It is an interesting observation that Russell and Saunders published their spectroscopy work in an astronomy journal.

Almost 100 years ago Dirac established a quantum theory of the electron that interacts with an electromagnetic field and writes the Hamiltonian as

$$F = \left(\frac{\Psi}{c} + \frac{e}{c}\boldsymbol{A}_0\right)^2 + \left(\boldsymbol{p} + \frac{e}{c}\boldsymbol{A}\right)^2 + m^2c^2 \tag{4.44}$$

The electromagnetic field is composed by a vector potential $\boldsymbol{A}$ and a scalar potential A_0, and $\boldsymbol{p}$ is the momentum vector; c is the speed of light, e the electron charge, and m the electron rest mass. Dirac chose as the wavefunction

$$\Psi = \mathrm{i}h\frac{\partial}{\partial t} \tag{4.45}$$

Recall that in quantum mechanics, the derivative of the wavefunction with respect to the time t is first order only and not second order.

The momentum vector is the derivative of the three components of the spatial coordinate vector $\boldsymbol{x}$

$$p_r = -\mathrm{i}h\frac{\partial}{\partial x_r} \tag{4.46}$$

With these transformations and operators, we obtain the wave equation

$$F\Psi = \left(\left(\mathrm{i}h\frac{\partial}{c\partial t} + \frac{e}{c}\boldsymbol{A}_0\right)^2 + \sum_r\left(-\mathrm{i}h\frac{\partial}{\partial x_r} + \frac{e}{c}\boldsymbol{A}_r\right)^2 + m^2c^2\right)\Psi = 0 \tag{4.47}$$

which needs to be invariant against a Lorentz transformation. As an exercise you can write down the Hamiltonian in the case of the *absence* of the electromagnetic field, which will then simplify the equation. Dirac rationalizes his approach by considering that the electron moves in empty space, and, therefore, the Hamiltonian would not require the time and space variables t and $\boldsymbol{x}$

When there is no field, the following equation emerges

$$\left(\beta + \sum_{i=0}^{3}\alpha_i p_i\right)\psi = 0 \tag{4.48}$$

and the operators β and α_i are independent of the components of the momentum vector $\boldsymbol{p}$, and they commute with t and the components of $\boldsymbol{x}$. The algebraic structure of the problem requires that four matrices are at hand with conditions $\sigma_r^2 = 1$, the anticommutator $\{\sigma_r\sigma_s\} = 0$, and $r \neq s$. The Pauli matrices, developed for accounting for the electron spin, are useful to solve this problem, but there are only three, as shown below. They are Hermitian and unitary:

$$\sigma_1 = \begin{bmatrix} 0 & 1 \\ 1 & 0 \end{bmatrix}$$

$$\sigma_2 = \begin{bmatrix} 0 & -\mathrm{i} \\ \mathrm{i} & 0 \end{bmatrix}$$

$$\sigma_3 = \begin{bmatrix} 1 & 0 \\ 0 & -1 \end{bmatrix}$$

To get the required fourth matrix, Dirac extends the dimension of the matrix from 2 to 4 across the diagonal; two exemplary such matrices are shown below.

$$\sigma_1 = \begin{bmatrix} 0 & 1 & 0 & 0 \\ 1 & 0 & 0 & 0 \\ 0 & 0 & 0 & 1 \\ 0 & 0 & 1 & 0 \end{bmatrix}$$

$$\rho_1 = \begin{bmatrix} 0 & 0 & 1 & 0 \\ 0 & 0 & 0 & 1 \\ 1 & 0 & 0 & 0 \\ 0 & 1 & 0 & 0 \end{bmatrix}$$

With the consideration of the electron in an electromagnetic field, Dirac extends the Hamiltonian by the scalar potential A_0 and the vector potential A, which reads, with the matrices ρ and σ,

$$\left[p_0 + \frac{e}{c}A_0 + \rho_1\left(\sigma, p + \frac{e}{c}A\right) + \rho_3 mc\right]\Psi = 0 \tag{4.49}$$

This eventually results in two extra terms, two extra energies from a new degree of freedom of the electron spin, the magnetic moment, which is being charged. There is an electric pendant in the energy, but this is only from the imaginary part of the Hamiltonian and does not exist.

4.14 Light absorption by organic molecules

When an atom absorbs a photon with the energy $h\nu$, the following relation holds according to Bohr's model for the atom: $E_f - E_i = h\nu$; E_i is the initial state of the atom, and E_f is the final state of the atom, after absorption of the photon. When light is absorbed by molecules, the absorption spectrum is not a sharp line but a somewhat broadened absorption band, mostly because some of the photon energy is converted into molecule vibrations and, thus, into thermal energy.

Theodor Förster pointed out in his paper (Forster 1938b, a) that light absorption is a common process for the ultraviolet short wavelength range. However, for the visible UV range and the visible wavelength range, absorption is a hardly observed process; only particular materials would absorb in this range. In carbohydrates you observe absorption only in specific unsaturated molecules. This holds particularly for the aromatic hydrocarbons (these are the molecules where the carbon atoms are arranged in a ring, and the hydrogen atoms are attached outside from each carbon, as figured out originally by Kekule). The carbon atoms are held together in the ring by carbon double bonds of the form C=C, and single bonds attach the hydrogen atoms to the carbon atoms: C–H. One carbon atom in an aromatic ring, therefore, has three neighbors: two carbon atoms to the left and right, and one hydrogen atom on the outside.

Förster starts out with his work by resorting to the early quantum mechanical concept by Slater and Pauling (Slater 1931; Pauling and Wheland 1933, 1934) and assumes for the time being that the carbon atoms are only linked by single bonds: C–C. Such a bond is considered as the bond made by a pair of electrons from both neighboring atoms, and these two electrons have antiparallel spin: ↑↓ or ↓↑. With the aforementioned three neighboring atoms for every carbon atom linked with single bonds only, only three valence electrons from the central carbon are required to form this aromatic cycle. The fourth valence electron per carbon atom is, thus, available for any further valence participation. When we pair two of these extra valence electrons, we notice that we can do this in an ambiguous conformation, as demonstrated in Figure 4.6[10].

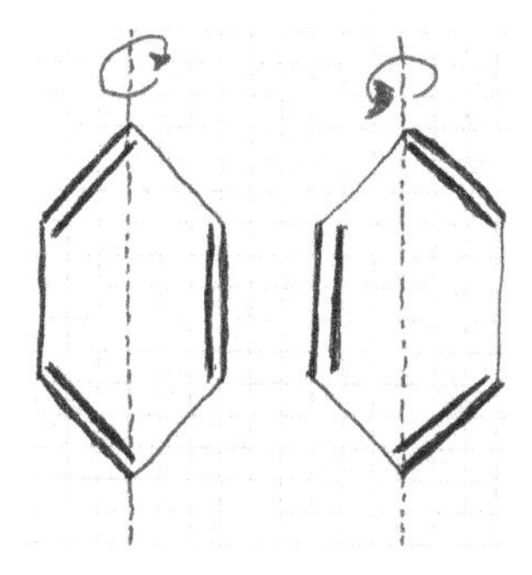

Fig. 4.6: Two alternative conformations of an aromatic ring (Forster 1938b, a), which can be transformed to each other by a simple rotation (symmetry operation) along an axis in the paper plane which goes through any pair of opposing corner points. The red lines indicate one of the three possible axes.

This unambiguity will not permit the system, the benzene ring, to assume as an electronic state either of both alternatives or both alternatives. Rather, the actual state will be a coupling of both states, the superposition of the aforementioned states. The physical origin for the coupling of both states is the existence of a symmetry operation, which transforms both states upon rotation by 180°. The mechanical analog for this superposition is given by a pair of pendula with identical properties, which are coupled with a spring coil. The oscillation of one pendulum with its eigenfrequency is not possible because it is coupled to the other pendulum. Only the frequencies under coupling are physically possible. The frequencies are given by the cases where the pendula are oscillating in phase and out of phase. The strength of the coupling (let us call it the oscillator strength; this is an important metrics in spectroscopy) determines by how much the frequencies differ from each other.

With this in mind, we can try to interpret the valence electrons and bonds in the benzene ring as the impossible electronic states (valence states) analog to the aforementioned impossible pendulum frequencies. Förster notes in his paper (Forster

10 Says Slater in his paper (Slater 1931): Other cases which we surely could not treat by the general method would be those like C6H6, the benzene ring, where an ambiguity between two ways of drawing the valence bonds seems to be an essential feature of the structure.

1938b, a) that it is not entirely correct to call the electronic configurations in the valence structures “states”. States in the quantum mechanical basis are only those that arise from superpositions of such valence structures. The superposition state with the lowest energy $E = h\nu$ is the ground state E_0 of the molecule, the “normal state”, as Förster puts it. The other states are excited states with energy E_i, which require a specific energy that they obtain by the absorption of a photon.

For the case of the benzene molecule, there will be no bond breaking. Moreover, the necessary energy for making excited states is relatively small. These states are made only by superpositions of the valence states from three double bonds in the aromatic ring molecule. It appears, therefore, that these molecules have excited energy levels that are only slightly above the ground-state levels. When the energy difference is small enough, then the low energy photons can be absorbed, right? This is the reason why the absorption band of benzene (aromatic carbon) is in the range of 250 nm, whereas the absorption bands of aliphatic carbons (chains of carbon, not rings) are in the wavelength range of 200 nm and lower. Note that smaller wavelengths mean higher excitation energy. It depends, therefore, on the structure, on the configuration of the atoms, in which wavelength range the molecules will absorb.

We recall that the possible spin orientations in paired electrons in the particular molecular arrangement are – because of the commutability – governed by statistical principles. However, also the neighboring atoms with their valence electrons interact with the atoms in their proximity. Accordingly, there is a particular statistical probability that the spins of the excess valence electrons are antiparallel as well. It is an exercise for the reader to derive that the probability for this case to happen is $1/2^{n-1}$. Förster outlines further in his paper how larger and more complex ring structures can be treated. Förster used the Pauling–Slater determinant method for the calculation of the energies of the relevant states of molecules.

4.15 Example of the calculation of absorption spectra

The fundamental considerations on cyclic hydrocarbon structures and their excess valence electrons is now being worked into the mathematical apparatus of quantum mechanics that was determined by J.C. Slater (Bader 2011) and L. Pauling (Pauling 1992). Förster considers here only those valence electrons that are in excess of those atoms that are used for building the molecule and not employed in normal single bonds.

These excess valence electrons – their number is by default 2n – are characterized by eigenfunctions that depend on the coordinates of the spin and the angular momentum of either valence electron. Förster defines the state as the superposition of valence eigenstates, a linear combination of eigenfunctions ψ_l like

$$\psi = \sum c_l \psi_l \tag{4.50}$$

Such an eigenfunction can, for example, have the form

$$\psi_1 = \frac{1}{\sqrt{(2^n)}} \frac{1}{(2n)!} \sum (-1)^P P\phi_1(1)\phi_2(2)[\alpha(1)\beta(2) - \beta(1)\alpha(2)] \cdot \phi_3(3)\phi_4(4)[\alpha(3)\beta(4) - \beta(3)\alpha(4)]\ldots \quad (4.51)$$

You will have to use pencil and paper in order to work yourself through the above equation. I am following here the exercise by Forster (1938a); P is any arbitrary permutation of the electron coordinates s and l (spin and angular momentum) over which we have to add up. Odd permutations have a negative sign (-), and even permutations have a positive sign (+), which is shown by the factor $(-1)^P$. The second electron (2) in the first atom (1) with *positive* spin has the eigenfunction $\phi_1(2)\alpha(2)$. The second electron (2) in the first atom (1) with *negative* spin has the eigenfunction $\phi_1(2)\beta(2)$.

When you carry out the multiplication with the terms in the n brackets, every resulting sum term is a state of the system, where every particular electron with a particular spin is at a particular atom.

With the understanding of these eigenfunctions ψ_j we can move on and write the Schrödinger equation with the aim of eventually determining the energy states E_j of the molecules, which manifest in the absorption spectra. The calculation of the energy values is laborious, quite laborious. Pauling discovered some rules that are based on considerations regarding the geometry and symmetry of molecules (Pauling and Wheland 1933, 1934), which considerably simplify the calculations.

$$(H - E_j)\psi_j = 0 \quad (4.52)$$

With the eigenfunction $\psi = \sum c_l \psi_l$ we can pull out the constant c_{jl} and obtain

$$\sum_l c_{jl}(H - E_j)\psi_l = 0 \quad (4.53)$$

The Hamilton operator is built from the kinetic energy $p^2/2m$ (we call the mass of the electron μ; in which we recognize the Laplace operator Δ as the second derivative for the spatial coordinates) and the potential energy V

$$H(1, 2, \ldots, 2n) = \frac{1}{2\mu} \sum_m \left(\frac{\partial^2}{\partial x_m^2} + \frac{\partial^2}{\partial y_m^2} + \frac{\partial^2}{\partial z_m^2} \right) + V(1, 2, \ldots 2n) \quad (4.54)$$

Now you multiply the Schrödinger equation one by one with the valence eigenfunctions ψ_k and integrate over the entire coordinate space, so that you arrive at a linear equation system

$$\sum_l c_{jl} \int \ldots \int \psi_k(H - E_j)\psi_l \,\mathrm{d}\tau_1 \ldots \mathrm{d}\tau_{2n} \quad (4.55)$$

which you simplify to

$$\sum_l c_{jl}(H_{kl} - d_{kl}E_j) = 0 \quad (4.56)$$

with the appropriate abbreviations d_{kl} and H_{kl}; the energies E_j are not functions or vectors but mere numbers, scalars. This helps with the visual simplification of the problem, which turns out to be a linear equation system. Because the right-hand side is 0, we are dealing here with a homogeneous equation system. The condition for the solution of this system is that the determinant of the system (in German: *Säkulargleichung*) equals 0. This condition then reads

$$H_{kl} - d_{kl} \cdot E_j = 0 \tag{4.57}$$

and the roots of the equation are the energy values E_j, which we experimentally find in the optical absorption spectra.

By superposition of alternative molecule structures a number of polygons, "islands", shows up. We call this number i_{kl}. The matrix element d_{kl} is calculated by the number of these islands according to

$$d_{kl} = \frac{1}{2^{n-ikl}} \tag{4.58}$$

In the superposition polygon, we may identify also a number κ_{kl} of pairs of atoms that are adjacent or connected with an odd number of dashes, and λ_{kl} is the number with even such dashes or links. The component H_{kl} for the Hamiltonian then reads

$$H_{kl} = -\frac{3}{2}\frac{\kappa_{kl} - \lambda_{kl}}{2^{n-i_{kl}}} \cdot a \tag{4.59}$$

The quantity a is the exchange integral, which depends on the atomic distance and the nature of the atomic state, and the overlapping areas of the atomic orbitals. By default we consider this integral as always positive $a \geq 0$. We can exercise this protocol for the r rings with six bonds, which include benzene ($r = 1$), naphthalene ($r = 2$), anthracene ($r = 3$), naphtacene ($r = 4$), pentacene ($r = 5$), and so on. So we are talking here about macromolecules that are built up from a number r of rings. They may be arranged in a line or in a two-dimensional structure[11].

11 I worked for a couple of years at the University of Kentucky for the Consortium of Fossil Fuel Sciences. There I developed soft x-ray spectroscopy methods for the analyses of the molecular structure of carbon and hydrocarbon materials in fossil fuels and biomass. The foundations are well explained in the book by Joachim Stöhr (Stöhr 1992). The x-ray core-level spectra of carbon compounds allow a very good analysis of their molecular structure. The spectrum of benzene, the benzene ring, has particular spectroscopic features. Naphthalene is made from two such connected rings, but the x-ray spectrum is not simply the double of the spectrum of one benzene ring. There are also three-dimensional carbon structures, such as carbon nanotubes and Buckminster fullerenes. Here we are no longer dealing with molecules but with materials. For those carbon structures, including graphene sheets, Nobel Prizes have been awarded. A while ago I came across a paper by my fellow colleagues at the University of Kentucky, who wrote that they had noticed carbon nanotubes in their samples before they were reported by the Nobel Prize winner in Japan. I believe this is true. This is why I fully agree with one of my PhD students' thesis advisors and the ETH President, who said "and as soon you have results, publish them."

In such a set of structures, the number of nonexcited valence electrons is $r + 1$, and r is the number of the benzene rings in the molecule. With these numbers, we can fill Equations 4.58 and 4.59 and obtain the following matrix elements, which depend on the parameters k and l:

$$d_{kl} = \frac{1}{4^{l-k}} \tag{4.60}$$

$$H_{kl} = -\frac{2}{3}\frac{2r + 3l - k + 1}{4^{l-k}} \cdot a \tag{4.61}$$

and are also dependent on the number r, which is the number of benzene rings

$$\begin{vmatrix} \frac{9}{2}a + E & \frac{9}{4}a + \frac{1}{4}E \\ \frac{9}{4}a + \frac{1}{4}E & \frac{9}{2}a + E \end{vmatrix} = 0 \tag{4.62}$$

For example, for naphthalene, we find a 3×3 matrix. The matrix elements on the side diagonal stand for the coupling of the valence structures. This coupling is of the order $1/2^{n-1} = 1/4$. Our aim is to diagonalize the entire matrix, so that the entries off the diagonal are all zero. This constitutes the transformation of principal axes in linear mappings and, thus, the orthogonalization of the solution space. With the mathematical rules for the manipulation of determinants[12] we can try to simplify the determinant by creating zeros in the entries of the determinant by multiplying a row with a constant and subtracting one row from another.

$$\begin{vmatrix} \frac{27}{2}a + \frac{5}{2}E & 0 \\ 0 & \frac{9}{2}a + \frac{3}{2}E \end{vmatrix} = 0 \tag{4.64}$$

Since we are dealing here with a 2×2-matrix, we expect two solutions, i.e., two roots of the equation. We can, of course, exercise this according to the definition for the determinant as shown in the footnote, which leads us to write

$$\left(\frac{27}{2}a + \frac{5}{2}E\right) \cdot \left(\frac{9}{2}a + \frac{3}{2}E\right) = 0\,. \tag{4.65}$$

Recall that when a product of the factors $A(x)$ and $B(x)$ are zero, $A(x) \cdot B(x) = 0$, then $A(x) = 0$ or $B(x) = 0$, or both. Thus, the solution of the above Equation 4.65 is given by the solution of the two equations $(27/2\,a + 5/2\,E) = 0$ and $(9/2\,a + 3/2\,E) = 0$. For the energies E, this yields the solutions $E_1 = -27/5\,a$ and $E_2 = -9/3\,a$. We can see that the wanted energies for the atomic state, E_1 and E_2 depend on the exchange integral a. With $a \geq 0$, we conclude that E_1 would be the ground state of the atom and E_2 the first excited state.

---

12 The determinant is a scalar. The determinant of a square matrix of dimension 2 is, for example, given as $\det A =$

$$\begin{vmatrix} a & b \\ c & d \end{vmatrix} = ad - bc\,. \tag{4.63}$$

The difference $E_2 - E_1 = (-9/3\,a) - (-27/5\,a) = 12/5\,a$ would be the range of the light absorption that causes the excitation of the atom. Now the question arises: how large or small is the quantity a? We continue to read in Förster's paper (Forster 1938a, b) and learn that a was determined by three different groups in the 1930s with different experimental methods, ranging from 1.55 over 1.92 to 2.72 eV. Förster had doubts about the accuracy of these values, given that they were based on assumptions that he felt were not necessarily justified. He, therefore, resorted to his own solution to the problem. Förster calculated the energies for a whole set of aromatic hydrocarbons with different sizes and structures, the results of which are listed in Table 4.1. The experimentally measured absorption wavelength for benzene is 2550 Å.

Tab. 4.1: Comparison of measured and calculated absorption maxima and calculated excitation energies in multiples of the exchange integral a (Forster 1938a, b). The two columns on the right show the wavelength of the absorption range maxima.

Molecule	Excitation energy in multiples of [a]	Wavelength calculated	Absorption range Å measured
Benzene	2.40 a	2450	2550
Naphthalene	1.97 a (3.34 a)	2950	2750
Anthracene	1.60 a (3.00 a, 3.80 a)	3650	3700
Napththacene	1.31 a (2.62 a, 3.56 a, 4.07 a)	4500	4600
Pentacene	1.08 a (2.20 a, 3.1 a, 4.0 a, 4.2 a)	5450	5800
Phenanthrene	1.94 a (...)	3000	2950
Pyrene	1.70 a (...)	3450	3300

As a rule of thumb, the larger the aromatic hydrocarbon molecules become, the larger the possibility, the number of excited states, becomes. You can now consider more complex molecules, formulate their Hamiltonian, and solve the Schrödinger equation accordingly to determine the energy levels of the ground state and the excited states. As an important side note, the determined energies are not accurate in absolute terms because there is no reference known as to where the energy would be 0. However, we consider here only the energy differences ΔE and can, therefore, neglect the missing reference energy.

Förster plotted the absorption maxima of the first absorption range of these molecules versus the calculated energy differences $\Delta E = E_2 - E_1$ in a double logarithmic representation. The energy difference ΔE is given in multiples of the exchange integral a, and the slope in this log-log plot (actually, the slope of the least-squares fit with a log-log function) will then yield an average value for the exchange integral, for which we find $a = 2.11$ eV.

With the calculated energy differences ΔE of more than a handful of different hydrocarbon molecules, you want to know where the systemic differences are, where the unknowns are, and where the constants are. As a researcher you should do this by intuition anyway, without any specific purpose.

The reader can consider this as a small exercise as follows in Figure 4.7. The energy of a photon is given by $E = h\nu$ according to Bohr's frequency condition. In our notation, the correct formulation is $\Delta E = h\nu$ or $\Delta E = h/\lambda$. We know the energy differences ΔE from our calculations (Slater) and we know the intensity maxima of the absorption ranges from the spectroscopy experiments. We can plot these data in linear axes in normal coordinates, such as $\Delta E/a = h/\lambda a$. You can try this and apply a least-squares fit to $1/\lambda a$, but this may turn out to be somewhat difficult. A convenient trick is to apply the logarithmic function to the equation, which will yield

$$\log\left(\frac{\Delta E}{a}\right) = \log\left(h\frac{1}{\lambda a}\right) = \log\left(\frac{h}{a}\right) + \log\left(\frac{1}{\lambda}\right) = \log\left(\frac{h}{a}\right) - \log\lambda\,. \tag{4.66}$$

Recall that this entire derivation was exercised by Forster (1938a).

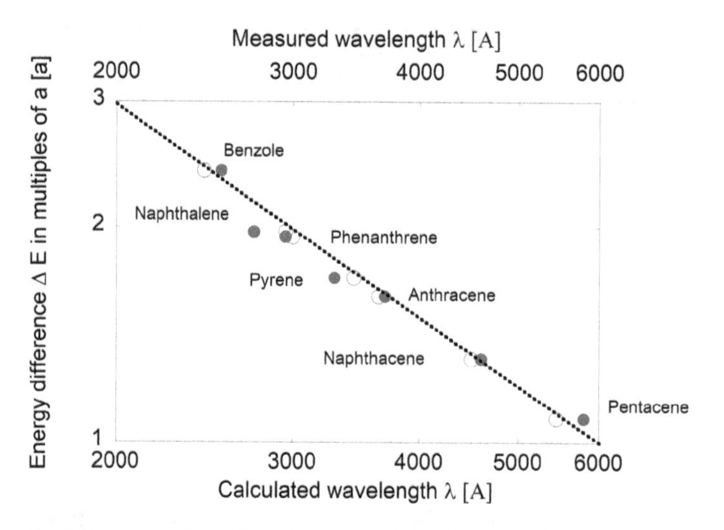

Fig. 4.7: Log-log plot of the calculated energy differences $E_2 - E_1$ in multiples of the unknown exchange integral a.

4.15.1 The relevance of mathematics in science and engineering

You realize now that the absorption spectra will be an application of linear algebra because the fundamental considerations around the valency states that absorb photons eventually merge into mathematical equations. Not everyone interested in physics and chemistry is also interested in mathematics, although many physicists are inclined towards mathematics. Moreover, a number of chemists do quite well in mathematics. This is in spite of the prejudice of younger physics students, who may feel some rivalry towards and superiority over those who do not study physics or mathematics. The legal framework (curriculum and syllabus and the like) around the physics studies at RWTH Aachen specifically mentioned that the physics student and future physicist was expected by the school to have serious interests in mathematics. Few physicists have problems with mathematics. I know that many students of mechanical engineering drop out from university in the first few semesters because they did not know that the undergraduate studies would be filled with very demanding mathematics classes and exams. I believe that at RWTH Aachen, 60 to 70% would not make it past the next semester.

Anyway, mathematics is not taught at university in order to tease you and keep you from progressing and graduating. The reason why mathematics is taught and required at university is a very practical one. When you find, as a researcher, that a problem you are working on, a system you are studying, has mathematical properties, then your physical (chemical, engineering, biological, physiological, economic, social, any such problem I will call a physical problem or system) problem will become a mathematical one. Then you can try to identify the mathematical solutions to that problem.

Here is a very simple example: all problems where you can count items, such as in stocking shelves in supermarkets and stores, adding and subtracting items when you buy stock or when you sell to customers. Even children can do that by intuition. These transactions of selling and buying obey the simple laws of arithmetics, and you can outsource the bookkeeping of these transactions to the accounting department.

In Chapter 2, we saw how the constellation of sun and Earth constitutes a geometrical (*geos* means Earth) system, and by resorting to fundamental geometrical principles in triangles, we are able to calculate the size of the Earth, see Erastothenes. Without seeing this astronomical constellation as a geometrical system and problem, we would not be able to estimate the size of the Earth.

Here in spectroscopy, we are dealing with algebra. The "mechanical" properties of the atoms held together by chemical bonds manifested by an equation of motion, which is written as a Hamilton function and a Schrödinger equation, the solutions of which yield eigenfrequencies that correlate with the optical absorption measured by experiment. The task of spectroscopy is to look for resonances. The absorption peak is a resonance effect where an equation of motion has an inhomogeneity with solution 0 in a denominator. All further complications, such as damping of an oscillation or multiciplity of a solution in an equation system, have their physical analogon in the real world.

When you know about population dynamics in biology, you may use it in social sciences. We will see in Chapter 8 how problems that we encounter in economy and banking have their analogons in physics. When after your university studies you know the full course on mathematics, then you have the minimum experience in finding the mathematical properties in the physical sciences problems you are working on. Then, for a long time, the physical problem will be a mathematical problem, and you can apply the whole mathematical apparatus to the problem. We have seen this above, where we arrive at a system of linear equations for which we seek solutions and for which simple manipulation of determinants yields these solutions.

I remember a class in metal physics (not part of the physics curriculum but part of the metallurgy and mining curriculum at RWTH Aachen) that I took because I chose it as a minor and also because I was enrolled in metallurgy and mining because of my particular interest in that field. In that class, we had an exercise where a determinant was given as equation 3.68, and we students were expected to determine the solutions. There were three physics students in this class, the majority of students were engineering majors. We physicists wrote down the solutions as $E_1 = -27/5\,a$ and $E_2 = -9/3\,a$. The numbers were different, and the actual example was from the strain tensor in mechanics, but the mathematical structure was the same. The teaching assistant was surprised how we calculated the solutions of the equation so quickly. Well, we did not calculate the solutions, we wrote them down – because this was a matrix in normal form; it had no off-diagonal entries, and, thus, the diagonal entries were identical to the solutions.

The teaching assistant, however, and the engineering students multiplied the entries and equated the product to zero $(27/2\,a + 5/2\,E) \cdot (9/2\,a + 3/2\,E) = 0$, and later factorized the quadratic (parabolic) function again. This certainly was not necessary at all because the solutions are given with the diagonalized matrix.

The manipulation of the matrices and determinants, where in the end you have a fully diagonalized matrix where only the principal diagonal has entries $\neq 0$ and all the other entries are 0, is a very good example of how a problem falls apart into a set of solutions. You only have to realize the solutions.

4.16 The calculation of bio-organic dyes

Now we can look at how the optical spectra of a natural dye can be treated. Here, I follow very closely the treatment summarized by Rienk van Grondelle and Bas Gobets in chapter 5 in Papageorgiou and Govindjee (2004)[13]. Chlorophyll is a natural dye in photosynthesis that contains magnesium Mg as the metal center. Photosystem I (PS I) contains around 96 of such chlorophylls, where the distance between the Mg atoms is around 1 nm. The electron clouds of these dyes, thus, strongly interact with each other to the extent that the optical absorption spectra show noticeable changes in the visible and infrared wavelength range. The Chlorophyll Q_y transition at 680 nm is, for example, dominated by Coulomb interaction. The interaction between two chromophores (1) and (2) with distance $\boldsymbol{R}_{12}$ is given in lowest-order approximation by the

13 A book edited by Papageorgiou and Govindjee, Chlorophyll a Fluorescence, which I strongly recommend to the reader.

interaction of their dipole moments $\boldsymbol{\mu}_1$ and $\boldsymbol{\mu}_2$ (their scalar products):

$$V_{12} = 5.04 \cdot \left(\frac{f^2}{\epsilon_r}\right) \cdot \left(\frac{\boldsymbol{\mu}_1 \cdot \boldsymbol{\mu}_2}{R_{12}^3} - \frac{(\boldsymbol{\mu}_1 \cdot \boldsymbol{R}_{12})(\boldsymbol{\mu}_2 \cdot \boldsymbol{R}_{12})}{R_{12}^5}\right) \tag{4.67}$$

The factor f^2/ϵ_r accounts for the electric properties of the medium (a continuum medium, a cavity field) in which the chlorophylls are embedded.

With the Coulomb potential energies V_{ij} and the site energies ϵ_{ij} known, we can build the complete Hamiltonian[14] H_0 in Equation 4.68

$$H_0 = \begin{bmatrix} \epsilon_1 & V_{12} & \cdots & V_{1i} & \cdots & V_{1(N-1)} & V_{1N} \\ V_{21} & \epsilon_2 & \cdots & V_{2i} & \cdots & V_{2(N-1)} & V_{2N} \\ \vdots & \vdots & \ddots & \vdots & \vdots & \vdots & \vdots \\ V_{i1} & V_{i2} & \cdots & \epsilon_i & \cdots & V_{i(N-1)} & V_{iN} \\ \vdots & \vdots & \vdots & \vdots & \ddots & \vdots & \vdots \\ V_{(N-1)1} & V_{(N-1)2} & \cdots & V_{(N-1)i} & \cdots & \epsilon_{N-1} & V_{(N-1)N} \\ V_{N1} & V_{N2} & \cdots & V_{Ni} & \cdots & V_{N(N-1)} & \epsilon_N \end{bmatrix} \tag{4.68}$$

To solve the corresponding Schrödinger equation we need a set of N quantum mechanical base functions ψ_i, which reads $|i\rangle = |\phi_1\phi_2 \ldots \phi_i^* \ldots \phi_N\rangle$ with $i = 1, 2, \ldots, N$.

When this arrangement of magnesium metal centers in the chlorophyll is hit by photons with energies higher than the difference between the highest occupied molecular orbital (HOMO) and the lowest unoccupied molecular orbital (LUMO), electrons are excited from the HOMO level to the LUMO level.

4.17 Förster resonance energy transfer (FRET)

It is physically possible that electromagnetic energy is passed on from a donor chromophore to an acceptor chromophore by an electromagnetic *field* without the involvement of an electromagnetic *wave*. This is a radiationless energy transfer and thus does not involve photons.

Figure 4.8 shows schematically two oscillators which are able to emit and absorb electromagnetic radiation. Above we see situation (a), where oscillator (1) is emitting radiation as illustrated by the arrows pointing away from the center of the oscillator. In some not further specified but remote distance is the other oscillator (2), which is not emitting any radiation, indicated by the lack of arrows.

Now look at situation (b) below where the two oscillators are approaching each other until they meet half way. Oscillator (2) is hit with a considerable portion of ra-

14 Hamiltonian for photosynthetic excitations after Rienk van Grondelle and Bas Gobets in Papageorgiou and Govindjee (2004).

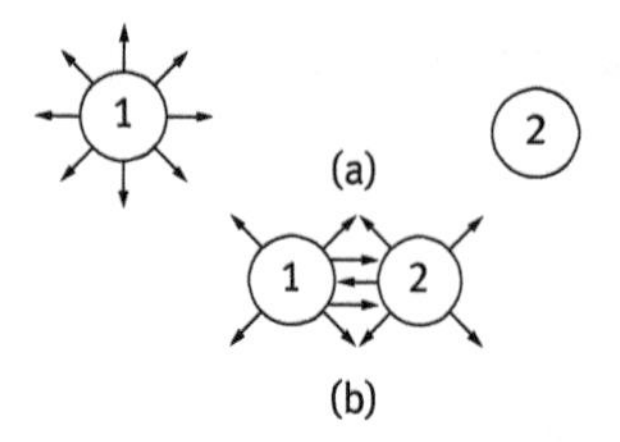

Fig. 4.8: The irradiation of two oscillators with a distance d in schematic view. a) $d \gg d_0 d \gg do$. Only oscillator (1) emits because it is somehow primarily stimulated. b) $d \ll do$. Gleiche Ausstrahlungswahrscheinlichkeit der Oszillatoren 1 und 2. Reprinted by permission from Springer Verlag. Naturwissenschaften 33 (6), 166–175, Energiewanderung und Fluoreszenz, Theodor Förster, (1946). https://doi.org/10.1007/BF00585226 (Forster 1946).

diation from oscillator (1)[15]. As oscillator (2) receives radiation from (1), (2) becomes excited from (1) and also begins emission of radiation. You can call this a resonance effect.

This resonance effect is not much different from a large musical instrument in your home, such as a piano – because, when you clap your hands or make a humming sound, you will notice that the interior of the piano somehow responds with a particular sound. Any other resonance body, be it a musical instrument or some other apparatus, will do the same. The sound waves from you are being absorbed, and a sound specific to the mechanical properties of the resonance body will be echoed.

The Jablonski diagram for such process is sketched in the top panel in Figure 4.9. The donor absorbs light and gets excited from state S_0 to, say, excited state S_2. This state may be discharged into state S_1 while releasing energy. Instead of releasing the rest of the energy by falling further down to level S_0 in the donor, the energy from S_1 is passed on to the acceptor level S_1, see the red horizontal arrow with label FRET. There, the acceptor may release light energy $h\nu$ by photoemission, and the final state is again S_0.

The efficiency of such transitions depends on the distance between donor and acceptor; the farther away the acceptor, the smaller the probability of energy transfer. Specifically, the relation between distance R and efficiency E is given by equation

$$E(R) = \frac{1}{1 + \left(\frac{R}{R_0}\right)^6} . \tag{4.69}$$

15 We remember that the flux through an area A that originates radially from a point source increases parabolically with the square of the reciprocal distance. This is why you learn in radiation safety class that you should increase your distance from a radiation source in order to keep your body away as much as possible from harm from radiation, such as γ-rays or x-rays. Currently, in early 2020 with the Corona virus crisis (Zhang and Liu 2020; Cohen and Kupferschmidt 2020), people are forced by the government under penalty law to maintain so-called “social distancing” in order to lower the probability of spreading the Corona virus and interpersonal infection.

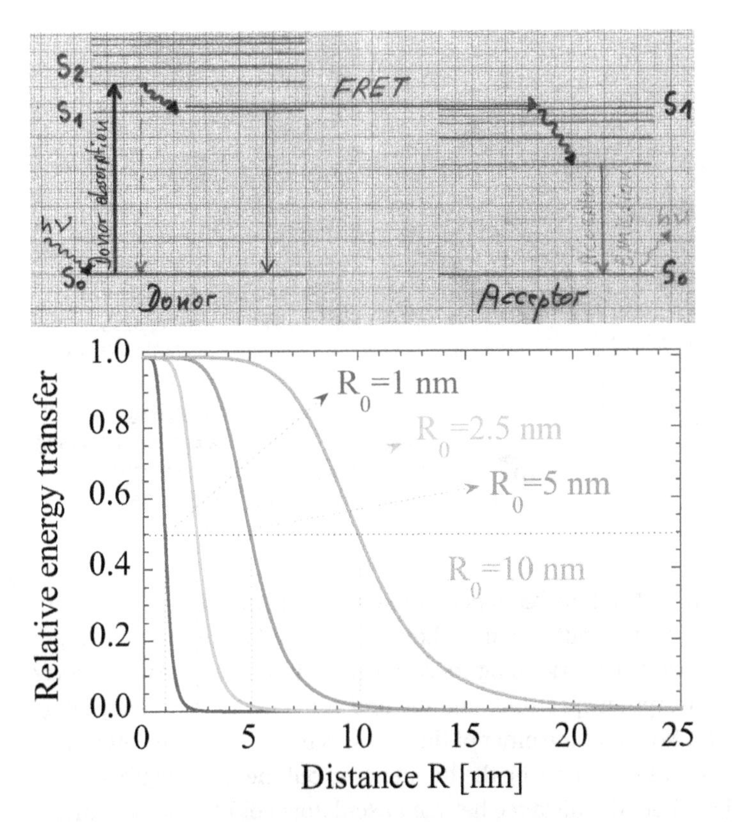

Fig. 4.9: Top: a Jablonski diagram representing Förster resonance energy transfer (FRET). Bottom: the dependence of the energy transfer from the distance of acceptor and donor chromophore, for four different Förster distances ranging from 1 to 10 nm.

The distance between acceptor and donor chromophore is plotted as the ordinate axis in the bottom panel in Figure 4.9 in multiples of 1 nm. The abscissa is the relative energy transfer, and 1 stands for 100% transfer. The distance at which the relative energy transfer is 1/2, i.e., 50%, is called the Förster distance. I have plotted this for several situations from 1 to 10 nm. The Förster distances are denoted by the vertical dotted colored lines. The horizontal dotted line denotes the 50% mark.

It may depend on the physical constellation of the "Förster arrangement" whether the Förster distance is 1 or 10 nm. However, the energy transfer scales with the sixth power of the distance. Thus, the chromophores must be close together to allow for Förster resonance energy transfer. Important is that there is an S_1 level in the acceptor available; this is the resonance condition for the transfer.

The condition for resonance between both oscillators is valid only for a short period, during which their eigenfrequencies are identical, or at least overlapping with respect to their coupling width, which depends on the dipole moments M and the distance d of the oscillator molecules; then the overlapping period or transmission time

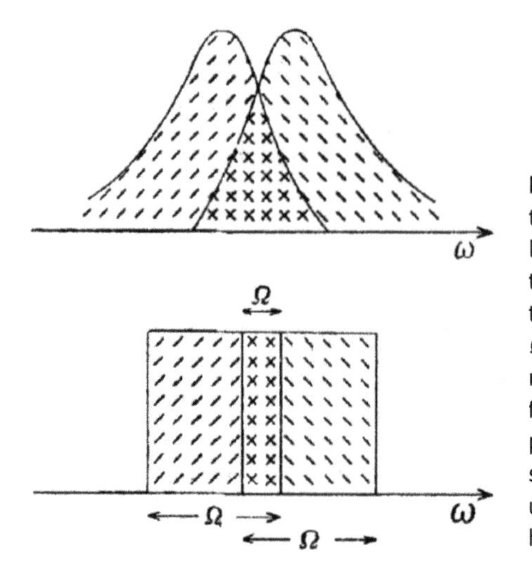

Fig. 4.10: Schematic of an absorption spectrum and fluorescence spectrum that overlap. Top: realistic intensity distribution. Bottom: simplified spectra with rectangular intensity distribution and an overlap width of Ω. The shaded area with increasing dashes means absorption, and the shaded area with falling dashes means emission. Reprinted by permission from Springer Verlag. Naturwissenschaften 33 (6), 166–175, Energiewanderung und Fluoreszenz, Theodor Foerster, (1946). https://doi.org/10.1007/BF00585226.

is $t_0 \sim \hbar\, d^3/M^2$. The corresponding frequency overlap Ω is illustrated by two overlapping rectangular intensity distributions in the bottom panel in Figure 4.10. The upper panel in Figure 4.10 shows two overlapping spectra (Forster 1946), the one at the lower frequency being an absorption spectrum, and the one at the higher frequency being an emission spectrum. Note that the order of the two spectra would be the other way around if the ordinate axis would not be the frequency ω but the wavelength λ.

You may wonder where the distance between oscillators could be a real parameter. In the laboratory, this is the case with dye molecules, where it was observed that very diluted dye solutions have a qualitatively different spectroscopy behavior than concentrated solutions, as was discovered and treated by Jean and Francis Perrin (Berberan-Santos 2001).

When you think of a solution in which dye molecules are solvatized, you are looking at more than two oscillators. One particular oscillator may have some six next-neighbor oscillator molecules with which it can interact. In his paper (Forster 1946), Förster notes that the absorption spectra of single molecules look different to those from solutions with high dye concentrations. In the case of two dye molecules, one can think of the similar excitation energy of the single molecule, where the one has a slightly higher energy than the single molecule and the other a slightly lower energy. This would correspond to a splitting of the primary energy term.

In Figure 4.11, you can see the sketch of a double molecule of methylene blue, which is known as a strong blue color dye in analytical chemistry. For photocatalysis experiments, you can use the blue solution as an optical marker. When the photocatalyst, such as TiO_2 is dispersed in the methylene blue solution, and the UV light is switched on, the photocatalyst will over time decompose the methylene blue, and the color change can be measured with a photometer. The transient of the photometer curve allows for the determination of the reaction rate of the photocatalyst.

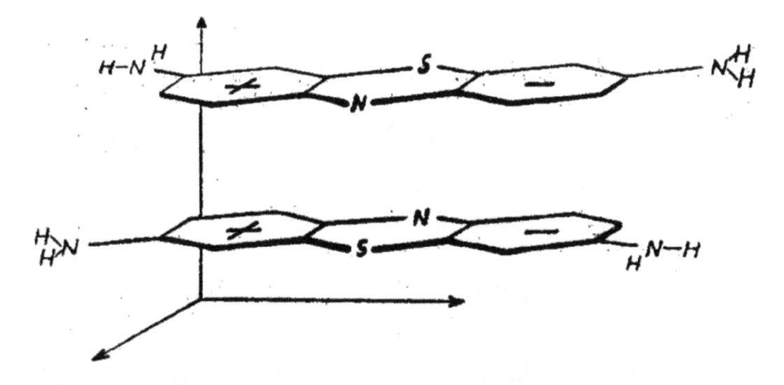

Fig. 4.11: Doppelmolekül von Methylenblau. Die eingezeichneten (willkürlich lokalisierten) Ladungen entsprechen Momentanwerten der Elektronenoszillation. Die stationären Ladungen der Moleküle sind nicht berücksichtigt. Reprinted by permission from Springer Verlag. Naturwissenschaften 33 (6), 166–175, Energiewanderung und Fluoreszenz, Theodor Förster, (1946). https://doi.org/10.1007/BF00585226.

One would expect a double absorption band for such stacked dye molecules, but in fact, only one band is observed. Förster argues that this lack of a second band can be explained with quantum mechanical selection rules, which would require particular geometrical arrangements of the molecules. The stacking arrangement as shown in Figure 4.11 satisfies the geometrical requirement for a selection rule that does not permit a second absorption energy term.

With respect to the Förster resonance energy transfer (FRET), it is worthwhile mentioning that noone lesser than Robert J. Oppenheimer presented a paper at the APS meeting in Pasadena, 18–21 June 1941, where he explained the energy transfer from a dye to chlorophyll, and how this amount of energy could not be explained by conventional energy transfer (Oppenheimer 1941; Bethe 1997).

When Förster studied photosynthesis in the 1930s, he found that his findings made little sense; or no sense within back then contemporary theory and understanding.[16] Förster measured the photosynthetic performance of a system that certainly should depend on the input, such as the number of photons, i.e., the light intensity,

16 Whenever you have experimental findings you should check whether they make sense, check them for plausibility. In German, we call this *Plausibilitätsbetrachtung*.
It is my opinion that graduate students in science and engineering should include in their experimental research theses also a theoretical or computational part. One reason for this is that when you go for an academic degree you should have exercised the mathematical body of your field, which you learnt at university. Thus, that you are able to master the mathematics distinguishes you from researchers who have no university degree (usually). The other reason is that you should always be able to check whether your experiments, your research data, and your findings make sense. You check what you measured for plausibility. You compare your numbers with numbers published elsewhere. You refine and combine your data and check whether they are in the ballpark of what is possible, or whether they are fully out of physical range.

the photon flux, and the number of photosynthetic reaction centers. When you do a quantitative analysis of the thylakoids in your sample, their concentration per unit volume, their number in a microliter, then you can calculate, or at least estimate, the number of reaction centers in that sample. With the known turnover frequency you can determine the amount of reaction product. These numbers did not add up properly in experiments on the photosynthetic apparatus, which 100 years ago were called *Zur Theorie der Assimilation* (On the theory of assimilation). This refers to the assimilation of carbon dioxide. The contradictory results were also measured in dye solutions.

In their paper on the theory of assimilation (Gaffron and Wohl 1936b, a), Gaffron and Wohl noted the results that around 1000 dye molecules would work around one CO_2 molecule in photosynthesis, which was called "assimilation". The chapter in which they discussed the theory was called *On the location and mechanism of assimilation*. Obviously, in 1936 researchers were able to count the dye molecules in a photosynthesis complex, but apparently there were still fundamental questions as to how photosynthesis and how assimilation would occur, and where.

Apparently, back then there was the opinion that in the CO_2 assimilation, i.e., in the photosynthesis, the individual carbon dioxide molecule would be associated with the individual chlorophyll molecule, and that the assimilation would be a process coherent in time and location. However, experimental observations would not support the hypothesis of such simple mechanism. Instead, the cell would be able to accumulate over time, and over an extended spatial range sufficient energy for the reduction of one CO_2 molecule. Such a process would require a high level of cooperation of different molecules.

Consider the volume of a leaf with a number N of reaction centers. When you have N photons arriving on this volume, at maximum every reaction center will absorb one photon and create one electron that works in the redox reaction. Then, all photons are absorbed and used, and none are left. When you increase the number of photons by a factor of 2, no more redox reactions will take place, because the redox centers are already occupied. Only beyond the turnover frequency can another photon be absorbed. With this rationale, it would not make much sense to increase the light flux above the leaf. However, the leaf does produce more redox reaction electrons empirically. Therefore, something in the theory that I just laid out must be wrong. The system does not scale. The explanation is that there are absorbers in the leaf that are not reaction centers. The photons are funneled by light harvesting proteins and pass on their energy not as photons but as electromagnetic nonradiative energy. You cannot come to such a groundbreaking conclusion unless you do the math, be it as simple as here.

You may have noticed in the mathematical outline I gave during the quantization of the electromagnetic field that I made a distinction between the electromagnetic field and the radiation field. This discrimination is necessary because in one case, the energy is stored in the electromagnetic field, and in the other instance, in the photons. According to Heitler (1947), the dipole radiation must have a minimum distance from the dipole matter before it will condense to a photon (see, also, the textbook by Robert M. Clegg, chapter 4, (Clegg 2004)).

Oppenheimer presented his work in the *Journal of General Physiology* (Arnold and Oppenheimer 1950). He was a physicist, and hence not known for any work in life sciences, and his discovery went virtually unnoticed by the photosynthesis community despite its groundbreaking nature. As an extensive review of the Förster resonance, consider the book by Govorov et al. (2016).

4.18 The porphyrins

We saw in the previous section how carbon atoms form cyclic structures (ring structures) and how their optical spectra change with the complexity of the arrangement of the carbon atoms[17].

There are also other ring structures, and these are known in living matter, in biology and biochemistry, and, finally, in macromolecular chemistry. These are the porphyrin rings. They are the base pigments of human and animal blood cells, the hemin and hemocyanin groups, and the plant pigments chlorophyll and bacteriochlorophyll, which are necessary for the photosynthesis and, thus, primary production. You can look at them as a carbon skeleton of unfinished aromatic rings, where the aromatic ring is not closed with a carbon atom but with a nitrogen group. The four nitrogens form a cavity within a tetradentate ligand, which can be filled with a divalent ion, such as Fe(II). Then, when iron is included, it is called a heme structure as it is found in blood. It is possible that other divalent ions, such as Mg^{2+} or Zn^{2+} can fill the cavity, such as in the case of porphyrin zinc, as shown in Figure 4.12 (PubChem 2019). Porphyrin zinc is strongly fluorescent and can be used in analytical chemistry as a dye for quantitative analyses.

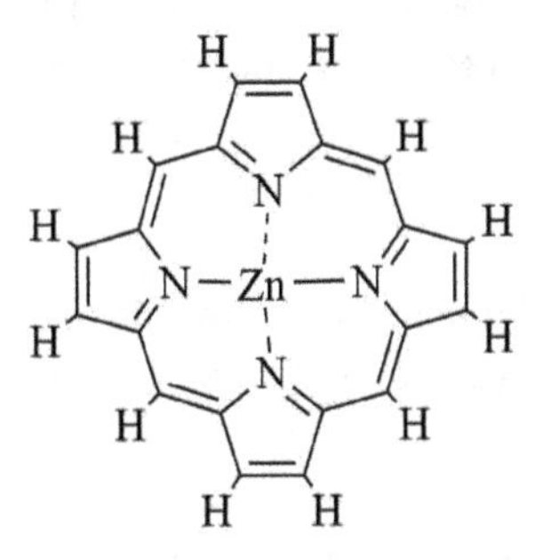

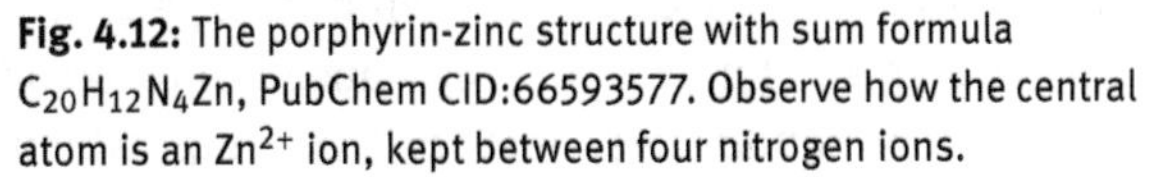

Fig. 4.12: The porphyrin-zinc structure with sum formula $C_{20}H_{12}N_4Zn$, PubChem CID:66593577. Observe how the central atom is an Zn^{2+} ion, kept between four nitrogen ions.

17 Carbon is an interesting element. It forms graphite structures from laterally extended graphene sheets, when you role them you get carbon nanotubes, and there are also closed spherical structures, such as the Buckminster fullerenes. Later, it was found that also other elements and compounds can form such roundish structures.

The macromolecules with such a cavity structure can be used as sensitizer molecules like, for example, in dye sensitized solar cells (DSSC), such as like the Grätzel cell. As long as 75 years ago, Rabinowitch provided a simple geometry for the porphyrin structure, which is reminiscent of origami patterns, see Figure 4.13 and 4.17. It is an exercise for the reader to count the number of conjugated double bonds in this geometric structure in the Figure. The color of this molecule and its derivatives is green, and, thus, they are called chlorophylls (in Greek, *chloros* means green).

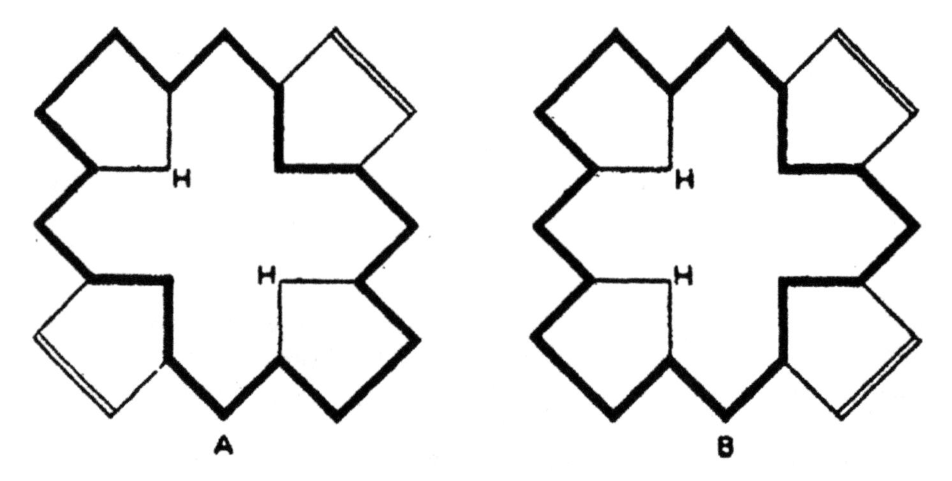

Fig. 4.13: The porphin system. A – diagonal form; B – lateral form. The heavy line represents an all-round conjugated system of nine single and nine double bonds. The four corner positions nearest to the center are occupied by nitrogen atoms; thus, the whole system can be derived from four pyrrole nuclei connected by four CH-bridges. http://dx.doi.org/{10.1103/RevModPhys.16.226}. Reprinted with permission from E. Rabinowitch, Reviews of Modern Physics, 16, 226–235, 1944. Copyright (2019) by the American Physical Society.

You can read from Figure 4.13 that the structure can have a diagonal arrangement with respect to an additional carbon double bond as shown in 4.13A, and a lateral form as shown in 4.13B. They have a reddish or purple color that is reminiscent of the color of blood. Hydrogenation of the molecule causes the color to change to green, like the color we know from the green chlorophyll in plants.

Figure 4.14 shows the optical transmission spectrum of the molecule porphin dissolved in ether. The spectrum is taken from a paper published by Albers and Knorr in 1936 (Albers and Knorr 1936). The transmitted intensity is plotted versus the wavelength given in nanometers, here written as mμ, which correctly stands for millimicron. Note that the ordinate begins on the left-hand side with the larger wavelengths, which is the lower photon energy and, thus, the red range. The peak labeled with A is at 490 nm and, thus, corresponds to absorption in the turquoise color range. Peak B has its maximum at 390 nm and is, thus, in the ultraviolet range (compare Figure 4.18). Albers and Knorr measured several different porphin molecules with different aliphatic side chains, where the porphin molecule is just the reference without

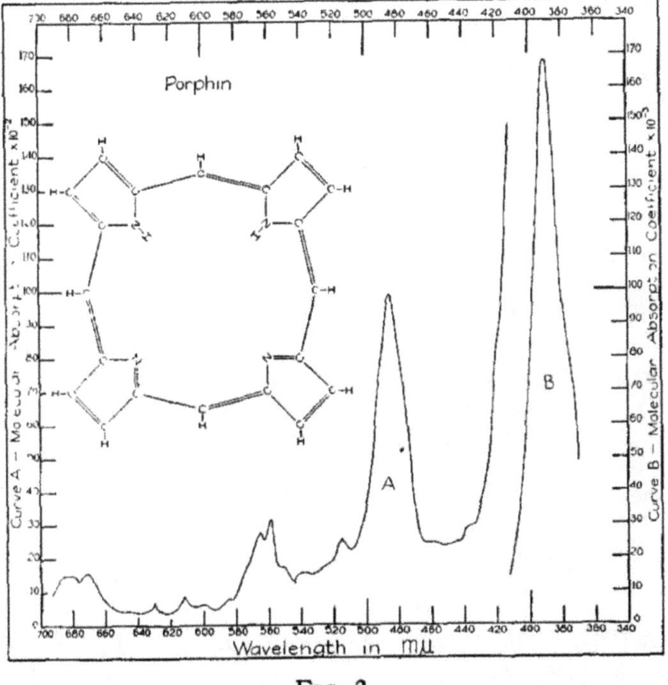

Fig. 4.14: The absorption spectrum ("absorption coefficient curve") of porphin, along with its structure formula. Note the two different abscissae for peaks A and B, where the scale differs by one order of magnitude. Reprinted from V.M. Albers and H.V. Knorr, Spectroscopic Studies of the Simpler Porphyrins I. The Absorption Spectra of Porphin, ms-Methyl Porphin, ms-Ethyl Porphin, ms-Propyl Porphin and ms-Phenyl Porphin J. Chem. Phys. 4, 422 (1936); https://doi.org/10.1063/1.1749873, with the permission of AIP Publishing. (Albers and Knorr 1936).

such a chain. They found that the presence of an aliphatic chain increases the absorption band intensity, but the amplitude of the peak becomes lower as the aliphatic chain becomes longer.

Back in 1936, light sources were so-called photoflood lamps. These were light bulbs with filaments that were operated at electric currents and, thus, at temperatures higher than those of conventional light bulbs. Therefore, the brilliance of the lamp is much higher with the tradeoff, however, that the lifetime of the lamp is much shorter. This is not a problem, if you want to do an important experiment. The light is then guided into a spectrograph (Figure 4.15), with the key optical element being a quartz prism that splits the white light from the photoflood lamp into its spectral components. By rotating the prism by small angular increments you can select the color and the wavelength and send it through the sample of which you want to measure the transmission.

A camera is placed behind the spectrometer, in this case a spectrograph from the Adam Hilger company, which will record the transmitted intensity for that particular

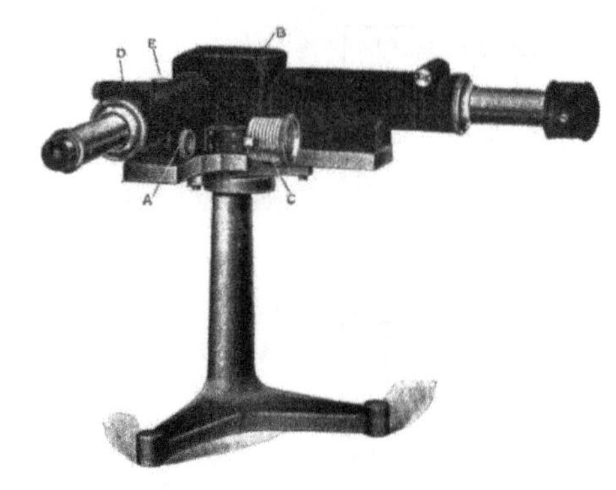

Fig. 4.15: The new model (1926) constant deviation wavelength spectrometer deserves special attention. As will be seen in Figure 3, the working parts of the instrument, as well as the prism, are protected and covered with cast aluminum shrouds, and the appearance of the instrument is aesthetically pleasing – not an unimportant consideration in design. Reprinted with permission from A.F.C. Pollard, Notes upon the mechanical design of some instruments shown at the exhibition of the physical and optical societies, Journal of Scientific Instruments, 4 (6), 184–190, 1927. Copyright IOP Publishing. Reproduced with permission. All rights reserved. (Pollard 1927)

wavelength. The camera (Figure 4.16) contains a photofilm, specifically I-L spectroscopic plates from Eastman (1935). The raw data that you obtain is then a photofilm, where the illuminated regions are darkened, and the developed film would show a white spectrum. You can see the experimental setup, which is shown in a museum today, in Carvalhal and Marques (2015). You can imagine that the emission spectrum of a gas would send a sharp line spectrum, whereas the spectrum of a solid would produce a broader spectral intensity distribution. This is why they are called absorption bands or emission bands; the darkened regions on the film are then not sharp but broad.

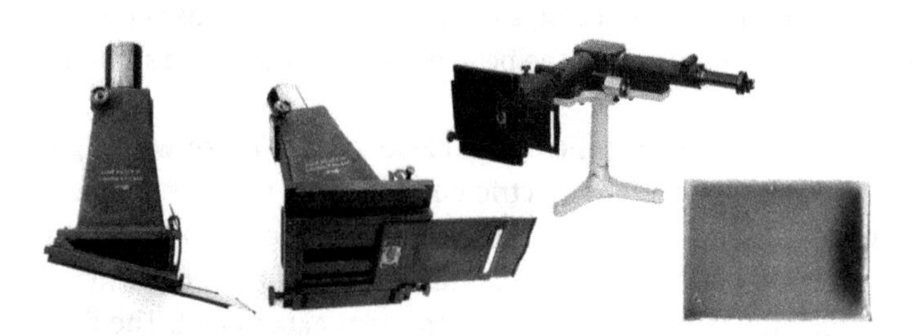

Fig. 4.16: The photographic camera and a metal dark slide, which takes 3 1/4 × 4 1/4 inch plates, known as quarter plates. Reproduced from M. Joao Carvalhal; Manuel B. Marques, Adam Hilger revisited: a museum instrument as a modern teaching tool, Proceedings Volume 9793, Education and Training in Optics and Photonics: ETOP 2015; 979328 (2015) https://doi.org/10.1117/12.2223202 Event: Education and Training in Optics and Photonics: ETOP 2015, 2015, Bordeaux, France. Published under Creative Commons (CC BY 4.0) license https://creativecommons.org/licenses/by/4.0/legalcode.

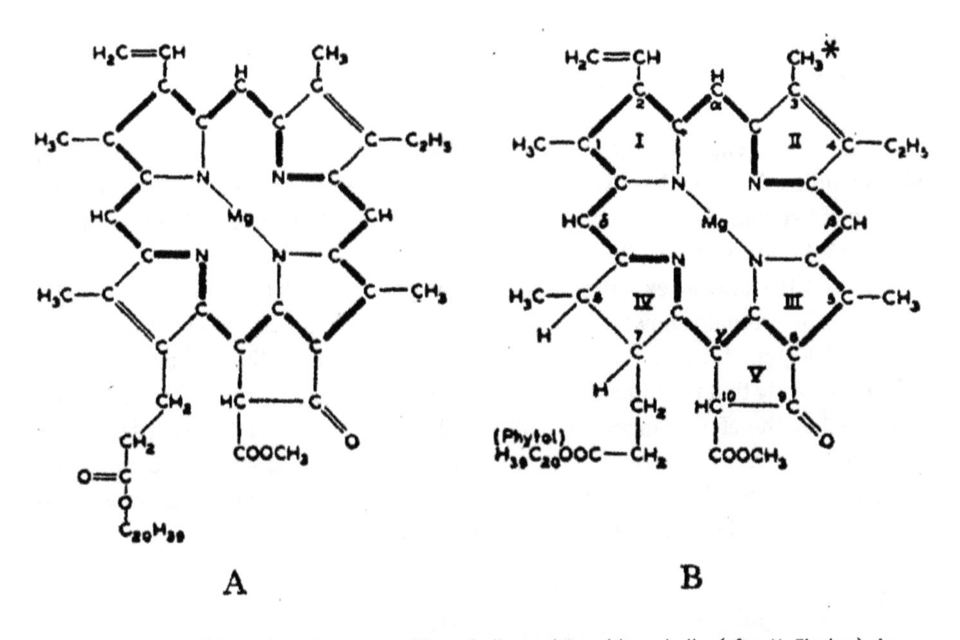

Fig. 4.17: Structural formulae of A – protochlorophyll a and B – chlorophyll a (after H. Fischer). An asterisk designates the CHa-group, which is replaced by CHO-group in chlorophyll b. http://dx.doi.org/{10.1103/RevModPhys.16.226}. Reprinted with permission from E. Rabinowitch, Reviews of Modern Physics, 16, 226–235, 1944. Copyright (2019) by the American Physical Society.

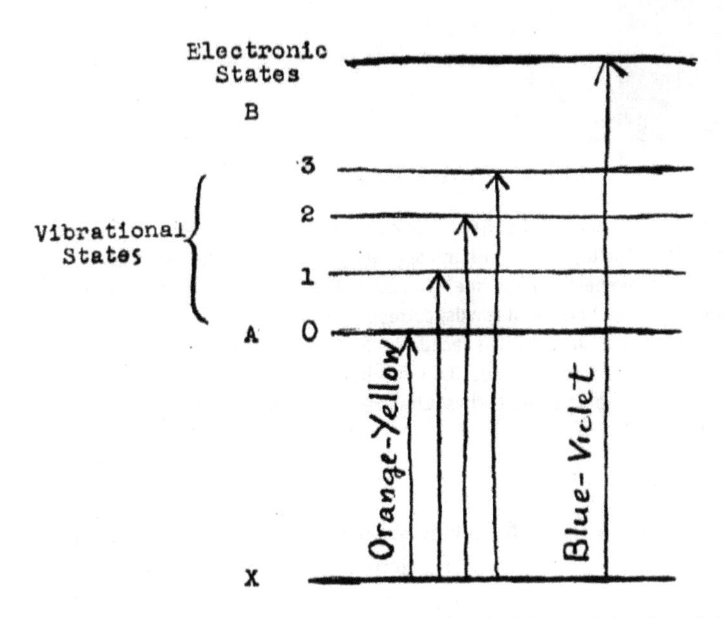

Fig. 4.18: The energy system of porphine. Reprinted with permission from E. Rabinowitch, Reviews of Modern Physics, 16, 226–235, 1944. Copyright (2019) by the American Physical Society. http://dx.doi.org/10.1103/RevModPhys.16.226.

4.18.1 Vibrational coherence in porphyrin by two-dimensional electronic spectroscopy

I present here an exercise that was authored by Camargo *et alii* and by Heisler *et alii*, which studies the so-called two-dimensional spectroscopy method. It identifies the validity of Feynman diagrams and is applied to a porphyrin chromophore (Camargo et al. 2015; Butkus et al. 2012).

Figure 4.19 shows an experimental spectrum graph of such two-dimensional spectroscopy. The ordinate shows an angular frequency range ω_1 divided by $2\pi c$ as multiples of $1000\,\mathrm{cm}^{-1}$. Since the angular frequency is 2π times the frequency f, or ν, we find correctly that the dimension of the ordinate is a (the) wavenumber 1/cm. The wavenumber, therefore, ranges from 14800 to 16400 1/cm, which translates to 675.68 to 609.76 nm. This is the red color range. The abscissa shows the same frequency number range, but labeled as ω_3. So we have a two-dimensional plot where one wavenumber range is plotted versus another wave number range.

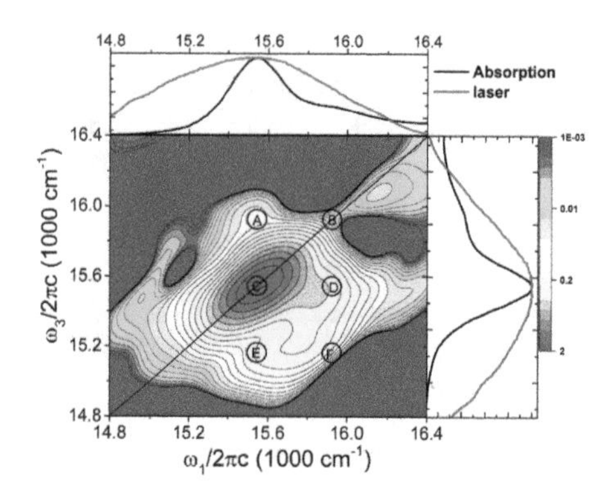

Fig. 4.19: Absorptive 2D emission spectrum of porphine for a population time equal to 120 fs (which is a maximum of the oscillation). The top and right-hand side graphs show the linear absorption spectrum (black line) and the laser spectrum (red line). The 2D plot vertical (y axis) corresponds to detection wavenumbers whereas the horizontal (x axis) corresponds to the excitation wavenumbers. The amplitude (z axis) has 21 contour lines evenly spaced on a log10 scale. Circles with letters inside mark positions where Feynman diagrams predict oscillation coherences during the population time. Reprinted with permission from Camargo et al., J. Phys. Chem. A 2015, 119, 1, 95–101, Copyright (2015) American Chemical Society.

The experimental procedure here is the following. A porphyrin compound is dissolved in an organic liquid matrix (toluene in this particular case) and then excited with a red laser at one particular wavelength λ_1 (wavenumber ω_1). Then the absorption is measured over the aforementioned same wavenumber range $14800/\mathrm{cm} \leq \omega_3 \leq 16400/\mathrm{cm}$. This gives one absorption spectrum over the wavenumber range ω_3. Then, the excitation energy (the excitation wavenumber) in the laser is changed by some increment, like $\omega_3 + \delta\omega$, and a new absorption spectrum is recorded. You then get a stack of absorption spectra of the same sample in the absorption range, but with the sample excited with photons with different wavenumbers[18].

Here it is worthwhile noting that the wavenumbers of the electron band structure of a dissolved molecule may depend on the solvent[19]. You may argue here that isolated molecules, be they in the gas phase or dissolved in some fluid or liquid, are not yet condensed enough for there to be some energy band formation. The general experimental observation is that the absorption and emission spectra of molecules show shifts in peak position, peak height, and changes in the curvature of peaks. This is not surprising because the molecular interactions, such as polarization and the dispersion between the solvent and the dissolved molecule, impact on the molecular orbitals and, thus, influence the relative energy levels.

Liptay (1965) bases his consideration on the polarizability of the dissolved molecule and carries out a classical calculation and a quantum mechanical calculation. In general, the polarization dependent part of absorption and emission from the solvent is a function of the dipole moment in the ground state g or excited state e, and by the change of that dipole moment as a result of the stimulation by the light. The difference between both calculations is that the quantum mechanical treatment delivers an additional term, which depends on the dispersion interaction, and this interaction causes a red shift for absorption and emission when the molecule is brought into a solvent. The shift turns out to be the sum of two terms, one dependent on the energy levels of the dissolved molecule, and the other dependent on the intensity of the energy "band".

$$H = H^u + \sum_{p=1}^{N} H^{v(p)} + H' \tag{4.70}$$

For the Hamiltonian of the system H, Liptay considers that all dissolved molecules have the same identical structure and can be treated equally with a corresponding Hamiltonian H^u that corresponds to a free molecule; $H^{v(p)}$ is the Hamiltonian of the solvent molecule number p in the ensemble of N such solvent molecules. These are also considered free molecules; H' is the operator describing the interaction.

---

18 In the course of my career, I have used many different kinds of spectroscopy methods. In x-ray spectroscopy, you typically use as parameter the energy of the x-ray photons in eV or keV. In optical spectroscopy with visible light and ultraviolet light (uv vis), you typically use the parameter of the wavelength of the photons, in nm. In older literature, you would sometimes read Å for the optical experiments and also for the x-ray experiments. You realize here the two different physical concepts of radiation, one resembling the wave nature (wavenumber, frequency, wavelength) and the other resembling the particle nature (photon energy).

19 This was pointed out, for example, by Liptay at the 2nd *Internationales Farbensymposium*, 21.–24. April 1964 in Schloss Elmau in Bavaria (Liptay 1965).

With some data plot programs you can then picture the spectra in a two-dimensional plot, in a contour plot. The amplitude of the absorption (this is the absorption) is then given as a color bar, as shown in Figure 4.19 with intensity value 1E-03 as the lowest absorption, given in deep blue color, and as intensity value 2 in red color as the highest absorption. We see now that the red color intensity distribution is right in the center of the plot, marked with an encircled "C". It is not a coincidence that the maximum absorption is in the center of the plot. Rather, the authors of the paper (Camargo et al. 2015) noticed where the highest absorption is and cut off the wavenumber ranges by 800/cm wavenumbers in either direction. The solid lines separating the color fringes from dark red, red to orange, yellow, turquoise, and blue are like the equipotential lines that you can see on the map of a landscape. Here, we see the absorption landscape given by the excitation wavenumbers and absorption wavenumbers. This is the two-dimensionality of the information we get from the contour plot.

Let us now look at one actual absorption spectrum, which is shown in Figure 4.20 by the blue solid line. It has a sharp maximum at around 15500/cm, a shoulder at around 15900/cm, and a broad peak with a center at 17000/cm. The red solid line is the emission spectrum from the laser. Inscribed in the graphic is also the structural formula of the porphyrin zinc. The two arrows near the molecule sketch indicate the transition dipole moments Q_x and Q_y in the x and y-directions.

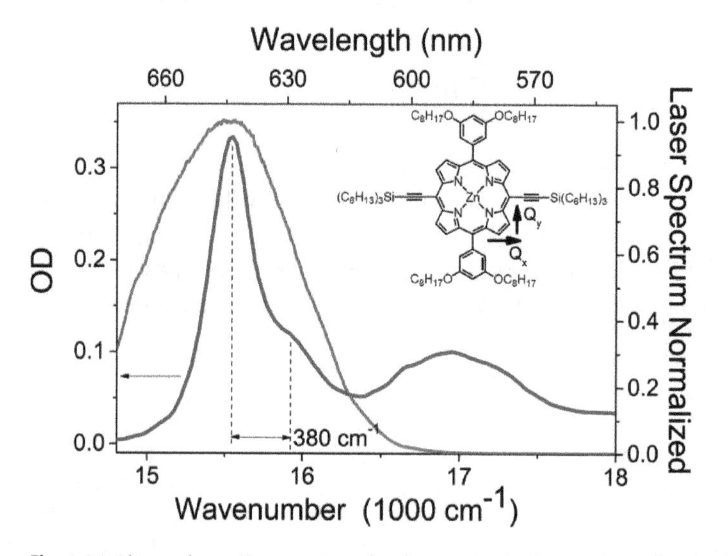

Fig. 4.20: Linear absorption spectrum for the porphyrin chromophore dissolved in toluene in the region of the Qy band (blue) along with the laser spectrum (red). The inset shows the molecular structure, and the arrows show the Qx and Qy transition dipole moment directions. Reprinted with permission from Camargo et al., J. Phys. Chem. A 2015, 119, 1, 95–101, Copyright (2015) American Chemical Society.

This laser-optical method is relatively complex and detailed in theory and technical specifications in Heisler et al. (2014). The light source is a laser system with a wavelength of 800 nm, which can deliver pulses of 120 femtoseconds (fs) at a rate of 10 kHz; 90% of the laser light power is used to pump an amplifier that allows tuning the light wavelength between 490 and 750 nm and even shorter pulses of 15 fs. For the experiment, three different pulses are needed, and this requires four phase coherent beams, which requires the use of two beam splitters. The four beams – after passing an entire range of optical elements – coincide at one place, and the sample is placed there. The underlying mathematics of the processes to be analyzed is taken from third-order perturbation theory (in German: *3. Ordnung Störungsrechnung*).

The scattered light signal is a result of several laser light pulses interacting with the porphyrin. The electric field E of the light signal (in Maxwell's theory, this is the light vector $\boldsymbol{E}$; this is considered the more active component because it already attacks on the electric charges when at rest, unlike the magnetic vector $\boldsymbol{B}$) is a convolution of the light vector from the laser excitation and the nonlinear response function of the porphyrin, and certainly the solution and any other matter in the light beam.

The first laser pulse $P1$ may come at some arbitrary time t_1, followed by the second pulse $P2$ at some time t_2 with the delay time, actually the coherence time τ. After some time T, the third pulse $P3$ follows at time stamp that is set $t_3 = 0$. This time T is the population time. The excitation pulses have the electric field envelopes E_n with phases ϕ_n and wavevectors k_n. Figure 4.22 illustrates when the four laser pulses are applied along the time axis. The electric vectors of the arranged laser pulses will interact with the sample. The sample is fully characterized by the response function R_n. This can be a tensor, such as the optical constants. This is what we actually want to quantify and qualify – the materials' properties.

The electric field detected after this interaction with the sample reads as shown in the equation below:

$$E(t) = \left(\frac{\mathrm{i}}{\hbar}\right)^3 \iiint_{t_i} \sum_n R_n(t_1, t_2, t_3) E_3(t - t_3) \mathrm{e}^{-\mathrm{i}\omega(t-t_3)} \cdot E_2(t - t_3 - t_2) \mathrm{e}^{\mathrm{i}\omega(t-t_3-t_2)} E_1^*(t - t_3 - t_2 - t_1) \mathrm{e}^{\mathrm{i}\omega(t-t_3-t_2-t_1)} \cdot \mathrm{e}^{\mathrm{i}(k_3+k_2-k_1)} \mathrm{e}^{\phi_3+\phi_2-\phi_1} \, \mathrm{d}^3 t_i \,. \tag{4.71}$$

We here carry out an integration over the three given time variables. The argument of the integral contains the electric field vectors of the laser light pulses before the interaction, and the phase relations and wave vectors. Using the three different laser pulses, we are looking at a physical problem that follows the third-order perturbation theory. Therefore, the response function will contain $2^3 = 8$ components.

As Camargo et al. point out in the Supporting Information section in their paper (Camargo et al. 2015), the number of components would increase if the electric vector would contain additional components, such as the polarization or the wavelength. Both certainly are possible and can deliver additional information on the structure of

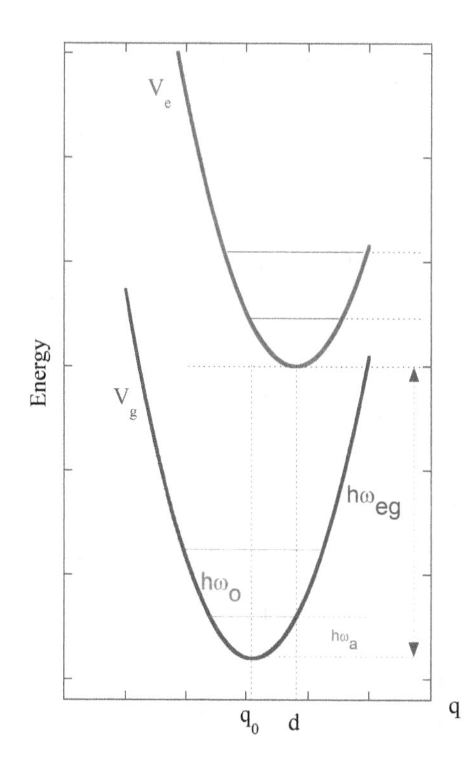

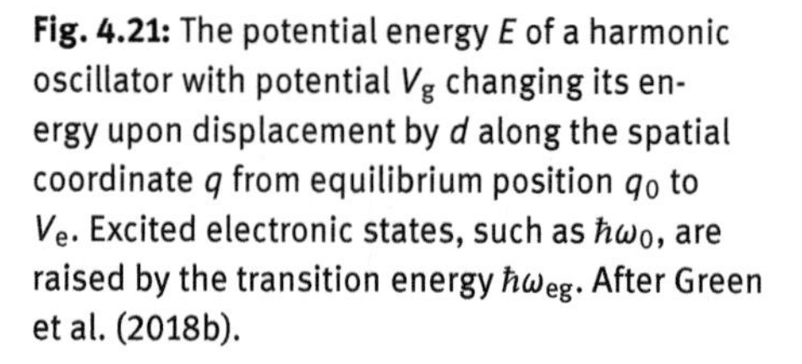

Fig. 4.21: The potential energy E of a harmonic oscillator with potential V_g changing its energy upon displacement by d along the spatial coordinate q from equilibrium position q_0 to V_e. Excited electronic states, such as $\hbar\omega_0$, are raised by the transition energy $\hbar\omega_\mathrm{eg}$. After Green et al. (2018b).

the sample. The spectra of porphyrins were discussed long time ago, for example in Rabinowitch (1944), Oliver and Rawlinson (1955), and Gouterman (1961).

The experimental key to the molecular structure is the optical technique. The conceptual key to the materials' structure or molecular structure and their interaction with photons or phonons is via the Feynman diagrams. The necessary parameter for this is the time axis, as shown in Figure 4.22. Hamm and Zanni use Feynman diagrams as shown in Figure 4.23 and explained them in their book on 2D spectroscopy, (Hamm and Zanni 2011).

Camargo et al. adopt this approach in their paper and resort to groups of non-rephasing (R1 and R4) and rephasing (R2 and R3) contributions. The connection to the response functions, Rn, can be found in the next section, where the calculation of 2D spectra is discussed. They (Green et al. 2018a, b) build the Hamiltonian by considering the ground states g and excited states e that are applied on harmonic oscillators, which reflect the intramolecular vibrations; for the ground state of the molecular system with spatial coordinate q_j, they write

$$h_\mathrm{g} = \sum_j \left[\frac{p_j^2}{2m_j} + \frac{1}{2} m_j \omega_j^2 q_j^2 \right] . \tag{4.72}$$

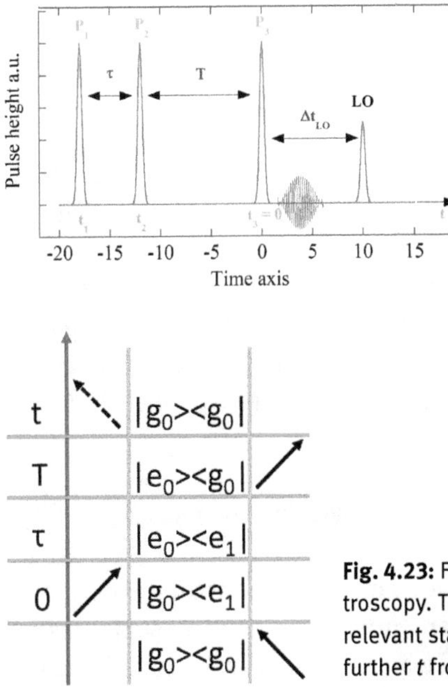

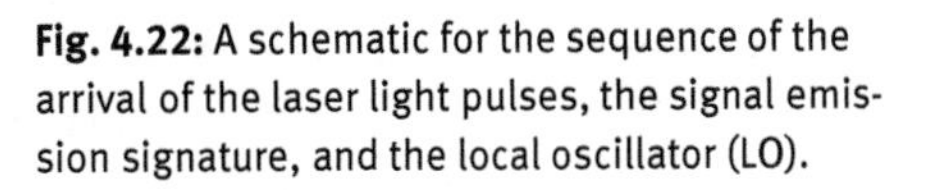

Fig. 4.22: A schematic for the sequence of the arrival of the laser light pulses, the signal emission signature, and the local oscillator (LO).

Fig. 4.23: Feynman diagram for optical transitions in 2D spectroscopy. The time axis is vertical and indicates the spectroscopic relevant stations of $t = 0, \tau$, and T according to Figure 4.22, and further t from the ground states g to excited states e.

When there is a d_j displacement along the coordinate q_j, the excited state reads

$$h_\mathrm{e} = \hbar\omega_\mathrm{eg}^0 \sum_j \left[\frac{p_j^2}{2m_j} + \frac{1}{2} m_j \omega_j^2 (q_j - d_j)^2 \right] . \tag{4.73}$$

This is illustrated in Figure 4.21, where the energy E of the harmonic oscillator with the parabolic potential V_g in the ground state is shifted by the displacement and raised along the energy axis to V_e. The Hamiltonian for the system then reads

$$H_\mathrm{S} = \langle g | \, h_\mathrm{g} \, | g \rangle + \langle e | \, h_\mathrm{e} \, | e \rangle \tag{4.74}$$

The thermal bath also produces a Hamiltonian H_B with an assumed infinite number of harmonic oscillators, which requires working with the infinite dimensional Hilbert space. The system sources its vibration energy from the thermal bath (Braun et al. 2019), and the corresponding Hamiltonian is

$$H_\mathrm{B} = \sum_n \sum_\alpha \hbar\omega_{n\alpha} \left(\alpha_{n\alpha}^\dagger \alpha_{n\alpha} + \frac{1}{2} \right) . \tag{4.75}$$

The total Hamiltonian of the system includes the contributions from the electronic transitions and the vibration transitions.

A graphic representation of the processes that occur in a 2D spectroscopy experiment is shown in the Appendix in the paper by Camargo et al. (2015). Figure 4.19, which I took from their paper, shows the letters A–F in the 2D spectrum map. Note that the system is rationalized by two electronic levels. The authors of the study included only the Q_y electronic transition, which was coupled to one *underdamped* vibrational mode with 380 cm^{-1}, with the result that up to 16,000 cm^{-1}, the absorption spectrum fits sufficiently. This is their spectral region of interest and can be well captured with the 2D spectroscopy method shown in this section.

The supporting information to figure 4.19 in the original paper (Camargo et al. 2015) presents a scheme showing where the different contributions locate on the 2D spectrum. Open symbols represent Feynman diagrams that oscillate (vibrational coherence) during the population time. Solid symbols, called static, represent Feynman diagrams, which are not on a vibrational coherence during population time and, therefore, account for electronic population relaxation. The circles A through F correspond to the same locations as in figure 2 in the manuscript (Camargo et al. 2015).

4.18.2 Vibration excitation energy transfer in light harvesting complexes

In the mid 1970s, Fenna and Matthews did an x-ray crystallography study on chlorophyll and found that bacteriochlorophyll is made from three identical subunits (the Fenna–Matthew–Olson complex, FMO), each of which contains seven bacteriochlorophylls that are irregularly arranged and enveloped by a protein scaffold (Fenna and Matthews 1975).

Oh et al. (2019) show that for bilin-based pigments, specifically the light harvesting complex phycocyanin 645 (Harrop et al. 2013), site-depending vibration excitations are important for the transport of excitation energy. In the 2D optical spectra, so-called long-lived oscillatory features have been observed, which extend unusually over hundreds of femtoseconds, which the discoverers, Engel et al. call "long-lived electronic quantum coherence" (Engel et al. 2007). The energy transfer after light absorption shows wavelike characteristics. The authors suspect that the FMO complexes are by this mechanism "to sample vast areas of phase space to find the most efficient path." This sounds very exciting, right?

Oh et al. (2019) find that particulary in the PC645, the wavelike characteristics have a drastic effect on the energy transfer between the chromophores in marine algae pigment protein complexes. Their Hamiltonian is composed of the electronic system contribution H_S, the vibronic bath contribution H_B, we remember both from the previous Section 4.18.1 by Camargo and Green (Camargo et al. 2015; Butkus et al. 2012; Green et al. 2018a, b). Additionally, Oh et al. bring in a third contribution H_SB, the coupling of the electronic system with the thermal bath:

$$H_\mathrm{tot} = H_\mathrm{S} + H_\mathrm{B} + H_\mathrm{SB} \tag{4.76}$$

where the latter contribution is expressed as

$$H_\mathrm{SB} = \sum_{i,m} u_{i,m} \left(b^{\dagger}_{i,m} + b_{i,m}\right) \langle m| \, |m\rangle \tag{4.77}$$

The information about the interaction between the molecular system and the thermal bath is given for each pigment m in the spectral density

$$J_m(\omega) = \pi \sum_i |u_{i,m}|^2 \delta(\omega - \omega_{i,m}) \tag{4.78}$$

In the course of energy transmission, the pigment proteins may reorganize, and this corresponds to a reorganization energy λ_m[20], which is a function of the spectral intensity $J_m(\omega)$ as follows:

$$\lambda_m = \frac{1}{\pi} \int_0^\infty \frac{J_m(\omega)}{\omega} \,\mathrm{d}\omega \tag{4.79}$$

For the as-good-as-possible description of this dynamic system, Oh et al. consider the intramolecular vibration modes in addition to the thermal bath, rather than including them integrally as part of the bath. They rationalize this with the caution that slow relaxation of the vibrations would not be accurate enough when treated by perturbation theory applied to the thermal bath. With a number of k vibronic modes included, the Hamiltonian by Oh et al. (2019) yields

$$\hat{H}_S^v = \sum_m^N \left(E_m + \hbar \sum_j \omega_{m,j}^{\text{vib}} v_{m,j} \right) |m, v_m\rangle \langle m, v_m| + \sum_{m \neq n}^N J_{m,v_m;n,v_n} |m, v_m\rangle \langle n, v_n| \,. \tag{4.80}$$

which they identify as a Holstein Hamiltonian. Theodore Holstein contributed in a major way to the theory of energy and electron transport in condensed matter, including electron–phonon coupling and polaron theory (Bederson and Stroke 2011). A polaron is a quasi particle that is produced by an electric charge elastically trapped in a crystal lattice, when the so dressed charge is kicked away from its equilibrium position by the vibration energy that the crystal sources from the thermal bath. The mass of the charge is in so far larger and the mobility restricted, but charge transport is warranted to some extent, and in some materials the only way for charge transport (Braun and Chen 2017). By definition, the polaron has a polaron mass, polaron radius, and polaron energy. The physical model is then that the point charge may jump from one site to a different site. Consider the first site as occupied with an electron, and the other site as not occupied. You then can think of a creation operator that makes that at the nonoccupied site an electron state is suddenly generated. At the first occupied site, you apply an annihilation operator that annihilates the electron from the occupied site.

The structure of a Holstein Hamiltonian for a two-site system with a hopping amplitude t and a coupling parameter g, which stands for the electron–phonon inter-

20 Having a λ, be it λ_m, standing for an energy is an unfortunate choice in optical experiments, because it may be easily mistaken with a wavelength. However, this here is the convention and nomenclature for the insider.

action, is given as a function of the phonon frequency ω (see the paper (Tayebi and Zelevinsky 2016))

$$H = t(c_L^\dagger c_R + c_R^\dagger c_L) + \omega(b_L^\dagger b_L + b_R^\dagger b_R) + \sqrt{2}g(c_L^\dagger c_L(b_L^\dagger + b_L) + c_R^\dagger c_R(b_R^\dagger + b_R)) \,. \quad (4.81)$$

We have here annihilation operators and creation operators for fermions (for the charge carriers, which can be electrons, holes, protons, and other ions) and for bosons, which are the phonons. With two arbitrary approximate positions in the lattice or in the molecule complex, we can define a left site and a right site. For example, $b_R^\dagger$ is the right site boson creation operator. Tayebi and Zelevinsky succeeded in finding an analytical solution for the Schrödinger equation to this Holstein Hamiltonian (Tayebi and Zelevinsky 2016).

Coming back to the work of Oh et al., Figure 4.24 shows the protein structure of the PC645 complex (phycocyanin, PC). Inscribed are eight functional complexes, which are explained in detail in the paper by Lee, Bravaya, and Coker (Lee et al. 2017), who made first-principles calculations on the light harvesting of algae. In the focus of their study are the phycobiliproteins (for a review on these, see Glazer 1994); these are the light harvesting antenna proteins such as phycoerythrin and phycocyanin. Oh et al. compared phycoerythrin PE545 with PC645. Both complexes have different chromophore compositions and also different π-bond lengths. This amounts, for example, in the shift of the absorption band by 100 nm, which lends the number to their names, PE545 with an absorption peak at 545 nm and PC645 at 645 nm.

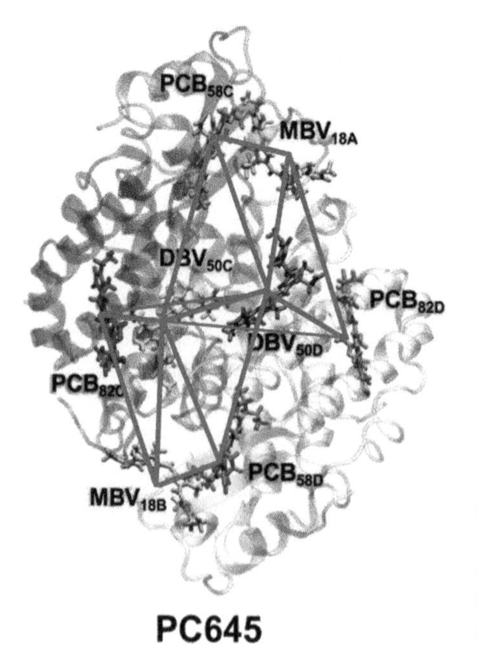

Fig. 4.24: Structure of pigment-protein complex PC645. Reprinted and adapted from with permission from First-Principles Models for Biological Light-Harvesting: Phycobiliprotein Complexes from Cryptophyte Algae, Mi Kyung Lee, Ksenia B. Bravaya, David F. Coker, J. Am. Chem. Soc. 2017, 139, 23, 7803–7814. Copyright (2017) American Chemical Society, (Lee et al. 2017).

The eight bilins as abbreviated in Table 4.2 are covalently bound to cysteine sites in the protein scaffold. The bilins interact in funneling energy to the chlorophyll. The numerals next to the abbreviations of the chromophores denote the site configuration, as is explained in the Supporting Information section in Lee et al. (2017). The DBV are at the dimer interface and constitute source states, whereas the PCB and MBV are intermediate states and sink states (Lee et al. 2017). The blue thick lines in Figure 4.24 are to aid in demonstrating the stereological configuration of the eight chromophores.

Tab. 4.2: Abbreviation of phycobiliprotein units in PC645.

Chromophores	
DBV	dihydrobiliverdin
PCB	phycocyanobilin
MBV	mesobiliverdin

In the paper by Oh et al. (2019), the chromophores are additionally labeled for their position, and the coupling strength constants have been determined. Table 4.3 shows the coupling constants for the 14 linkages in increasing order, the smallest being an arbitrary 17 and the strongest being 212. This is a factor of 12 to 13 in coupling strength. This will give you an impression of how strong and diverse a framework the chromophores build and what kind of vibration structure can emerge from this.

Tab. 4.3: Coupling constants of phycobiliprotein units according to (Oh et al. 2019).

Coupling constants between chromophores	
DBV_{50C} – PCB_{82D}	17
PCB_{82C} – DBV_{50D}	20
PCB_{58C} – DBV_{50D}	24
PCB_{58D} – DBV_{50D}	34
PCB_{58D} – DBV_{50C}	37
PCB_{82C} – DBV_{50C}	43
DBV_{50D} – MBV_{18A}	49
DBV_{50D} – PCB_{82D}	50
DBV_{50C} – MBV_{18B}	53
PCB_{82D} – MBV_{18A}	67
PCB_{82C} – MBV_{18B}	69
PCB_{58C} – MBV_{18A}	78
PCB_{58D} – MBV_{18B}	78
DBV_{50C} – DBV_{50D}	212

Bukartė et al. (2020) investigated chlorophyll *c1*, a molecule that looks similar to the chlorophyll in Figure 4.17, with a Mg^{2+} central ion. They used 2D optical spectroscopy with linear and circular polarized light with the samples cooled down to 77 K. The spectra of chlorophylls and porphyrins show a strong Soret band (Soret 1883) at around 400 to 500 nm wavelength, and the so-called Q_x and Q_y (Gouterman 1961) bands at larger wavelengths, which originate from transitions with perpendicular transition dipole moments.

The vibronic coupling causes a mixing of the states, and it is suggested that the coupling could also speed up the energy transfer in chromophore complexes, but it is not a trivial task to quantify the coupling (Bukartė et al. 2020). The selected chlorophyll *c1* molecule has two close-ranging Q_x and Q_y states, which can be covered with one single laser pulse.

The measurement geometries for the optical experiments were done in an all-parallel arrangement, where the four laser pulses had the identical linear polarization with all set at 0°, and a so-called double-crossed geometry with angles 45°, −45°; 90°, and 0°, allowing for the detection of vibronic coupling.

4.19 Chromophores in green fluorescent proteins

The recording of the optical spectra of chromophores is relatively trivial. It is important to isolate them from their environment first. Otherwise you have a composition spectrum of all components in the sample. It is, therefore, important to be able to calculate a spectrum of a photosynthesis component, like a chromophore or a hydrogenase or oxygenase, based on first principles. Agreement or disagreement between the calculated spectrum and the experimental spectrum can then help you to resolve and understand the molecular structure and functionality.

Chromophores in green fluorescent protein were studied by Voityuk et al. (1998). The green fluorescent protein glows green when stimulated with UV light. It is well known in jellyfish. In 2008, Shimomura, Chalfie, and Tsien were awarded with the Nobel Prize in Chemistry "for the discovery and development of the green fluorescent protein, GFP."

Voityuk et al. (1998) investigated native and mutant green fluorescent protein in various solvation environments. The chromophore molecule is embedded in an infrastructure of amino acids. The GFP has 238 amino acids. For the calculation of the absorption spectra of such large organic molecules, the INDO method was established in the 1970s. INDO stands for intermediate neglect of differential overlap and is a semi-empirical quantum mechanical calculation procedure (Voityuk and Rosch 2004).

They used the INDO in order to model and calculate the chromophore in an aqueous environment; the method is well suited to also include protonation effects on molecules. Figure 4.25 is a simple sketch of a chromophore molecule hydrated by a

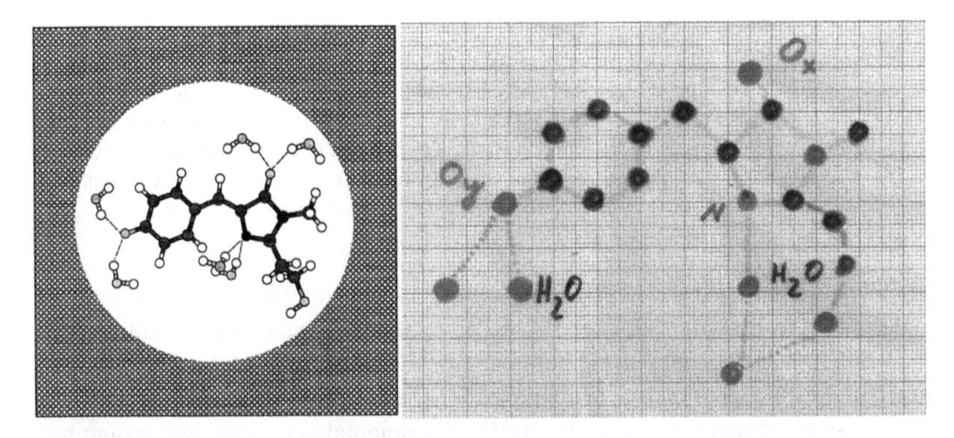

Fig. 4.25: Left: model for the hydrated anionic chromophore. Reprinted from Chemical Physics, Alexander A. Voityuk, Maria-Elisabeth Michel-Beyerle, Notker Rösch, Quantum chemical modeling of structure and absorption spectra of the chromophore in green fluorescent proteins, 13–25, Copyright (1998), with permission from Elsevier, (Voityuk et al. 1998). Right: a sketch of the structure with labeled oxygen O_x and O_y and nitrogen N atoms which are used by Voityuk in their modeling.

number of water H_2O molecules, an anionic chromophore (Voityuk et al. 1998). The structure was determined by Yang et al. (1996). For various conformations of the molecule, different energies were obtained by the calculations. The conformation can be changed by putting the chromophore into different chemical environments with different protonation.

The purpose for doing spectroscopy is to learn about the structure and function of molecules, materials, components, devices, and systems. The purpose for doing spectroscopy here on the green fluorescent proteins could be similar to the question of what the influence of protonation on the chromophores is. So, you try to induce various protonation to the chromophores and measure them and then test whether the hypotheses, if any, make sense. What you find from the calculations, where the protonation is a structural parameter, are the different transition energies. Based on that, you can try to determine the proton affinity *PA*.

The approach here is that the difference in the proton affinity of the chromophore system in the ground state and the excited state would be the same as the difference in the transition energies between the protonated form and the nonprotonated form. Let the chromophore have the ground state M and excited state M^*, and the protonated state HM^+, then we can write (Voityuk et al. 1998)

$$PA(\mathrm{M}^*) = \mathrm{PA}(\mathrm{M}) + h\nu(\mathrm{M}) - h\nu(\mathrm{HM}^+) \tag{4.82}$$

and then the changes in the states and their energies upon protonation can be sketched in a Förster cycle (Verhoeven 1996; Förster 1950), as shown in Figure 4.26.

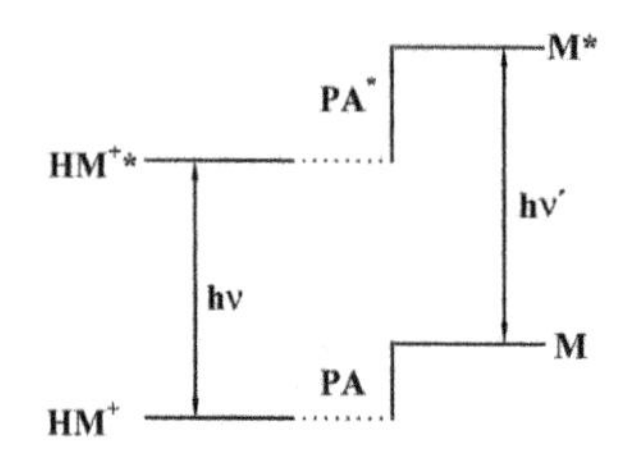

Fig. 4.26: Förster cycle for deriving the proton affinity *PA** of an excited state. Reprinted from Chemical Physics, Alexander A. Voityuk, Maria-Elisabeth Michel-Beyerle, Notker Rösch, Quantum chemical modeling of structure and absorption spectra of the chromophore in green fluorescent proteins, 13–25, Copyright (1998), with permission from Elsevier (Voityuk et al. 1998).

Förster found (Förster 1950) that the emission spectra of some fluorescent molecules would change upon electrochemical dissociation, whereas the absorption spectra would not change. Moreover, there was no *shift* of the peaks of the emission spectra, but the disappearance of one peak and the appearance of another peak. He rationalized his observations: the pH was the experimental parameter, with a quantum mechanical energy term scheme, as he found that the dissociation of the molecule would be promoted when it was excited by long wave ultraviolet light.

The right-hand panel in Figure 4.25 is a simpler representation of the Tyr66 chromophore in the GFP along with its intimate molecular environment, with the two oxygen O_x, O_y and N. It is a so-called zwitterion because it has a positively charged and negatively charged functional group attached – and thus is electrically neutral: (O_y, HN, O_x). This, and the corresponding cation $(HO_y, HN, O_x)^+$, turned out to be the spectroscopically relevant forms of the chromophore. The difference in the molecular environment, produced by the different protonation treatment, causes changes in the spectroscopic features when the chromophore is optically excited.

The results by Voityuk et al. are graphically sketched in a Jablonski diagram in Figure 4.27, which is also known as the Förster cycle. The chromophore in its cation form $(HO_y, HN)^+$ can be slowly (langsam) hydrated to form an H-bonded complex $[-H^+ \ldots O_y^-, HN^+]$. The experimentally determined emission line of the cation is at 397 nm[21].

I have adopted the Förster cycle concept in Figure 4.27 from Voityuk et al., where I show on the ordinate the changes of the structure upon hydration and complexion and deprotonation, with the black thick horizonal lines the ground state, and on the abscissa the excited states with the corresponding transition wavelengths in nm and light color RGB codes. Depending on the chemical state of the green fluorescent protein, as indicated in the structure model in Figure 4.25, the emission lines differ, as mentioned above.

21 In Figure 4.27, I have put the RGB color code for the three colors of the green fluorescent protein. You can find a calculator for the conversion of wavelength into RGB color code, for example, on the webpage of John D. Cook Consulting firm in Houston TX, www.johndcook.com.

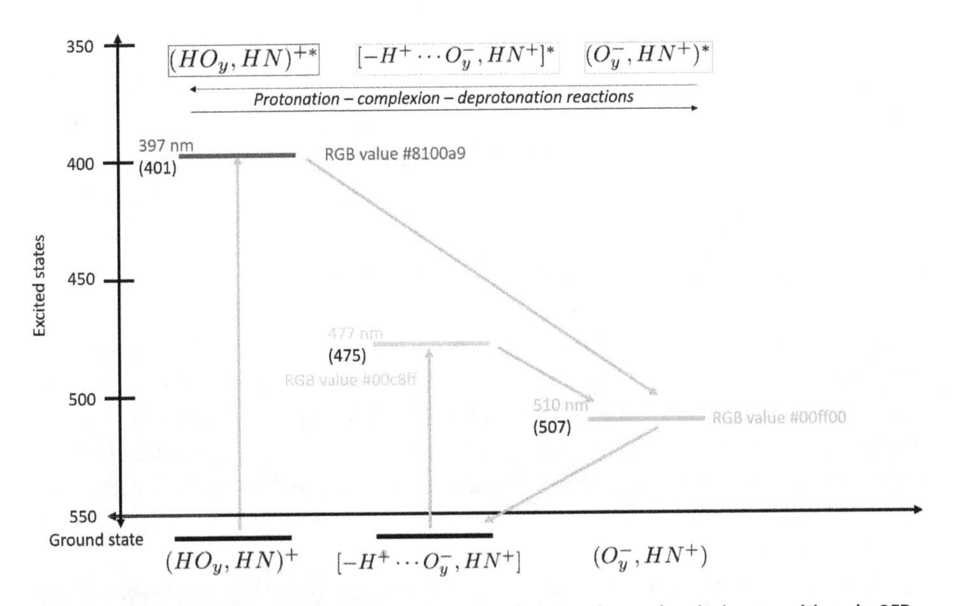

Fig. 4.27: Experimental and calculated (in parentheses) absorption and emission transitions in GFP. $[H^+ \dots O_y, HN]$ designates an H-bonded complex between the zwitterionic form of the chromophore and its environment. Inspired by the graph from Chemical Physics, Alexander A. Voityuk, Maria-Elisabeth Michel-Beyerle, Notker Rösch, Quantum chemical modeling of structure and absorption spectra of the chromophore in green fluorescent proteins, 13–25, Copyright (1998), with permission from Elsevier (Voityuk et al. 1998).

4.20 Example: combination of ring and chain of chlorophyll for excitonic wave

We characterize two chromophores by two dipole transition vectors with notation $\boldsymbol{\mu}_i$ and $\boldsymbol{\mu}_j$ (μ for the dipole moment) at the positions $\boldsymbol{r}_i$ and $\boldsymbol{r}_j$. The distance r_{ij} between the chromophores is given by the conventional modulus of the distance vector; $|\boldsymbol{r}_j - \boldsymbol{r}_i|$. The two chromophores will then experience the Coulomb interaction, specifically the dipole–dipole coupling. For the determination of the coupling energy, we must consider the dielectric properties of the medium in which the two chromophores rest, this is, for example, a medium with dielectric constant ϵ.

Depending on the orientation, the dipole energy then reads

$$J_{ij} = \frac{\mu_i \mu_j}{4\pi\varepsilon r_{ij}^3} \kappa_{ij} \tag{4.83}$$

The factor κ_{ij} is the scalar product (normal vector $\hat{n}$) that accounts for the orientation of the chromophores and is given by

$$\kappa_{ij} = \hat{\mu}_i \cdot \hat{\mu}_j - 3(\hat{n} \cdot \hat{\mu}_i) \cdot (\hat{n} \cdot \hat{\mu}_j) \tag{4.84}$$

The coupling strength of the chromophores depends on the orientation of their dipoles. The chain built from the chromophores is represented by the following Hamiltonian

$$\hat{H} = \frac{1}{2}\sum_{i} \omega_i \sigma_i^z + \sum_{i \neq j} \left(J_{ij} \sigma_i^\dagger \sigma_j + H.c. \right) \tag{4.85}$$

4.21 Light scattering by moth-eye structures

When not all of the incoming light can be absorbed by plants or animals, nature has developed some tricks during evolution of life to scatter the light in a controlled way and keep the light in its vicinity of control for subsequent absorption, rather than reflecting it away from the plant or animal where the light, the photons, would be lost for further use. One such example is found in the moth eye, but not for energy conversion, but for sensing purposes of the moth, more specifically, for it to detecting enemies without being detected.

We[22] employed this principle for energy efficiency purposes in photo electrochemical cells, which are found in the PhD thesis of Florent Boudoire (Boudoire et al. 2014; Boudoire 2015) and which triggered a considerable media echo (Greenemeier 2014; Rothkopf 2014; Economist 2014; Lemonde 2014).

The photoelectrodes were prepared by a synthesis route, which is graphically sketched in Figure 4.28. Ammonium tungstate is dissolved in water and along with polystyrene sulfonate this will form vesicles that can be spin coated on an electrode support. For PEC experiments, such electrode supports are glass slides coated with a transparent conducting oxide (TCO), such as fluorine-doped tin oxide (FTO). The electrode is then baked in a furnace, and the tungsten salt turns into yellowish tungsten oxide WO_3. In the next step, iron nitrate solution is sprayed on the electrode, which is then baked again, and then an iron oxide layer of α-Fe_2O_3 grows on the tungsten oxide structure.

Figure 4.29 shows electron micrographs of these structured electrodes. Particularly the lower left-hand image resembles the structure of the eyes of moths. The top of the right-hand panel shows three so-called STXM images, where STXM stands for scanning transmission x-ray microscpectroscopy. This methods helps us to show where on a sample which chemical element is present. Moreover, the method shows not only the position of the element but also its molecular structure, such as the oxidation state or molecular environment. Consequently, the right-hand bottom panel shows two x-ray spectra recorded from inside a vesicle droplet and outside such a

22 Mainly funded from the following two projects by the Swiss National Science Foundation http://p3.snf.ch/Project-137868, from which PhD student Florent Boudoire was funded, and http://p3.snf.ch/Project-139698, which was a personal Grant by the Marie Heim-Vögtlin Foundation for Dr Rita Toth.

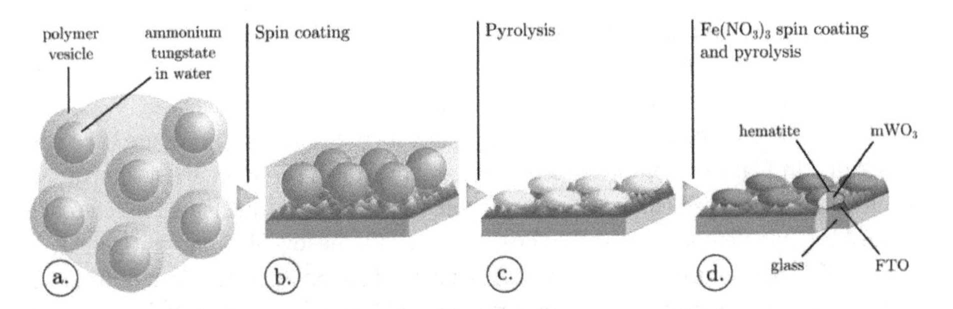

Fig. 4.28: Flow sketch of the mWO_3 microspheroid self-organization and hematite coating. (a) PSS vesicle suspension in an ammonium tungstate solution for spin coating. (b) Polymer film enclosing ammonium tungstate, after spin coating on FTO-coated glass. (c) mWO_3 spheroids after the first pyrolysis. (d) Finalized films after $Fe(NO_3)_3$/ethanol spin coating and pyrolysis. Courtesy of Florent Boudoire (Boudoire 2015).

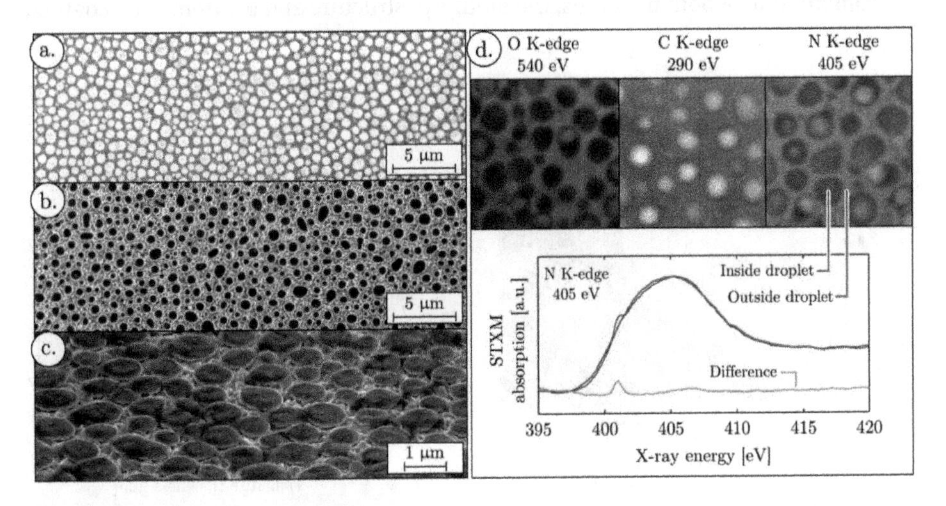

Fig. 4.29: Ammonium tungstate/PSS film surface (spin coating speed 3000 rpm). (a) SEM picture before pyrolysis. (b and c) SEM picture after pyrolysis.(d) Scanning X-ray absorption microscopy (STXM) pictures at the O K-edge, the C K-edge, and the N K-edge, before pyrolysis. Courtesy of Florent Boudoire (Boudoire 2015).

droplet. The spectra were recorded at the nitrogen N-edge at around 400 eV, and there is a small peak at 401 eV present inside the droplet and not outside the droplet, which is experimental proof that the N atom is still inside the vesicle.

However, this was only a side information, showing you which kinds of experiments sometimes assist in studies on artificial photosynthesis. The photons with energy in the range of $h\nu \sim 400$ eV are not visible and also not in the ultraviolet range. They are soft x-rays and allow for core-level x-ray spectroscopy. This method can be used for to count the hole states in the metal ion centers of proteins from photosynthesis, such as Ni, Fe, and Mn. Also on a side note, as the peak is relatively small, we here used a trick, where we subtract the two spectra from each other, and the result, the difference spectrum, shows one new prominent peak in the spectrum, which is characteristic for the molecular structure of nitrogen in the electrode layer.

What is the practical benefit of employing such a moth-eye structure? We can check this in Figure 4.30. On the left-hand side, you can see the current density in an electrochemical cell equipped with two different electrodes, measured in reference to the Ag/AgCl reference electrode. The red curve with the low photocurrent density is from a thin unstructured α-Fe_2O_3 layer. The onset of the photocurrent is at 600 mV. At 1000 mV, the photocurrent is a maximum value of 150 μA cm^{-2}. The yellow photocurrent curve is from the WO_3 moth-eye material. The photocurrent onset is at 150 mV, so there is a substantial benefit: we need a 450 mV lower DC bias then with the iron oxide electrode, but the maximum photocurrent is not higher than with iron oxide. You can see both curves, red and yellow, intersect at around 1000 mV bias. However, the combination of both materials, the moth-eye structure and the iron oxide coating,

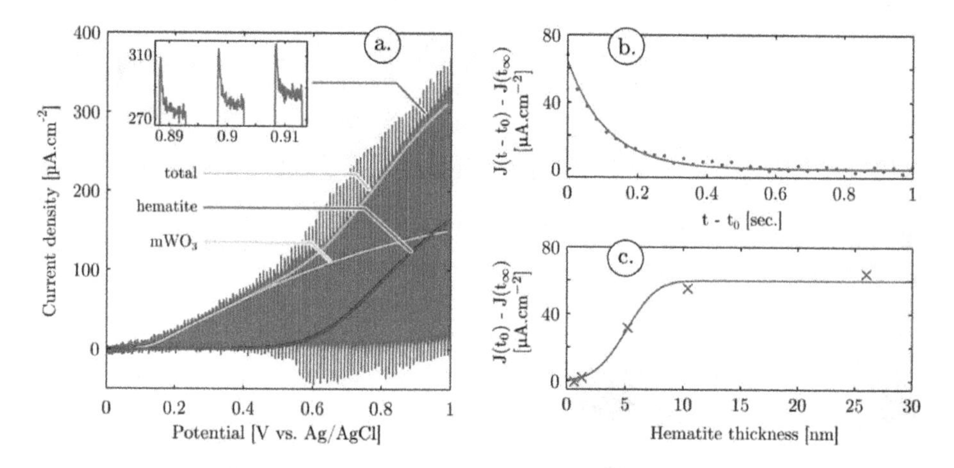

Fig. 4.30: (a) Typical curve obtained when chopping the light at 1 Hz while recording the photocurrent at different potentials versus the Ag/AgCl reference electrode in PBS electrolyte (pH ~ 7). The blue curve is the result of such a measurement on an mWO_3 film made of spheroids (processed at 2000 rpm) with a 5 nm thick[23] hematite overlayer. The yellow and red curves represent the two sigmoids that can be fitted to the photocurrent at equilibrium; they correspond to mWO_3 and hematite respectively. (b) Example of decay observed for a 30 nm hematite thin film and the corresponding exponential fitting according to Equation(2). (c) Dependency of the charging amplitude over the film thickness. Courtesy of Florent Boudoire (Boudoire 2015).

23 Some people may think that 5 nm is very thin. I am a surface scientist and did my physics diploma thesis on ultrathin films, which were films of a fraction of a monoatomic layer. So it is a matter of perspective as to what we consider thick and thin. It is, therefore, good when you can always provide the thickness of the films you are working with. Also, it is not always trivial to determine the thickness of a film.

yields a photocurrent density double as high as the photocurrent density of both electrode structures when applied individually. This is shown by the light blue curve, with a maximum photocurrent density of 300 μA cm^{-2}.

The blue shaded area in Figure 4.30 is actually the densely situated data points for the measurements. Moreover, the photocurrent measurement was modulated by regularly repeated light on/light off switching by an automatic and computer-controlled light shutter mounted in front of the xenon lamp (solar light simulator, see figure 8.4 in my book (Braun 2019a)), which is used for the illumination of the PEC cell. Switching the light on and off causes sharp current peaks in the photocurrent curves, which can be used for further analyses. Three such current peaks are shown in the inset in the left-hand panel in Figure 4.30. They were done at potentials of 890, 900, and 910 mV. The peaks have a fine structure, which can be mathematically modeled with exponential functions, as illustrated in the upper right-hand panel in Figure 4.30. The invariant in such exponential behavior is given by the time constant τ (not shown here).

Florent Boudoire carried out these measurements with many samples, while he also varied the thickness of the iron oxide layer. Some were so thin that he could barely measure the thickness. The bottom right-hand panel in Figure 4.30 shows the peak height photo currents measured on a handful of such films with various thicknesses. When a film is extremely thin, the peak height is very small. With increasing film thickness, the current increases in a sigmoidal shape and has a plateau value of 60 μA cm^{-2}, which appears to remain constant for larger thicknesses. The peak current is a result of the electric charging that takes place in the α-Fe_2O_3 film when illuminated. The impinging photons create electron–hole pairs, and we want the electrons go in one direction and the holes in the other. Specifically, we want the holes to diffuse to the electrode surface, where it meets the water molecules that the holes can oxidize and evolve oxygen O_2. The problem is that the holes, on average, diffuse only 5 nm before they recombine with electrons and become annihilated. Figure 4.30 shows that the current density no longer increases when the thickness is larger than 10 nm (Boudoire et al. 2014).

However, such thin films cannot absorb enough light. The absorption length of α-Fe_2O_3 is around 500 nm, which is almost two orders of magnitude larger than the hole diffusion length. So there is a fundamental problem, at least with this material, that the charge transport properties require a thin material, whereas the light absorbing properties demand a thick material. The way out of this dilemma is to orthogonalize the light absorption and the charge transport. This can be achieved by making thin but long absorbers (Kayes et al. 2005). We could not make the structures long in one direction, but in two. These were the moth-eye structures.

Figure 4.31 compares the attenuation of light with wavelengths from 300 to 650 nm into the nanostructures grown by the PhD student. The top graph shows the experimental UV-visible spectra for six different samples with six different sizes of nanoparticles. The PhD student was able to control the size of the nanoparticles with the speed of the spin coater that he used to make the nanoparticle films. With a speed of 1000

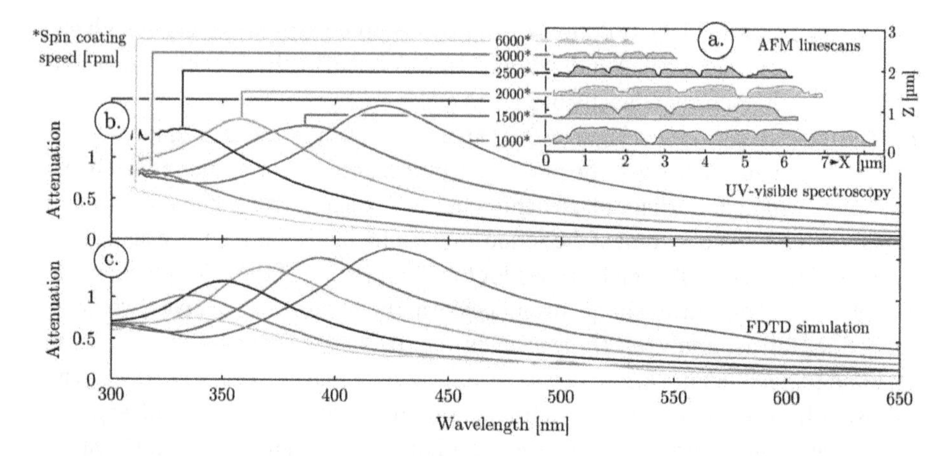

Fig. 4.31: Correlation between the tungsten oxide microstructure and the far-field scattering attenuation peak of the mWO_3 films composed of spheroids, processed at different spin coating speeds. (a) AFM line scans of the surfaces, (b) UV-visible absorption spectroscopy performed in water on the different films, (c) simulated attenuation spectra using the finite-difference time-domain method. Courtesy of Florent Boudoire (Boudoire 2015).

rounds per minutes, the particles were large, as indicated in the inset in Figure 4.31, and at 6000 rounds per minutes, the particles were very small. The PhD student used an atomic force microscope (AFM) to measure the size of the nanoparticles. On close inspection you see that the particles are actually microparticles, and not nanoparticles, given that their size ranges from 2 to 0.5 micrometers.

The optical absorption peak of tungsten oxide WO_3 is at around 310 nm. This absorption is due to the electronic structure of the material. However, in Figure 4.31, we notice broad intensity maxima in the spectra, which means that there is an additional attenuation of light at the wavelengths, which are not characteristic for WO_3. For the 1000 rpm sample, the additional attenuation maximum is at around 420 nm. The smaller particles made by 1500 rpm yield the attenuation the maximum at 390 nm. With decreasing thickness, the maxima shift to 360, 330, and 320 nm. Obviously, the size of the attenuator particles has an effect on the additional absorption of the electrode material, which should be beneficial for the performance of the electrode. The bottom panel in Figure 4.31 shows the simulated spectra. Observe how well the experimental spectra and calculated spectra match.

This light scattering process can be modeled with the classical electromagnetic light theory envisioned by James Clark Maxwell and exercised, for example, in the optics book on electromagnetic light theory by Max Born (Born 1933). The computational photonics modeling was done with commercial software from Lumerical (Riaz et al. 2017; Radhakrishnan and Murugesan 2014), which uses the finite-difference time-domain (FDTD) numerical method.

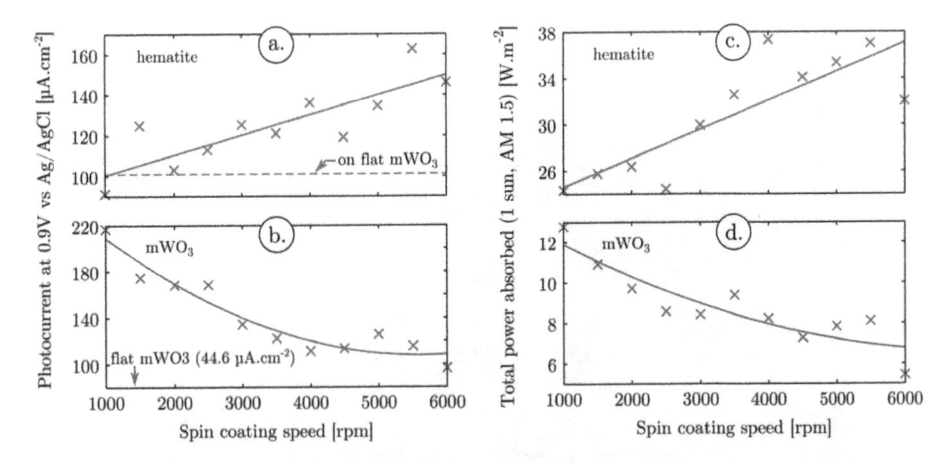

Fig. 4.32: Photocurrent densities at 0.9 V vs. Ag/AgCl reference (current density under illumination – dark current density) as a function of the spin coating speed. Contributions from: (a) hematite and (b) mWO_3; deconvoluted from the photocurrent presented in the ESI, Figure VIII.† Calculated total power absorbed under simulated solar light conditions in: (c) hematite and (d) mWO_3. Courtesy of Florent Boudoire (Boudoire 2015).

In Figure 4.32, you can see the photocurrent densities of electrodes as a function of the spin-coated speed in rpm. Panel (a) shows the photocurrent density of hematite increasing linear with the speed, i.e., with larger moth-eye shape particles. The photocurrent density of the tungsten oxide decreases concavely with increasing coater speed. The total absorbed power for the series of electrode samples is shown in the two right-hand panels in Figure 4.32. It was determined (Boudoire 2015) by the divergence of the Poynting vector as

$$P_{\text{abs}} = -\frac{1}{2}\omega|E|^2 k\,. \tag{4.86}$$

The observation that the absorbed power increases in the iron oxide and decreases in the tungsten oxide can be explained by changes in the interaction of light with the particles depending on the spheroid size distribution, as Florent Boudoire explains in his PhD thesis (Boudoire 2015). Finally, he investigates the distribution of the optical intensity of the films as a function of the light wavelength λ and the radius R of the moth-eye type spheroids in the photoelectrodes. The intensity is given by a color code from blue to red. The nine panels in Figure 4.33 cover the λ, R parameter space, which is divided into three classes: red frames for $\lambda < 0.75R$; yellow frame for the intermediate range $0.75R < \lambda < 1.25R$; and the wavelength larger than the radius $\lambda > 1.25R$.

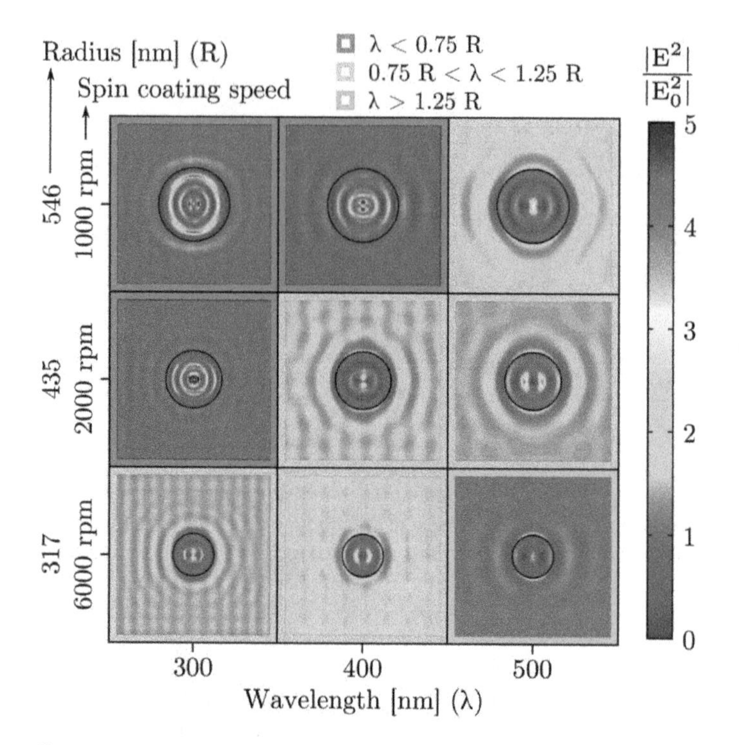

Fig. 4.33: Optical intensity distribution in the near field as a function of spheroid size for different incident light wavelengths. The simulations where the light wavelength is lower than the spheroid radius ($l < 0.75R$) are highlighted with a red frame, and the simulations where the light wavelength is comparable to the spheroid radius ($0.75R < l < 1.25R$) are highlighted with a yellow frame. Courtesy of Florent Boudoire (Boudoire 2015). This light scattering process can be modeled with the classical electromagnetic light theory envisioned by James Clark Maxwell and exercised, for example, in the book by Max Born (Born 1933).

4.22 Emission and fluorescence spectroscopy

In photosynthesis research it is the fluorescence spectra that basically constitute the photoemission spectra. We excite the material, be it a plant, a component of a plant or an inorganic material, with light of an energy E_0 (or wavelength λ_0) and we notice that the material responds with the emission of a light with energy $E_F < E_0$ (or wavelength $\lambda_E > \lambda_0$). As a matter of fact, we typically will not notice one sharp emission line but rather an extended emission spectrum or fluorescence spectrum. The electronic terms and transitions are shown, for example, in the Jablonski diagram in Figure 3.22.

Let us look at two emission spectra obtained from bean leaves that were grown in the dark, and greened for a short time by subjecting them to white light flashes. The spectra are shown in the left-hand panel in Figure 4.34, which I took from the paper by Strasser and Butler (1977b). The upper spectrum was obtained by exciting the leaf

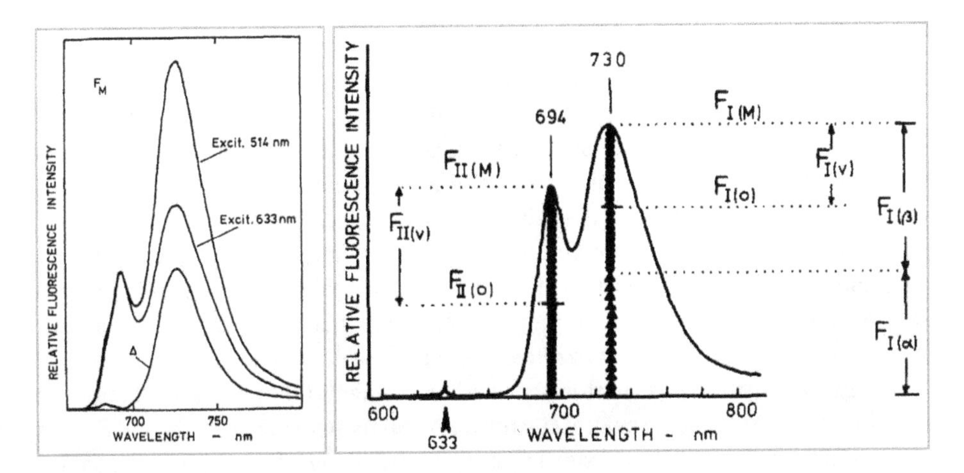

Fig. 4.34: Left: emission spectra of a flashed leaf at −196 °C excited at 514 and at 633 nm; *Δ* is the difference spectrum. Reprinted from Biochimica et Biophysica Acta (BBA) – Bioenergetics, 462 (2), Reto J. Strasser,Warren L. Butler, Fluorescence emission spectra of photosystem I, photosystem II, and the light harvesting chlorophyll a/b complex of higher plants, 307–313, Copyright (1977), with permission from Elsevier, (Strasser and Butler 1977b). Right: emission spectrum of a flashed leaf at −196 °C at the F_M level excited by 633 nm light. Intensities of fluorescence at the F_0 level at 694 and 730 nm are also indicated. The categories of fluorescence excited by photosystem II are indicated by •, the category of fluorescence excited by photosystem I is indicated by ▲. Reprinted from Biochimica et Biophysica Acta (BBA) – Bioenergetics, 462 (2), Strasser, Reto J. Butler, Warren L., Fluorescence emission spectra of photosystem I, photosystem II, and the light harvesting chlorophyll a/b complex of higher plants, 307–313, Copyright (1977), (Strasser and Butler 1977c).

with 514 nm, which is nice green light. The resulting emission spectrum extends from 650 to 800 nm and shows two clear fluorescence peaks at 680 and 725 nm; 680 nm corresponds to a dark red, and 725 nm is already considered infrared.

The spectrum in the middle was obtained by exciting the lead with 633 nm light, which is red light. The overall intensity of this spectrum is lower than the one excited with higher energy light. The authors normalized the spectra by matching the intensity for both spectra for their first emission peak at 680 nm (actually, at 694 nm). Then, they built the difference of both spectra, so that the resulting intensity at the matching peak height becomes zero (0).

The right-hand panel in Figure 4.34 shows how one can handle such a spectrum for further quantitative analysis. It is important to note that the spectra were recorded when the samples were cooled down, actually frozen down to −196 °C. We can relate this Celsius temperature to the Kelvin temperature scale, where 0 °C is 273 K. The difference is 273 K − 196 °C = 77 K; 77 Kelvin is the temperature reached by using liquid nitrogen as coolant. By cooling down the sample, the lattice and molecule vibrations become smaller because there is less thermal excitation. Often, you cannot identify spectroscopic signatures unless you cool the sample down. Then, the peaks of spectra become much narrower. Also, the low temperature may sometimes prevent what is known as radiation damage in the sample.

The two emission bands (left, Figure 4.34) originate from the antenna chlorophyll a, specifically photosystem I at 730 nm and photosystem II at 694 nm, with the centers indicated by strong vertical lines from • and ▲. The peak height from PSII has the value $F_{\text{II(M)}}$, and that of PSI has the value $F_{\text{I(M)}}$. With a keen eye, you will notice a very weak intensity shoulder at around 685 nm, which originates from low-temperature emission of mature chloroplasts with light harvesting chlorophyll a/b protein. This sample apparently had little chlorophyll b formed, which why this structure is lacking in the spectrum. Note the bean has been grown in the dark and thus is not green. It was only green with a few short flashes with white light. Continuing irradiation, or illumination makes that the peak at 694 nm from PSII increases from the height $F_{\text{II(0)}}$ to $F_{\text{II(M)}}$, and the difference in emission intensity is labeled $F_{\text{II(V)}}$ because of fluorescence from *variable* (v) yield.

The intensity of the fluorescence peak from PSI increases in the same course from $F_{\text{I(0)}}$ to $F_{\text{I(M)}}$, and again the difference in intensity is the *variable* yield. The emission peak at 730 nm is the spectroscopic signature of chlorophyll from PSI. However, the cause of the excitation of PSI can be in the immediate excitation of PSI, or be some initial excitation of PSII, which passes the energy on – to PSI. Strasser and Butler extended a physical model for the energy exchange between PSI and PSII (Strasser and Butler 1977a), which translates into a relatively simple linear mathematical relation of the form

$$F_{\text{I}} = (\alpha + \beta \cdot \phi_{T(\text{II}\to\text{I})}) I_{\text{a}} \cdot \phi_{F_{\text{I}}}\,, \tag{4.87}$$

where F_{I} is the emission intensity of PSI. The portion n of light $(n \cdot h\nu)$, which is given to PSI is given to α, and the remainder $1 - \alpha = \beta$ is given to PSII. The transfer of the energy from PSII to PSI is termed $\phi_{T(II\to I)}$. The entire photon flux that is absorbed by the overall photosynthetic apparatus is given as I_{a}. Finally, $\phi_{F_{\text{I}}}$ is the probability that PSI, when excited, will, indeed, perform fluorescence.

With Equation 4.87 we now understand the meaning of the two fluorescence contributions $F_{\text{I}(\alpha)}$ and $F_{\text{I}(\beta)}$, which add up to the fluorescence intensity $F_{\text{I(M)}}$, as indicated in Figure 4.34. The fluorescence from PSI, therefore, has two sources as described above (Strasser and Butler 1977c). When you alter the concentration of chlorophyll, and you can do that by exposure of the plant to light and to dark, the fluorescence spectra will change, as you can notice from the five bean leaf spectra in Figure 4.35. The spectra are normalized with respect to the fluorescence intensity of PSII at 694 nm. The peak of the PSI signature at 730 nm increases when the chlorophyll content increases. We know there are two sources contributing to the PSI peak. One can consider the measured intensity as the sum of two contributions as follows

$$\frac{F_{730(\text{M})}}{F_{694(\text{M})}} = \frac{F_{\text{I(M)}(730)}}{F_{694(\text{M})}} + \frac{F_{\text{II(M)}(730)}}{F_{694(\text{M})}} \tag{4.88}$$

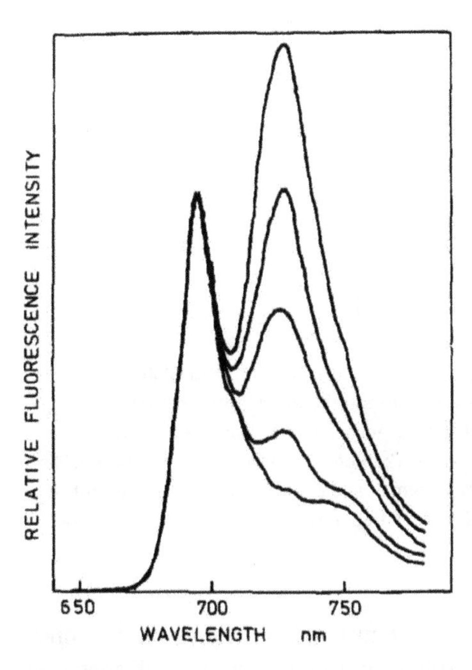

Fig. 4.35: Emission spectra of different flashed leaves at −196 °C excited at 633 nm. The leaves were the same age and had been subjected to the same number of flashes but had different chlorophyll content. The spectra were normalized at 694 nm. Reprinted from Biochimica et Biophysica Acta (BBA) – Bioenergetics, 462 (2), Reto J. Strasser, Warren L. Butler, Fluorescence emission spectra of photosystem I, photosystem II and the light-harvesting chlorophyll a/b complex of higher plants, 307–313, Copyright (1977), with permission from Elsevier, (Strasser and Butler 1977b).

and then, based on a parameterized experiment like the one shown in Figure 4.35, carry out an "extrapolation to zero"[24], which is shown in Figure 4.36. There you plot the intensity ratio $F_{730(M)}/F_{694(M)}$ versus the ratio $F_{I(\alpha)}/F_{694(M)}$, and you notice the data points do not follow a linear relation. However, you observe a trend that allows the extrapolation. When the authors of (Strasser and Butler 1977b) mention a "straight line", they do not necessarily imply a linear line. However, the line approaches the value 0.12 on the y-axis, which is the mentioned 12%.

The young researcher learns from the Figures 4.34, 4.35, and 4.36, that science is not only about showing and interpreting spectra. Instead, here it is shown how you can use a ruler and pencil[25] to extract particular quantitative data, such as in photo-

24 The extrapolation to zero, to infinity, or to any other specific value is an experimental trick that you carry out in the data analysis. The experimental system may contain hidden or not direct accessible information, and by the variation of a specific parameter you can pinpoint the wanted information. A simple example may be a physical system where a quantity y is linearly related with a quantity x, so that $y(x) = a \cdot x + b$. Your experiment delivers you the data points (x_1, y_1), (x_2, y_2), and so on. Two such data points would allow for the determination of the slope a. We may not have enough data points to immediately read the value for b, but when we realize that the data points follow a straight line, we can extrapolate the line, so that we have the y-value for $x = 0$. This is the extrapolation to zero, and the cross section between the straight extrapolated line and the x-axis gives us the value for b.

25 When you begin from scratch, for the analysis of which you have not yet developed a routine, you typically start with pen and paper and maybe a ruler, and then play around with the spectrum or other data. A long time ago as student I had to work on electron diffraction data. There I used a pen

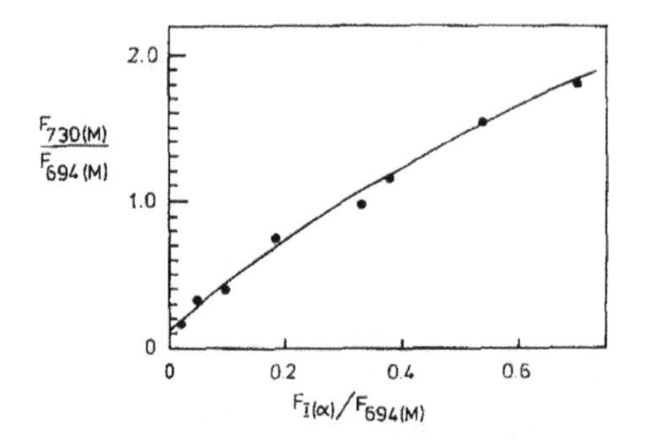

Fig. 4.36: Extrapolation of F730(M)/F694(M) vs. FI(α)/F694(M)· FI(α), F694(M), and F730(M) was measured on each of the leaves indicated in Figure 4.35. The extrapolation indicates that photosystem II has emission at 730 nm, which is 12% of the peak emission at 694 nm. Reprinted from Biochimica et Biophysica Acta (BBA) – Bioenergetics, 462 (2), Reto J. Strasser, Warren L. Butler, Fluorescence emission spectra of photosystem I, photosystem II, and the light harvesting chlorophyll a/b complex of higher plants, 307–313, Copyright (1977), with permission from Elsevier, (Strasser and Butler 1977b).

synthesis. For some studies, it may be sufficient to just record the maximum fluorescence intensity F_M at some wavelength. For other studies, you record a full spectrum and identify peaks and shoulders, and maybe, depending on the changes of the samples, redistribution of spectral weight. For other studies, you make a full analysis of the transient of the spectral intensity.

For example, Wang et al. studied two macroalgal species (*U. prolifera* and *U. intestinalis*), which coexisted freely floating in green tides in the Chinese Sea (Wang et al. 2016). They performed a fluorescence scanning with a Hitachi F-4500 Fluorescence Spectrophotometer (HITACHI, Japan) and found the intensity maxima for chlorophyll a were at 436 and 663 nm, and those for the chlorophyll b were at 463 and 645 nm. The *U. prolifera* turned out to have a larger photosynthetic capacity than the *U. intestinalis*. They found, for example, that this better photosynthesis means that *U. prolifera* can float better during green tides, so that it does not sink out of what is known as the euphotic zone, the surface layer of water where enough light can still reach the photosynthesis life to continue. By the way, it is possible to monitor such green tides and algal blooms with remote sensing (Klemas 2012).

The fluorescence of chlorophyll is strongly time dependent, and its transient characteristic shows variations over several orders of magnitude in time ranging from milliseconds to minutes upon illumination. This is demonstrated in Figure 4.37, which

and a ruler and measured distances over the data image, which showed the diffraction spots (Bragg reflections) in reciprocal space. I exactly described this way in a manuscript to be published. However, my supervisor suggested that I should avoid the words pencil, ruler, and instead use wording like "determined with high precision [...]".

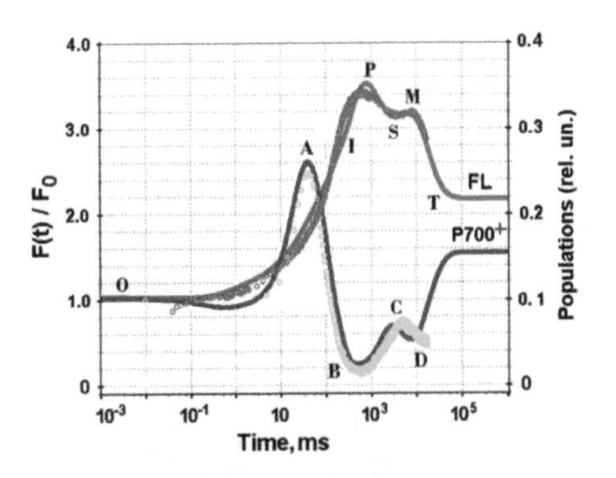

Fig. 4.37: Parallel fittings of the experimental chlorophyll a fluorescence induction (FL) and ΔA_{810}, absorbance data ($P700^{+}$) (shown by beige and green circles) with theoretical curves (magenta and violet) calculated with the thylakoid membrane model. Reprinted by permission from Springer, Photosynthesis Research, 130(1-3), 491–515, Thylakoid membrane model of the Chl a fluorescence transient and P700 induction kinetics in plant leaves, Belyaeva, N.E., Bulychev, A.A., Riznichenko, G.Y., Rubin, A.B., Copyright (2016), (Belyaeva et al. 2016).

shows the maximum intensity of fluorescence $F_M(t)$ of pea leaves (*Pisum sativum*) as a function of time on the millisecond scale. The phenomenon of the fluorescence transient is known as fluorescence induction (Lazár 1999).

Let us begin briefly with the *experimental* details (Belyaeva et al. 2016). The pea leaves were first kept in the dark for about 15 minutes at ambient temperature. The intensity F_0 is generally determined with a light intensity much lower than 1 μmol $cm^{-2}s^{-1}$ photons. It is a simple arithmetic exercise to determine the number of photons per mm^2 per second from this number. Still, such a low photon flux still constitutes some detectable photosynthetically active photon flux density (PPFD). After 15 minutes in dark, the plant is illuminated with 200 μmol $cm^{-2}s^{-1}$ PPFD. Figure 4.37 shows two curves with green circles and brown circles. The green dots represent the absorbance of the pigment P70* from the reaction center chlorophyll a in PSI. The brown data points come from the chlorophyll a fluorescence (FL).

Belyaeva et al. also *calculated* the $F(t)$ intensity based on a biophysical model of the thylakoid function and processes, shown in Figure 4.38, which yields the formula

$$F(t) = \frac{k_F}{k_L} \times (x_2(t) + g_2(t) + y_2(t) + z_2(t) + x_6(t) + g_6(t) + y_6(t) + z_6(t)) \,. \tag{4.89}$$

The rate constant for the fluorescence emission is k_F. At high light intensities, the rate constant changes to k_L. The variables x, g, y, z are clarified in the thylakoid model in Figure 4.38. This model includes a number of energy dissipative processes, which are illustrated in the geometrical model of the thylakoid shown in Figure 4.39, the thylakoid membrane model. In short, the thylakoid is a closed container, the membrane being the container and the lumen being the content.

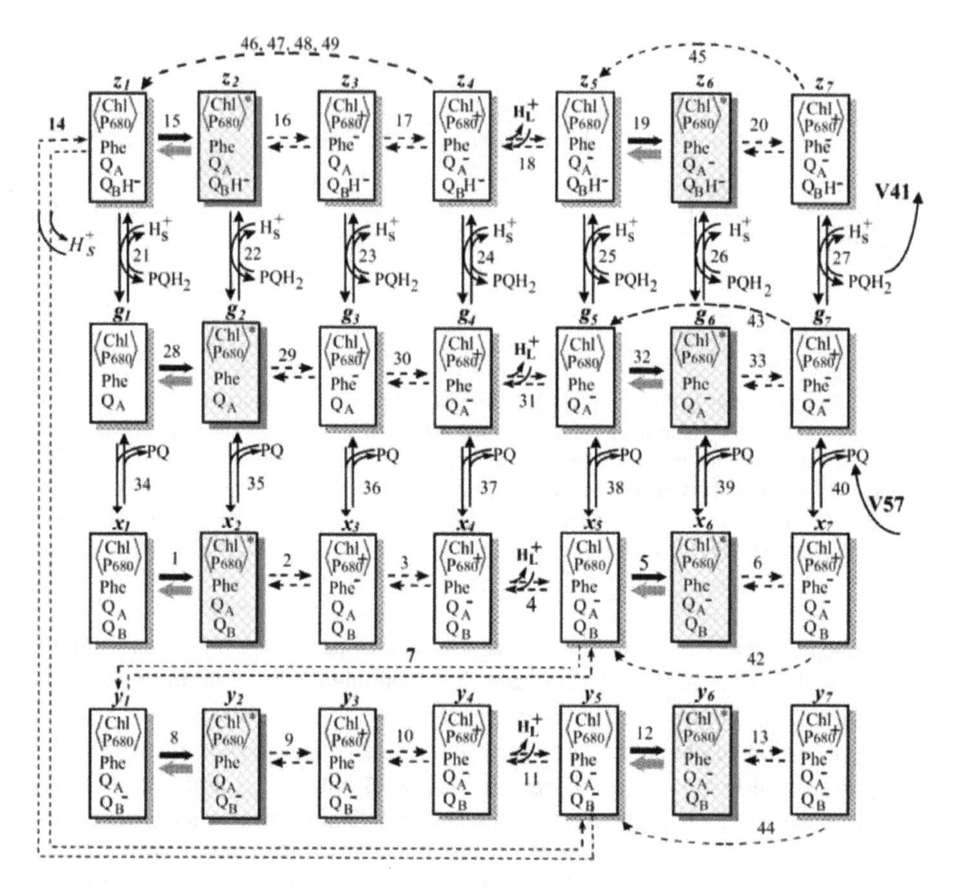

Fig. 4.38: Scheme of the catalytic cycle of photosystem II. One rectangle denotes a certain state form of PSII, determined by the transient redox states of the PSII cofactors as electron carriers: <Chl P680>-chlorophyll pigments of antenna and RC P680 (singlet excited states $^1Chl^*$ being delocalized on all pigments); phe-pheophytin, the primary electron acceptor of PSII; Q_A and Q_B primary and secondary quinone acceptors; PQ-plastoquinone; PQH_2-plastoquinol; H_L^+ or H_S^+-protons in lumen or stroma. The model variables (x_i, y_i, z_i, g_i, $i = 1, \ldots, 7$) are defined above the rectangles. Reaction numbers are denoted above the arrows. Dashed arrows show fast (< 1 ms) transitions. Dashed arcs show irreversible reactions of nonradiative recombination of Phe^- with $P680^+$ (42–45) and Q_A^- with $P680^+$ (46–49). States capable of emitting fluorescence and quanta are shaded. The gray backward arrows specify the sum of all deactivation processes of $^1Chl^*$ (except the photochemical quenching) given by Equation (10). Reprinted by permission from Springer, Photosynthesis Research, 130(1-3), 491–515, Thylakoid membrane model of the Chl a fluorescence transient and P700 induction kinetics in plant leaves, Belyaeva, N.E., Bulychev, A.A., Riznichenko, G.Y., Rubin, A.B., Copyright (2016), (Belyaeva et al. 2016).

The functional components are in the membrane, such as the chlorophyll, which receives the light energy from the outside and gives the energy to PSII for water oxidation and proton production, and PSI for the NADP production. Protons and potassium ions are pumped outside into the stroma, and chlorine anions are dragged from the

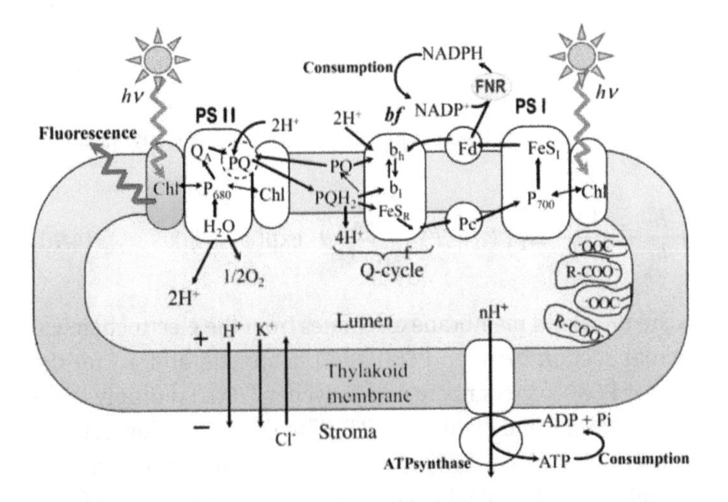

Fig. 4.39: Thylakoid membrane model. The diagram comprises the presented components of the whole electron transfer chain (ETC) in leaves or alga: the thylakoid compartments and charge fluxes induced by light. PSII, PSI—photosystems II and I; PQ(PQH_2)—plastoquinone (quinole) pool; bf—Cyt b_6f complex; mobile Fd—ferredoxin; Pc—plastocyanin; and R–COO^- – buffer groups. The stromal phase components: $NADP^+$—nicotinamide adenine dinucleotide phosphate, oxidized form; FNR—ferredoxin-NADP-oxidoreductase; and ATP synthase—CF_1–CF_0 ATPase complex responsible for the synthesis of ATP from ADP and inorganic phosphate (P_i). All components and processes are discussed in the main text. Reprinted by permission Springer, Photosynthesis Research, 130(1-3), 491–515, Thylakoid membrane model of the Chl a fluorescence transient and P700 induction kinetics in plant leaves, Belyaeva, N.E., Bulychev, A.A., Riznichenko, G.Y., Rubin, A.B., Copyright (2016) (Belyaeva et al. 2016).

stroma to the lumen. The kinetic processes that occur among the multitude of components shown in the Figure 4.39 act back on the structure of the chlorophyll, which can be detected in the fluorescence signal.

The 4 × 7 rectangles in Figure 4.38 stand for the model, and each rectangle contains electron carriers, specifically the antenna and reaction center in chlorophyll 680, pheophytin (this is chlorophyll a without the Mg^{2+} central ion), the primary plastoquinone acceptor Q_A, and the secondary plastoquinone acceptor Q_B. They are all involved in electron transport while their redox state changes as part of charge transport. The transport is reflected by the particular reaction rates. They come in reduced, neutral, excited, and oxidized states, which yields an overall of 48 different states to include in the model of 4.38.

Due to some constraints, the number of states is further reduced to 24. This is because an electron cannot be stabilized on the oxidized pheophytin cofactor at the subnanosecond timescale. This section does not have the scope to go into further details of the model, but we understand that the transient of the fluorescence intensity is affected by the light intensity and the multitude of reaction rates.

Vredenberg and Bulychev came up with a hypothesis that the chlorophyll fluorescence yield Φ is dependent on the electric field, which is a result of charge carrier generation (electron–hole pairs), when light hits the photosynthetic reaction centers. The charges are equilibrated by currents that propagate throughout the thylakoid lumen:

$$1/\Phi = 1 + \frac{k_t}{k_f} + \frac{k_d}{N \cdot k_f} \exp(\Psi_0 - \Psi) + \frac{k_e}{N \cdot k_f} \theta \cdot \exp(\psi_0 - \psi) \tag{4.90}$$

The electric field across the thylakoid membrane originates from the electrochemical potentials ϕ for a particular section between P680 and pheophytin and ϕ_0 for the midpoint redox potential for P680*. Vredenberg and Bulychev (2002) multiply these potentials with Faraday's constant F and divide it by RT. With this dimensionless constant, the potential can be written as $\psi = \phi F/RT$ in multiples of 25 mV, as you can see in Figure 4.40; N is the number of chlorophyll molecules per reaction center (in PSII), and θ the fraction of semi-open reaction centers to closed reaction centers. Note that Figure 4.40 does not show experimental data but a graph to Equation 4.90.

The various constants in Equation 4.90 denote rate constants for nonradiative losses, k_t; the sum of the rate constant of nonradiative recombination of the

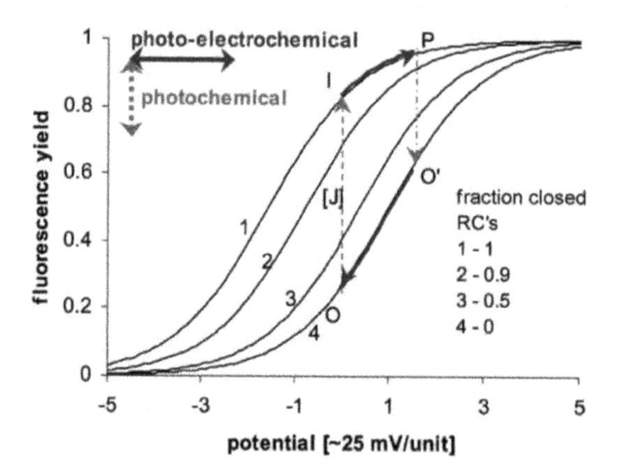

Fig. 4.40: The effect of potential ($\psi = \phi F/RT$) on fluorescence yield (Φ) under conditions differing in the degree of RC closure (θ) is shown [Equation (2); for a quantitative description, see [4])] Fluorescence and photocurrent measurements (not shown) in chloroplasts indicate a lower time constant for lateral (photo-) potential propagation (~ $2\,s^{-1}$) than for full closure of RCs (~ $30\,s^{-1}$). As a consequence, release of photochemical fluorescence quenching (O-J-I phase) and photo-electrochemical stimulation (I-P phase) become distinguishable in the $F(t)$ curve. I and P are fluorescence levels at saturated photochemistry in absence and presence of a photo-electrochemical effect, respectively; O and O' are at ceased photochemistry (dark) in absence and presence of a photo-electrochemical effect, respectively. Reprinted from Bioelectrochemistry, 57 (2), Vredenberg, W.J. Bulychev, A.A., Photo-electrochemical control of photosystem II chlorophyll fluorescence in vivo, 123–128, Copyright (2002), with permission from Elsevier, (Vredenberg and Bulychev 2002).

P680+Pheo state and formation of the P680$^+$ triplet state, k_d; the rate constant for fluorescence is k_f; and k_e is the rate for electron transfer. The variation of the fluorescence yield with the electrochemical potential follows a sigmoidal profile and resembles a hysteresis curve. Curve number 1 stands for 100% of the reaction centers being closed. When 10% of the reaction centers are open, the fluorescence curve shifts by some 6 mV towards higher potentials. With 50% of the reaction centers open, another 10 mV shift is observed. Vredenberg and Bulychev used a letter code in the graph that is connected with arrows and resembles a Carnot cycle.

The edge points that define the cycle are labeled O-J-I-P-O' (this is a common notation in this context). This brings us back to the fluorescence induction. You may have seen these letters already in Figure 4.37. These letters are pronounced stages during the fluorescence inception and induction. Let us compare this with Figure 4.41, which shows the fluorescence induction measured from *Chenopodium album*. O stands for origin and means the dark fluorescence yield $\Phi_{f,0}$; you can see this in Figure 4.41.

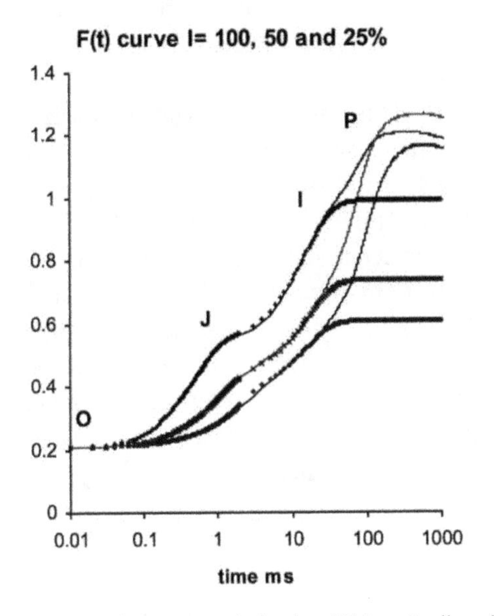

Fig. 4.41: Fluorescence induction $F(t)$ in a *C. album* leaf upon illumination with a 1-s light pulse of 100% (~ 600 W m^{-2}), 50% and 25% intensity (from top to bottom). Symbol curves are calculated with a three-state trapping model (TSTM), with fixed rate constants for electron transport at donor and acceptor sites, and excitation rate k_L proportionally variable with intensity [20]. The close correspondence of simulated and experimental curves indicates that the release of photochemical quenching (O-J-I) is completed after about 30 ms, with approximately 100, 67, and 51% full closure of RCs at the intensities used. The final rise (I-P) in the 30–300 ms time range is attributed to photoelectrochemical stimulation of the fluorescence, which in confirmation with theory [4] is supplementary to the release of photochemical quenching. Reprinted from Bioelectrochemistry, 57 (2), Vredenberg, W.J. Bulychev, A.A., Photo-electrochemical control of photosystem II chlorophyll fluorescence in vivo, 123–128, Copyright (2002), with permission from Elsevier, (Vredenberg and Bulychev 2002).

I means intermediate, and J is also an intermediate level. D stands for dip, which is sometimes observed between the intermediate stages I and J, when a very high light intensity is used. P means peak. S is the semisteady state. M means a maximum, and T is the terminal steady state (for the nomenclature and its references, see Lazár 1999). The "dip" is observed in Figure 4.41 for the fluorescence curve obtained at 100% illumination but is not noted with a *D*. Note how well the mathematical model can reproduce the experimentally measured fluorescence transient (Vredenberg and Bulychev 2002).

Remember that the processes during fluorescence are rationalized along the time axis. The leaf is exposed to 1 second light pulse of 600 Watt per m^2 for the 100% flash. For 50 and 25% it is 300 and 150 Watt, respectively. This is the same duration of the entire fluorescence detection, which ends at 1000 milliseconds. When looking at the stages O-J-I we can see how well the mathematical model reproduces the experimental data for three different intensities of light stimulation. For the 100% experiment, after around 30 ms at stage I, the plateau for the completion of photochemical quenching is reached, which means that the photosystem II reaction centers are all closed. This is where the model ends, and $F(t) = 1$, by normalization and the experimental data, the fluorescence transient continues to rise to around 1.2. This extra fluorescence is stimulated by photoelectrochemical processes.

With light pulses we can probe the dynamics of the photochemical and photoelectrochemical processes in photosynthetic components and systems such as leaves. We can do similarly with electric pulses when we switch a current on or off with the photosynthesis component assembled on an electrode. Another approach is to "pulse" the gas concentration over a leaf. Since leaves perform CO_2 assimilation, carbon dioxide is a suitable gas. Kocks et al. (1995) exposed leaves from *Arbutus menziesii*, a tree native to coastal California[26] – to CO_2 concentration ranging between 5% (high dose) and the natural concentration in ambient air (low dose). The leaf is initially illuminated for some period, and then the gas concentration is periodically varied out-of-phase or in-phase with the periodic illumination. Under in-phase conditions, oscillations with a resonant photosynthetic effect were observed (Kocks et al. 1995).

4.23 Ultrafast spectroscopy

When you record a spectrum of a sample, it may take a while. The light source and the spectrometer may need some time so that a spectroscopy scan for a necessary wavelength range can be performed. Meanwhile, the sample you want to probe may be altering its condition, its structure, and so on. When you want to monitor processes that take place in the specimen, such as a chemical redox process, it is important that the measurement does not last longer than the chemical process. For example, if there is a

---

26 The study was done by a Stanford research group.

chemical process in a lithium battery with a manganese containing electrode that oxidizes and reduces the manganese from oxidation state Mn^{3+} to Mn^{4+} back and forth, and this might occur within minutes, whereas the recording of a spectrum may last an hour, then you may not be able to pinpoint the reaction spectroscopically. The same may hold for the cycling of manganese in photosystem II (PSII). So, you either want to speed up the measurement or slow down the redox reaction.

Sometimes, the only solution to this dialectic problem is to come up with an entirely better spectroscopy method, for example by using high-frequency pulsed apparatus, such as short light pulses, or short x-ray and neutron pulses. I am lucky in that I can present you an example from photography here. I have been working since 2012, with great colleagues in Europe, towards a European flagship program on artificial photosynthesis, which was headed by Professor de Groot at Leiden University. In 2018, we were successful and were granted 1 million Euros to prepare a large grant proposal for a billion Euros (Abbott 2019; Kupferschmidt 2019).

To the meeting that followed the announcement of the successful application, at Leiden University in the Netherlands, I brought a bottle of champagne that I had received as a gift from my colleague Professor Emil Roduner at the University of Pretoria in South Africa. At about that time and period, I always brought my camera to meetings – we had many –, as I was doing my business trips in Europe with Empa's new hydrogen fuel cell car. I wrote a book about all the trips I made with the fuel cell car (Braun 2019b). So, when I opened my champagne bottle for the celebration of the contract we had won with the European Union for our flagship proposal, our host and *spiritus rector*, Huub de Groot, took a photo of the *moment* when I opened the bottle. You can see the photo in Figure 4.42.

The camera was set to automatic mode. Hence there was no manual settings by the photographer in play. The automatic mode had sensed it was not enough lighting in the room and thus the inbuilt flashlight would be used. When the trigger button was pushed on the camera, sensing light from an LED was emitted forward and reflected by the cork flying in the air. In that moment the flashlight was automatically triggered, which illuminated the entire space forward, including the cork. This was an incidental snapshot[27]. When the flashlight was triggered, the camera aperture was open for 1/60 seconds – the exposure time (0.0167 seconds). The flash however has a duration of 1/1000 seconds (0.001 seconds) or less, 1/10000 seconds in the extreme case. During this very short time of the flash the cork was illuminated and its motion virtually freezed in the position where it flew during the flash. When the flash was discharged, exhausted, extinguished, the cork certainly was still flying and there was still the illumination from the lighting in the room, the background light intensity. From the 0.0167 seconds of aperture opening there were still 0.0167 s – 0.001 s = 0.0166

27 Professional photographers use the play with the flashlights in order to get special effects, see, for example, https://neilvn.com/tangents/flash-photography-faq-questions-answers/.

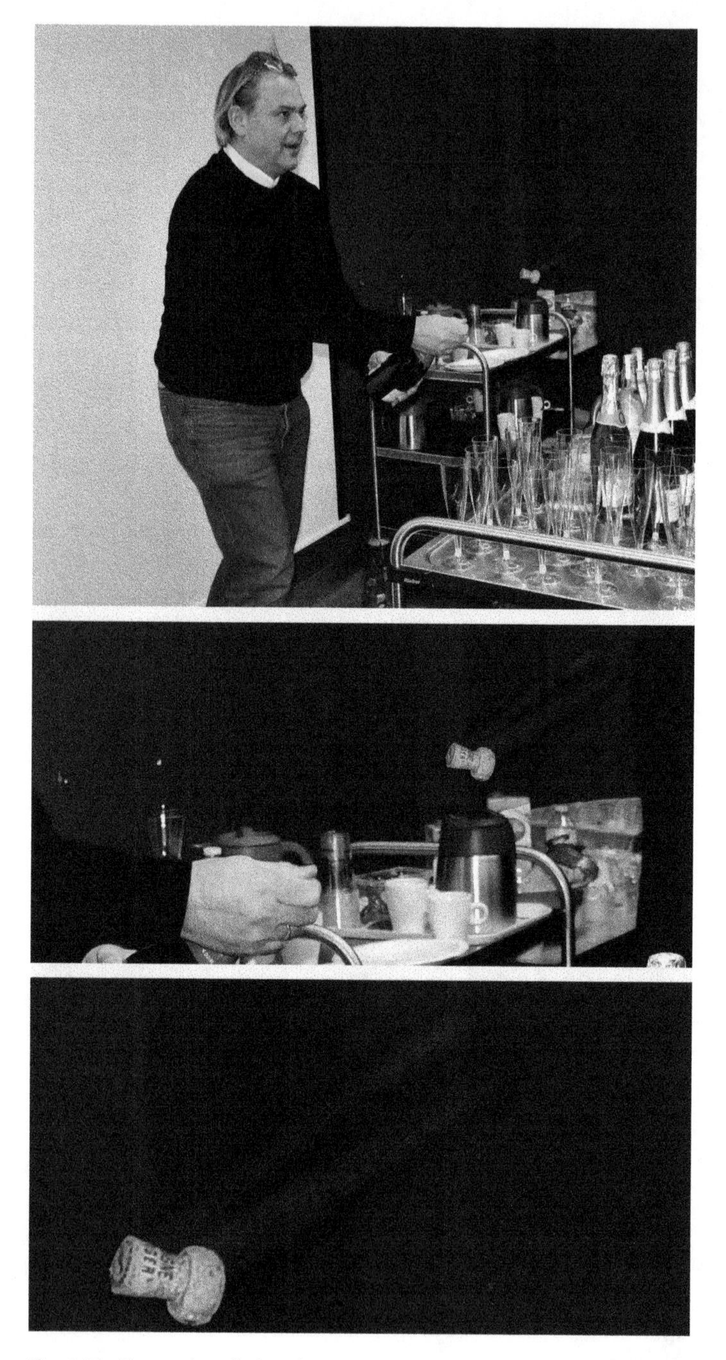

Fig. 4.42: Photo taken during the ceremony for the successful SUNRISE CSA project (Aro et al. 2017). Top: the author of this book has just opened a bottle of champagne. Middle: magnification of the photo. You can see the cork around 20 cm away from the bottle flying to the right, upwards. (Bottom) further magnification of the photo. You can see the sharp shape of the cork and also two diffuse red broad stripes in front of the cork, highlighting the trajectory of the cork after it was hit with the light from the camera flash. Photo taken by Prof. Huub de Groot, Leiden University. 17, January 2019. Nikon D7500 camera settings: exposure time 1/60 sec; flash auto, strobe return.

seconds to go. During that remaining time everything in front of the lense and the camera CCD was recorded. As the cork was a moving target, its contours are distributed, or smeared, along its trajectors. This smeared intensity is the arrangement of red stripes forward to the cork.

We can understand this by looking at Figure 4.43. Let's say the aperture of the camera is open for 12 milliseconds. This means that for 12 milliseconds, the film or the detector will collect light through the lens and form a picture. During these 12 milliseconds, we are "shooting" at a moving target, an object in motion with the camera. In the beginning, we have a short and high-intensity light pulse from the flash of light. During this short pulse, the object is moving but it has not passed a long way. Therefore, the image of the object on the film is relatively sharp and also relatively bright – because the light intensity was high. After 2 milliseconds, the flash is gone, and the only light intensity that can help make an image on the detector or film is the light from the ambient. Its intensity is constant, but rather low. Also, the object is still moving. Therefore, the entire motion of the object is being recorded on the film. We, therefore, do not have a sharp feature of the object but an object with "smeared" contours.

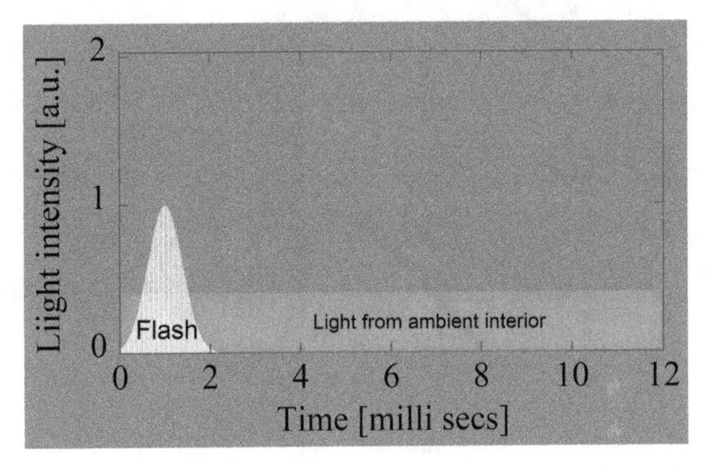

Fig. 4.43: Schematic visualization of how the short high-intensity light pulse from the flash burst and the long-time low-intensity background from the ambient add their information to the photo or film.

We are interested not only in the femtosecond and picosecond processes in photosynthesis. We want to know about the entire timescale, as we can assess it. Of course, it is much easier to study slow processes in photosynthesis, the slowest being the growth of a plant, for example, the growth of a tree. Figure 4.44 from Tsimilli-Michael et al. (2000) shows the fluorescence intensity of chlorophyll a over a timescale from 50 µs to 1 s. We know already that the fluorescence transient has a fine structure. We already know the O-J-I-P structure of the fluorescence transient, which happens at particular time stamps along which the system can be rationalized.

The O-J-I-P are the time stamp positions, and the corresponding fluorescence values are, thus, F_O at, for example, 50 μs, where all reaction centers of photosystem II can be considered open and the quinone acceptor fully oxidized. The inset in Figure 4.44 shows a construction around the 300 μs time stamp for the relative variable fluorescence kinetics. Note that the time in this inset is plotted on the linear scale, whereas the common O-J-I-P is plotted on the logarithmic timescale. This is why you see the fluorescence growing linear during onset of fluorescence. The slope of the construction yields M_O, which is one of the various quantities to be derived from the transient. The specific energy fluxes, the phenomenological fluxes, and yields can be calculated according a protocol laid out in the paper by Tsimilli-Michael et al. (2000).

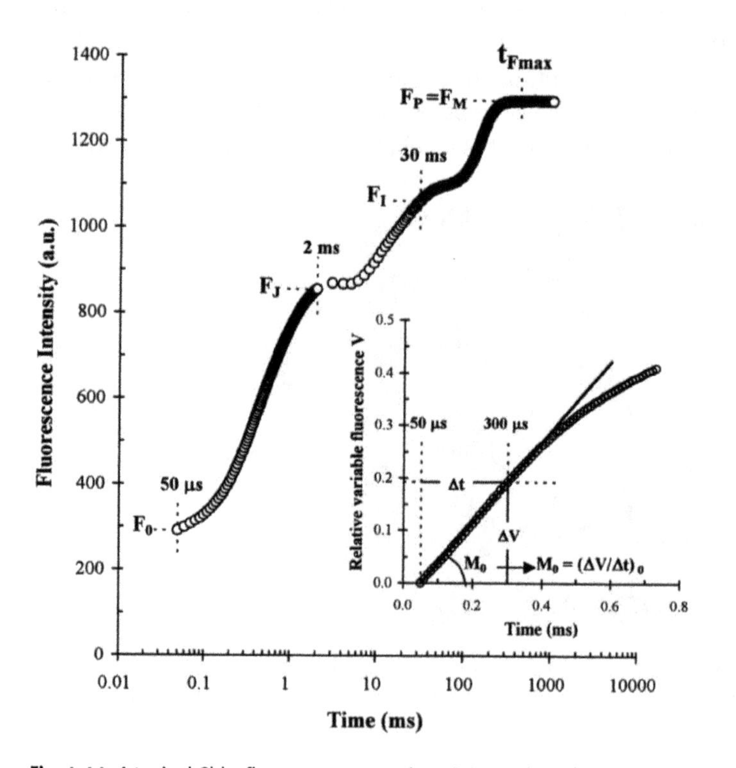

Fig. 4.44: A typical Chl a fluorescence transient O-J-I-P, plotted on a logarithmic timescale from 50 ms to 1 s. The marks refer to the fluorescence intensities at the selected times used by the JIP-test: the fluorescence intensity F0 (at 50 ms); the fluorescence intensities FJ (at 2 ms) and FI (at 30 ms); the maximal fluorescence intensity FPDFM (at tFmax). The insert presents the relative variable fluorescence on a linear timescale, from 50 ms to 0.8 ms, demonstrating as well how the initial slope is calculated: M0D(dV/dt)0 D (1 V/1t)0DV300 ms/250 ms. Reprinted from Applied Soil Ecology, 15, M. Tsimilli-Michael, P. Eggenberg, B. Biro, K. Köves-Pechy, I. Vörös, R.J. Strasser, Synergistic and antagonistic effects of arbuscular mycorrhizal fungi and Azospirillum and Rhizobium nitrogen fixers on the photosynthetic activity of alfalfa, probed by the polyphasic chlorophyll a fluorescence transient O-J-I-P,169–182, Copyright (2000), with permission from Elsevier.

4.23.1 Streak camera

For the detection of a light transient, such as the fluorescence signal stimulated by a light impulse from a light source, a so-called streak camera was developed. The camera records the light intensity's variation versus time and maps it into a light intensity variation over a space. The timescale is mapped to a spatial scale. This is the way to *memorize* the temporal light pattern. It depends, then, on the specifications of the components of the streak camera as to how fast it can detect, how high a temporal resolution of processes it can provide.

When you look briefly at Figure 4.41 you see that a time resolution of 0.1 ms of the spectroscopy apparatus would be just good enough to reproduce the data profile, the fluorescence transient curve; 0.01 ms resolution would be better; this is 10 µs. Ito et al. (1991) developed an absorption spectrometer for the picosecond resolution, which is 10^{-12} seconds. It uses a probe light with a continuous light spectrum covering the full range from infrared to ultraviolet with a 50 ns pulse duration. With a monochromator and a streak camera, they can record absorption spectra with 50 picosecond time resolution. One advantage is the primary light source laser pulse, which is sufficient to produce the entire result. This helps to limit radiation damage from the light source. A similar concept is nowadays employed with free-electron lasers (FEL).

An example of how such ultrafast spectra are recorded is given in Komura and Itoh (2009), with the experimental setup sketched in Figure 4.45. The light pulse is created with a light-emitting diode (LED), which has a device-specific wavelength but a considerable broadening. This light pulse excites the sample, which is the green solution in the glass (quartz or other) or plastic vial. All necessary information is now in the multiwavelength fluorescent emission that passes through two lenses and one optical slit and is then guided onto a grating, which acts as a monochromator.

The grating splits the multiwavelength information from the sample into a "rainbow" distributed in horizontal direction. This "streak" rainbow hits a photocathode and, thus, produces electric charge carriers that are accelerated onto a phosphor screen. This screen produces an image. The horizontal image is extended in vertical direction by an extra electric field, which is produced by a couple of deflection plates. This is basically a cathode ray tube as invented by K.F. Braun in 1897 (Braun 1909). The time of arrival of photons on the photocathode will create the vertical position on the phosphor screen. A trigger mechanism between the laser excitation and the voltage applied to the deflection plates warrants that the fluorescence signal is recorded as transient with highest resolution.

Behind the phosphor screen is a CCD (charge-coupled device, CCD chip) camera. In the end, you get a two-dimensional image which spans the wavelength information over the time. Thus, you get an image of time-resolved absorption spectra. Representative data images are shown in Figure 4.46 of the study by Komura and Itoh (2009). They were recorded from photosystem I (PSI) from spinach at 4 K, excited

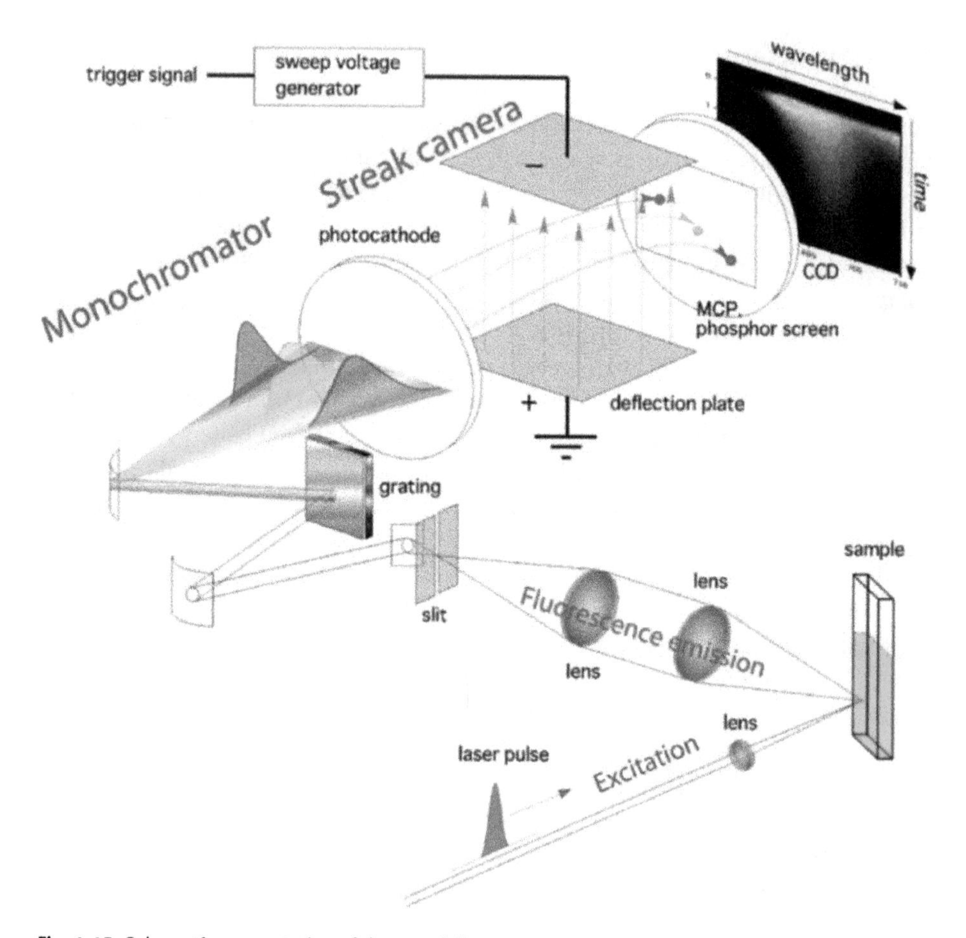

Fig. 4.45: Schematic presentation of the streak fluorescence camera measurement system. See text for the details. Reprinted by permission from Springer Nature, Photosynthesis Research 101,119–133, Fluorescence measurement by a streak camera in a single photon-counting mode, Masayuki Komura and Shigeru Itoh (2009).

with 405 nm in the upper left image. The other three images are from three other samples measured at 77 K or ambient temperature, including one membrane from purple bacteria.

Komura and Itoh made a detailed study on photosystem II (PSII) from spinach with their streak camera spectrometer at 4 K and at 77 K[28]. They measured the fluorescence decay with picosecond resolution. Their results are summarized in a scheme shown in Figure 4.47. It shows the estimated flow path of the excitation energy. The flu-

28 These are the temperatures that you can achieve easily with liquid helium and liquid nitrogen cooling.

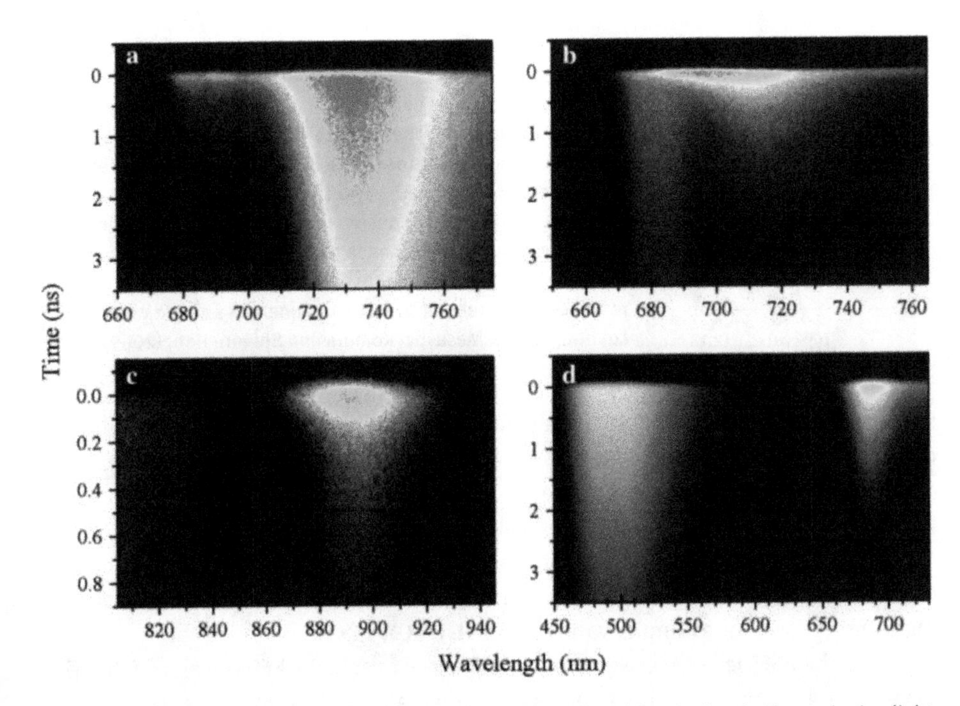

Fig. 4.46: a. 2D fluorescence image at 4 K of PS I particles isolated from spinach. The excitation light source was a 405-nm laser diode operated at 1 MHz. b. 2D fluorescence image at 77 K of the FCP-PS I supercomplex isolated from a marine centric diatom, *C. gracilis*. The fluorescence measurement was performed with a 460-nm, 120-femtosecond laser flash that excited mainly Chl c. c. 2D fluorescence image at 77 K of isolated membranes from a purple photosynthetic bacterium, *A. rubrum*, which has Zn-bacteriochlorophyll a in contrast to all the other photosynthetic bacteria that use Mg-bacteriochlorophyll a. The excitation light source was a 405-nm laser diode operated at 1 MHz. d. 2D fluorescence image at room temperature in a living coral with an intrinsic green fluorescent protein (GFP) and an endosymbiont, Zooxanthella. The long-decay fluorescence at 400–500 nm comes from GFP, and the fast-decay fluorescence at around 680 nm comes from Chls in Zooxanthella. The excitation light source was a 405-nm laser diode operated at 1 MHz. Details of the experimental conditions for the measurements are given in the text. Reprinted by permission from Springer Nature, Photosynthesis Research 101,119–133, Fluorescence measurement by a streak camera in a single photon-counting mode, Masayuki Komura and Shigeru Itoh (2009).

orescence band F677 has the smallest wavelength and, thus, the highest energy and passes on energy to F685. The time constant τ for this process is smaller than 5 ps at the "high" temperature of 77 K and increases to 28 ps upon cooling to helium temperature.

Spectroscopy and scattering are the tools that we physicists and chemists use for the determination of the structure of matter. To some extent also microscopy is important. At the molecular scale, microscopy becomes difficult; and even when you employ electron microscopy for higher spatial resolution, the dissipated energy from the elec-

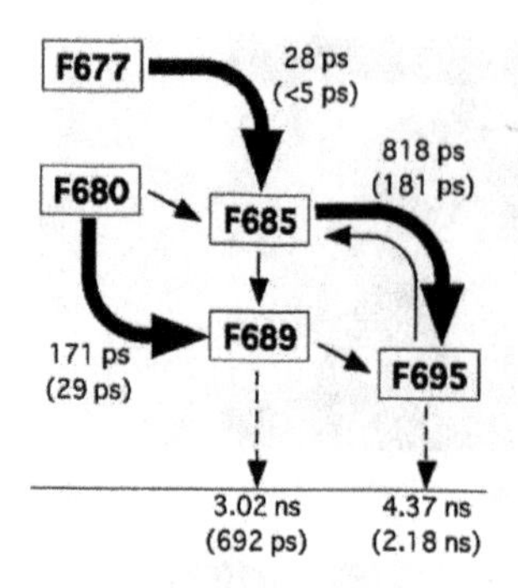

Fig. 4.47: Estimated scheme of excitation energy transfer among the fluorescence component bands in PS II based on the 2D fluorescence measurements at 4–77 K. The bold arrows represent the major energy transfer pathways. The estimated time constants of the excitation energy transfer/relaxation at 4 K; 77 K time constants are shown in parentheses. Reprinted with permission from Springer Nature, Photosynthesis Research 101,119–133, Fluorescence measurement by a streak camera in a single photon-counting mode, Masayuki Komura and Shigeru Itoh, (2009).

tron beam may incur radiation damage to the specimen. This holds in particular for organic samples, which we always have when we study photosynthesis.

Van Oijen et al. spent several years on developing a method for the detection of fluorescence spectra from the light harvesting complex LH2 in purple bacteria. For this method to succeed, it was necessary to cool down the sample to 1.2 K – for which you need cooling with liquid helium. For information on the structure of the light harvesting complex, see, for example, van Oijen et al. (2001, 1999).

From the F685 level, excitation energy is passed over to the F695 level, which is the lowest energy band here. The time constants increase from 181 to 818 ps. The authors found the hitherto unknown new fluorescence band F689, which is fed by the band. From a blueshift of the steady-state spectrum upon cooling, the authors concluded that the fluorescence from the shorter wavelengths increases, because low temperatures slow down the excitation energy transfer.

On the occasion of a project meeting for the SUNRISE flagship at Consiglio Nazionale delle Ricerche (CNR) in Bologna I could visit the laboratory being used by my colleague Dr Andrea Barbieri. Figures 4.48 and 4.49 show a streak camera system used for photochemical experiments at this laboratory.

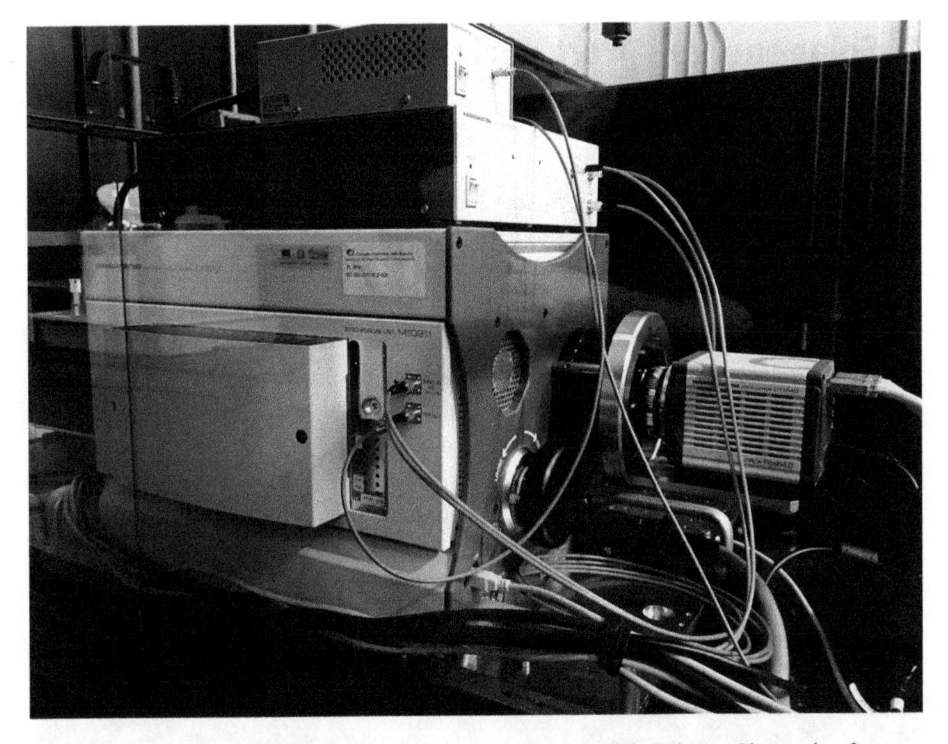

Fig. 4.48: A streak camera system at the photochemistry lab at CNR in Bologna. Photo taken from side view by the author, 16 May 2019.

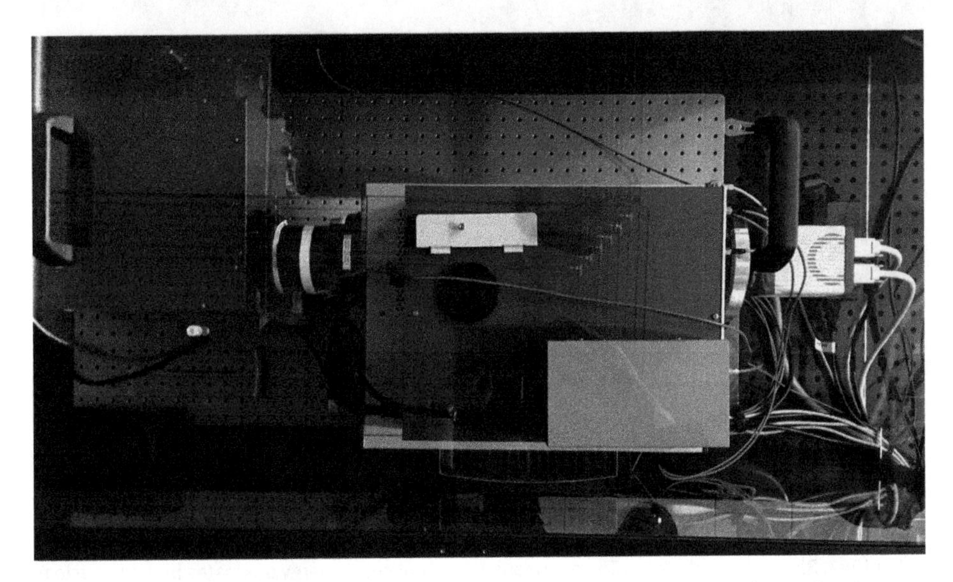

Fig. 4.49: A streak camera system at the photochemistry lab at CNR in Bologna. Photo taken from top view by the author, 16 May 2019.

5 Solar energy – from nuclear forces

5.1 The power of the sun

Planet Earth and the sun were formed 4.5 billion years ago. By gravitation, the sun keeps our Earth on its trajectory. The time that Earth needs for one cycle around the sun is what we call 1 year. The energy from our sun created life on Earth. After 1 year, the cycle of life repeats itself. Its four seasons have a major impact on life in nature. The Earth rotates around its own axis. This means that every point on Earth faces the sun once per cycle and, thus, becomes illuminated. In the changes from night to day, from winter to summer, the sun is the almighty and reliable constant upon which life depends.

Every second, the Sun emits 20,000 times more energy from its 6,000 °C hot surface than was used for the entire industrialization of mankind. 200 years of modern industrial revolution required just 200 hours of sunshine. One full month of solar power was sufficient to build our modern civilization. The amount of energy that we receive from our sun, is to human imagination unmeasurable and inexhaustible. The photo shown in Figure 5.1 was taken in Death Valley. In the forward direction, on the right-

Fig. 5.1: Sunrise over Death Valley, California, one of the hottest spots on Earth (Lingenfelter 1986; Kahle 1987; Bjorkman et al. 1972). The enormous amount of solar energy originates from the nuclear fusion of hydrogen isotopes. For 4 billion years the sun has delivered this energy to Earth. The hydrogen pool of the sun will be exhausted in the next 4 billion years to come. Photo taken 15. April 2015 by Artur Braun.

https://doi.org/10.1515/9783110629941-005

hand side, you can see the sun shortly after sunrise with very high white intensity, surrounded by a yellowish ring, which has a longer wavelength than the blue light, and which is not concentrated on the disk of the sun but is visible from all directions with respect to the position of the sun. The reason why we can, thus, see the sky in blue color is the λ^4-law, which is found in the Rayleigh scattering of light.

100 million miles away from the sun, at a safe distance, in 1 hour Earth receives enough energy to meet mankind's energy demand for an entire year. The Sun has now completed half of its lifetime. Half of its fuel – hydrogen – has already been burnt to helium by nuclear fusion. There is more hydrogen left for us – for another 5 billion years.

5.2 Origin of sunlight

The sunlight that arrives on Earth is produced inside the sun by nuclear fusion. The sun is still a huge fire ball of hydrogen plasma, protons that have a high speed and that upon collision fuse with each other and form reaction products, chemical elements with a heavier weight, such as helium, plus some excess reaction energy that is irradiated into the cosmos.

We assume that each proton π with mass m is part of an ideal gas and has an individual speed, velocity v_i. The kinetic energy of such a proton is then $E_{\text{kin}} = m/2v^2$. Moreover, we assume that the entire ensemble of protons is characterized by a distribution of velocities that follows the Maxwell distribution. This implies that the probability that the equilibrium state is occupied is given by a Boltzmann factor

$$W = W(E) \sim \exp\left(-\frac{e_{\text{kin}}}{k_B T}\right) \tag{5.1}$$

The Maxwell–Boltzmann $p(v)$ then distribution reads

$$p(v) = 4\pi\left(\frac{m}{2\pi k_B T}\right)^{3/2} v^2 \exp\left(-\frac{mv^2}{2k_B T}\right) \tag{5.2}$$

We can calculate that the momentum of two high-energy protons colliding with each other is not sufficient for them to overcome the nuclear Coulomb potential, which is modeled by a Yukawa potential. The classical statistical mechanics and physics, therefore, cannot explain the effect of nuclear fusion. The only way we know that allows for nuclear fusion as it happens in the sun is by virtue of the tunnel effect. The probability for this tunnel effect to happen for protons is incredibly small. However, the number of protons in the sun is incredibly large. This keeps the nuclear fusion in the sun going. Figure 5.2 shows the Maxwell–Boltzmann distribution for three different temperatures $T_3 < T_2 < T_1$ on an arbitrary unit scale.

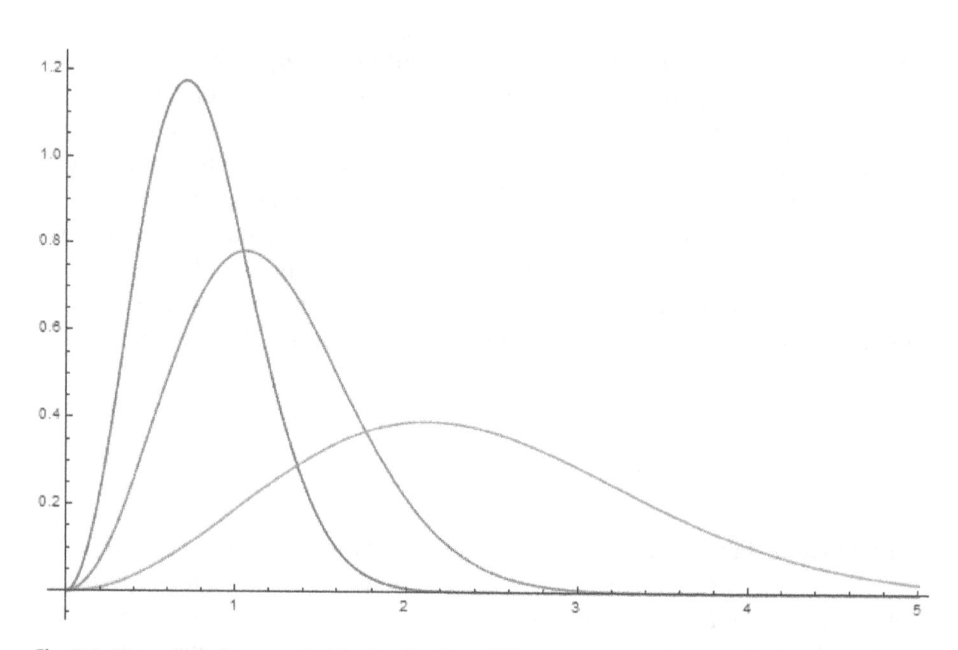

Fig. 5.2: Maxwell–Boltzmann distribution for three different temperatures. As an exercise, label the axes correctly and find out which of the three curves have the temperature $T_3 < T_2 < T_1$.

5.3 Nuclear fusion reactions

As electrochemists, chemists, and condensed matter physicists, we are typically dealing with electron binding energies that arise from the relations between the atoms or ions and within molecules and condensed matter. Their energies range in the order of 1 eV and below. We pay no attention to the nucleus of the atom, with the exception that its relatively heavy mass as compared to the mass of the electrons plays a role in the total mass of the matter we are dealing with. The nuclear power, however, originates from reactions among the particles that constitute the nuclei, and these are protons and neutrons. We see its actions in the light from the sun during the day and from the stars at night. In simple words, it is the hydrogen that makes the beginning of what our universe looks like (von Ditfurth 2015). This is why particle physics is in the in-between of astronomy and cosmology, and nuclear physics.

Nuclear science is a field of its own and beyond the scope of this book. However, we need to deal with some very simple nuclear models here in order to understand the working principle and scope of fusion and cold fusion. This is notwithstanding that nuclear science is still a further developing field. We restrict ourselves here to the picture that the nucleus of the atom is built from protons with mass 1 and charge +1, and neutrons with mass 1 and 0 charge. The chemical element is defined by the number of protons that it has in its nucleus. Elements may have a varying number of neutrons, which is schematized in a nuclide chart. At a later point in this book we also want to consider the Feynman diagram for the neutron reaction with a neutrino, where it decays into a proton plus a muon. The neutron is built from three quarks named up quarks and down quarks, and the proton is built up from three quarks, up quarks and down quarks where the middle quark transforms from a down quark to an up quark.

Let us consider two protons that will be fused together so as to build a helium nucleus. Protons are positively charged, and their Coulomb interaction (Coulomb force) will make them repel each other. However, there is another force acting as well, and this is the nuclear force. The nuclear force is an attractive force and has a very short range and follows a different law to Coulomb's law. The nuclear force is also stronger than the Coulomb force. Relevant is here the short-range attractive Yukawa potential

$$U_{\text{Yukawa}}(r) = -g^2 \frac{e^{-amr}}{r} , \tag{5.3}$$

which adds to the long-range repulsive Coulomb potential and yields a superposition of potentials with a minimum that defines the distance of the nuclei.

Both forces act on the proton. When an additional force comes into play and pushes the two protons beyond the minimum of the nuclear force, which is achieved at around 1.3 fm (femtometer, 10^{-15} m), the protons can fuse and form a new nucleus, such as deuterium ^{2}H, which contains one proton and one neutron. Typically, it is considered that the kinetic energy of protons follows a Maxwellian velocity distribution , and then the protons collide. This situation is illustrated in the proton–proton chain (cycle) in Figure 5.3. Consider a pool of protons ^{1}H, which we can, somehow, force to approach each other beyond the aforementioned minimum of the potential located at around 1.3 fm.

This process seems simple here in the book, but in reality, it requires a tunneling process that is extremely rare and occurs in the sun only once every $1.4 \cdot 10^{10}$ years. This is the reason why the sun has been burning for such a long time and is not yet exhausted. The reaction product is a deuterium nucleus ^{2}H (which is built from a proton and a neutron), one positron p^+, and one electron neutrino ν_{e^-}. The energy released from this process step is 0.42 MeV per proton[1]. The neutrino carries away 0.267 MeV of this energy. Since neutrinos hardly interact with any matter, this energy is carried away from the sun and is basically lost in the universe. The positron p^+ will immediately annihilate with an electron e^- and release 1.022 MeV in the form of two γ quanta. The deuterium ^{2}H will find further protons from the pool and then fuse to a helium nucleus ^{3}He and release one γ quant of 5.493 MeV. With 1.4 s, the lifetime of the deuterium is very short. The helium nuclei ^{3}He can now fuse to a heavy ^{4}He nucleus with

1 It is an exercise for the reader to convert 1 MeV to watt-hours Wh or milliampere hours mAh. These are the metrics used in the heating of a residential home or in the capacity of batteries.

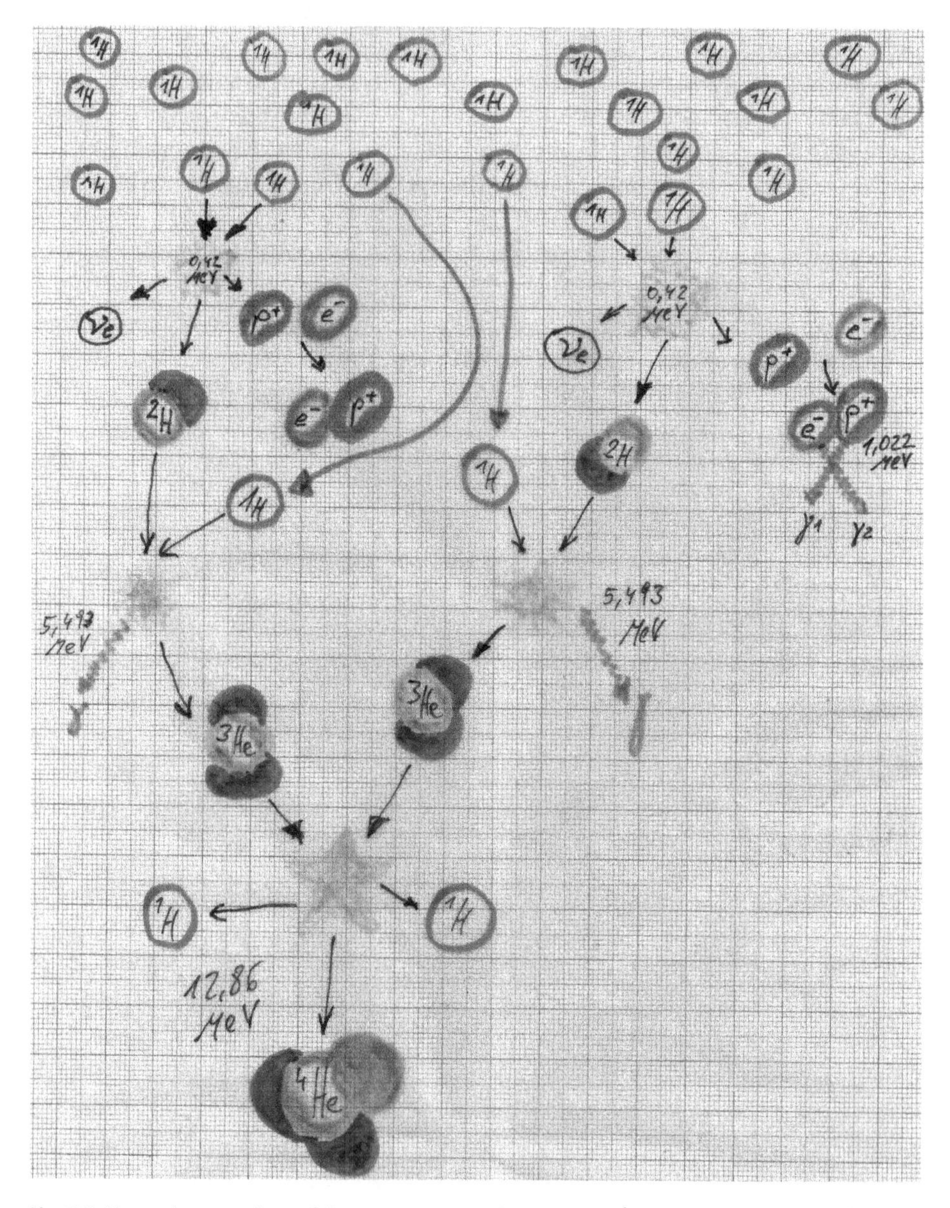

Fig. 5.3: The nuclear reactions of the proton–proton chain towards ^{4}He production in a so-called hydrogen burner.

the release of two further protons, which will add to the proton pool. The released energy is 12.86 MeV. This nuclear reaction takes place at temperatures between 10 million and 14 million Kelvin.

There are two alternative and competing protons – proton processes that form boron, lithium, and beryllium as reaction products, in addition to ^{4}He. These are of lesser probability and will be ignored here. The energy released from the chain is 26.196 MeV. I leave it as an exercise for the reader to figure out from Figure 5.3 how this value is obtained by addition. It is also interesting to estimate how much energy is released in total when we begin the chain with 1 mol of nuclei. This compares, then, with 1 mol of hydrogen converted in a fuel cell, and with 1 mol of carbon representative of coal (fossil fuel) or wood (biomass). This, too, is left as an exercise for the reader. The huge amount of energy released by nuclear fusion is a strong motivation for research and technology in this field.

To the physicist, nuclear fusion is a scattering problem like any other mechanical problem dealing with collisions of objects (Figure 5.3) with an elastic and inelastic contribution. The scattering cross section of the nuclei and the kind of interaction determines whether a nuclear reaction such as nuclear fusion can take place. Quite early on, scientists were able to determine by simple calculations how likely it was that a proton would enter an atom when it was in a proton cloud with a kinetic energy distribution according to Boltzmann statistics (Atkinson and Houtermans 1929; Gamow 1938). The reader is referred to the literature for details on nuclear physics, specifically (Adelberger et al. 2011, 1998). What we need to remember here is that the conditions for a nuclear fusion are very extreme and not all are easily possible on Earth, specifically in a laboratory.

As soon as scientists (for example, see Bethe 1939, Gamow 1938, von Weizsäcker 1937, and von Weizsäcker 1938) had figured out which nuclear reaction would cause fusion of nuclei with the corresponding release of huge amounts of energy, they thought about how to harness these nuclear reactions in order to make them useful for mankind on Earth. It is certainly ironic that the first application of nuclear fusion was realized in the hydrogen bomb (Bernstein 2010; Gorelik 2009)[2].

---

2 During my time in California I had a colleague who, as life would have it, was suddenly faced with a difficulty in continuing his scientific career in the United States, despite being an excellent scientist on an excellent career path. It was an issue with visa and the immigration authorities, founded in a ridiculous travel detail from the time when he was an undergraduate student, the later consequences of which had been not predictable. The only way out for my colleague was the issuing of a so-called O-visa. Such an O-visa is "For persons with extraordinary ability or achievement in the sciences, arts, education, business, athletics, or extraordinary recognized achievements in the motion picture and television fields, demonstrated by sustained national or international acclaim, to work in their field of expertise. Includes persons providing essential services in support of the above individual." So it reads in the regulations of the U.S. Department of State. Before the visa problem occurred, my young colleague had met Professor Edward Teller at a conference in Stanford, approached him and had a nice chat with him. Teller had studied in Karslruhe and Leipzig and had worked in Göttingen, so he spoke German very well. So my German colleague spoke with Teller in German. When he needed the O-visa, he asked Professor Teller if he could furnish him a letter of recommendation. Edward Teller wrote my colleague the recommendation letter; and then he got the O-visa. My German colleague's career in USA could continue.

Today, it is established which of the various nuclear reactions release the most energy in the sun and in the universe (Siegel 2017). In a hydrogen bomb, deuterium ^{2}D and tritium ^{3}T are brought to an extremely high temperature and pressure in a condition where they can react, fuse, and release energy as electromagnetic radiation in an amount that corresponds to their mass loss of 0.3%. This extreme condition is realized by the ignition of a nuclear bomb or by the activity inside the sun and in other stars.

Note: this is the origin of sunlight and of the light from the stars we see at night. Particles fuse and transform into other particles, including photons (electromagnetic radiation). We are here at the core of quantum electrodynamics because the nuclear fusion process provides a situation with an electric field and charged particles.

5.4 Quantum mechanical formulation of nuclear fusion

Edwin Ernest Salpeter worked in nuclear science and elaborated nuclear reactions that would take place in the sun and in other stars. In a note from 1952 Salpeter presented calculations and results that were aimed at being suitable for astrophysical applications to main sequence stars (Salpeter 1952b).

The main sequence is a term in astronomy and means the continuous band of stars when these are plotted on a map of stellar color (a function of temperature) and their brightness. This plot is called a Hertzsprung–Russell diagram according to its inventors Ejnar Hertzsprung and Henry Norris Russell. For an interpretation of these diagrams the reader is referred to (Bondi 1950).

Hertzsprung and Russell invented these diagram independently around 1910. Stars on this band are known as main sequence stars or dwarf stars. These are the most numerous true stars in the universe, and include the Earth's sun.

In Figure 5.4, you can see a Hertzsprung–Russell diagram that was produced from data (Babusiaux et al. 2018) very recently measured by the Gaia astronomic observatory, which is a satellite orbiting on the L_2 Lagrangian point. On the top axis, the diagram shows the temperature of the stars as determined form their color; note that the values increase from right to left. The right abscissa shows the luminosity of the stars on a logarithmic scale. This image was made by around 65 million stars observed by Gaia. The bright diagonal is virtually the main sequence stars. Our sun is part of this main sequence.

The proton–proton chain involves predominantly the conversion of (in the end) four hydrogen atoms to one helium atom, plus two neutrinos (Salpeter 1952a):

$$\mathrm{H}^1 + \mathrm{H}^1 \rightarrow \mathrm{D}^2 + \mathrm{e}^+ + \nu + 0.42\,\mathrm{MeV} \tag{5.4}$$

This reaction takes place once every 8×10^9 years, which by human standards we would call unlikely, or even impossible. The following reaction takes place once every

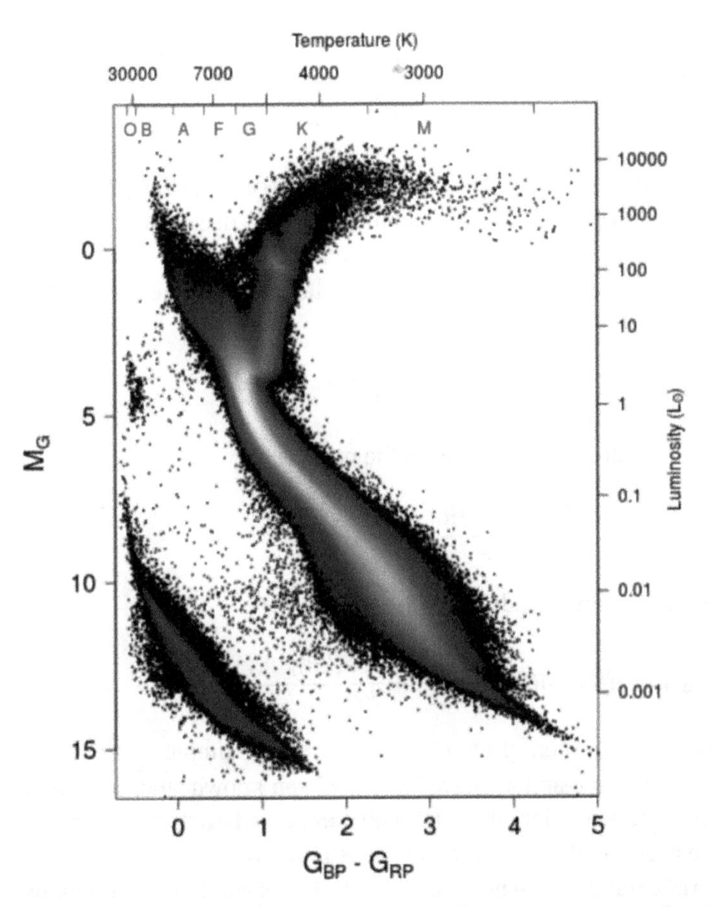

Fig. 5.4: Gaia HRD of sources with low extinction ($E(B-V) < 0.015$ mag) satisfying the filters described in Sect. 2.1 (4 276 690 stars). The color scale represents the square root of the density of stars. Approximate temperature and luminosity equivalents for main-sequence stars are provided at the top and right axis, respectively, to guide the eye. Reprinted from Astronomy & Astrophysics Special issue 616, Gaia Data Release 2 – Observational Hertzsprung–Russell diagrams, A10 (2018), Open Access article, published by EDP Sciences, under the terms of the Creative Commons Attribution License (http://creativecommons.org/licenses/by/4.0), which permits unrestricted use, distribution, and reproduction in any medium, provided the original work is properly cited. (Babusiaux et al. 2018).

four seconds (Salpeter 1952b):

$$H^1 + D^2 \rightarrow He^3 + \gamma + 5.5\,\text{MeV} \tag{5.5}$$

$$He^3 + He^3 \rightarrow He^4 + 2H^1 + 13\,\text{MeV} \tag{5.6}$$

The reaction rate p for the deuterium production in the reaction Equation 5.4 is given by Bethe (Salpeter 1952a) as

$$p = 16\pi \times 3^{-5/2} gf(W)|M_{\text{sp}}|^2 \gamma^{-3} \Lambda^2 N_0 A_{\text{H}}^{-1} \rho x_{\text{H}} \tau^2 \text{e}^{-\tau} F_\tau \tag{5.7}$$

Salpeter concluded from the proton reaction rates on the amount of deuterium in the main sequence stars, which was considerably smaller than the deuterium found on Earth, which led him to suggest in his paper (Salpeter 1952a) that the Earth's material was not from a main-sequence star.

There are a number of constants in Equation 5.7 that we can find easily, such as the hydrogen concentration x_H, the β-decay constant g and the β-decay function $f(W)$, γ^{-1} the deuteron radius, F_τ a correction factor not much different from unity (1), and M_{sp} the spin component of the matrix element. The orbital matrix element Λ has the general form

$$\Lambda = \left(\frac{\gamma^3}{8\pi C_0^2}\right)^{\frac{1}{2}} \int \psi_D^* \psi_P \, d^3 r \tag{5.8}$$

C_0 is a measure for the Coulomb barrier between the protons,

$$C_0 = \left(\frac{2\pi\eta}{e^{2\pi\eta} - 1}\right)^{\frac{1}{2}} \tag{5.9}$$

The wavefunction for the proton ψ reads

$$\bar{\psi}_P(y) = e^{i\delta} \sin\delta C_0^{-1} \frac{\Theta(y) + C_0^2 \cot\delta(kR) y \Phi(y)}{kr} \tag{5.10}$$

The exact calculation of the proton–proton reaction is not as simple as exercised above. Although these nuclear reactions have basically been known since long ago, the formulation of the underlying Hamiltonian is being improved even these days, for example for the more accurate determination of the reaction rates by mathematical methods where experimental data are not available for the relevant energy domains (Acharya et al. 2016).

5.5 Absorption of γ-rays by neutrons

In the early 1930s, Chadwick and Goldhaber wondered whether it could be possible that a nucleus of an atom could be excited or ionized when it was excited with suitable energy in the range of γ-rays (Chadwick and Goldhaber 1934, 1935). This is an analogy to the excitation of the electron configuration of the atom with x-rays and is, thus, a natural idea for a physicist. They had γ-rays of thorium available, $h\nu = 2.62$ GeV and suspected that the target nucleus could emit a particle like a proton or a neutron, and thus have a nuclear photoelectric effect, or photonucleonic effect.

Chadwick and Goldhaber expected this effect to be some disintegration according to the nuclear reaction

$$_1D^2 + h\nu \rightarrow {}_1H^1 + {}_0n^1 \tag{5.11}$$

They carried out an experiment and filled an ionization chamber with heavy hydrogen ^2H, or $_1D^2$. Upon exposure to a thorium source with γ-rays, the oscillograph recorded what the experimenters call “kicks”. They believe this was an indication for proton production upon heavy hydrogen decay. They also used a radium γ-radiation source, which produced only a few such “kicks”.

5.6 The TOKAMAK fusion reactor

Engineers have tried to simulate the high temperatures and pressures as they occur in the sun and stars in a controlled environment, which is provided by high temperature and strong magnetic fields in a fusion reactor. This reactor keeps a plasma at temperatures of 150 million °C. This is high enough to provide the thermal environment for the proton–proton chain, for example.

Well-known Russian physicists Igor Tamm, who established the term of surface states (Tamm 1932), and his illustrious doctoral student, later dissident and Nobel Peace Prize Laureate Andrej Sacharow (Sakharov 1975; Gorelik 2013), were among the first to figure out that one could harness such plasma in a toroidal chamber with magnetic coils. The acronym built from the Russian term adds up to TOKAMAK; this kind of fusion reactor has been called TOKAMAK since then. The first reports on the TOKAMAK design appeared in the early 1960s (Matveev and Sokolov 1961).

In the mid to late 1960s, TOKAMAKs were considered for nuclear reactions (Artsimov et al. 1969; Gashev et al. 1965; Holcomb 1969; Hubert 1969). Several national and multinational projects have been aiming at demonstrating a controlled nuclear fusion for future energy production. Energy production for mankind by nuclear fusion in TOKAMAKs is not being considered a realistic solution for the midterm. Maybe we must wait for another 50 to 100 years for this to become realized. It appears to be technically too difficult at this time to control a very hot proton and deuterium plasma for that purpose.

When Fleischmann and Pons (1989), and Hawkins, whose name was omitted in the original paper (Fleischmann et al. 1989), discovered that nuclear fusion can be forced when protons and deuterons were in or on palladium electrodes, the excitement and response by the scientific community was big. It seemed then that nuclear fusion was possible not only under extremely hot conditions but also under cold conditions. Cold fusion: the “mother load” of energy sources. Fleischmann later wrote that he had figured out this concept long before, but he had to conduct this project as a “hidden agenda” (Fleischmann 2006a).

5.7 Cold fusion: electrochemists go nuclear

In the late 1980s, electrochemists Fleischmann and Pons discovered that their electrochemical cells, operated under conventional ambient conditions ($T \sim 300$ K, $p \sim 1$ bar) warmed up slightly when they experimented with electrocatalysis on noble metal electrodes in deuterated electrolytes (Fleischmann and Pons 1989; Fleischmann et al. 1989). The amount of heat released from the cell was small but noticeable and reproducible. They found no other explanation for this observation than that of the nuclear fusion of protons and deuterons, mediated by the electrochemical environment. This implies the very strong claim that nuclear fusion was possible without the harsh, hot conditions that you have in sun or in a TOKAMAK.

Consequently, their discovery, published on 10 April 1989 as a "Preliminary Note", was called *Cold Fusion*. This sounds unbelievable. The scientific community was, therefore, caught between excitement, scepticism, and doubt, and ultimately by rejection. When a great discovery has been reasonably claimed, soon there will be followers, critical or not, who try to reproduce the results claimed. It is interesting how some researchers put great efforts in disproving claims made by other researchers.

By 10 May 1989 (1 month after the first cold fusion paper appeared), *Physical Review Letters* received a theoretical work from Sun and Tomanek (1989), who calculated with density functional theory (DFT) the structure of a hypothetical PdD_2 crystal and compared it with Pd and PdD structures. Even at very high loading of Pd with deuterium, the distance between the deuterium nuclei would be so far apart that a fusion of the deuterium nuclei was "very improbable": "Our results show that the intramolecular distance d(H2) is expanded to 0.94 Å in the Pd lattice even at very high H concentrations in the hypothetical crystal PdH_2. At this large distance, cold nuclear fusion is even less probable than in the deuterium gas phase."

This sounds like an early and final verdict against the possibility of cold fusion. Gittus and Bockris were less critical and sought possible theoretical explanations for the hypothesized cold fusion, which they published in May 1989 in Nature (Tien and Ottova-Leitmannova 2000). The aforementioned DFT study does not take into account potential dynamic effects when the lattice is externally excited. Protons in ceramic proton conductors can move along with the thermal excited lattice vibrations and, thus, behave as proton polarons (Braun and Chen 2017). Resonant excitation of crystal lattices with infrared radiation can cause dramatic enhancement of proton transport (Samgin and Ezin 2014; Spahr et al. 2010). It would certainly be interesting to see how such experimental conditions could affect the old experiment by Pons and Fleischmann. They also argue that the protons and deuterons could act as oscillators and delocalized species in the crystal lattice.

It was reported in 2002 that very strong acoustic excitation of organic compounds containing deuterium have shown neutron emission (Taleyarkhan et al. 2002). The acoustic excitation caused extremely hot bubble implosion conditions, similar to the ones required for nuclear fusion.

5.8 The experiment by Pons and Fleischmann

Pons and Fleischmann worked with a specifically designed electrolysis cell that was integrated in a Dewar calorimeter, which is sketched in their paper (Fleischmann and Pons 1993). They felt a calorimeter was necessary for their study in order to use the suspected Joule heat evolution in the cell as a proof of nuclear fusion. Some readers may wonder know why such a delicate analytical instrument like a calorimeter is used, which is designed to detect minute amounts of heat in a reaction, whereas cold fusion should produce so much heat that the experimenter would notice it immediately on their own body or maybe with a thermometer.

However, I am here reminded of my own research work, specifically hydrogen production by photo electrochemical water oxidation. In the beginning of my projects, I felt we would need a gas chromatograph for the precise quantification of the hydrogen and oxygen produced. Only later, when we were working with quite successful metal oxide semiconductor photoanodes, could we see the gas bubbles evolve and catch them with simple glassware and measure the evolved gas volume with a simple ruler – for every bystander to see.

I guess that Fleischmann and Pons, too, wished they had been be able to produce such pretty obvious results to the general public, but they were bound using the calorimeter. Moreover, they and their followers, for example Melvin Miles (Miles 2000; Miles et al. 1994, 2001, 1990a, b; Fleischmann 2006b), became experts in calorimetry and instrument technology.

The cathode was a thick palladium sheet of 2 mm thickness and 8 cm length and 8 cm width. The palladium was surrounded by a large platinum counter electrode. These electrodes were inserted into a large Dewar so that the electrochemical experiment could take place under precise temperature control. The electrolyte was a 0.1 molar solution of LiOD (you may know LiOH) dissolved in 99.5% D_2O with 0.5% H_2O. Let us go one step back and think about what this means. Lithium hydroxide LiOH can be dissolved in water H_2O, and this could be a conventional aqueous electrolyte. Both substances deliver the protons $^1H^+$ in the electrolyte. Fleischmann and Pons, however, wanted to deuterate their system; they wanted to have as nuclear reaction products $^2H^+$ (this can also be written as $^2D^+$) instead of $^1H^+$. We recall that the deuteron contains one proton and one neutron and, thus, has the mass number 2.

When you look at the section on ceramic proton conductors in my book (Braun 2019a), you will read that we hydrated ceramic slabs with vapor either from water H_2O or from heavy water D_2O. There may be various reasons for doing so, but it was always for analytical purposes. You may, for example, see an isotope effect in the diffusion constants of protons $^1H^+$ or deuterons $^2H^+$. In my experiments, I protonated the ceramic slabs with water H_2O when we did quasi elastic neutron scattering (QENS) (Braun and Chen 2017; Chan et al. 2016; Chen et al. 2013, 2012). When we carried out the neutron diffraction (ND) experiments, I deuterated the ceramic slabs by using heavy water D_2O (Braun et al. 2009); ^1H and ^2D have different coherent and incoherent scattering cross sections for neutrons. The proper choice of H or D for scattering methods where we are interested in the coherent of incoherent response yields optimized results.

Pons and Fleischmann calculate that the pressure necessary to push deuterons together would be in excess of 1029 bar. They use the term “astronomically high” with respect to such high pressure. Fleischmann and Pons write they “compressed” the D^+ ions ($^2H^+$) from the electrolyte into the palladium cathode by using a galvanostatic method with a moderate current density of $1.6\,\mathrm{mA/cm^2}$. This means that they applied a negative potential to the cathode so that the D^+ ions would adsorb at its surface and potentially enter the palladium crystal lattice and become intercalated. This is the same principle that is used in the lithium intercalation battery.

The electrolysis steps are as follows (Fleischmann and Pons 1989): the heavy water molecule D_2O is oxidized at the cathode (palladium), and the resulting deuterons are adsorbed at the electrode surface. Instead of the corresponding hydroxyl group OH^-, we now have a deuteroxyl group OD^-:

$$D_2O + e^- \rightarrow D_{ads} + OD^- \tag{5.12}$$

With more heavy water available from the electrolyte, more deuterons will be produced, which will bind to deuterium gas molecules at the electrode surface.

$$D_{ads} + D_2O + e^- \rightarrow D_2 + D_{ads} + OD^- \tag{5.13}$$

Deuterons adsorbed at the electrode surface will diffuse into the crystal lattice of the palladium electrode

$$D_{ads} \rightarrow D_{lattice} \tag{5.14}$$

$$D_{ads} + D_{ads} \rightarrow D_2 \tag{5.15}$$

Fleischmann was of the opinion that the anticipated cold fusion was a bulk effect (Fleischmann et al. 1994). They, therefore, used electrodes of various geometry and size, that is, rods, sheets, and cubes of palladium. They carefully monitored the evolution of the Joule heat, and it is interesting to note that they used current densities up to $512\,\mathrm{mA/cm^2}$. The size of electrodes and the evolved heat are listed in table 1 in their paper. I have not found any study that looked into the change of the molecular and electronic structure of palladium before, during, and after deuteration.

By doubling the diameter of the palladium rods (0.1 to 0.2 to 0.4 cm), the excess rate of heating increased from 0.0075 W by a factor of 4.8, and then again by a factor of 4.25, to finally reach 0.153 W, when the current density was $8\,\mathrm{mA/cm^2}$. The increase by roughly a factor of 4 corresponds to the parabolic increase of the mass of the Pd cylinder due to doubling the diameter. When the experimenters increased the size of the cylinders and the current density, however, the excess specific heating rate increased disproportionately.

I have multiplied the values for the excess specific rate of heating at 8 mA/cm^2 by the factors 8 and 64 to check how much these values differ from the actually measured rates. The bulky rod with 0.4 cm diameter produces almost three times more heat when it is heavily loaded with deuterons due to the very high current density. Pons and Fleischmann attribute this considerable gain in heat due to the cold fusion of the deuterons in the palladium, which they "compressed" galvanostatically into the palladium.

The large palladium sheet shows an enhancement by the factor 3 (0.0021 → 0.0061) when the current density is increased by 33% from a low 1.2 mA/cm^2 to a still low 1.6 mA/cm^2. This is a huge effect. However, the *catastrophic effect* sets in when the experiments used a bulky palladium cube of 1 cm × 1 cm × 1 cm size; in their paper, the experimenters issue the warning that ignition sets in at 125 mA/cm^2 with palladium in cube geometry (Fleischmann and Pons 1989).

The primary energy that is released from the fusion of deuterium and protons comes in the form of electromagnetic waves with extremely short wavelengths and very high energy in the MeV range. These are the γ-rays that cause severe harm to the human body. The collisions of the particles produced, such as deuterium, will cause a warming up of the entire matter, which will raise the temperature. This is the excess heat measured during cold fusion. Pons and Fleischmann also claimed that they detected γ-rays and provided the evidence for this (Fleischmann et al. 1989).

$$^{2}\mathrm{D} + {}^{2}\mathrm{D} \rightarrow {}^{3}\mathrm{T}\,(1.01\,\mathrm{MeV}) + {}^{1}\mathrm{H}\,(3.02\,\mathrm{MeV}) \tag{5.16}$$

$$^{2}\mathrm{D} + {}^{2}\mathrm{D} \rightarrow {}^{3}\mathrm{He}\,(0.82\,\mathrm{MeV}) + n\,(2.45\,\mathrm{MeV}) \tag{5.17}$$

5.8.1 Other electrochemist's aid to help

Soon after the claim by Pons and Fleischmann was published, Lin et al. of the Bockris group at Texas A&M University, submitted a as a Preliminary Note a speculative explanation about the potential mechanisms that lead to cold fusion observed by Fleischmann and colleagues (Lin et al. 1990). Lin et al. speculated that during the extended electrochemical treatment of palladium, dendrites would grow with sharp tips, which would allow for extraordinary large electric fields, over which a dielectric breakdown of the water molecules could take place, along with formation of a fluctuating gas volume over the dendrite tip. Then, the strong electric field could accelerate one deuteron from the gas phase to a deuteron present at the tip with a collision of a very high kinetic energy of 2000 eV.

The aforementioned process of nuclear tunneling, which we know is a statistically extremely rare event, was determined by Gamow to be probable as (Gamow 1928)

$$G = \exp\left\{-\pi e_0^2 \sqrt{\frac{M_\mathrm{D} 4\pi^2}{h^2 E}}\right\} \tag{5.18}$$

with M_D the rest mass of the deuteron and E the energy that the deuteron has when it transits from the electrolyte solution into the electrode. Lin et al. consider that deuterium gas can evolve electrochemically at the palladium surface until it is fully covered, depending on the current density i. The rate of collision between two deuterium atoms is, then, the current density divided by Faraday's constant, i/F, in mol per cm^2 per seconds.

Lin et al. then speculate further and write that if there is a fraction Γ of the electrode surface with an abnormally high field strength, then the energy at this location could be equal to the energy necessary for a D-D collision; then the rate for the penetration (this is the fusion rate f) of the Gamow barrier would read:

$$f = \Gamma(i/F) \exp\left(-\frac{2\pi^2 \epsilon^2}{h} \frac{\sqrt{M}}{\sqrt{E}}\right) \tag{5.19}$$

At some point, the electrolyte would start boiling and the electrolyte would evaporate rapidly. In order to monitor the time properly where this would happen, they used a video camera and recorded time lapse images, so that they could later inspect the condition of all cells in the photos and assign the state of cell to the time from the time stamp.

Also other researchers commented on the possible mechanism for the reported cold fusion. Around 5 years after their discovery, Fleischmann and Pons, and Preparata speculated on possible theories for the cold fusion (Fleischmann et al. 1994). Today, research on cold fusion is known as LENR – low-energy nuclear reactions. A small group of researchers is active in the field. Said Ludwik Kowalski in a commentary (Kowalski 2010) in Physics Today: *I believe that reports made by recognized scientists should be taken seriously, even when their results conflict with what is expected.* Moreover, Stanford physics student Chung Chonwoo wrote in his 2015 term paper (Chung 2015): *Further study of the properties of deuterium in palladium may well be warranted as part of chemical physics, but deeming any phenomenon in that system to be cold fusion is not well-substantiated at present.*

5.9 Rectangular potential for cold fusion

We already saw in Section 3.7.2, which is about Fourier series, that the presence of a rectangular potential between protons could, in principle, allow for cold fusion (Schneider 1989) (see, also the review paper (Storms 1991)). There, Schneider refers to experiments of Jones et al. (1989) on cold nuclear fusion, which are explained by the quantum mechanical tunneling effect. He derives a transmission coefficient with Schrödinger's equation in one dimension, using a rectangular form of the potential. This special form of the Coulomb barrier for the deuteron–deuteron reaction does not prevent a significant fraction of nuclei from penetrating the barrier.

The Schrödinger equation for the particle with mass m in the potential well $V(x)$ (Figure 3.14) is written as

$$\frac{\hbar^2}{2m}\frac{\mathrm{d}^2\Phi(x)}{\mathrm{d}x^2} + V\Phi(x) - E\Phi(x) = 0\,. \tag{5.20}$$

You can introduce the abbreviation $W^2 = 2m(V - E)/\hbar^2$, which makes that the Schrödinger equation

$$\frac{\mathrm{d}^2\Phi(x)}{\mathrm{d}x^2} + W^2\Phi(x) = 0\,. \tag{5.21}$$

The potential $V(x)$ can be defined similarly to the Heaviside function in Equation 3.28

$$V(x) = \begin{cases} 0\,, & \text{if } x > \frac{L}{2} \\ V\,, & \text{if } \frac{-L}{2} < x < \frac{L}{2} \\ 0, & \text{if } x < \frac{L}{2} \end{cases} \tag{5.22}$$

Let us consider here that the potential is given as one-dimensional potential well with the length L, which in two and three dimensions would be the radius R; L, or R, is here the interaction length (radius), the thickness of the barrier; $L = 2 \cdot R_\mathrm{n} \cdot \sqrt{Z}$, and $R_\mathrm{n} = 1.2 \times 10^{-15}$ m is the radius of the hydrogen atom; Z is the number of nuclei that interact.

By solving the Schrödinger equation you will obtain one solution for each of the three definition ranges of the potential $V(x)$. For the regions where $V(x) = 0$, the solutions are identical by the function

$$\Phi(x) = C_1 \mathrm{e}^{\mathrm{i}kx} + C_2 \mathrm{e}^{-\mathrm{i}kx} \tag{5.23}$$

and k is the wavenumber (wave vector) $k = \sqrt{2mE/\hbar^2}$. This solution is the plane wave and has exponential solutions for the case that the energy is smaller than the potential: $E < V$. For the region inside the potential well and $E > V$, the solution is oscillatory and reads

$$\Phi(x) = C_3 \cdot \mathrm{e}^{W \cdot x} + C_4 \cdot \mathrm{e}^{-W \cdot x} \tag{5.24}$$

With the boundary conditions (consult the Appendix in Schneider (1989), it is possible to derive from the solutions of the Schrödinger equation for the potential well a transmission coefficient T, which depends on the characteristic of the potential

$$T = \frac{V^2 - (2E - V)^2}{(V^2 - (2E - V)^2) + V^2 \sinh^2(WR)} \tag{5.25}$$

and which is a measure of the probability that the Coulomb barrier between two nuclei can be overcome and the nuclei can be joined as nuclear fusion.

5.10 Research funding for energy production in black holes

Instead of detailing here the energy production in black holes, I want to make a general statement about research funding. On 10 April 2019, the National Science Foundation (of the United States) held a press conference for the first imaging of a black hole, and I am citing here from their transcript after I saw the news video in the Internet (Flagg 2019). While the speculation of the existence of the black hole dates back to over 200 years ago, it was a difficult task to actually picture it. It required a multinational team. Yesterday was the announcement of their great success. For sure this could be Physics Nobel Prize material. Says Shep Doeleman EHT Director Center for Astrophysics, Harvard & Smithsonian (35:10) at the press conference:

"I just wanted to point out that when we first started the Event Horizon Telescope project, the group was small .." Our SUNRISE project group started out small too – many years ago. An experienced scientist and researcher, Doeleman concludes his address with

> I also want so say something in particular about funding and support. This has been a high risk but high pay-off endeavour. So you have to kiss a lot of frogs before you get the prince, before you get the black hole image. And you need supporters, you need funders who will stand by you for long periods of time, who take the long view, who understand that basic science never goes out of style and who also understand that in basic science you never know when its going to pay off but ultimately it usually does and that you have to play the long game. And we have wonderful partners with the National Science Foundation and with the international funding agencies and foundations and our head is off to them for sticking by us for so long ...

6 Foundations of photosynthesis

Photosynthesis is an essential process that is possible only with living matter or by artificial photosynthesis. From the perspective of the evolution of life, it is a very old process. Yet what do we mean by old? In Figure 6.1, I have listed a whole range of processes and stages that are important in the development and evolution of the universe, matter, nature, and life. It is basically a timeline of nature, which you read along the time axis from the bottom to the top.

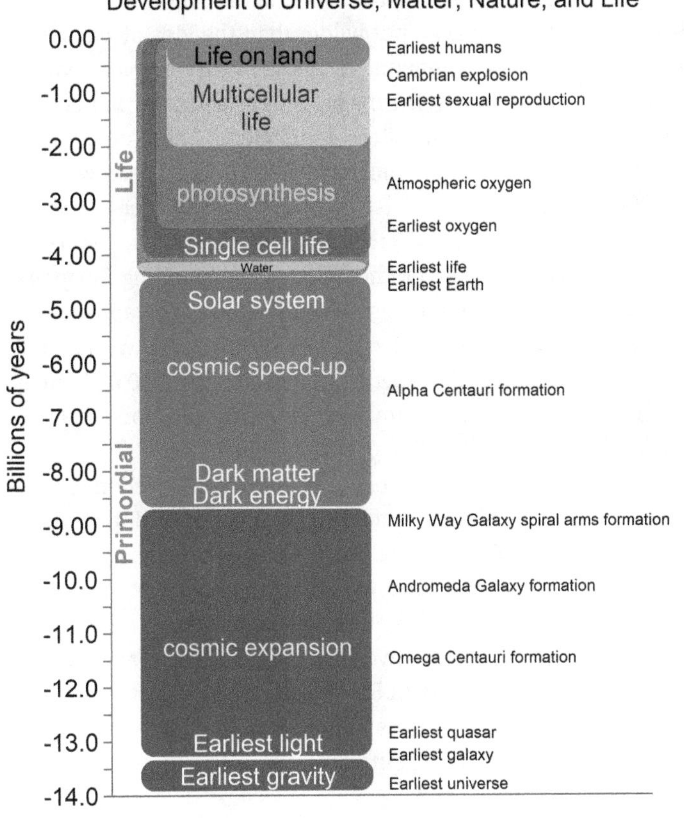

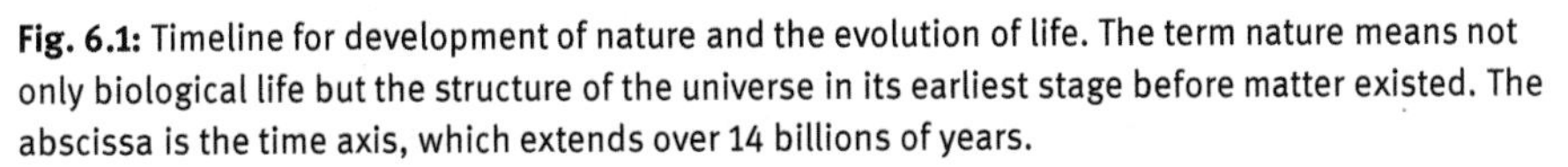
Fig. 6.1: Timeline for development of nature and the evolution of life. The term nature means not only biological life but the structure of the universe in its earliest stage before matter existed. The abscissa is the time axis, which extends over 14 billions of years.

14 billion years ago, which means today (minus) – 14 billion years, the signs of the earliest universe were detected, the time right after the Big Bang, the *Urknall*. At that time, also the first gravitation developed because matter (with mass) developed, as

https://doi.org/10.1515/9783110629941-006

well as light. We have known since Albert Einstein about the equivalence of matter and energy via the relation $E = mc^2$. Since, the universe has been expanding. Consider the universe as a sphere that has an inside but no outside. This is the true meaning of the word universe. Like a Möbius band, it has only one side, the inside. That the universe is expanding implies that there is a radius of that sphere that is increasing.

The idea of a growing universe is rather new. Nikolaus von Kues (Gestrich 1997) in the fifteenth century was of the belief, and this was the belief of the leading minds back then, that the universe was endless and infinite. Nikolaus von Kues concluded by mathematical principles that a universe of infinite expansion could not have a center. This means that he, a Christian bishop and assistant to the Pope, by the way, did not believe that the Earth, or even the Sun, could be the center of the universe.

That idea the universe is expanding is an not wilder than the idea of a universe of infinite size. However, such wild ideas remain wild, unless they are verified and proven by empirical analysis, by experiment. During my time as a postdoc in Berkeley National Laboratory (Ernest Orlando Lawrence Berkeley National Laboratory, LBNL, or just LBL), every other week, or month, the homepage at LBL would open with another wild story by Saul Perlmutter, astrophysicist in the Physical Sciences Division. He looked at supernovas and their electromagnetic spectra and red shifts and, eventually, it emerged to him and his coworkers that the universe is expanding. His paper *Measurements of Omega and Lambda from 42 high-redshift supernovae* appeared in June 1999 in the *Astrophysical Journal* (Perlmutter et al. 1999), a journal with a moderate impact factor of below 6. Since, their paper has been cited over 10,000 times. Perlmutter, along with Brian Schmidt and Adam Riess, received the Nobel Prize in Physics in 2011 for their work on the acceleration of the universe[1].

During the expansion phase, which extends over the time from 13 billion to 9 billion years, the Andromeda galaxy was formed and also Omega Centauri, the largest accumulation of stars in the universe. About 8.5 billion years ago, the crab arms of our Milky Way galaxy formed. Four billion years later, our solar system formed. Moreover, some four billion years ago, the earliest form of our globe was shaped and, not very long after that, water was formed and started covering the globe. Then, also the earliest, most primitive forms of life were formed on Earth, the single cells. Half a billion year later, that is 3.5 billion years ago, the process of photosynthesis was established by life, and this process has been working ever since.

1 On my return from a synchrotron beamtime in Berkeley I happened to meet Saul Perlmutter in the airplane on my way back to Zürich. I spotted him already at San Francisco Airport but only when I approached him at the gangway at Zürich International Airport was I sure that it really was him. He was on his way to the World Economic Forum in Davos and promised to come by at Empa next time when he came to Switzerland – to speak with the students. Maybe I should remind him and fix a day when he is in Switzerland next time. Saul Perlmutter is the prime example of a genius who is still a regular guy, just watch the Ice Bucket Challenge with Paul Alivisatos (Alivisatos et al. 2014).

For photosynthesis to work, it certainly requires the light as an energy source, which comes from the sun. At that time, also the atmosphere on Earth had an increasing concentration of oxygen. Oxygen is formed by the oxygen evolving complex in photosystem II by the oxidation of the water.

One and a half billion of years later, multicellular life evolved, including the animals that feed on the plants, and including the predators that feed on animal prey, all of which need oxygen to live. For 500 million years, there has been life established on "dry" land, as opposed to life in sea and water. At about this time, the so-called Cambrian explosion took place, which means that in a relative short period of only 5 to 10 million years, as compared to the timeline of 14 billion years, an incredible variety and diversity of life evolved.

Around 2 million years ago, human life evolved as the *homo erectus*, the upright walking human. The first *homo neandertalensis* in Germany was found in the village of Ochtendung near the Rhein river, in an extinct volcano cave in the Vulkaneifel between Koblenz and Nürburgring and is estimated to be over 150,000 years old (Flohr et al. 2004). The neanderthalens are considered to have been primitive humans, but they had a social life and they would burry their deceased family members – and, thus, had a sense for reflection. Possibly they may have looked up at the stars at night and found regularities that they worked into a mystic view of life and the environment.

As humans we are free to chose whether we believe in creation in the biblical sense, which is outlined in the book of Genesis, which I have copied in the Appendix, or whether we believe in the evolution theory, which is, *nota bene*, a scientific theory. Even if the scientific mind rejects the creationist theory, it is appalling how the timeline of evolution of matter and nature and life in Figure 6.1 is chronologically similar to the book of Genesis. Nature has developed by increasing its complexity, beginning from a homogeneous cosmos in the very early stage 14 billions of years ago, up to the life that exists only under mild thermodynamic conditions because of fragile and sensitive protein molecules. Also, technology has developed along the axis of complexity (Solée et al. 2013), so we can compare the technological evolution with biological evolution. We physicists are used to applying a reductionist stance on the things that concern us, whereas functionality in animate and inanimate matter is always a result of complexity, and not simplicity, as we try to achieve by simplifying structures and models (Anderson 1972).

Over seven billion of human beings live on Earth. Their food, and the food of animals, plants, is made from primary production via photosynthesis from water, CO_2 and solar energy. Plants produce hydrocarbons as sugar, then larger molecular aggregates like starch, and finally cellulose.

The oxygen O_2 that we need to breathe comes from the dissociation of the H_2O water molecules, which is a product of photosynthesis. Oxygen is produced by the oxygen evolving complex (OEC) in photosystem II (PS II). Carbon dioxide CO_2 is also dissociated by plants. The protons from the water dissociation are combined with the carbon atoms from the CO_2-reduction to hydrocarbons, sugars, actually, of the form

C_nH_m, which are then combined to long-chained starch molecules or aromatic cellulose molecules. The chemical reaction reads

$$CO_2 + H_2O \xrightarrow[\text{chlorophyll}]{h\nu} (CH_2O)_n + O_2 \tag{6.1}$$

A simplistic sketch of this important process for life is shown in Figure 6.2, where, Melvin Calvin is shown explaining the basic steps of the photosynthesis that he and co-workers have discovered, in an oral presentation.

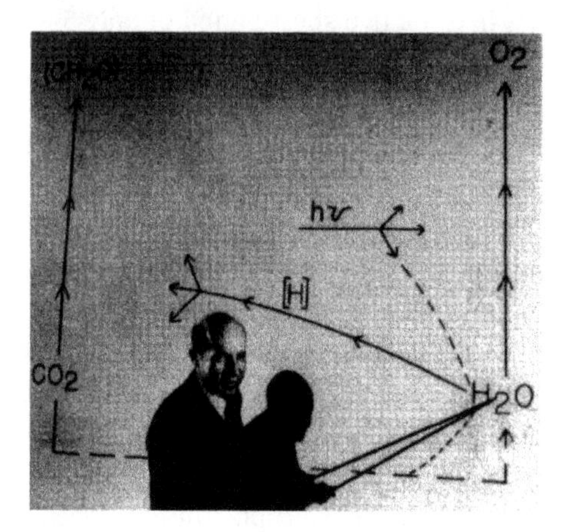

Fig. 6.2: Melvin Calvin explaining a simplified version of the photosynthetic pathways that he discovered. Reprinted with permission from Journal of Chemical Education, 73 (5), 412–416, Kauffmann, George B., Mayo, Isaac, Multidisciplinary scientist – Melvin Calvin – His life and work. Copyright (1996) American Chemical Society, (Kauffmann and Mayo 1996).

It may be an interesting side note that the research on photosynthesis was a national security interest for the United States, particularly in relation to the Soviet Union. There is a declassified report in CIA (1954) by the Central Intelligence Agency with an assessment done by a scientist, whom I suspect to have been working at what is today known as LBNL (Bio-organic Chemistry group of University of California Radiation Laboratory). The file has been declassified as a result of the MuckRock lawsuit by Emma Best. I do not dare to speculate about who the author of the report was.

Said 1961 Chemistry Nobel Prize winner Melvin Calvin in an interview with David Ridgeway in 1973 (Ridgway 1973): “It was a remark by the physics teacher that because of the speed with which I tended to try and answer his questions and occasionally got them wrong, that I would never make a scientist.”

6.1 Algae, plants, and plankton

Photosynthesis means that biological material, organic material, is synthesized by virtue of light energy. Synthesis means synthesis in the chemical sense – by chemical reaction. The chemical reactions that take place in the photosynthetic apparatus take carbon dioxide from the atmosphere and convert it with the protons H^+ (p, p^+) from water H_2O into hydrocarbons C_nH_m, first sugars, then starch, and, eventually, cellulose, which makes up wood. We can see this by the growth of plants like grass, flowers, trees, and so on. Over a season a farmer can grow hay stacks on a field or corn and crops, and over the decades a huge forest can grow[2]. It is not soil which makes trees, but CO_2 and water. This is why you can grow plants in air and in water. The oceans are full of plankton; these do not produce starch and cellulose. This biological material, once you consider it a material, is called biomass.

Biomass is the result of photosynthesis and is known as primary production (Reynolds and Lund 1988; Papageorgiou and Govindjee 2004). Primary production includes everything produced by photosynthesis. This is what grows on the agricultural crop fields, the plants that grow in nature, such as flowers and grass, bushes and trees, and also the algae and plants that grow in the oceans. Figure 6.3 shows a photo that I took during a walk through a local forest between two villages. The photo was taken in fall when tree loggers had already cleaned up the forest and piled up the wood ready for transport to consumers. It is necessary to take care of the forest by first planting it, then making sure it has enough water and nutrients to grow, and preventing small trees from being damaged or eaten by animals, and when the forest

Fig. 6.3: A pile of cut wood in a Swiss village. This amount of wood can substitute 75,000 liters of heating oil. Photo taken 14 October 2018 by Artur Braun.

2 "Growth has its seasons", says simple-minded gardener Chauncey Gardner (played by Peter Sellers) in the movie *Being there*, (Ashby 1979). You notice this from tree rings.

has grown to some size and density, cutting out a number of trees to make space for the remaining trees, so that they can grow further. Wood waste should not be left in the forest, because in a drought it will be a fuel that can fan the flames in a wildfire[3].

Figure 6.4 shows the cross section of a tree that was cut and that I found during my hikes in the local forest in Switzerland. Part of the tree is covered with green microbes or fungi due to light exposure. This helps to increase the optical contrast and, therefore, you can better identify and count the annual tree rings, according to which I estimate the tree having an age of 25 years. The log below is from a beech tree that has dried over an extended period in the woods. You will notice, when you burn the two types of logs, that the dry one will cause less vapor and also less particulate matter; it makes a big difference in the molecular structure of the soot what the type of wood it is and how long it has dried (Braun et al. 2008).

The wood pile shown in Figure 6.3 will be burnt very soon. The logs in the pile will be cut to small pieces, dried for 2 or 3 years in the ambient and will then be burnt in ovens in residential homes. The home owners are, thus, using a local energy source, instead of foreign mineral oil, which happens to also be a fossil fuel. Mineral oil fuel is not a renewable fuel, unlike wood and other biomass. The smaller part of the wood might be shredded and pressed into wood pellets and sold for central heating in Minergy homes, one of which I live in. In either way, the wood will be burnt very soon and thus reacted to CO_2, water, and ash. However, it may have taken up to 20 years for the trees to reach the size you see in the photo.

Historically, working in the forest and cutting trees is very laborious and very dangerous work. Today, wood loggers use heavy equipment to handle the timber logs, as is shown in Figure 6.5. The work also depends on the season. There are particular seasons for cutting trees and particular seasons for planting new trees. When winter is cold and humid, snow may pile up on branches of the trees and bend them until they break. Heavy storms can break trees. The situation is then that you have a “mikado of trees” piled up in disorder and under mechanical tension. Then, it is particulary difficult and dangerous for the wood loggers to cut the trees free from the mikado. In this case, it is very helpful and almost mandatory to use heavy equipment, which also increases the protection for the operator. I grew up in a wooded area and I knew several cases where workers died because they had accidents in such dangerous situations.

To become independent of foreign oil by using domestic wood is not a viable solution, given that it takes ages before a full-sized forest is grown, while it can burn so fast. I have exercised this with some calculations in chapter 9 of my book from 2018

3 The forests in Germany were cultivated 300 years ago for economical reasons. In the late 1970s, a movement to allow again for more nature in the forests was started. Wood is a valuable goods for construction and civil engineering, in addition to being fire-wood fuel. New policies were to be established, like not cleaning up the forests anymore, but to let the scrap wood back in the forests. These, however, attract harmful bugs and also constitute an additional and unwelcome fuel in the case of wildfires in a very hot and dry summer.

Fig. 6.4: Cross section of two logs of wood I found in the forest. The green one on the top was a piece of fir tree showing the tree rings (photo taken 28 December 2018). You can count them as an exercise and estimate the age of the tree. The log in the bottom photo was from a piece of beech (4 August 2018). It has dried for over 1 year, and the loss of water has caused changes in the structure of the wood.

(Braun 2017). In California, there are national parks that feature sequoia trees, also known as redwood trees or mammoth trees. They can reach a height of 100 meters, and some of them are 2000 years old. They are huge. It would be irresponsible to cut and burn these trees to obtain energy. However, wildfires, which are a part of nature, occasionally destroy such trees.

You can see the author of this book standing in front of a huge sequoia tree in Yosemite National Park in California. I do not know the height of the tree, but you can

Fig. 6.5: Heavy equipment for timber logging in forest. Pine trees and some beech trees have been cut. The machine can drive around and grab a stem and cut it into pieces. Photo taken 30 September 2018 by Artur Braun.

see that it must be a huge tree, given how small my body appears in comparison to the diameter of the tree, which I estimate to 130 cm. The height of a tree is an important parameter for the estimation of the biomass production in a forest (Mugasha et al. 2013). The relationship between height and diameter and width of annual tree rings is still a matter of scientific research (Carroll et al. 2018; Sumida et al. 2013). Based on the data in Carroll et al. (2018), I assume that the height of the tree in Figure 6.6 is around 55 meters.

It is very important for a country to grow its own food to feed the people. For this, it needs land with nutrients, water, sun, CO_2, and also some workforce and technology if agriculture is to be done. Some countries have a very efficient agriculture, and the efficiency can be ranked as tons of grains per hectare, for example. The statistics of which countries grow how much and at which efficiency is a good indicator for the power of the country and its technological advancement. Therefore, primary production of biomass is certainly of interest for intelligence agencies and firms.

It is, therefore, not so surprising that a study was prepared for the European Energy Commission (EEC), titled *Attempt to formulate common European energy policy*,

Fig. 6.6: The author standing at a large sequoia tree in Yosemite Valley National Park in California. The diameter of the tree is estimated to 130 cm (see yellow vertical lines) based on my body height, as indicated by the distance of the two horizontal green lines. The two vertical green lines indicate my waist. Photo taken 2 May 2011 by Artur Braun.

published in *Paris Revue de l'Energie* (March 1980, pp 153–159), or a study like *Sweden's energy policy*, Paris, *Le Progres Scientifique* (November–December 79, pp 33–40).

"The biomass energy potential is available in two forms: that of a biomass cultivated or produced solely for its energy properties, and that of waste rejected in an activity other than energy production. As a complement to energy-producing forests other rapid-growth plants are to be studied from the viewpoint of energy production. The sunflower is an annual whose growth reaches 79–104 g/m daily with a photosynthesis efficiency of 75 percent."

Such studies are then, for example, collected and compiled by the (United States) Foreign Broadcast Information Service under the name "West Europe Report' – Science and Technology" (For Official Use Only FUOU 7/80), and recorded by the Central Intelligence Agency as file CIA-RDP82-00850R000200090033-4. (CIA 1980). For example, in 1951, the CIA compiled a report on the capability of Russian scientists in the Soviet Union with respect to photosynthesis research (CIA 1954).

6.2 The photosynthetic apparatus

6.2.1 The complexity of living systems

We will find that the cells of plants, which we might consider the building blocks of the plant, are actually small but highly complex biological factories, even biological industrial complexes at the mesoscopic scale. Complexity is the precondition for functionality, and this stands in contrast to what we physicists like to see in order to make problems transparent and analytically treatable, i.e., simplicity, as may be found, for example, in the high symmetry of crystals, which is a low degree of complexity. Physics Nobel Laureate Anderson explained this in a position paper in Science in 1972 (Anderson 1972). The high complexity of systems also brings about their vulnerability to damage and destruction. This principle extends throughout life and also extends to socially complex structures, as has been illustrated by Tainter and Taylor in their paper on sustainability and resilience in buildings and ancient, extinct societies (Taylor and Tainter 2016; Tainter and Taylor 2013).

6.2.2 Lessons for sustainability

Tainter and Taylor, based on their own work and on references in Tainter and Taylor (2013), come to a set of conclusions, or lessons, about what sustainability means, how much sustainability will cost to a society or system.

- Sustainability is a function of success at solving problems. It does not emerge, as is commonly thought, as a passive consequence of consuming less.
- Complexity in human societies grows through the mundane process of solving problems, as discussed, including problems of sustainability. This can be seen not only at the societal level but also, as discussed above, in the design and building of cities.
- Complexity is an economic function, with benefits and costs, and can reach diminishing returns.
- Since sustainability depends on solving problems, it promotes the growth of complexity and complexity's associated costs. Sustainability may, therefore, require greater consumption of resources, not less.
- Under diminishing returns, complexity in problem-solving causes damage subtly, unpredictably, and cumulatively.
- A society or other institution can be destroyed by the cost of sustaining itself.

6.3 The leaf and its components

Figure 6.7 shows a photo I took in a sunflower field in Switzerland in July 2018. I take the information for granted, see[4], that "The sunflower is an annual whose growth reaches 79–104 g/m daily with a photosynthesis efficiency of 75 percent."

When you have the pleasure to live near a place with sunflowers, you may notice that the sunflowers are typically directed to the position of the sun: southwards. The ability of a plant to point the flower in the direction of the sun for growth promotion is called heliotropism and is based on the action of a plant hormone called auxine, which causes some parts in the plant to grow faster than other parts, and this difference in growth speed causes a motoric action of the plant, so that the direction is changed. Note that in the southern hemisphere, the direction towards the sun is not southward, but northward. I was reminded of this fact during my recent trip to South Africa in July/ August 2019, which was a hot summer in Europe and a mild winter in South Africa.

In the Southern Hemisphere, July is winter time. I was driving from Pretoria to Pilanesberg, and I had a navigation system in my car with an electronic compass, but I was confused because on a sunny day, the sun was in a different direction from what I expected from the compass. The compass was pointing to the north, of course, but the high-noon sun was also in the north direction. When you grow up in the Northern Hemisphere, you learn that at noon the sun is in the south. Actually, we take it so much for granted that we do not even discriminate between Northern and Southern Hemisphere. Then, when you program a solar PV tracking system, you imply that it is working in the Northern Hemisphere. Then, when the system is shipped to South Africa or Australia or Argentina, it will not work properly because the PV panels face away from the sun.

The magnified image of the green sunflower leaf shows that the leaf is built up of or sectioned into prismatic-shaped compartments of the size of 1 cm^2. The sunlight

4 "The biomass energy potential is available in two forms: that of a biomass cultivated or produced solely for its energy properties, and that of waste rejected in an activity other than energy production. As a complement to energy-producing forests other rapid-growth plants are to be studied from the viewpoint of energy production. The sunflower is an annual whose growth reaches 79–104 g/m daily with a photosynthesis efficiency of 75 percent.". This is how it reads in a study by the European Economy Commission EEC:

ATTEMPT TO FORMULATE COMMON EUROPEAN ENERGY POLICY, Paris REVUE DE L'ENERGIE in French Mar 80 pp 153–159, specifically a STUDY MADE OF SWEDEN'S BIOENERGY POLICY, Paris LE PROGRES SCIENTIFIQUE in French Nov–Dec 79 pp 33–40, collected by the

West Europe Report
SCIENCE AND TECHNOLOGY
(FOUO 7 /80)
[FBIS] FOREIGN BROADCAST INFORMATION SERVICE
FOR OFFICIAL USE ONLY

APPROVED FOR RELEASE: 2007/02/08: CIA-RDP82-00850R000200090033-4

Fig. 6.7: Top: a field full of sunflowers (*Helianthus annuus*) in Switzerland. Middle: part of a leaf from the sunflower. Bottom: magnified area of a leaf from a sunflower. Photo taken July 2018 by Artur Braun.

is shining through from the back, so that you can follow the structure better. Each of these compartments is further sectioned into smaller mosaics with a size somewhat smaller than 1 mm^2. The compartments are all separated by a hierarchical network of plant stems, like a road system or pipe system that can provide nutrients from the main plant stem that is connected by the roots with the soil.

We can call this network of plant stems the cytoskeleton, which holds the plant cells. The cells do not look isometric and round, but rather flattened with a sixfold geometry or a hexagonal close packing density for efficient utilization of available space.

When you further investigate the plant cell with a microscope, you will find a number of subunits, which I have illustrated in the next Figure 6.8. The red dot in the center is the cell nucleus. It is connected with the endoplasmatic reticulum (ER) plus the ribosomes. The lengthy blue shapes are the cell Mitochondriae. There is typically one

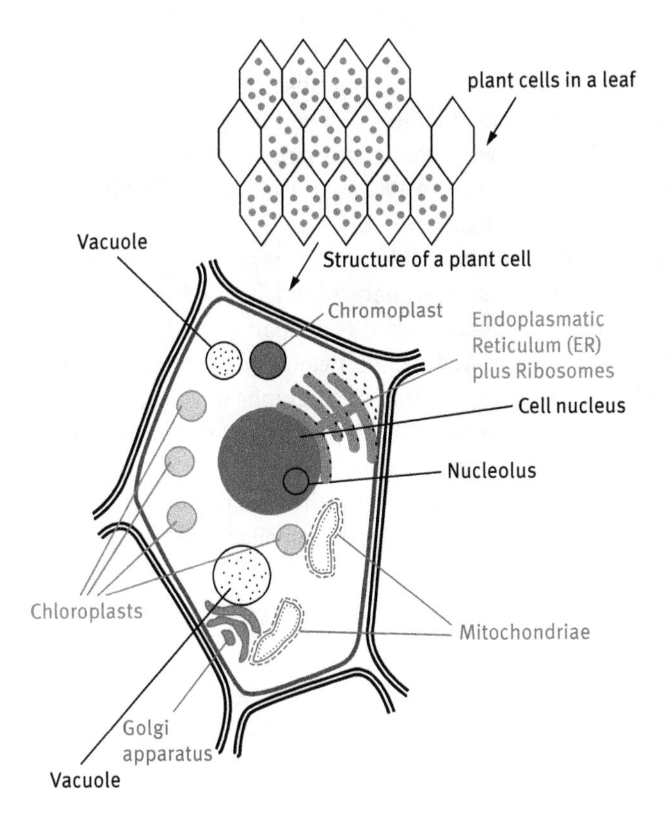

Fig. 6.8: Sketch of a plant cell with its functional components. The upper right-hand side shows 15 plant cells arranged in a leaf; note the pentagon or hexagon geometry of the cells. The center shows a magnified plant cell with the large red cell nucleus, which includes the small red nucleolus and the green endoplasmatic reticulum (ER) plus the surrounding ribosomes. The blue lengthy ellipsoids are the mitochondriae, close to the grayish Golgi apparatus with a large round vacuole. The red ball on the top is the chromoplast. The green round bodies are the chloroplasts. Artur Braun.

or more handfuls of greenish, literally greenish, round bodies, which are the chloroplasts. These are the working machines of photosynthesis and contain all relevant functions for the solar energy conversion of the plant.

When you cut the leaf into its components you will eventually come to the size of the plant cells. The cells, however, are also made of subunits, all of which have particular functions in the plant's growth and subsidy and in photosynthesis. Important for the photosynthesis function are the chloroplasts.

6.3.1 Chloroplasts

Figure 6.9 shows a selection of three electron micrographs of chloroplasts, which were published in the 1950s (Leyon 1954; Elbers et al. 1957). The size of chloroplasts ranges from 1 μm to 5 μm. The micrographs show the chloroplasts from two different plants. The chloroplast on the upper left-hand side is from the *Synechococcus cedrorum* cyanobacterium. The image on the bottom left-hand side is from the always-green cast iron plant *Aspidistra elatior* from the inner leaf of a young shoot. The image on the right-hand side shows a chloroplast with starch accumulation (food production). Compare these photos with those published in 1961 in a PNAS paper (Rabinowitch 1961) by photosynthesis pioneer Eugene Rabinowitch.

When you further inspect the chloroplasts in Figure 6.9 (see my color sketch in Figure 6.11), you will notice lamellar substructures made from lengthy lines and tapes, which extend and fill the entire chloroplasts (see the comparison with connecting stairways in Figure 6.12). Albertsson and Andreasson determined the relative abundance of grana and stroma lamellae using electron micrographs of chloroplasts (Albertsson and Andreasson 2004). You will also find the single-cell nucleus, which is dark or black in the image, and some of the vacuoles, which look round and white. Albertsson developed a quantitative physical model for the domain structure of thylakoids (Albertsson 2001). We will stick, however, with the lamellar substructures built up from the thylakoids.

Figure 6.10 shows Nobel Laureate and photosynthesis pioneer Melvin Calvin in the laboratory with his algae-culturing flasks.

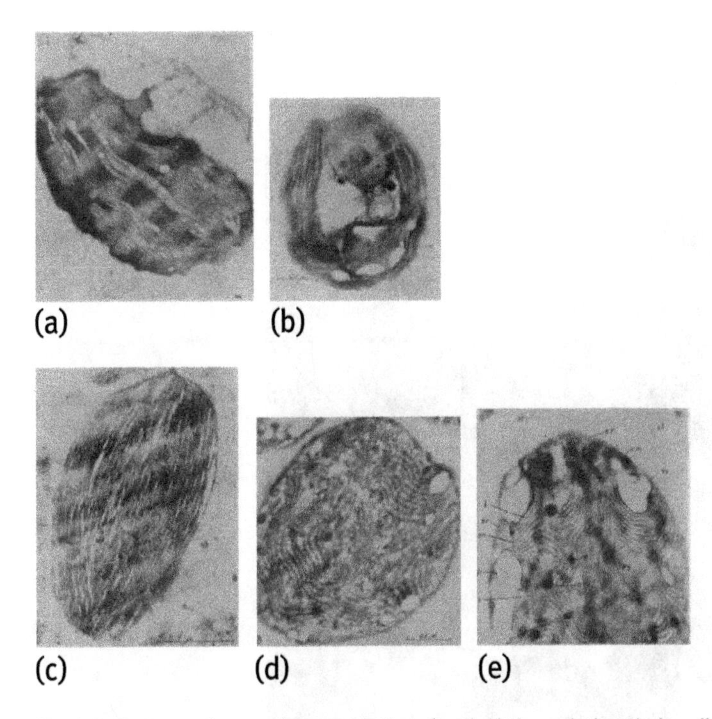

Fig. 6.9: Electron micrograph cross sections (top left) through the whole cell of Synechococcus cedrorum, after (Elbers et al. 1957); (bottom left) chloroplast from a light green part of the inner leaf of a young shoot, after (Leyon 1954). Reprinted with permission from Acta Botanica Neerlandica, 6 (3), Elbers, P. F., Minnaert, K., Thomas, J. B., Submicroscopic Structure of Some Chloroplasts, 345–350, Copyright (1957), with permission from Wiley & Sons. Right: chloroplast with starch accumulation, after (Leyon 1954). Reprinted from Experimental Cell Research, 7 (1), H. Leyon, The structure of chloroplasts IV. The development and structure of the Aspidistra chloroplast, 265–273, Copyright (1954), with permission from Elsevier.

Fig. 6.10: Melvin Calvin in the old Radiation Laboratory on the UC, Berkeley campus, with the algae-culturing flasks used in his work on photosynthesis. Reprinted with permission from Journal of Chemical Education, 73 (5), 412–416, Kauffmann, George B., Mayo, Isaac, Multidisciplinary scientist – Melvin Calvin – His life and work. Copyright (1996) American Chemical Society, (Kauffmann and Mayo 1996).

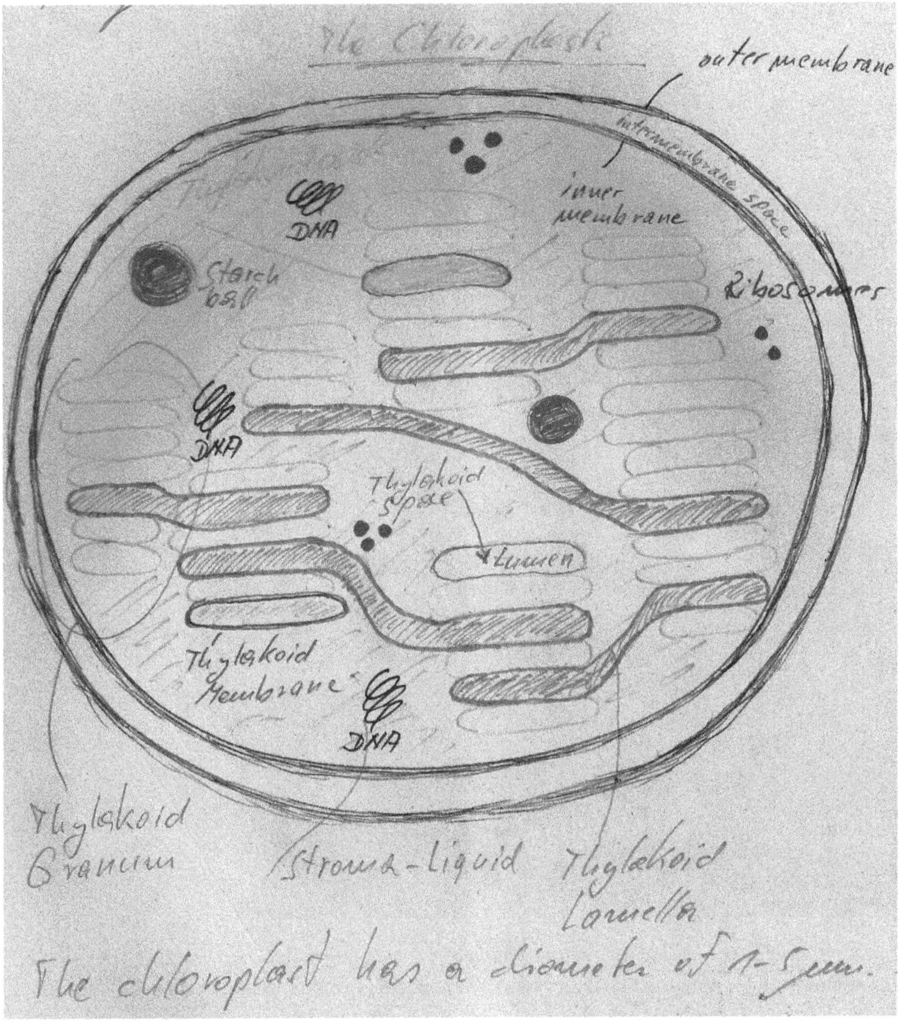

Fig. 6.11: Sketch of the structure of a chloroplast. Consider the size of the chloroplast of 1–5 μm. The two black rings are the inner and outer chloroplast membranes. The numerous green flat objects are the thylakoids, the outer boundaries of which are called thylakoid membranes. Some of the thylakoids are connected by so-called lamellae.

Fig. 6.12: Apartment building with external stairways connecting the apartments. This arrangement resembles the links between thylakoid plates, the *lamellae*. Photo taken in Seoul by Artur Braun.

6.3.2 Thylakoids

Figure 6.13 shows the sketch of a pair of thylakoids that I found in a paper by Stingaciu et al. (2016). It is so nicely sketched that it really complements the scanning electron migrographs in Figure 6.14; those are experimental data, and the other sketches, including mine in Figure 6.20, which are more of a conceptional kind. The two thylakoids in Stingaciu's paper look like folded carpet, artificial grass carpet from a plastic soccer field, folded, enclosing a seemingly empty space.

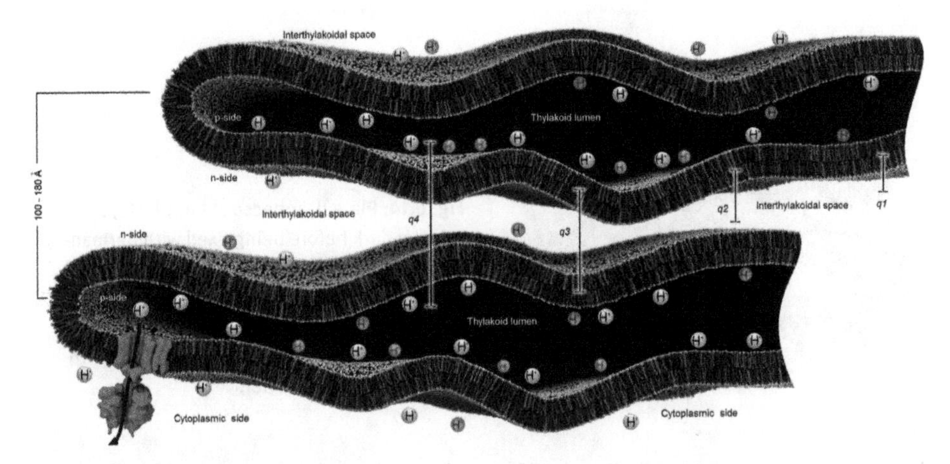

Fig. 6.13: Thylakoid membrane sheet-like structure dynamics. A small region of two adjacent closely appressed membranes that cannot accommodate phycobilisomes is presented undulating freely in the cytoplasm and the corresponding interthylakoidal distances. From $q1$ to $q4$ NSE samples a larger distance (57 Å to 175 Å) within the interthylakoidal space and different relaxation behavior is observed. The excess of protons (H^+) in the thylakoid lumen results in restricted membrane dynamics observed during light. At dark, H^+ pressure is alleviated by chemiosmosis and the membranes undulate freely resulting in higher relaxation rates and excess dynamics. Purple feature is the ATP-Synthase responsible for moving down the electrochemical gradient. The center-to-center distance of a membrane pair corresponds to SANS correlation peaks for closely appressed membranes in Figure A 2 (100 Å to 180 Å). Reproduced from Stingaciu et al. (2016) based on a Creative Commons Attribution 4.0 International License http://creativecommons.org/licenses/by/4.0/.

It is characteristic for thylakoids that they look like flat stacked platelets, which are called granum, and which are connected with each other by bridges and stages, as is shown in the chloroplast cell of a tobacco plant in Figure 6.14. The types of plants that have such chloroplasts with thylakoids inside are certainly countless. It depends on which plant serves best for particular studies. Sometimes it is just important to obtain a visual image of the cell and its contents with electron microscopy. They have to be prepared as extremely thin slices and sections using microtome technology (the method as such is over 250 years old, see Smith 1915).

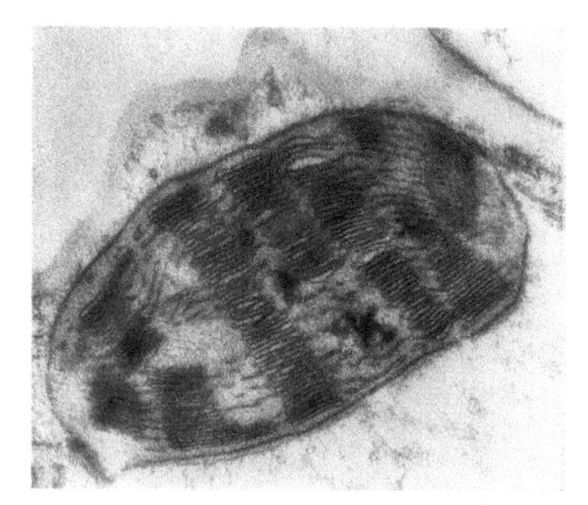

Fig. 6.14: Plate II. Tobacco chloroplasts; 24–36 h in the dark before being fixed with permanganate (Weier, University of California, Davis). (Calvin 1961b). Reprinted from J. Theoret. Biol. 2, Melvin Calvin, Quantum Conversion in Photosynthesis, 258–287, Copyright (1961), with permission from Elsevier.

However, when you cut and slice the plants, they will not live any more, and the studies that you carry out on them are called *post mortem* analyses. There is nothing wrong with studies on dead material, if you are just interested in structural studies. It is often more interesting, however, to carry out the study when the object is still alive and under action and performance. In biology and physiology, such studies are called *in vitro* studies.

Stingaciu et al. carried out such an *in vitro* study with small-angle neutron scattering (SANS) on living cyanobacteria in solution under dark and light conditions. They carried out this experiment at the neutron spallation source at Oakridge National Laboratory (ORNL)[5].

Small-angle scattering (SAS), with x-rays (SAXS) or with neutrons (SANS), can be used for the determination of the morphology of objects with a size on the mesoscale.

5 The nuclear science division at the research center in Jülich, Germany, has an outpost in Oakridge in Tennessee. It also has an outpost in Garching, near Munich in Bavaria. Jülich has a strong neutron research community, but its neutron reactor is decommissioned and no longer in operation, and is, thus, not available for their neutron community. I remember that I once visited the Kernforschungsanlage Jülich in the early 1990s when I was Physics student in Aachen. Back then, they were already shutting down the reactor and its instruments. It sounds quite odd that the neutron experiments are outsourced to nuclear-friendly states like Bavaria or USA.

The methods are well known for the determination of particle size, pore size, also particle shape, fractal dimension, internal surface area, and pore volume. When the structure factor or atomic form factor (Nelms and Oppenheim 1955) of the object is known, one can simulate the SAS patterns and carry out modeling with least-squares fitting on experimental SAS curves. The authors of the paper (Stingaciu et al. 2016) found that the thylakoids apparently undergo a stratification or rectification from their originally wobbly structure when they are illuminated.

It is not possible (or at least not trivial) to carry out such experiments with electron microscopy, because electron microscopy requires high vacuum, and the thylakoids' integrity will suffer in vacuum. It would be difficult to study a thylakoid in such conditions *in vivo*. With SAS, however, this is possible. Neutrons are preferred over x-rays, because the latter cause instantaneous radiation damage, cell damage actually, whereas neutrons are less harmful during the experiment. Also, with SAS it is easy to carry out *in vivo* experiments on living systems, done in the work in Stingaciu et al. (2016).

Since thylakoids provide a SAS pattern that suggests that they are stratified under illumination, and because photosynthesis is performed under illumination, the authors conclude that stratification is connected with the photosynthesis activity of the cyanobacteria or their thylakoids.

One important aspect in this experiment is that thylakoids have a proton gradient. Neutrons are very good probes for hydrogen and protons. It is also possible to label experiments by using deuterium instead of hydrogen, or heavy water D_2O instead of water H_2O, and, thus, have a large scattering contrast that originates from the different neutron scattering cross section for H and D.

As the thylakoids have a proton gradient, you may wonder about their electrical properties in general. In the 1950s, the question about the electric properties of porphyrine and related organic molecules emerged. Those had been used for some time as analogs for chlorophyll. It became known in the 1950s that these molecules behave like organic semiconductors. This is why Melvin Calvin, who used to work on porphyrin structures in the 1930s, decided to investigate the electric properties of chlorophyll (Calvin 1961b):

> Our first measurements were purely of conductivity: Could these layers carry an electronic current in the dark? What would happen to the conductivity of such a system if one put donor or acceptor layers together in such a configuration?

A sketch of the experiment setup by Calvin and colleagues is shown in Figure 6.15. The mechanical support for the sample is a thin glass slide – because of its transparency to visible light. The glass is then painted with a graphite ink (colloidal graphite solution), which was known as Aquadag (Acheson 1910). Specifically, they drew interdigited electrode terminals – with a spacing of 0.1 mm – on the glass to be able to spread the chlorophyll sample, actually a thylakoid and chloroplast solution by sub-

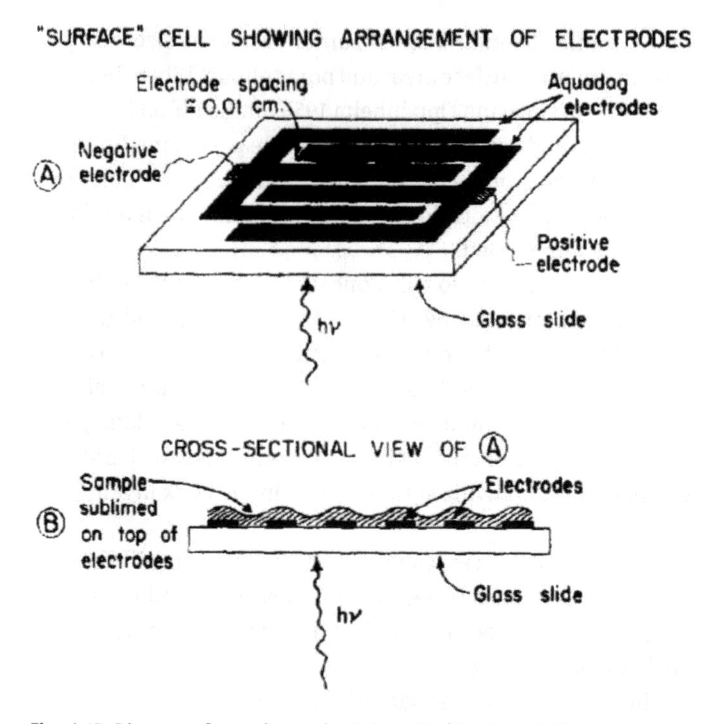

Fig. 6.15: Diagram of sample conductivity cells (Calvin 1961b). Reprinted from J. Theoret. Biol. 2, Melvin Calvin, Quantum Conversion in Photosynthesis, 258–287, Copyright (1961), with permission from Elsevier.

limation over the electrode fingers. Prior to measuring biological samples, they put layers of phtalocyanine on the electrodes (Calvin 1961b).

With this arrangement it is possible to record optical absorption spectra and so-called action spectra. The action spectrum is the variation of the photocurrent as a function of the light wavelength. This is also known as IPCE measurement, which stands for incident photon to current efficiency, also known as quantum efficiency. Naturally, the ordinate for both experiments is the wavelength λ of the light. You need to have a monochromatic light source for this kind of experiment.

We may wonder who, or what, provides for the nice stacking of the thylakoids in the chloroplasts that you can see in the electron micrographs. Wood et al. (2018) have conducted a study and found that upon changing from dark to light, the phosphorylation in the light harvesting complex in photosystem II (LHCII) does change the architecture of the thylakoid membrane and the organization of the photosynthetic complexes. The diameter of the grana and the number of membrane layers per granum are decreased under illumination, whereas the total number of grana inside the chloroplast increases. This causes the geometric contact area between granum and a stromal lamella to also increase. Overall, this process is a rearranging of the stacking of the thylakoids.

This rearrangement has an electric effect in the chloroplasts and influences the balancing of what is known as linear electron transfer (LET) and cyclic electron transfer (CET). When the granum is smaller, the travel distance for the electron shuttles plastocyanin and plastoquinone becomes shorter, and this means that the electric resistance for linear electron transfer becomes smaller.

When the granum becomes larger, then also the reservoirs for the granal and stromal lamellae plastoquinone becomes larger. This increases the efficiency of the cyclic electron transport. The CET produces a change in the pH, which prevents the “overreduction” of the PSI and protects the PSII from “overexcitation”. In the end, there is a dynamic sizing and stacking of the thylakoids in response to the change from dark to light in order to regulate the electron transfer for photosynthesis, see Wood et al. (2018).

In biology, there is another stacking phenomenon, and this is the stacking of the red blood cells, also known as *erythrocytes*. These stacks then extend “long” in one direction, and the aggregates are called “rouleaux”. The blood cells carry a negative charge on their surface and, therefore, they should repel each other and not attract each other to form a long, tall stack. Bradonjic et al. claim, or suggest that this counterintuitive effect is caused by the Casimir effect, where the attractive force of the Casimir effect overcompensates the electrostatic repulsion (Bradonjić et al. 2009).

The Casimir effect is a quantum mechanical effect and was derived in the work of H.B.G. Casimir and D. Polder, which was published[6] in 1948 in Physical Review (Casimir and Polder 1948). The authors call it the influence of retardation on London and van der Waals forces and find that interaction energy terms between neutral atoms and planes are of the order of distance with the 7th power.

Casimir and Polder actually only predicted the effect and did not provide any empirical results or verification. This took until 1957, when Spaarnaay provided experimental proof for the prediction (Sparnaay 1957). Since then the Casimir effect has been verified in many experiments. The force between two plates, or other objects, is a result of the quantization of the electromagnetic field between the objects and does not depend on any material parameters but only on their distance and on the constants c and h.

---

6 Casimir and Polder worked at the Philips Electronics Company in Eindhoven, The Netherlands. A long time ago, companies like Philips, and many more worldwide, had research laboratories with researchers and scientists who could publish high-profile papers like this one. Many of these companies had their own research journals. The well-known Van der Pauw method for measuring the electric conductivity of arbitrarily shaped thin films by applying four probe spots on the film surface is an original Philips research work (van der Pauw 1958). There is another Van der Paauw, with double a, who reported in 1935 that four red light photons are required to produce one oxygen molecule from intermediates in the CO_2 reduction process in photosynthesis (van der Paauw 1935; Gaffron and Wohl 1936b).

The attractive Casimir force F between two electrically conducting plane plates per geometric area A is $-(\hbar c\pi^2)/(240d^4)$ (Greiner 1989). We know that force per area is a pressure p. As there is only a dependency of the Casimir pressure from the nature constants c and $\hbar$, but not from the electron charge e, there is apparently no mechanistic coupling between the matter of the plates, or atoms, and the electromagnetic field. This is why we can say that the Casimir force is some sort of zero-point pressure of the zero-point oscillations of the photon vacuum. The origin of the force is the vacuum, but it is mediated by the two plates in that very vacuum.

One may wonder, then, whether this stacking of blood cells could be another manifestation of the Casimir effect (Bradonjić et al. 2009). Similarly to Calvin in the 1950s, who used porphyrins as synthetic models for the chlorophyll, Bradonjic et al. model the blood cells with charged dielectric plates in an ionic solution, where the repulsive Coulomb force is compared with the attractive Casimir force, and the resulting free energy per geometric area yields, as a function of distance d,

$$F(d)/A = \frac{\sigma^2 \Lambda}{2\epsilon_2}\left\{e^{-d/\Lambda} - \left(\frac{\pi^2 \hbar v \sqrt{\epsilon_0 \epsilon_2}}{360\sigma^2 \Lambda}\right)\frac{1}{d^3}\right\} . \tag{6.2}$$

Here, Λ is the Debye screening length of the liquid, or fluid, and ϵ_0 and ϵ_2 the dielectric constant in vacuum and in the fluid, respectively. The constant v is the ratio of the dielectric constants for the dielectric disks, ϵ_1, and the fluid between them, ϵ_2:

$$\frac{v}{c} = \left(\frac{\epsilon_1 - \epsilon_2}{\epsilon_1 + \epsilon_2}\right)^2 . \tag{6.3}$$

Thylakoids may carry positive charges when protons follow the gradient to their surface. They then repel each other, but we experimentally observe by electron micrographs like those in Figure 6.14 that they stack together, depending on the situation. The Casimir effect like the vacuum pressure should act everywhere and always, and provided that the geometry and dielectric constants allow for it, competing other forces like Coulomb forces may be overcompensated. Possibly this is the case between the thylakoids as well. Maybe someone will design an experiment and verify or falsify the suggestion for blood cells and for thylakoids.

There is another such suggestion where the Casimir effect may play a role in the ordering of biological systems. This is the stacking of the molecules in the lipid bilayer membrane. Here, it is important that the molecules are not penetrated by each other but slipped up against each other. This warrants, for example, the necessary rheological properties of the lipid membrane (Pawlowski and Zielenkiewicz 2013).

6.3.3 Method for the extraction of thylakoids from spinach

During my time as a postdoc at Berkeley I worked in the battery research group of Prof. Elton J. Cairns. As he held synchrotron radiation methods in admiration, he collaborated with the biophysics group of Prof. Stephen P. Cramer from UC Davis. As I spent many synchrotron beam times with the Cramer group, I became involved in their photosynthesis research (for a representative PhD thesis, see, for example, Gu 2003). We carried out protein spectroscopy experiments with hard and soft x-rays and with other spectroscopy methods (Mössbauer, EPR, ...) on photosystem II and on hydrogenases.

The protein samples were typically delivered by our collaborators in Germany, France, or from the Calvin Lab on the UC Berkeley campus. Back then was the first time that I heard that such samples could be basically made from spinach, from shredded spinach. This sounds very trivial, and I had not previously associated our fancy synchrotron radiation-based experiments with something as vulgar as spinach. What I learnt from my colleague Dr Daulat Patil was that proteins, particularly for protein crystallography, needed to be concentrated and also purified. These were some of the lengthy steps towards protein crystallography and spectroscopy. But it started with spinach.

It would take almost 20 more years before I personally had the opportunity to see how such samples were made – during my sabbatical at Yonsei University in Seoul, Korea. First, you buy spinach in a grocery store (see top left image in Figure 6.16), or you pick it from the field if you have access to a field. When you want to excel in photosynthesis research, there is nothing wrong with spending time in a grocery store. You are actually in good companionship. Said Melvin Calvin in an interview that was conducted with him in 1973 (Ridgway 1973):

> At that time, I worked in a retail grocer's shop as a delivery boy. I lived in an environment in which the earning of the daily bread was a concern on a daily basis.

Wash the spinach immediately and store it cool and dark. Depending on the case, you may also try leaves from other plants (remember Figure 6.14, which shows chloroplast from tobacco plants), or from cyanobacteria. You tear the leaves apart so as to remove all branches and tough fibres from the leaves (called deveining). Then you wash the leaves in clean, cold water and put them in a kitchen blender, which is filled with cold water from ice cubes, see Figure 6.16 top middle, and contains a buffer saline containing, in weight % per volume 5.96% HEPES (4-(2-hydroxyethyl)-1-piperazineethanesulfonic acid), 30% D-sorbitol, 0.29% NaCl, 0.37% EDTA (ethylenediaminetetraacetic acid disodium dehydrate), 0.5% $MgCl_2$, and 0.1% bovine serum albumin (BSA); you can look up such procedure in a paper by Sjoholm et al. (2012).

The cold environment from the ice cubes prevents the spinach from overheating in the blender, where it is shredded to a green "smoothie"[7], see Figure 6.17 middle, left. The smoothie is then poured through a filter into a glass beaker sitting in an ice cube environment. Eventually, the dark filtrate is collected in a plastic vial and stored in ice.

7 I call it *smoothie* because my wife sometimes uses the kitchen blender to make such smoothies from spinach and fruit.

Fig. 6.16: Method for the extraction of thylakoids from spinach. A grocery package with spinach from the supermarket. Preparation of cooling bath with ice cubes. Measuring cold water to add to deveined leaves in the blender. Sieving of the green spinach blend and collecting it in a beaker with cold water. Further sieving of the sieved liquid. Collecting the green chloroplast solution in a plastic container. Storing the chloroplast solution in cold ice water. Magnesium chloride tablets for osmotic shock. Two plastic containers with blue stoppers and chloroplast solution inside, ready in the centrifuge for separation.

Fig. 6.17: Plastic container with green spinach solution containing primarily the thylakoid solution and the whitish solid pellet at the bottom, all obtained after centrifugation.

The filtration warrants that the crude parts and components of the leaves are kept back and only the fine matter is dispersed in the filtrate. Next, you add two magnesium chloride pellets to the spinach solution. The effect of this is that the chloroplasts is subject to osmotic pressure by picking up the $MgCl_2$ solution and will eventually "explode" because the cell walls cannot sustain the osmotic pressure. This is called osmotic shock. As a result, all components in the chloroplasts will be liberated and accessible for further processing.

The mixture is then centrifuged for a few minutes at $200 \times g$ to remove the cellular debris; "$200 \times g$" means that the centrifugal force is 200 times the gravitational force g. Note that here I am using the term force incorrectly; the correct term is acceleration. The supernatant is then isolated from the debris pellet. Then, the supernatant fluid is centrifuged at $1000 \times g$ and carefully placed onto a 40% Percoll® density gradient and centrifuged for 6 minutes at $1700 \times g$. The resulting pellet is then redissolved using 10 mM $MgCl_2$ 2 for 30 seconds and resuspended in a buffer containing 330 mM sorbitol, 50 mM HEPES, and 2 mM $MgCl_2$ (pH 7.8).

You can see the result of such ultracentrifugation in Figure 6.17, where a vial with thylakoid solution is held by a graduate student from the research group of Professor Ryu at Yonsei University. This was the protocol of how to extract and separate thylakoids from chloroplasts. For a condensed matter physicist like myself, such biophysics experiments with algae and thylakoids look quite complex and difficult, but you can read the paper by Rob Dean and Ewa Miskiewicz who consider such experiments with thylakoid electron transfer a laboratory exercise (Dean and Miskiewicz 2003) for students.

Maybe you have a salad centrifuge at home in the kitchen, a technically simple plastic pot with a grid inset that is fixed to a gear system on the top of the first pot with a string attached, which you pull. Then, the grid inset will be driven into rotation. When you put wet salad leaves, after you have rinsed them, into the inset, close the pot and pull the string, the inset rotates at high speed, and the water drops from the salad leaves feel the centrifugal force and leave the salad leaves and fly to the outside. This is how the salad becomes dry.

Laboratory centrifuges basically work on the same principle, but they have stronger motors, better gears, and mechanically more stable material because the centrifugal forces are very large. These forces are counted in multiples of Earth's gravitation *g*, or equivalent, in rounds per minute, rpm. To some extent, centrifugation is some sort of chromatography. The difference is that centrifugation has the purpose of separating materials in order to accumulate them for further processing, whereas chromatography is a purely analytical method. As a side note, separation techniques are important for many technologies, also nuclear technologies. Ultracentrifuges are used for the separation of the isotopes of heavy elements for nuclear fuel and weapon production, and, thus, sensitive technologies and products subject to export limitations for several minutes. After that, the resulting pellet can be resuspended in one μL of the aforementioned buffer saline[8].

The ultracentrifuge is an invention reported by Theodor Svedberg and Herman Rinde in Uppsala in the year 1924 (Svedberg and Rinde 1924), so around 100 years ago, in the *Journal of the American Chemical Society*. This was a purely technical paper, and it has since been cited around 150 times. It could be used to determine the size distribution of particles in a colloidal or mesoscopic solution, and eventually also the molecular weight of proteins (Svedberg and Fåhraeus 1926). For his work on disperse systems, Theodor Svedberg received the Nobel Prize in Chemistry in 1926. His work on these systems also provided experimental evidence (the existence of molecules) for theories on the Brownian motion.

6.3.4 The lipid membrane

The lipid membrane is an essential part of the architecture of thylakoids and separates the lumen from the stroma. For the in-depth study of membrane lipids, I recommend the review on *haloarchaeal bioenergetics* by Kellermann et al. (2016).

I begin here with a general overview of the membranes as shown in Figure 6.18. The two reddish "tapes" in horizontal direction in the figure are two such membranes, also known as cytoplasmic membranes. Close inspection shows that each membrane tape is made from two layers attached together. These are the two aforementioned lipid layers, the bilayer that separates the lumen (inside the thylakoid) from the stroma (outside the thylakoid, the inside structure of the chloroplast).

These membranes are not entirely homogeneous. Instead, they have regions where proteins are built in, such as the PSII or bacteriorhodopsin bR, and are, thus, highly functional. The membranes also have regions with lipid layers of different per-

---

8 PBS is phosphate buffer *saline* and sometimes wrongly called phosphate buffer *solution*.

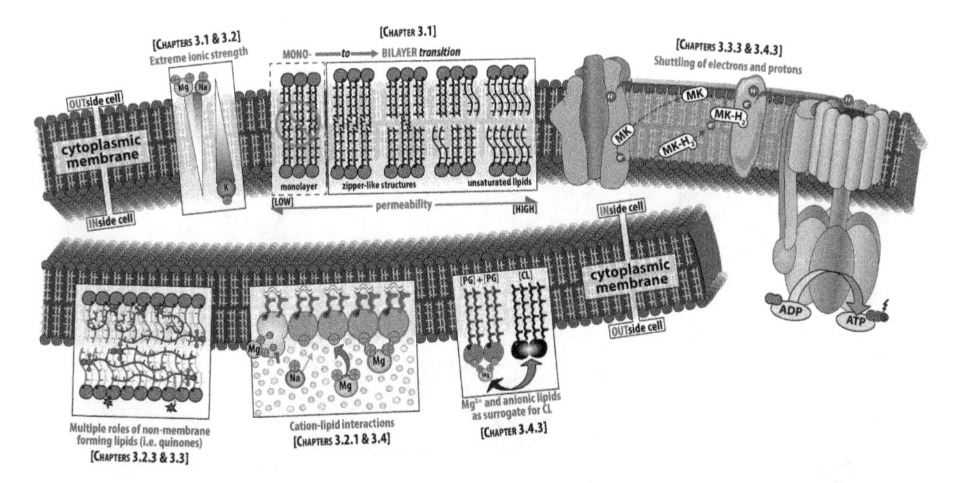

Fig. 6.18: Overview of major membrane lipid adaptations that allowed haloarchaea to thrive in an oxygenated environment. The here highlighted adaptations will be discussed in detail throughout this manuscript (Kellermann et al. 2016). Reprinted from Biochimica et Biophysica Acta, 1858, Matthias Y. Kellermann, Marcos Y. Yoshinaga, Raymond C. Valentine, Lars Wörmer, David L. Valentine, Important roles for membrane lipids in haloarchaeal bioenergetics, 2940–2956, Copyright (2016), with permission from Elsevier.

meability, which is illustrated in Figure 6.19. These regions may be open for charge transport by electrons and cations, positive charge carriers like protons, K^+, Na^+, Mg^{2+}.

Kellermann et al. point out that the models for the charge transport along the membrane are still hypothetical (Kellermann et al. 2016)[9]. It is not trivial to decide which processes take place at which locations at the molecular scale in such a complex region like the thylakoid membrane.

One question is how the transport of electrons and protons takes place along the lipid membrane. "Along" basically means "azimuthal" with respect to the thylakoid and not radially. This is illustrated in Figure 6.19. Three different models are shown in the figure. The "liquid model" assumes that protonated aromatic quinone groups carry electrons and protons and diffuse freely through the mid-plane in the membrane from one respiratory complex to another. The flip-flop model says the charges are soon transported to the heads of the lipids and then passed on along the membrane surface.

9 I have met people, colleagues, who believe if something like a model is hypothetical, then it is irrelevant. I have a different opinion on this. One of the core qualifications of a scientist is to be able to make a hypothesis, and then turn it into a model, or the other way around. Further studies, empirical studies, may then show whether or not the hypothesis reflects reality. A scientist has to have the necessary intellectual synthesis skills to come up with a hypothesis and a model.

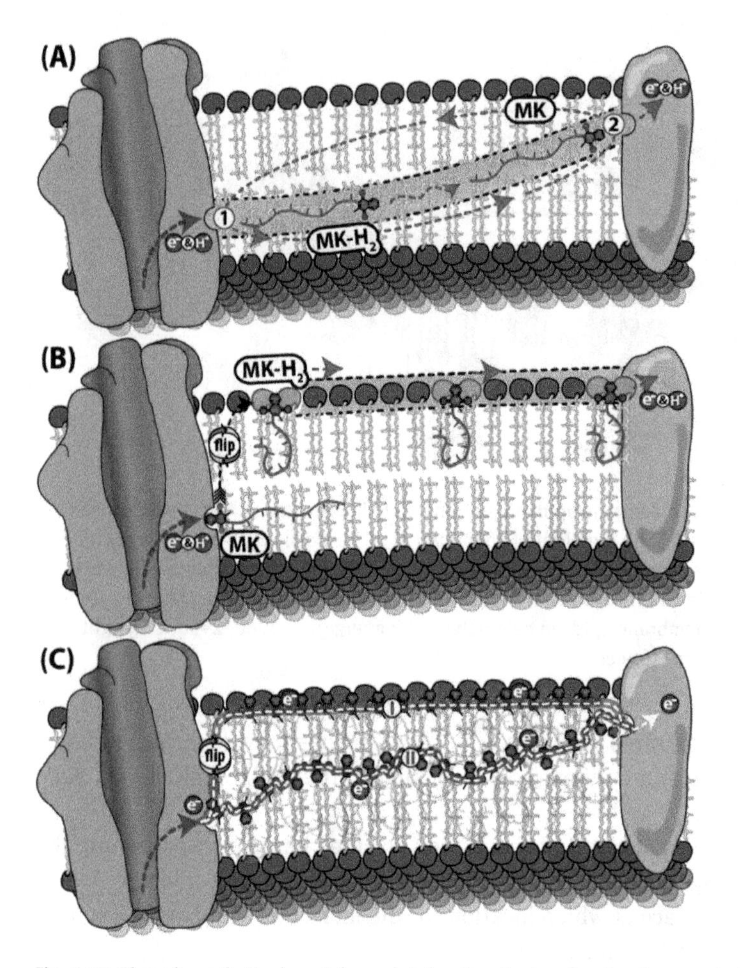

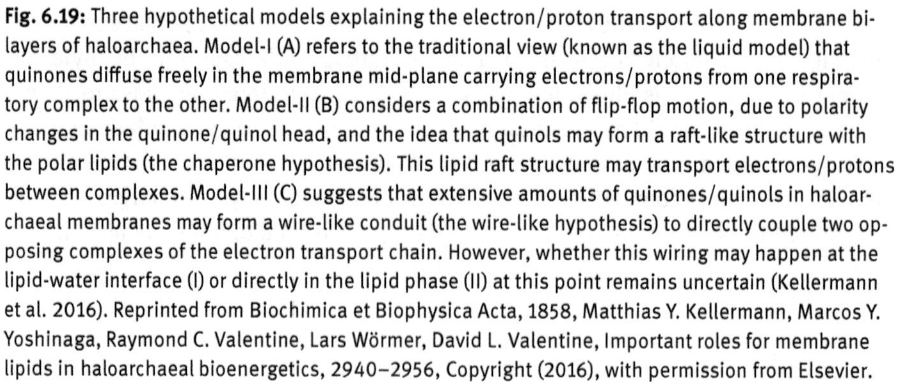

Fig. 6.19: Three hypothetical models explaining the electron/proton transport along membrane bilayers of haloarchaea. Model-I (A) refers to the traditional view (known as the liquid model) that quinones diffuse freely in the membrane mid-plane carrying electrons/protons from one respiratory complex to the other. Model-II (B) considers a combination of flip-flop motion, due to polarity changes in the quinone/quinol head, and the idea that quinols may form a raft-like structure with the polar lipids (the chaperone hypothesis). This lipid raft structure may transport electrons/protons between complexes. Model-III (C) suggests that extensive amounts of quinones/quinols in haloarchaeal membranes may form a wire-like conduit (the wire-like hypothesis) to directly couple two opposing complexes of the electron transport chain. However, whether this wiring may happen at the lipid-water interface (I) or directly in the lipid phase (II) at this point remains uncertain (Kellermann et al. 2016). Reprinted from Biochimica et Biophysica Acta, 1858, Matthias Y. Kellermann, Marcos Y. Yoshinaga, Raymond C. Valentine, Lars Wörmer, David L. Valentine, Important roles for membrane lipids in haloarchaeal bioenergetics, 2940–2956, Copyright (2016), with permission from Elsevier.

6.3.5 The light harvesting complex (LHC)

We have now progressed step by step to the components in the leaf, down to the thylakoids. However, we are not at the end yet. Thylakoids are built up from finer components such as the thylakoid membrane and the chlorophylls (Table 6.1) and the light harvesting complexes LHC and other proteins functional for photosynthesis.

Tab. 6.1: The characteristics of chlorophyll a and chlorophyll b.

Chlorophyll a	Chlorophyll b
Principal pigment capturing sunlight for photosynthesis	Accessory pigment collecting sunlight and passing it on to chlorophyll a
Absorbs light in range of 430–660 nm	Absorbs light in the range of 450–650 nm
Around 1–2 µm size particles are ingested	Around 0.1–0.2 µm size liquid droplets are ingested
430 and 662 nm light easy absorbed	470 nm wavelength easy absorbed
Absorbs violet-blue and orange-red light from the solar spectrum	Absorbs orange-red light from the solar spectrum
Reflects blue-green color	Reflects yellow-green color
Contains methyl group in third position of its Cl ring	Contains aldehyde group in third position of its Cl ring
Molecular weight 839.51 g/mol	Molecular weight 907.49 g/mol
Found in all plants, algae, and cyanobacteria	Found in all plants and green algae
Accounts for 3/4 of chlorophyll in plants	Accounts for 1/4 of chlorophyll in plants
Solubility low in polar solvents	Solubility high in polar solvents
Present at reaction center of antenna array	Regulates size of antenna

The photosynthetic light reaction in cyanobacteria, algae, and plants are performed by two photosystems. Photosystem I is plastocyanin-ferredoxin oxidoreductase and an integral thylakoid membrane protein complex that absorbs light for the catalytic transport of electrons across the thylakoid membrane from plastocyanin to ferredoxin. These electrons are used for the production of the chemical energy carrier NADPH (energy coupling and ATP synthase (Haraux and de Kouchkovsky 1998)).

Light-dependent oxygenic photosynthesis is performed by photosystem II. Electrons are transported for the reduction of plastoquinone to plastoquinol. PSII contains the light harvesting complex LHC for the oxygen evolution reaction OER. For a recent review on the function and structure of PSII, see Allakhverdiev (2008). LHC contains a manganese-oxygen cluster, along with a calcium ion. Researchers have tried to syn-

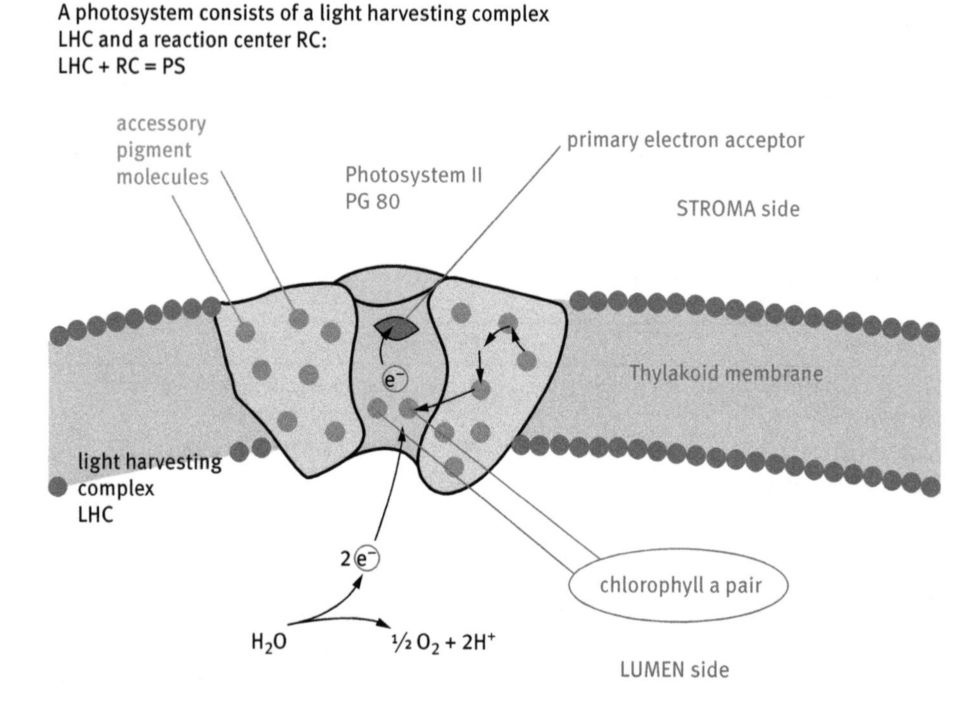

Fig. 6.20: Sketch of thylakoid membrane with photo system included. Artur Braun.

thesize inorganic photoelectrodes of these elements with the hope of obtaining better photoelectrodes; with little success, I believe. It is the molecular complex in its more complex protein environment, which performs very well the water oxidation. Its computational treatment is virtually impossible because of the incredible number of degrees of freedom. However, Kurashige et al. have calculated the wavefunctions for the Mn_4CaO_5 cluster in the oxygen evolving complex of photosystem II (PSII) (Kurashige et al. 2013) and taking care of 10E18 degrees of freedom.

Figure 6.20 shows a sketch with a section of a thylakoid membrane, which includes a photosystem II (PSII). The two red lines named "thylakoid membrane" are actually the lipid bilayer. This bilayer encloses the lumen; this is the inside space of the thylakoid. The red dot stands for a carboxylic group, which is hydrophilic. The two legs (alkyl chain) connected to the red dot are fatty acids, which are hydrophobic. The thickness of such a bilayer is around 6–10 nm.

Integrated in the thylakoid membrane are the quantasomes, such as you see on PSII. A photosystem is the combination of a light harvesting complex (LHC) and a reaction center (RC). The two green components are a chlorophyll pair, which contains a collection of accessory pigment molecules.

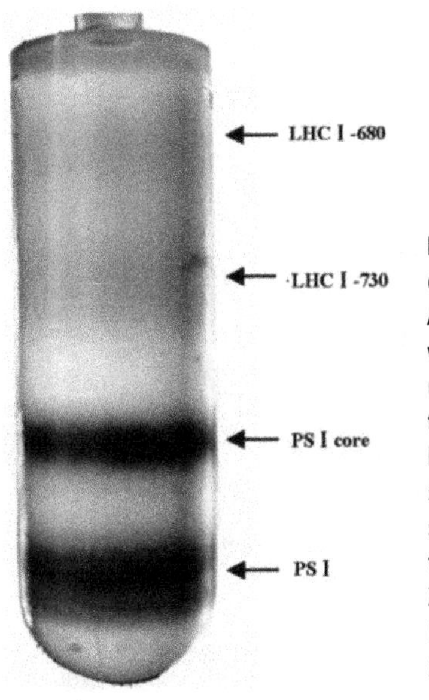

Fig. 6.21: Separation of PSI core and LHCI by sucrose gradient ultracentrifugation in a Beckman VTi 50 rotor at 4 °C. After ultracentrifugation at 200,000 × *g* for 4 h, four bands were distinguished: LHCI-680, LHCI-730, PSI core, and undissociated PSI, respectively, from the top to the bottom. Arrows indicate these different bands. Reprinted by permission from Springer Nature. Photosynthesis Research 2006, 90, 195–204, Rapid purification of photosystem I chlorophyll-binding proteins by differential centrifugation and vertical rotor, Xiaochun Qin, Kebin Wang, Xiaobo Chen, Yuangang Qu, Liangbi Li, Tingyun Kuang, https://doi.org/10.1007/s11120-006-9104-6, (Copyright 2007).

A detailed protocol for the separation of the various proteins inside the thylakoid membrane was reported by D'Amici, Huber, and Zolla. This requires not only (ultra) centrifugation, but also electrophoresis (D'Amici et al. 2009). After centrifugation with extremely high speed, corresponding to 200,000 × *g*, the fractions of the thylakoid membranes become separated, as demonstrated in Figure 6.21.

In Figure 6.21 you can see an upright vial filled with a greenish-yellowish solution with regular ring patterns. In every ring there is a concentration of a particular species of material. At the bottom, there is a starch pellet. The very dark ring at the bottom contains photosystem I (PSI). The second dark ring above the first one contains the PSI core fraction. Further above is a broad intense yellow ring that contains the light harvesting complex LHC-I 730. The 730 stands for the absorption maximum at 730 nm light wavelength. The broad, dim green ring at the top contains the LHC-I 680.

Figure 6.22 shows three absorption spectra from chlorophyll a, chlorophyll b, and carotenoids. We can easily distinguish the three types of pigment molecules by the position of their absorption maxima. Chlorophyll a has a distinct absorption maximum at 680 nm, and this is the spectroscopic feature we know from LCH-I 680. Many of the components in light harvesting complexes are classified by their spectroscopic features, such as the position of the absorption maxima.

Figure 6.23 shows fluorescence emission spectra of light harvesting complexes LHC-680 and LHC-730, and PSI. These samples were excited with a particular monochromatic wavelength light. The absorption spectra have their maxima at different positions to the emission spectra.

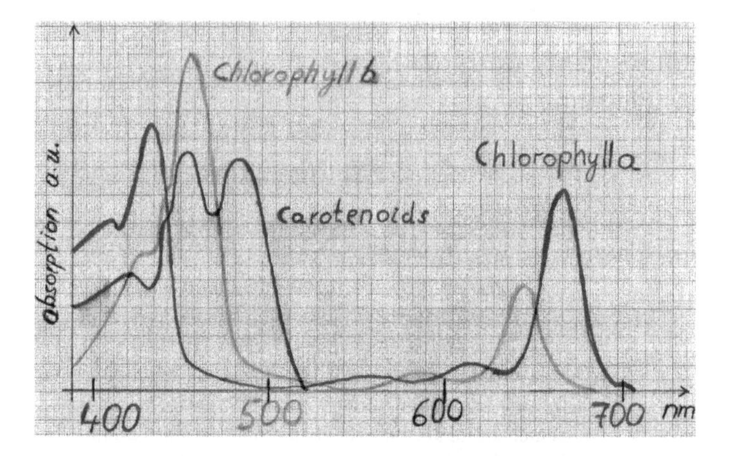

Fig. 6.22: Absorption spectra of chlorophyll a (blue) and b (green) and carotenoids (red).

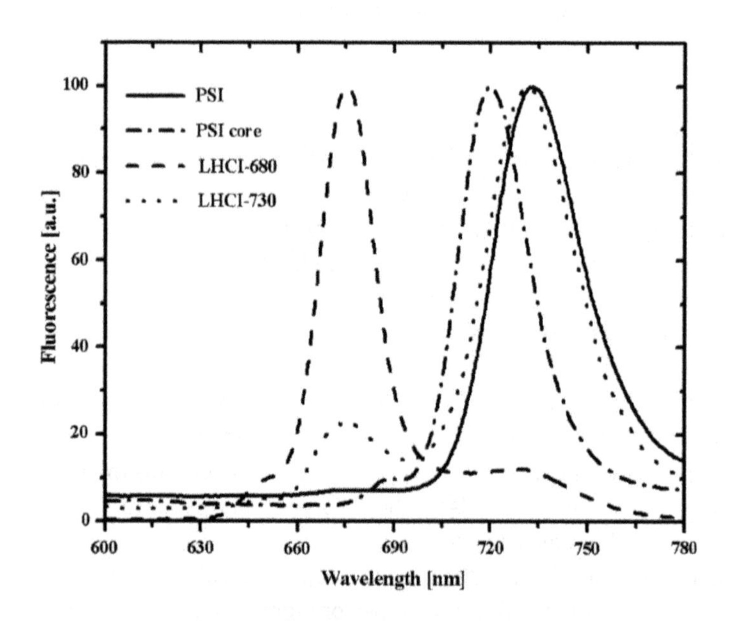

Fig. 6.23: Low-temperature (77 K) fluorescence emission spectra (excited at 440 nm) of PSI, PSI core, LHCI-680, and LHCI-730. The spectra were normalized with respect to their wavelength of maximal emission. Reprinted by permission from Springer Nature. Photosynthesis Research 2006, 90, 195–204, Rapid purification of photosystem I chlorophyll-binding proteins by differential centrifugation and vertical rotor, Xiaochun Qin, Kebin Wang, Xiaobo Chen, Yuangang Qu, Liangbi Li, Tingyun Kuang, https://doi.org/10.1007/s11120-006-9104-6, (Copyright 2007).

6.3.6 Cytochromes

Cytochromes are color proteins (dyes) and are typically found in the mitochondriae of living matter. They are omnipresent throughout all biological classes, but not necessarily in every species. Their function is typically the transport of charge, and they are classified by their heme, which has redox functions, and their light absorption properties. The well-known *Escherichia coli* bacterium has no such cytochromes.

The cytochromes were discovered in 1884 by McMunn. In the 1920s, David Keilin was able to interpret the optical spectra recorded from living cells as the three important cytochromes a, b, and c. Cytochromes have as linkages one or several heme molecules, the differences of which cause differences in the absorption spectra, which makes it possible to sort them as cytochromes a, b, c, and d. Cytochromes are proteins and as such are subject to conformational changes depending on their thermodynamic environment (Zheng et al. 1991). A structural representation is shown in Figure 6.24.

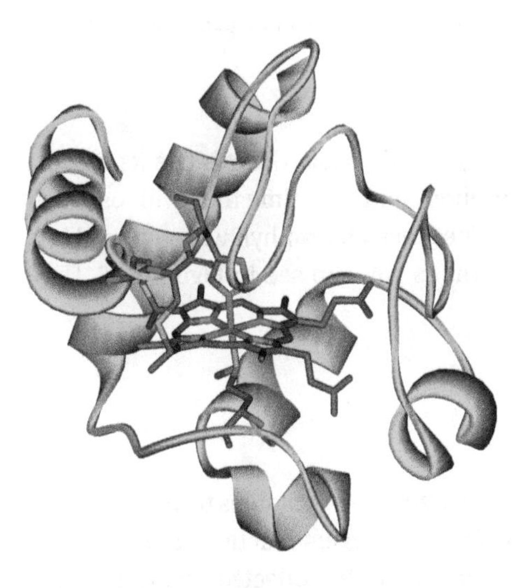

Fig. 6.24: Structural model of cytochrome c. Courtesy Klaus Hoffmeier, Universität Frankfurt.

The cytochrome P450 is actually an entire family of metaloproteins that contain a heme, an iron cofactor that works as an oxygenase. They come in mammals and in plants and oxidize and synthesize fatty acids and hormones, respectively. The suffix P450 stands for the observation that the enzyme has an optical absorption peak at 450 nm (this is blue in color), provided that it is ion the reduced state. For this, it is complexed with carbon monoxide CO. It is quite common in proteon research that you have a bottle of toxic carbon monoxide gas in your lab to cause a chemical gas phase reduction with CO. It is a reducing agent. As CO is toxic, particular cautions and safety procedures are necessary.

Fig. 6.25: Two beech tree leaves beginning to rot – photographed on 8 February (left) and 2 September (right) 2019 in Switzerland. Parts of the leaves are still green and contain chlorophyll, which keeps performing photosynthesis. The yellow part is mainly starch. The brown part is degrading severely. Photo taken by Artur Braun.

We recollect the Prologue in the beginning of this book. Growth has its season. In fall, the leaves of the trees fall onto the ground, and over time they will lose their green color and become brown and rot away. The green chlorophyll will stop working, colors will turn from green to yellow to brown, as you can see in Figure 6.25. Until spring comes again.

6.3.7 Photosynthesis in caves

Those readers who know about potatoes may have seen how potatoes turn green when you leave them outside exposed to the sun. The sun triggers reactions in the potato, which produces chlorophyll, so that more sunlight can be collected and turned into biomass. In Figure 6.26 you can see a photo I took in the Damlataş Cave (also known as the asthma cave) on the occasion of my visit to Mersin University in Turkey. To my knowledge, the case has no hole where light from the outside can come in. The cave is dark. The only light comes from strong electric lamps positioned every couple of 10 or 20 meters throughout the cave, as far as the locations are accessible to the public (Alakavuk and Yağmuroğlu 2010). I took many photos in this cave, which has lots of stalagmites, that is, long sticks growing downwards from the "ceiling" of the cave from recrystallized minerals that drop down in water droplets from the ceiling. Those spots

Fig. 6.26: Photo taken in the Damlataş Cave (Alakavuk and Yağmuroğlu 2010) in southern Turkey. The green regions on the lower left-hand side are green because they are exposed to high-intensity light from the lamps necessary to bring light into the dark cave. Photo taken 14 June 2019 by Artur Braun.

in the cave that are illuminated by lamps had a green color, obviously from microbes or fungi, which produce chlorophyll under illumination (Montechiaro and Giordano 2019; Mulec et al. 2008), which is known as lampenflora. The regions where light did not shine directly were brownish and showed only the mineral surface.

6.4 Thylakoid films

It is possible to extract electrons from single algal cells, as was, for example, shown in Ryu et al. (2010). You can think about distributing thylakoids over a substrate and, thus, make a thylakoid film. The goal would be to expose the thylakoid film to sunlight and then extract an electric thylakoid current from the substrate, after the thylakoids produce electrons under illumination, and inject them into the current collector on the substrate, so that the large amount of thylakoids produces a larger electron current. Hamidi et al. (2015), for example, embedded thylakoid membranes in an osmium-redox-polymer and used this as a photoelectrode for solar energy conversion. Agostiano et al. dispersed thylakoids in an electrolyte and then conducted cyclic voltametry on them (Agostiano et al. 1992, 1993); they call it thylakoid membrane preparations.

Figure 6.27 shows a series of thylakoid films that were deposited on gold films, coated on float glass substrates. The gold film is the electric current collector, which needs to be connected with an electric wire. The five electrodes at the bottom of Figure 6.27 have different thicknesses, as you can see from the saturatedness of the green color. The film on the left/hand side was prepared with a more dilute solution of thy-

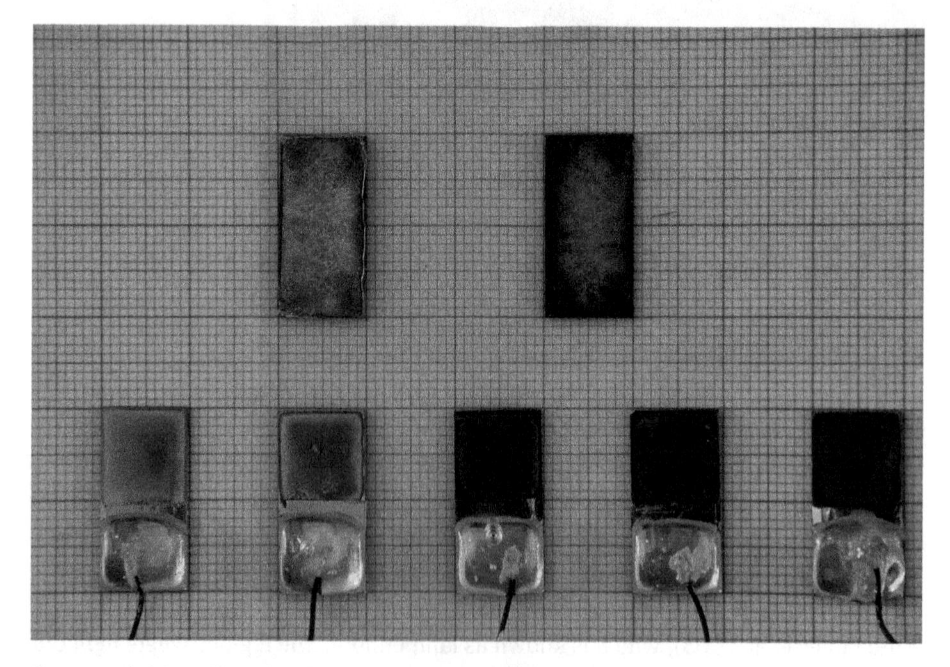

Fig. 6.27: Thylakoid films on gold with different thicknesses. The two upper samples are not in contact with electric wires. The five lower samples are in contact with wires and can be used as photoelectrodes. The electrodes were prepared by students of Prof. Wonhyoung Ryu, Yonsei University. Photo taken by Artur Braun.

lakoids, and the film on the right/hand side with a concentrated thylakoid solution. The bottom part of the five electrodes is soldered with a wire on the gold layer and then coated with rubber glue to provide firm mechanical contact with between the wire and gold.

The two electrodes on the top have no wires because those will be used for optical spectroscopy experiments where wires, and the solder contacts would be disturbing. As the samples are no longer transparent, the optical experiments must be done in the reflection mode. You can notice on the bottom samples that there is a pattern on every sample that shows how the solvent from the thylakoid solution has evaporated and the thylakoids were moved by the solvent recession line towards the center of the electrodes. This indicates that the films are not mechanically stable and the thylakoids might move on the electrode surface while exposed to the electrolyte in the assembled electrochemical cell.

The strategy was then to fix the thylakoid film with a sugar solution. Such a sample is shown in Figure 6.28, from the back and front sides. We want to have the thylakoids

Fig. 6.28: Thylakoid film in dried sugar solution deposited on nickel film. The left-hand side dark sample shows the back side of one electrode, the color coming from the titanium coating on the glass substrate. The right-hand side green sample shows the front side of a thylakoid electrode, the thylakoids embedded in a sugar solution. The electrodes were prepared by students of Prof. Wonhyoung Ryu, Yonsei University. Photo taken by Artur Braun.

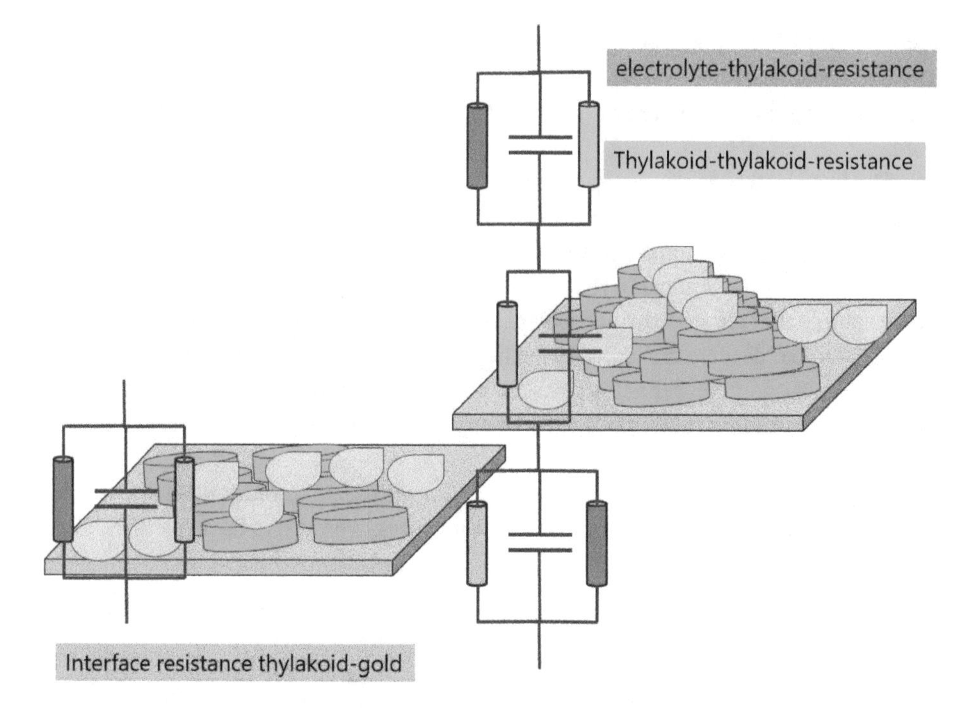

Fig. 6.29: Schematic of wet thylakoid film and electric circuit.

fixed in order to warrant a firm mechanical and, thus, good electric contact between thylakoid and current collector. This situation is illustrated in the sketch in Figure 6.29, which on the lower left-hand side shows a substrate with a yellow gold film over which several green thylakoid disks are distributed. The liquid electrolyte layer is indicated by the bluish drops.

Along with this sketch you see part of an electric circuit made from two parallel ohmic resistances and one capacitor. The blue resistor stands for the electric resistance between the thylakoid and the electrolyte. The yellow resistor stands for the electric resistance between thylakoid and gold layer.

The top right-hand side of the sketch shows the situation where several layers of thylakoids pile up in the vertical direction. The corresponding electric circuit, thus, becomes somewhat more complex because now an additional resistance among the thylakoids is to be considered. In parallel to the resistances we consider the capacitive contributions as well. We can place the thylakoid electrodes in an electrochemical cell and then run the electroanalytical techniques on them, such as cyclic voltametry and impedance spectroscopy. With the latter, one can check whether the experimental impedance spectrum can be modeled by the suggested electric circuit in Figure 6.29.

It is then a further technical question how to measure such a sample. Figure 6.30 shows two photos of a setup that I had recently purchased in Switzerland and used

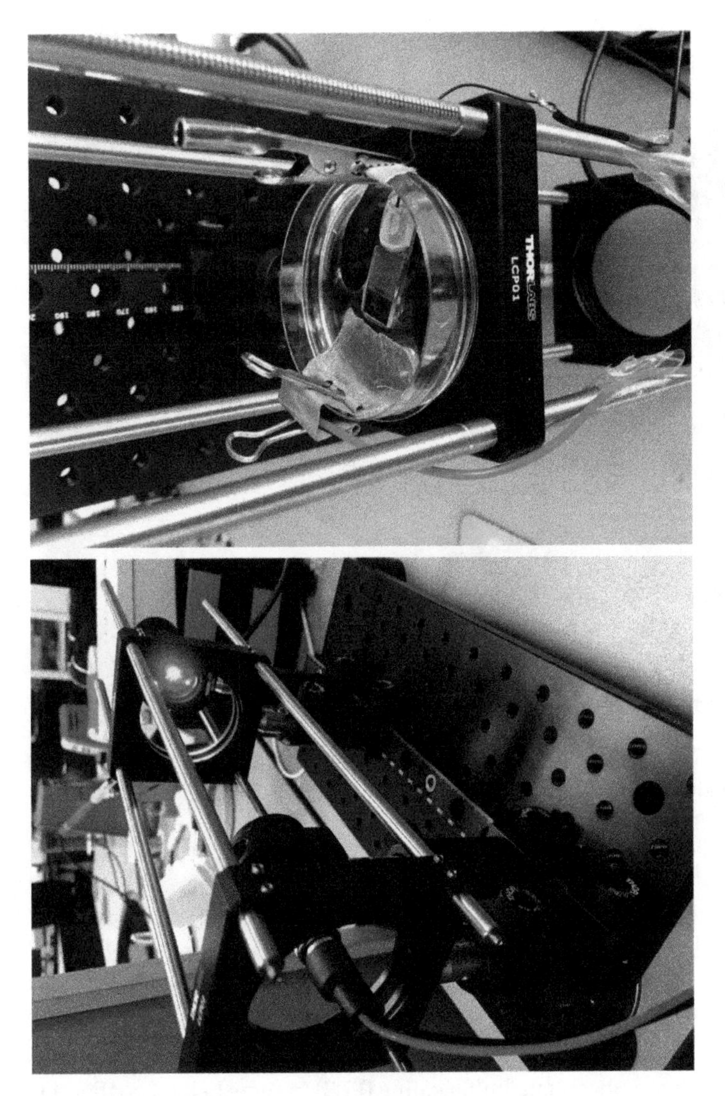

Fig. 6.30: Red LED shining on a photodiode. Photo taken by Artur Braun.

during my time at Yonsei University in Korea. On the left-hand side you can see a small optical bench with a red light-emitting diode (LED) fixed to it. The LED is switched on, as you can see. The red and black cable on the left-hand side are from a photo diode that is just receiving part of the red light from the LED and delivering a photovoltage to a meter. The optical bench allows us to fix the distance of the photodiode from the LED and, thus, to tell the photovoltage as a function of the distance from the LED. The light flux decreases with $1/R^2$, when R is the distance between the photodiode and the LED.

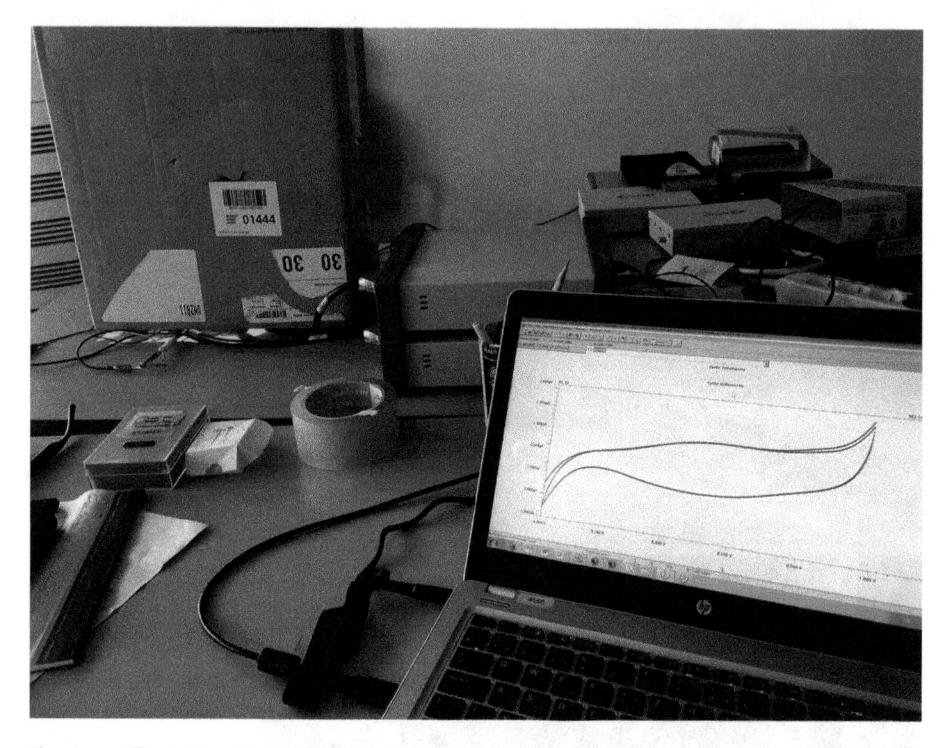

Fig. 6.31: Office table with potentiostat, laptop, and the photoelectrochemical experiment with the thylakoid sample under the cardboard box, from which some of the green light from the LED shines through. Photo taken by Artur Braun.

On the right-hand side you can see the very same optical bench, but it is turned up by 90°. This enables the LED to shine downwards on a plastic cap that contains thylakoid electrode and perpendicularly point the light onto it. The photodiode is removed because it is no longer necessary. The red cable is connected to a platinum mesh as a counter electrode. The black wire is connected with the thylakoid electrode. The plastic cap is filled with phosphate buffer saline (PBS) to some height, so that the thylakoid film is fully covered with electrolyte.

As the sample is thus exposed to the light from the LED and connected to a potentiostat, it is necessary to control the illumination properly. This also requires that carries out the electrochemical experiment under dark conditions and light conditions. Figure 6.31 shows a photo of my office where I had the equipment assembled to do such an experiment. On the upper left-hand side of the photo, you can see a cardboard box, which I put over the optical bench. At the bottom of this box, you can see the green light from the green LED shining through.

When the box covers the entire experiment, and the LED is not switched on, then you can record electrochemical data in the dark. You need the box for darkening the environment, because otherwise the light coming in through the window, or the light from the room lamps, will make a photocurrent in the thylakoids. You will notice the effect of light on the currents by checking out every situation and reading the current signal.

This experiment is certainly not optimized in any respect, but you can see for yourself how one can set up such experiments to get started. The entire experiment is computer controlled, as the thylakoid electrode and the LED are operated by two linked potentiostats.

As an example, I can show you the photocurrent transients obtained for four different wavelengths. For this experiment, we used two potentiostats. One potentiostat is used to power one LED with a current of 600 mA. The potentiostat is computer controlled, so that the LED is switched on and off every, say, 60 seconds. The other potentiostat is used to run the electrochemical cell with the thylakoid electrode in a chronoamperometry mode, say, with a DC bias of 400 mV for a time of 10 minutes.

The four data panels are shown in Figure 6.32. The top figure with the red curve was obtained with the red LED with a 625 nm wavelength. The LED was driven with 600 mA from one of the two potentiostats as power source for the LED, which was programmed to switch the LED off and on every 80 seconds. The current through the electrochemical cell was recorded, and the thylakoid electrode was manufactured from a thylakoid solution with 10 μL concentration. The electrolyte was PBS, and the DC bias imposed by the second potentiostat was 400 mV.

We are considering the red curve. At $t = 0$, the current is very high, exceeding 1×10^{-6} ampere, but it drops significantly in the first 60 seconds. After the first 60-second interval is over, the red LED is switched on, and there you can see that the current takes a small step upwards. This is the onset of the photocurrent. When, after another 60 seconds, the LED is automatically switched off, the current drops by a small step but continues to flow over time. So, with an interval of 60 seconds, the dark current and the photocurrent take turns, so to speak. After around 400 seconds, the overall current flattens out, but the pattern of switching off/on the LED is reproduced by the current response.

Below the red curve, you can see the grayish curve, this is the current transient of an electrode with no thylakoids on it, i.e., the blank sample. Other than the sharp decay of the current versus time, we see no fine structure, unlike with the thylakoid electrode. We can see, however, that even in the dark, the current by the thylakoid electrode is larger than with the blank only.

Now we move to the panel with the orange curve. This is the thylakoid current when the orange LED with 590 nm was used. The overall current is lower than with the red LED, but we can still see the step-fine structure between dark current and photocurrent. The current of the blank seems to be the same as with the red LED.

The green curve obtained with the LED with 530 nm shows a large and noticeable difference between the dark current and the photo current. It is interesting, however, that also the current for the blank, which contains no thylakoids, shows the fine structure originating from the light switching on and off. This effect is particularly strong

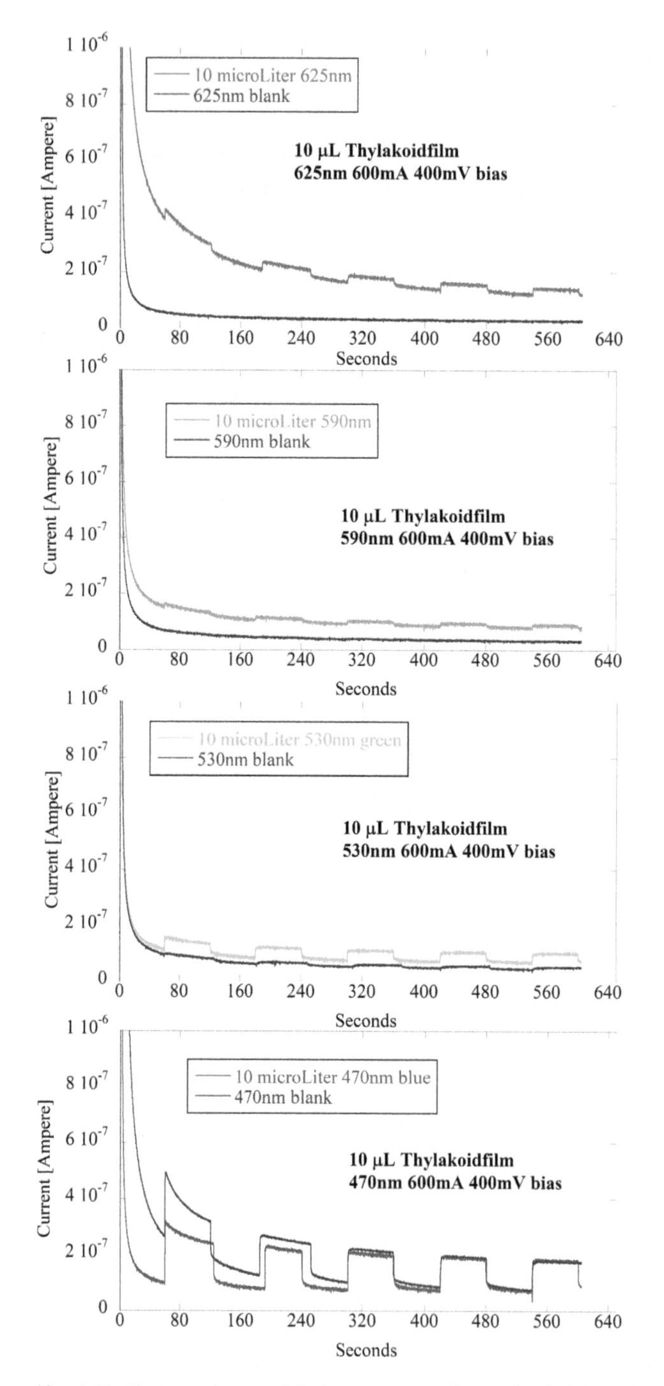

Fig. 6.32: Photocurrents and dark current transients of a thylakoid film grown from 10 μL concentration, recorded under illumination by LED wavelengths of 625, 590, 530, and 470 nm. The gray curve is the dark current and the photocurrent of the gold film on the substrate without thylakoids as the reference (blank). Photo talem by Artur Braun.

when the blue light with 470 nm is applied. Both the thylakoid current and the blank current are very high when the light is on. You may wonder why the blanks show a photocurrent, but we know that the photoelectric effect is measured in metals. The high-energy photons from green and blue light produce a correspondingly high current from the inorganic photoelectric effect. The lower-energy photons from orange and red light cannot produce such a photoelectric current effect.

While the thylakoid current shows an overall exponential decay, we are interested in the difference between the dark current and the photocurrent. Practically, you then apply an exponential fit function to the thylakoid curve and later subtract this very fit function from the experimental data. This is shown in Figure 6.33, which in the top panel shows the thylakoid current of a thick sample that was produced from a

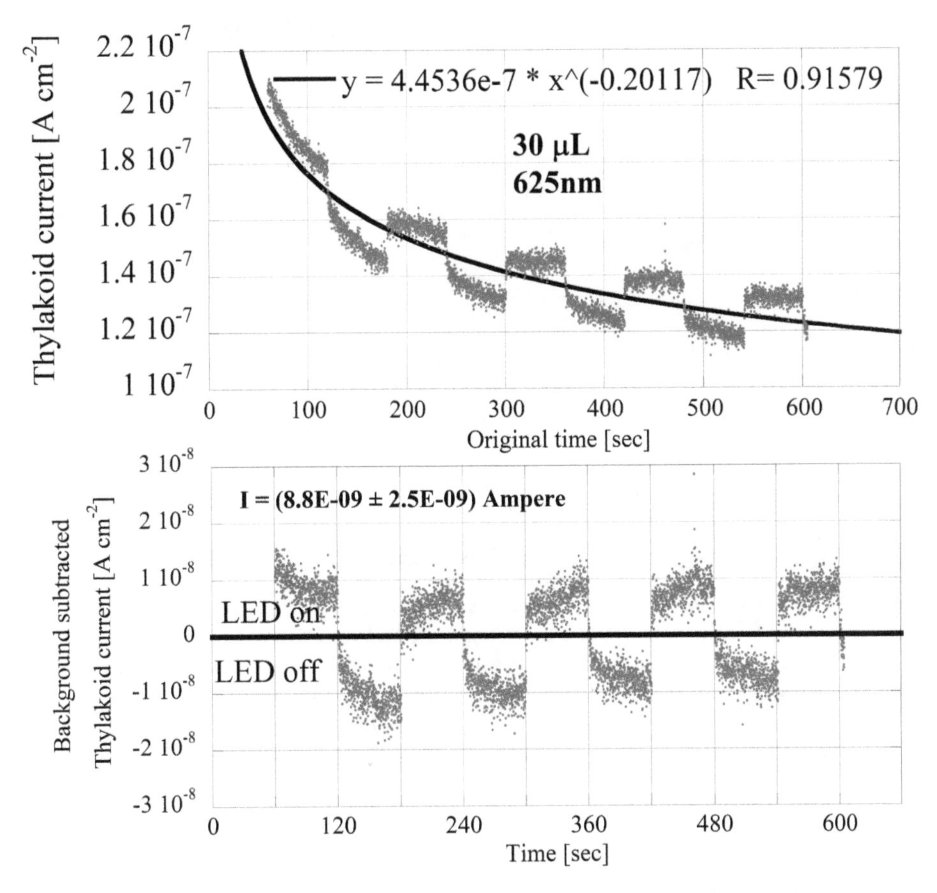

Fig. 6.33: Photocurrent and dark current transients of a thylakoid film grown from 10 µL concentration, recorded under illumination by LED wavelengths of 625, 590, 530, and 470 nm. The gray curve is the dark current and the photocurrent of the gold film on the substrate without thylakoids as the reference (blank). Photo taken by Artur Braun.

concentration with 30 μL thylakoid solution. The black solid line is a least-squares fit with a transient of $I(t) = 4.45\text{ E}{-07} \times t^{-0.2}$, which will be subtracted from the red data points.

The bottom panel in Figure 6.33 shows the result of this subtraction of the thylakoid current background. The top five segments denote the light current, because these were recorded with the LED on. The bottom four segments denote the dark current. Their absolute difference is, then, the photocurrent and yields roughly $2 \times$ E–08 ampere, around 20 nA.

When you plot the thus obtained photocurrents of the thylakoid films over the wavelength of the LEDs we used, you will get a graph like in Figure 6.34. Our approach to obtain these results is quite laborious. What we measure here is the incident photon to the current efficiency (IPCE). I used four wavelengths from four LEDs. Typically, one uses a monochromatic light source where the wavelength can be scanned while the photocurrent is recorded. The result is known as the quantum efficiency.

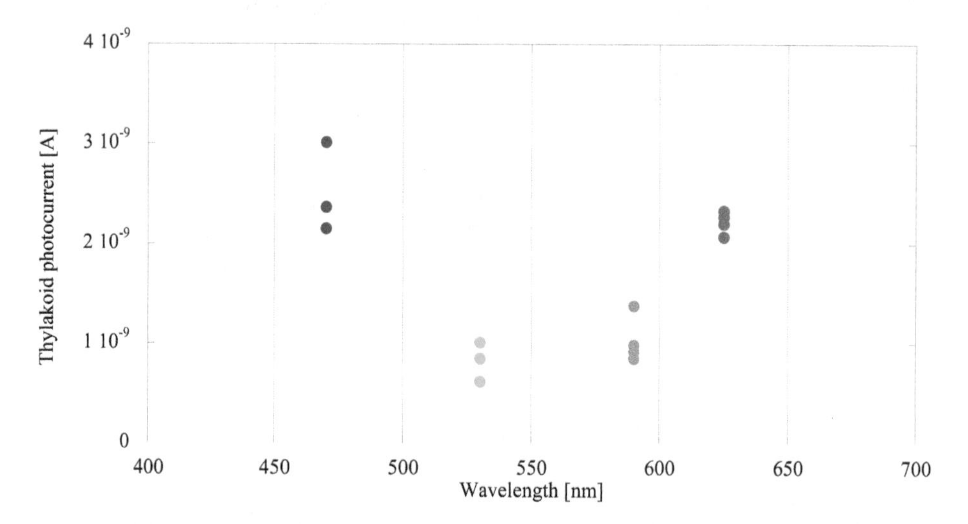

Fig. 6.34: Photocurrent and dark current transients of a thylakoid film grown from 10 μL concentration, recorded under illumination by LED wavelengths of 625, 590, 530, and 470 nm. The gray curve is the dark current and the photocurrent of the gold film on the substrate without thylakoids as the reference (blank). Photo taken by Artur Braun.

6.5 The mesoscopic aspect of quantum electrodynamics

We realize now that the photosynthetic apparatus has a finite size on the submicrometer scale. The fluorescence resonance energy transfer FRET is an effect that takes place at a larger distance than the electron transfer. Electron transfer, including tunneling, takes place on the molecular scale in few Ångstroms only. FRET takes place in the multinanometer, at the mesoscopic range, which, however, is still so short that no electromagnetic wave may evolve and a photon becomes generated (read the remark about Heitler (Heitler 1947) in Papageorgiou and Govindjee (2004): "Electromagnetic energy departing from a donor cannot be considered as a real photon until the distance from the emitting species is approximately the length of a wavelength of light." Speaking of tunneling, would the same principle apply and we can consider a charge an electron, or a proton, until the distance from the donor or emitting species is approximately some relevant de Broglie wavelength? This is an exercise for the reader.

We can wonder now how the electron –photon interaction can be described in the mesoscopic range in the framework of quantum electrodynamics. Cottet et al. (2015) worked out a scheme for this problem with a view on photonic device technology. They consider the electrons in a cavity and not in an electronic conductor ("medium"). The technical arrangement is sketched in Figure 6.35, with the "housing" being a cavity conductor for photons. It allows photons to pass through and interact with an electronically conducting arrangement of electrodes. The photons are indicated as a pale yellow light in the cavity.

The conductor itself is electronically conducting, of course, and connected with the electric ground. Inserted into the cavity is a Y-shaped "nanocircuit", which is built from a finite number of atoms, or quantum dots. This is basically a Y-shaped nanoparticle. This is electrically contacted by three electrodes inserted into the cavity, which operate as the source (the blue pin on the left at potential V_1) and two drain electrodes (two blue pins middle and right). Two electrostatic gates (green pins at potentials V_2 and V_3) are inserted into the cavity between the source and the drain electrodes. This arrangement is reminiscent of an electron vacuum tube (a triode, a pentode actually).

Directly opposite to these two gate pins are what Cottet et al. call "protrusions", barriers (purple pins coming from the cavity conductor) that serve as inhomogeneities to perturbate the photonic field to be sent into the cavity. The authors (Cottet et al. 2015) indicated this by a stronger yellow color, in proximity to the nanocircuit. The nanocircuit is the operating core in this mesoscopic device. At such a small size range, tunneling of charges becomes possible and may be considered as in the "concert" of interactions of photons with matter. As the nanocircuit is not much more than a nanoparticle, the number of atoms is relatively small, the electron density is, thus, rather small, and we cannot expect conventional bulk condensed matter physics to be dominant.

There may be some tendency by the nanocircuit to develop plasmon modes from conduction electrons, in addition to the tunneling of electrons. As the nanocircuit is

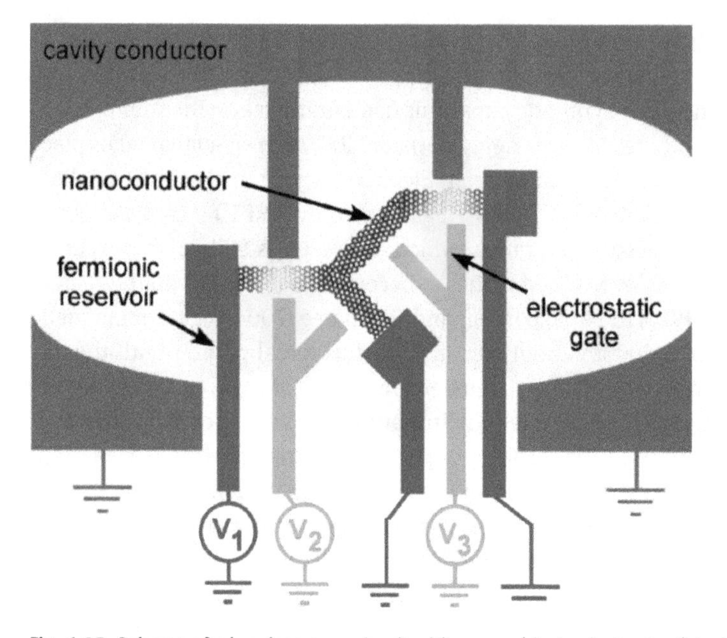

Fig. 6.35: Scheme of a loopless nanocircuit with source/drain electrodes (blue) and electrostatic gates (green) embedded in a photonic cavity (purple). The yellow cloud represents the photonic field. The cavity presents some protrusions (purple stripes) fabricated to increase the photonic field inhomogeneities (darker yellow areas). Reprinted figure with permission from A. Cottet, T. Kontos, B. Doucot, Electron-photon coupling in mesoscopic quantum electrodynamics, Physical Review B 91, 205417. http://dx.doi.org/10.1103/PhysRevB.91.205417. Copyright (2015) by the American Physical Society, (Cottet et al. 2015).

electrically connected to metallic current collectors, Cottet et al. (2015) assume it is reasonable that the dynamics of the plasmonic modes is faster than the dynamics of the tunneling electrodes and modes in the cavity. This difference in speed is accounted for by a capacitive element C in parallel to an ohmic element R, so that we basically consider this nanocircuit and its ambient as a quantum RC circuit in a cavity. In the extreme case, the nanocircuit is made just by a quantum dot coupled to the cavity.

The total Hamiltonian $H_{RC+\mathrm{cav}}$ of such a quantum dot circuit in a cavity is built by the contributions from the quantum dot, H_{e}, the electron charge reservoir/bath H_{res} in the metal current collectors, a coupling term h_{int} between circuit and cavity, and a radio frequency term H_{RF} when a microwave ωRF is fed into the cavity. Finally, the cavity has a renormalized electromagnetic mode $\hbar\omega_0\hat{a}^\dagger\hat{a}$. The interaction term h_{int} depends on coupling constants γ, which might contain signatures visible in the response of the cavity device (Cottet et al. 2015).

Such a kind of response is mathematically formulated as a response function. In the physical world, a sensor or a transducer represent such a response maker. When you address or excite a system with an information input, you get a response as information output. The signal may be a signature of the transducer or sensor. You can use this approach for the investigation of the interior functional components in a complex system by measuring it – by introducing a signal and recording the response.

Impedance spectroscopy is such a method for measuring dynamic systems, such as electric circuits and mechanic structures, including the human body, such as the lungs and breathing system. The mathematical description of such a complex system is given by the response function. Important are responses that are linear, and, thus, we are looking for a linear response function. Let us consider, therefore, a harmonic oscillator with a damping element, which we know from classical mechanics. Let $x(t)$ be the output variable and a force $F(t)$ be the input variable:

$$\ddot{x}(t) + \gamma\dot{x}(t) + \omega_0^2 x(t) = F(t)/m \tag{6.4}$$

The Fourier transform of the linear response function is given as

$$\widetilde{\chi}(\omega) = \frac{\widetilde{x}(\omega)}{\widetilde{F}(\omega)} = \frac{1}{\omega_0^2 - \omega^2 + \mathrm{i}\gamma\omega} \tag{6.5}$$

and is particularly useful when the input signal $F(t)$ is given as a wavefunction, such as a sine or cosine, or an exponential with complex argument. This is the reason why a sine signal is used for the excitation of the system in impedance spectroscopy. A recently developed instrument for solar cell studies is the IMPS[10] method, which modulates the intensity of the light with a sine perturbation. I had been wondering for quite some time whether an additional dynamic excitation could be applied in parallel to impedance spectroscopy, but did not come up with one[11]. A mathematical solver of the signal transduction problem is Green's function. Green's function for the problem shown in this section is given as the exact relation, (Cottet et al. 2015), by

$$\widetilde{G}_{\hat{a},\hat{a}^\dagger}(\omega) = \widetilde{G}_0\omega + \frac{1}{\hbar^2} G_0(\omega)\widetilde{G}_{\hat{h}_{\mathrm{int}}\hat{h}_{\mathrm{int}}}(\omega_{\mathrm{RF}} G_0(\omega)\,. \tag{6.6}$$

Cottet et al. (2015) find, after a number of considerations and approximations

$$\widetilde{G}_{\hat{a},\hat{a}^\dagger}(\omega_{\mathrm{RF}}) \simeq \frac{1}{\omega_{\mathrm{RF}} - \omega_0 + \mathrm{i}\Lambda_0 - \frac{G_{\hat{h}_{\mathrm{int}}\hat{h}_{\mathrm{int}}}(\omega_{\mathrm{RF}})}{\hbar^2}} \tag{6.7}$$

---

10 intensity modulated photocurrent spectroscopy and intensity modulated photovoltage spectroscopy.

11 I once had the pleasure to speak at the Ertl Conference in Stuttgart, where I presented the first *operando* NEXAFS study on PEC water splitting, which we had done at the Advanced Light Source in Berkeley in 2011 (Braun et al. 2012). Professor Ertl asked me whether it would be possible to conduct such study with different light intensity. Today, I know that he referred to IMPS, a method first published, to the best of my knowledge, in 1983 by Li, Peat, and Laurence "Laurie" M. Peter at University of Southampton (Li et al. 1984). To date Peter is still developing this method further (Peter 2019). The photoelectrochemical equipment that I used during my sabbatical at Yonsei University is such an IMPS setup, which uses one potentiostat to control and modulate the light intensity, and the other potentiostat is for the impedance spectroscopy.

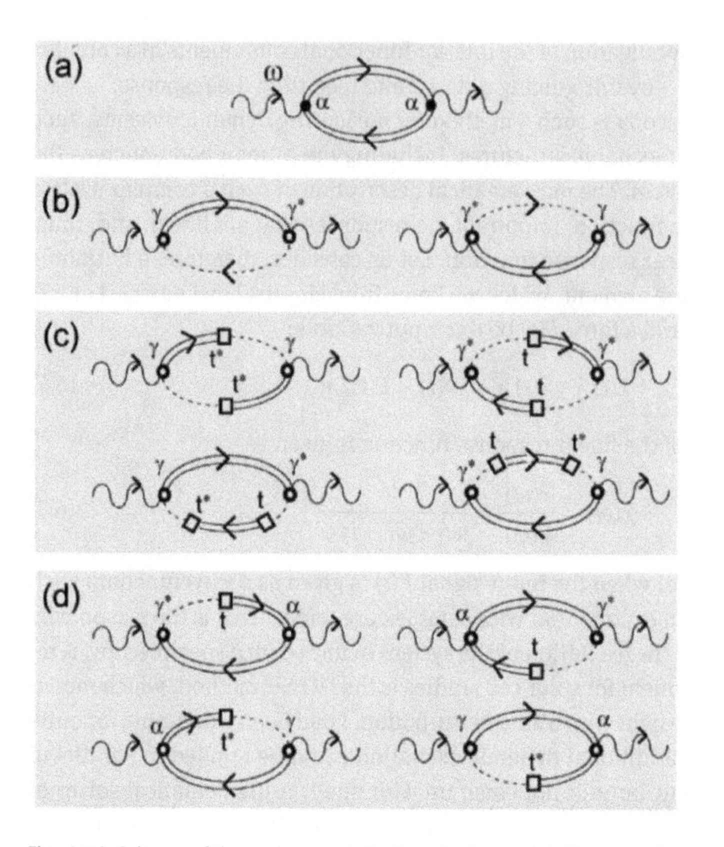

Fig. 6.36: Scheme of the various contributions to $G_{h_{int}h_{int}}(t)$. The wavy lines correspond to photonic propagators, the simple dashed lines to bare electronic propagators in the normal metal reservoir, and the double full lines to electronic propagators in the quantum dot dressed by the dot/reservoir tunneling processes. Reprinted figure with permission from A. Cottet, T. Kontos, B. Doucot, Electron-photon coupling in mesoscopic quantum electrodynamics, Physical Review B 91, 205417. http://dx.doi.org/10.1103/PhysRevB.91.205417. Copyright (2015) by the American Physical Society, (Cottet et al. 2015).

The above expression is valid under the conditions that τ and n_0 do not depend on the energy and that both constitute purely cavity damping. Moreover, the constant is $\Lambda_0 = \pi\tau^2 n_0/\hbar \ll \omega_0$. The propagator algebra then provides that the quantity $G_{\hat{h}_{int}\hat{h}_{int}}(\omega_{RF})$ falls apart into 11 contributions, which are shown as Feynman diagrams in Figure 6.36. The elements in these Feynman diagrams are termed as follows. The wavy lines that look like photons are the photonic propagators. They are defined by the frequency ω. The dashed lines that look like fermions are the electronic propagators in the fermion metal reservoir, the current collectors. The double solid lines stand for the electronic propagators (fermionic) that are in the quantum dot, and those are "dressed" by the tunneling processes between the quantum dot and the fermion reservoir. These three types of processes combine in the cavity device.

You can try to write the Feynman diagram for the fist case in Figure 6.36, (a), by using the TikZ Feynman package (Ellis 2017) in LaTeXwith the command

```
\feynmandiagram [layered layout, horizontal=b to c]{
a -- [photon, edge label=\(\omega\)] b
[dot] -- [fermion, edge label=\(\alpha\), half left] c
[dot] -- [fermion, half left] b,
c -- [photon] d,
};
```

Cottet et al. use a double arrow in their Feynman diagrams in Figure 6.36, which I believe is not necessary. It is not clear to me whether such a double arrow can be reproduced in the TikZ Feynman package (Ellis 2017), but we can "fake" a double arrow and also "squeeze" the circle a little, so that the Feynman diagram better resembles the original in the paper by Cottet et al. (2015) by using an alternative command (the *momentum*) from the TikZ package

```
\feynmandiagram [layered layout, horizontal=b to c] {
a [particle=\(\omega\)] -- [photon] b
[dot] -- [fermion, half left, looseness=1, momentum=\(\alpha\)] c
[dot] -- [fermion, half left, looseness=1, momentum=\(\alpha\)] b,
c -- [photon] d,
};
```

as you will see in Figure 6.37.

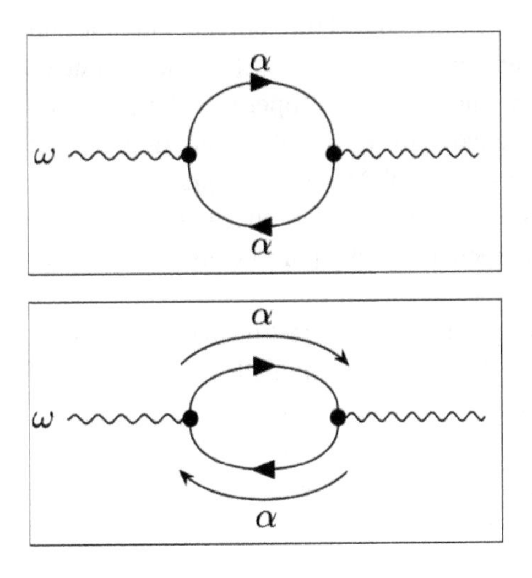

Fig. 6.37: TikZ Feynman diagrams of contributions to the response function of electronic propagators, inspired by Cottet et al. (2015).

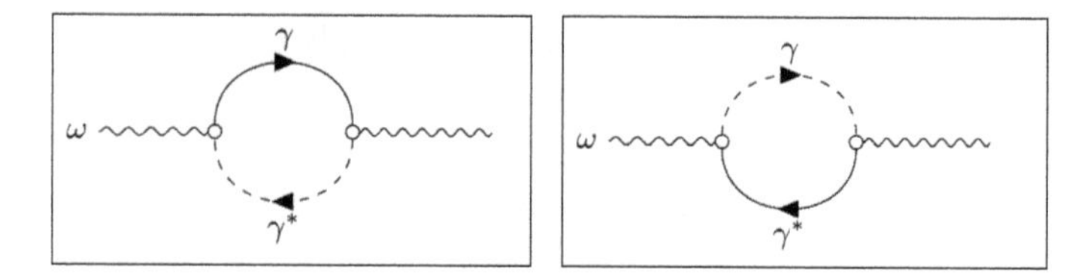

Fig. 6.38: TikZ Feynman diagrams of contributions to the response function of electronic propagators.

The TikZ-code for Figure 6.38 for the up and down versions reads

```
\feynmandiagram [layered layout, horizontal=b to c]{
a [particle=\(\omega\)] -- [photon] b
 [empty dot] -- [charged scalar, edge label=\(\gamma\), half left] c
[empty dot] -- [fermion, half left, edge label=\(\gamma^{\ast}\)] b,
c -- [photon] d,
};
```

and for the upside-down version

```
\feynmandiagram [layered layout, horizontal=b to c]{
a [particle=\(\omega\)] -- [photon] b
 [empty dot] -- [fermion, edge label=\(\gamma\), half left] c
[empty dot] -- [charged scalar, half left, edge label=\(\gamma^{\ast}\)] b,
c -- [photon] d,
};
```

The response function in Equation 6.7 allows the analysis of the coupled photonic and electrodynamic properties of the arrangement shown in Figure 6.35, the “mechanism” of the cavity nanocircuit device. One such “mechanistic” property is the damping. When we consider the limit $\omega_0 \ll \Gamma, D$, we can use $G_{\hat{h}_{\text{int}}\hat{h}_{\text{int}}}(0)$ in Equation 6.6 and calculate what is known as the cavity frequency pull $\Delta\omega_0$ in laser physics and the corresponding cavity damping pull $\Delta\Lambda_0$. Cottet et al. calculate those, and the result is plotted for three different choices for nanocircuit cavity parameters, as shown in Figure 6.39.

We should be aware that such a nanocircuit cavity has a technical or even a technological functionality. As Cottet et al. correctly state, it is technically possible to manufacture such circuits, but to the best of my knowledge, there exists no technology yet where such devices are utilized. The purpose of dedicating a section to this nanocircuit cavity device was to show how photonics and electrodynamics can interact at the mesoscopic scale, which is also relevant for the photosynthetic apparatus.

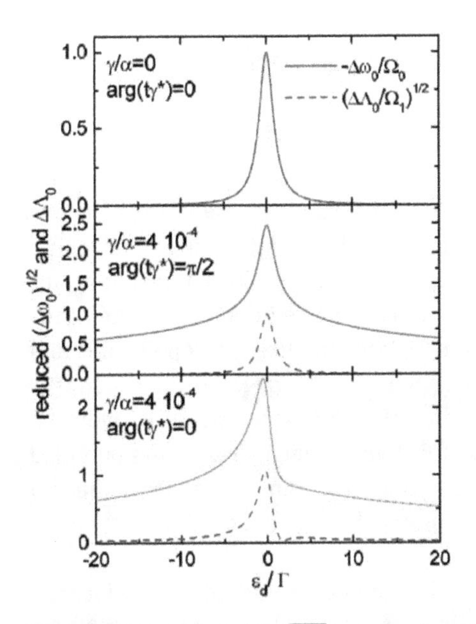

Fig. 6.39: Reduced $\Delta\omega_0$ and $\sqrt{\Delta\Lambda_0}$ versus ϵ_d for different values of γ. We used $t/\Gamma = 0.001$, $D = 20\Gamma$, and the pulsation scales $\Omega_0 = 2\alpha^2/\pi\Gamma\hbar$ and $\Omega_1 = \hbar\omega_0\Omega_0/\Gamma$. Reprinted figure with permission from A. Cottet, T. Kontos, B. Doucot, Electron-photon coupling in mesoscopic quantum electrodynamics, Physical Review B 91, 205417. http://dx.doi.org/10.1103/PhysRevB.91.205417. Copyright (2015) by the American Physical Society.

7 The light-driven proton pump: bacteriorhodopsin

7.1 The rhodopsins

The rhodopsins are a class of proteins that absorb light and use the light energy to carry out photochemical reactions. The rhodopsins fall into two groups, one of which are the visual rhodopsins and the other of which are bacterial rhodopsins. The visual rhodopsins can convert light into electric nerve impulses and are, thus, important for the vision apparatus in primitive animals like arthropods and mollusks, and more highly developed animals, such as vertebrates. Visual rhodopsins thus have an organic function. Bacterial rhodopsins also have a function in photosynthesis, as they convert light energy into electrochemical energy via proton pumping. Bacteriorhodpsin (bR) is the most studied bacterial rhodopsin. For details on the photophysics and molecular electronic applications of rhodopsins, see the review paper by Robert R. Birge (Birge 1990).

The number of different rhodopsin protein complexes is large, but they all share some common structural components. The light absorber molecule is a chromophore, which is called a retinal. It is large enough so that it can isomerize upon optical excitation. The retinal chromophore is located in the center of seven DNA helices, which are called transmembrane helices.

One of the most primitive ways in nature to convert light to chemical energy is done by archea, such as the *Halobacter halobium* and *Halobacter salinarum*. They contain bacteriorhodopsin which has a retinal group. When the retinal absorbs light, it will change its shape, which is known as photoimerization from trans-retinal to cis-retinal. This change in shape will cause the retinal to release a proton H^+ to the outside of the cell. This will cause the retinal to switch back and pick up a proton from the cell interior. One absorbed photon $h\nu$ works one proton H^+ like a proton pump (Lanyi 1993). The photofunction in bacteriorhodopsin is, thus, not a result of photon interaction with protons.

The so described bacteriorhodopsin thus has optical and electrical properties that can be considered as electro-optical elements and analyzed according to conventional electrostatics and electrodynamics (Porschke 1996). The structural changes due to the photoisomerization in the proton pumping are considerable and involve switching processes that extend over 10 Å (Varo et al. 1992).

Ahmadi et al. even suggested using bacteriorhodopsin – sandwiched in polyimide between polyethylene and an indium tin oxide layer – as an x-ray sensor for in-vivo x-ray measurements on patients (Ahmadi et al. 2012). Bacteriorhodopsins are in the outer of the plasma membrane of the cell. The membrane has a purple color from the bacteriorhodopsin, and this color causes ponds that are rich in such bacteria to have a reddish color. The bacterial colonies grow particularly well in warm, salty and sulfur-rich waters (see also Wang et al. 2017).

https://doi.org/10.1515/9783110629941-007

Fig. 7.1: Owens Lake in California Mojave Desert. Photo taken 28 April 2014 by Artur Braun.

Figure 7.1 shows Owens Lake in California, which I photographed on 28 April 2014 on my way to Death Valley. Owens Lake contains a lot of salt and other minerals, and sulfur, and this is used by bacteria (Pikuta et al. 2003; Ryu et al. 2006) that make it a reddish or purple color. You can identify the lake by the purple color when you pass by along the California State Route 190.

There are plenty of waters that allow for growth of purple bacteria populations. Figure 7.2 shows an aerial view of the San Francisco South Bay with the Dunbarton Bridge, which connects East Palo Alto with Fremont. The purple color in some of the water originates from the purple color of the membrane of *Halobacter halobium*. This is a purple archeon and lives in the salt marshes in the San Francisco Bay area.

The dynamics of hydrogen pumping can be measured at the molecular level by employing quasi elastic neutron scattering (QENS). The scattering contrast between neutrons and protons is large, and thus, neutrons are very good probes for protons and hydrogen. Disorder caused by diffusion and migration of protons can thus be very well resolved in the quasi-elastic broadening of neutron spectra, a neutron method that is very well established (Hempelmann 2000).

Pieper et al. developed the QENS method further over several years, so that increasingly sophisticated experiments on bacteriorhodopsin became feasible and also fruitful for deriving biophysical relevant information, the results of which were published in a series of papers. One important step (Combet et al. 2008) was to measure the light harvesting dynamics of an antenna protein phycocyanin, where the optical

Fig. 7.2: Aerial view of the San Francisco South Bay and San Mateo. The purple color in some of the water originates from the purple color of the membrane of *Halobacter halobium*. The white buildings in the lower left corner are the Facebook company. Photo taken 20 April 2017 by Artur Braun.

stimulation with laser light was synchronized with the inelastic neutron scattering (INS). This is possible when the laser is pulsed and the neutron beam is pulsed. The incoherent INS scattering reflects the density of states G of the vibration modes of the protons, which can be written as

$$G_{\text{incoherent}}(Q,\omega) = \frac{\omega^2}{Q^2} S_{\text{incoherent}}(Q,\omega) \tag{7.1}$$

with the incoherent dynamical structure factor S, which can be determined by the INS (Combet et al. 2008).

Bacteriorhodopsin (bR) is a membrane protein, which, by absorption of light, can build up a proton concentration gradient and, thus, produce energy for the production of ATP-synthase (adenosin tri-phosphate, ATP). Bacteriorhodopsin contains chromophores that absorb the green light in the wavelength range of 500 to 650 nm, with the absorption maximum at 568 nm, and can be considered a genuine proton pump (Lanyi 1993; Balasubramanian et al. 2013). Prior to light absorption, the chromophores are present as an isomer mixture of cis and transconfiguration

Upon light absorption, the trans-configurations become converted into cis-configurations, which cause morphological changes in the protein with an energetic imbalance of protons, which, in turn, influences the protonation of functional Schiff bases[1] in the protein. The proton is moved in the extracellular direction to regain energetic balance.

Further proton shifts follow before the protein regains its initial configuration (= morphological) and energetic state, ready for another such proton pump action. These steps are characterized by protein states that can be determined and distinguished by optical spectroscopy, and, thus, are called the "photocycle", see Figure 7.3. The photocycle is indicated by the corresponding color changes of the bR. The bR complex is built in the thylakoid membrane, like many other photosynthesis proteins.

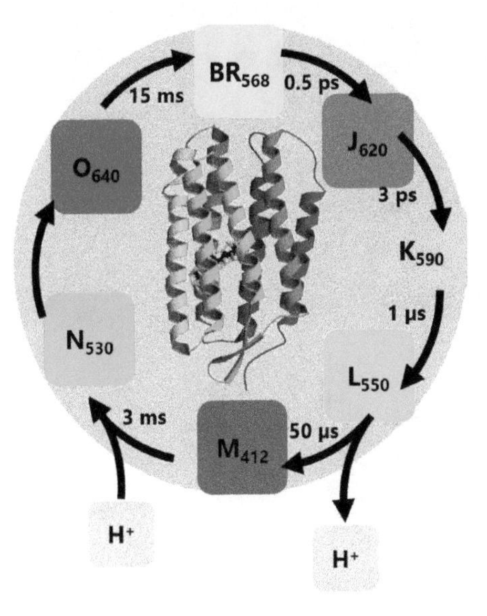

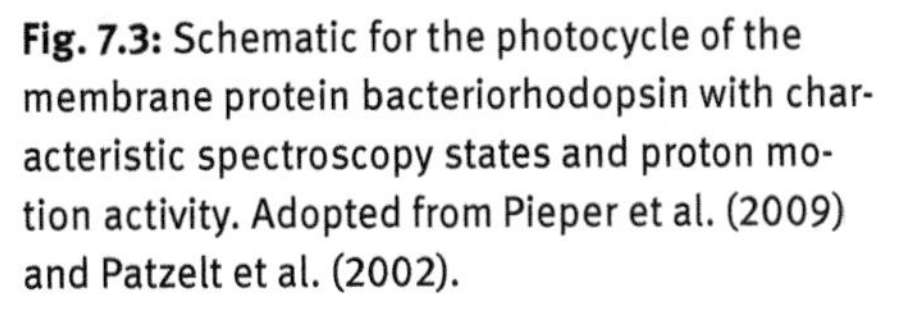

Fig. 7.3: Schematic for the photocycle of the membrane protein bacteriorhodopsin with characteristic spectroscopy states and proton motion activity. Adopted from Pieper et al. (2009) and Patzelt et al. (2002).

The bR is built from seven α-helices, which contain one retinal protein $C_{20}H_{28}O$. The retinal is a chromophore and belongs to the carotinoids, the A vitamins, which we know are important for a healthy vision apparatus. bR is a transmembrane protein, which means it is embedded in the lipid double-layer membrane (the purple mem-

1 A Schiff base, named after Hugo Schiff, has the general structure R2C=NR′, where R′ ≠ H. Note the C=N double bond characteristic for the Schiff bases (Schiff 1864).

brane PM, which has a characteristic purple color when extracted and dried as powder.) and allows for transport between lumen and stroma. The PM also has per one bR unit ten haloarcheal lipids with side chains, which have an identical saturation but different polarity heads.

bR is present in many biological systems and is the only protein in the purple membrane. In halobacteria, bR is a light harvesting protein (compare, for example, the other light harvesting protein phycocyanin). The net effect of the bR reaction is that one photon $h\nu$ comes in, and one proton p^+ goes out and leaves the cell. The proton transport in the bR can be understood as a proton wire[2], which is quantum mechanically treated in Section 7.2.4. A biomimetic analogon in a fuel cell is shown in the paper by Leem et al. (2008).

The structure of the retinals in Figures 7.4 and 7.5 can be found, for example, in the database[3] of the European Bioinformatics Institute, which is part of the EMBL, the European Molecular Biology Laboratory (Brooksbank et al. 2005). ChEBI stands for chemical entities of biological interest.

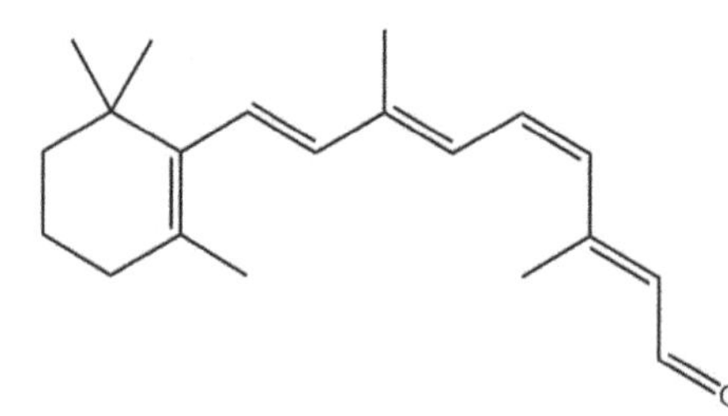

Fig. 7.4: Structure formula of the 11-cis-retinal.

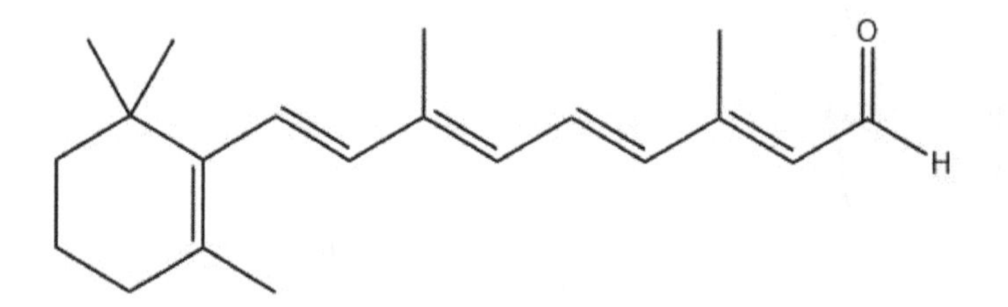

Fig. 7.5: Structure formula of the 11-trans-retinal.

2 At the astronomical scale, there exists a stream of neutral hydrogen H_2 between the two Magellan clouds. The stream is also called the Magellan current, although here we are dealing with neutral hydrogen atoms and not protons. The two clouds approach each other with a speed of 55 km/s. The distribution of matter in the universe can be studied with spectroscopy, see, for example, Prochaska et al. (2015) and Hindman (1967).

3 www.ebi.ac.uk/chebi/.

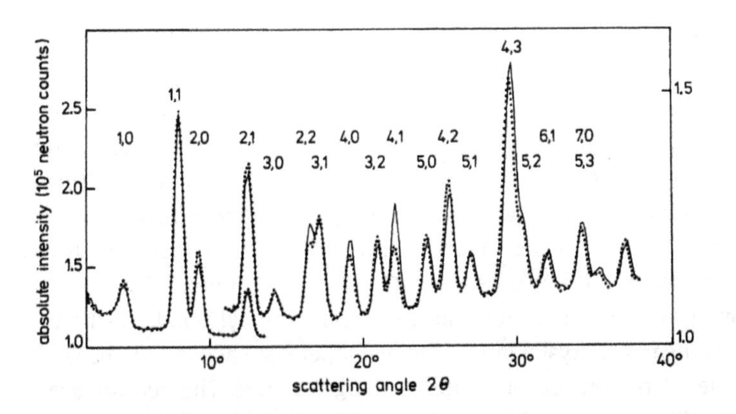

Fig. 7.6: Neutron intensities of purple membrane stacks in D_2O of the light-adapted ground state BR-568 (-) and of the *M* intermediate of the photocycle (. . .) at 93 K. Neutron counts as a function of the detector angle 2θ are shown. Reflections are indexed according to a hexagonal lattice (reproduced from Dencher et al. 1989 with kind permission from the Prof. Dencher).

The aforementioned conformational changes in proteins occur during exposure to light (which we term light activation) and cause changes in the proton transport properties. Such conformational changes can be better detected by neutron diffraction rather than by x-ray diffraction, because of the large scattering contrast between proteins and associated water. Using deuterated water D_2O allows for enhanced scattering contrast in neutron diffraction. Figure 7.6 shows neutron diffractograms of bacteriorhodopsin recorded at 93 K low temperature[4] in two different photo excited states, i.e., bR-568 (568 is the wavelength in nm, bacteriorhodopsin is the solid line) and at an intermediate position in the photocycle, referred to as state M. The Bragg reflections are indexed versus a hexagonal lattice. We notice minute but clearly distinguishable changes in the peak heights for the (2,0), (2,1), (2,2), (3,1), (4,0), (4,1), and (4,2) indexed Bragg peaks, which are interpreted as structural signatures for conformational changes (Dencher et al. 1989).

Since protons are being moved throughout the bacteriorhodopsin transport channels, it is possible to record protonic "photocurrents" in bacteriorhodopsin electrode assemblies with electroanalytical techniques (Horn and Steinem 2005). Using electrochemical impedance spectroscopy, it should, therefore, be possible to study the proton dynamics and charge transfer in bacteriorhodpsin electrode assemblies (Horn and Steinem 2005). Electrode assemblies have been made, such as bacteriorhodopsin/TiO_2 nanotube-array hybrid systems' (Allam et al. 2011, see also Balasubramanian et al. 2013) for solar photoelectrochemical water splitting.

4 Low temperature. High temperature. It is always good to disclose the actual temperature of an experiment, especially in the absolute temperature scale Kelvin. If a temperature is low, or high, depends on the observer. A colleague of mine used to work in a low-temperature institute, and he told me that 1 K was for them almost infinitely high.

For the investigation of ionic charge carrier dynamics at the molecular scale, quasi elastic neutron scattering has been proven to be a suitable method (Chen et al. 2013, 2012). I show here an example of a comparable system, that is, ceramic proton conducting electrolytes for intermediate temperature solid oxide fuel cells (Kaur 2019). In this particular case, we have a ceramic membrane of yttrium-substituted barium cerate, such as $BaCe_{0.9}Y_{0.1}O_3$ with a ABO_3-type perovskite structure, which contains engineered oxygen vacancies, which will be filled with oxygen ions from water molecules when the membrane is exposed to water vapor or humidity.

The two protons from the water molecule will form hydroxyl bonds O–H with the adjacent oxygen ions in the crystal lattice. Upon temperature activation, the O–H bonds break, and the H^+ protons become mobile charge carriers. The neutron scattering spectrum typically shows a peak with a finite width, which originates from the elastically scattered neutrons from the atoms in the crystalline structure of the probed sample. The temperature activated mobility of the protons causes a diffusive proton motion, which is visible in the neutron spectrum as a broadening of the peaks, the quasi elastic scattering, as demonstrated in Figure 7.7. The width of this diffuse scat-

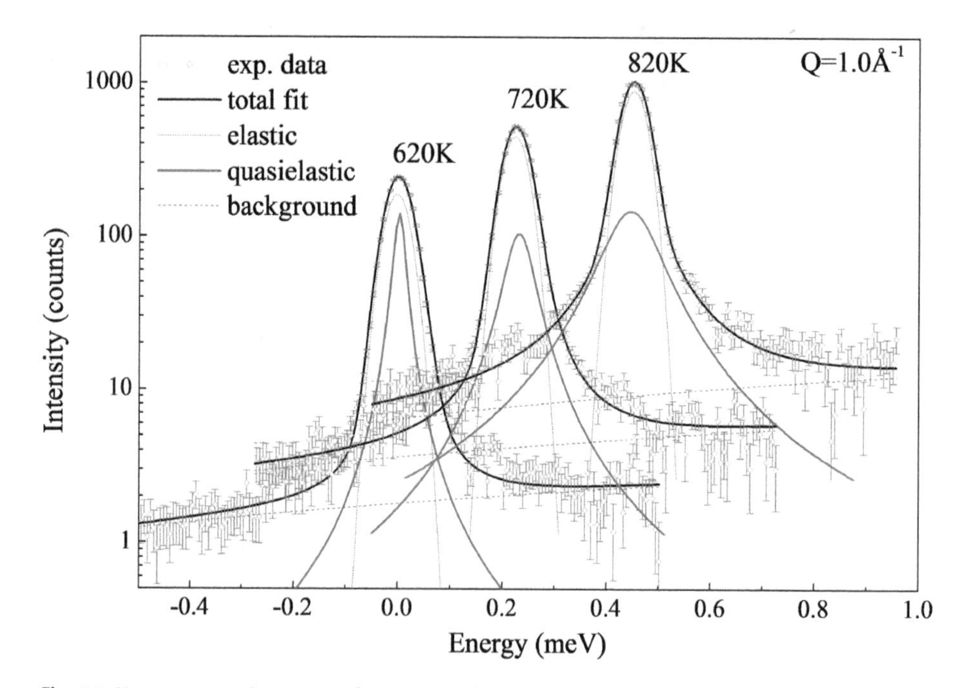

Fig. 7.7: Neutron scattering spectra from a ceramic proton conductor recorded at 620, 720, and 820 K, deconvoluted into elastic and quasi elastic contributions. The quasi elastic lines broaden significantly with increasing temperature. Reproduced from [Chen 2013]. "Reprinted from Q. Chen, J. Banyte, X. Zhang, J.P. Embs, A. Braun; Proton Diffusivity in Spark Plasma Sintered $BaCe_{0.8}Y_{0.2}O_3$: in-situ combination of quasi-elastic neutron scattering and impedance spectroscopy; Solid State Ionics 252 (2013) 2–6" with permission from Elsevier.

tered intensity is given by the diffusivity D of the protons. By deconvolution of the neutron spectrum into elastic and quasi elastic contributions we are, thus, able to determine the diffusivity and, via the Nernst–Einstein relation, the proton conductivity on the molecular scale.

In order to study the proton dynamics in bacteriorhodopsin, such as with QENS, it is necessary to be sure about the particular spectroscopic state in the photocycle. Since these photoactivated states have a limited lifetime, it is necessary to synchronize the photoactivation steps with the QENS probing steps. Practically, this amounts then to a typical combined spectroscopic pump probe experiment. This has been done recently with success (Pieper and Renger 2009), see Figure 7.8.

At this point it should be stated that the right sample preparation becomes more important the more sophisticated an experiment becomes. For example, in the aforementioned x-ray absorption studies, particularly for quantitative EXAFS studies in transmission sampling geometry, it is necessary to optimize the sample thickness to "1 absorption length", which is the sample thickness weighted with its mass density,

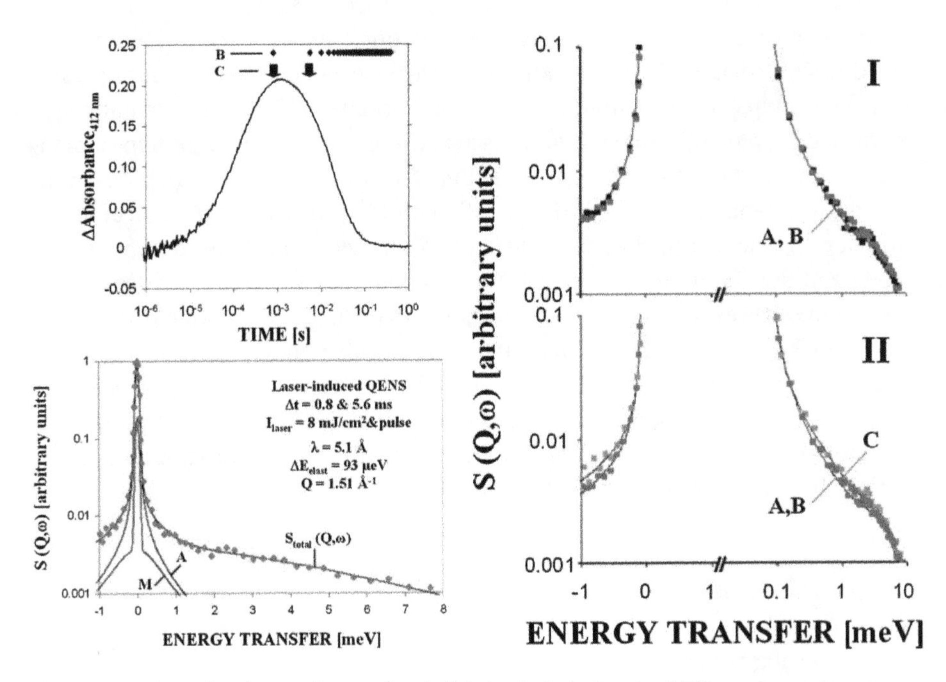

Fig. 7.8: Transient absorbance changes (top left) in bacteriorhodopsin at 412 nm at room temperature after laser excitation at 532 nm and $t = 0$ obtained using a laser pulse energy of 8 mJ cm^2 and a laser pulse repetition time of 400 ms, i.e., exactly those conditions used for the light-excited QENS experiments. Black diamonds and arrows indicate the arrival times of the neutron pulses at the sample relative to the laser pulse in two different measurement modes, i.e., experiments without (B, black diamonds) and with time selection (C, arrows), respectively. Here, the letters B and C correspond to the subscripts of the fit functions to the respective QENS spectra (right). Reprinted with permission from Pieper et al. (2009). Copyright 2009, John Wiley and Sons

to avoid loss of photons by extensive absorption and to avoid spectroscopic artefacts by multiple scattering. This holds particularly for operando and in-situ experiments, where several methods are applied on the very same sample. In a heterogeneous sample with different elements, the optimum sample thickness depends, then, also on which particular element is being investigated (Koningsberger and Prins 1988).

Soft x-ray spectroscopy experiments (NEXAFS) are typically done in the reflection mode, where the requirements for the optimum sample thickness are more relaxed. In the in-situ NEXAFS studies on photoelectrochemical water splitting, however, the sample thickness needs to be chosen carefully when specific spectral information is required on the electrode bulk, depletion layer range, electrode–electrolyte interface, and the electrolyte region. With a 100 nm thick iron oxide film, no information on the bulk and depletion layer can be extracted, and no signature from the electrolyte is visible in the measurement geometry shown in figure 8.49 in Braun (2017). A 100 nm film would be too thick already. It should be noted that exposure to x-rays can cause radiation damage to the sample, although x-rays are typically not as damaging to the sample as, for example, the electron beam in an electron microscope.

Neutrons, in contrast, hardly cause any radiation damage to the samples shown here. One shortcoming of neutron experiments is the need for a large amount of sample, which is typically a result of the small neutron flux. A QENS experiment needs around 300 mg protein materials, which is a very large number in the protein community. The optical absorption of such an amount of material is huge, so that virtually all light from excitation is absorbed. Then another problem arises. The fraction of bacteriorhodopsin reaching the intermediate state “M” depends on the absorbance of the protein and the quantum efficiency of the entire photocycle. To allow for combined optical and QENS experiments, it was, thus, necessary to lower the sample amount to 30 mg, which required an accordingly larger data acquisition time (Pieper 2010).

Protonic energy storage and conversion mechanisms appear to be more tolerant to structural and morphological deficiencies than materials with electronic properties. Such proton technology does not yet exist, but according to Helmut Tributsch, this is worthwhile considering (Tributsch 2000).

7.2 Exercise: proton wire

7.2.1 Molecular structure of liquid water

In this section, we will learn how to model the molecular structure of water. Before doing so, we need to look into one fundamental approach in theoretical physics, which helps us formulate a system. We can formulate the dynamics of the water molecule, for example, by using Lagrange formalism. The Lagrange function is a scalar function and reads in general form as

$$\mathcal{L} = \mathcal{T} - \mathcal{V} \tag{7.2}$$

where $\mathcal{T}$ stands for the kinetic energy, and $\mathcal{V}$ stands for the potential energy. The second kind of Lagrange equation is given as

$$\frac{\mathrm{d}}{\mathrm{d}t}\frac{\partial \mathcal{L}}{\partial \dot{q}_i} - \frac{\partial \mathcal{L}}{\partial q_i} = 0 \tag{7.3}$$

and contains the generalized coordinates q_i of the system along with its first derivatives $\dot{q}_i$, i.e., the momentum.

This second kind of Lagrange equation is a so-called Euler–Lagrange equation and can be derived by the first variation of the action integral in so-called Hamilton formalism[5]. Hamilton formalism and Lagrange formalism are global descriptions of a physical system that does not include, for example, the forces acting on the objects in the system. It is, however, possible to derive the forces from the two aforementioned formalisms by deriving the first kind of Lagrange equations, which will yield the forces that we typically derive from the d'Alembert principle (derivation of forces in molecules by wave mechanics was demonstrated by Feynman in Feynman 1939).

Here, we now write the water molecules in the system by their spatial coordinates of the ions, $\boldsymbol{R}_\mathrm{I}$ and their electronic wavefunctions (orbitals) $\phi_i(\boldsymbol{R})$:

$$\mathcal{L} = \mu \sum_i \int \mathrm{d}\boldsymbol{r} |\dot{\phi}_i(\boldsymbol{r})|^2 + \sum_i \frac{1}{2} M_\mathrm{I} \dot{\boldsymbol{R}}_\mathrm{I}^2 - E_\mathrm{KS}(\{\phi_i, \boldsymbol{R}_i\}) \tag{7.4}$$

This approach is outlined in Laasonen et al. (1993) and is based on the cooking recipe detailed in Remler and Madden (1990), which I recommend for the reader.

The wavefunctions are expressed in an orthonormal base system and, thus, must hold the condition

$$\langle \phi_i | \phi_j \rangle = \delta_{ij} \tag{7.5}$$

which means the scalar product of two wavefunctions is 0, when the index of the wavefunctions is not identical. If the wavefunction indices are identical, like $i = j$, then the scalar product is 1 (unity), as expressed by the Kronecker delta δ_{ij}.

For the electronic degrees of freedom a mass parameter μ is introduced, whereas M_I is the mass of the ions. The total energy of the system is abbreviated with E_KS. When we apply the formalism of the second kind of Lagrange equation on $\mathcal{L}$, we obtain the equations of motion with respect to the spatial coordinate,

$$N_\mathrm{I} \ddot{R}_\mathrm{I} = -\frac{\partial E_\mathrm{KS}}{\partial R_\mathrm{I}} \tag{7.6}$$

and with respect to the wavefunction (coordinate)

$$\mu \ddot{\phi}_i = -\frac{\partial E_\mathrm{KS}}{\partial \phi_i^*} + \sum_j \Lambda_{ij} \phi j \tag{7.7}$$

5 When I graduated as physicist from RWTH Aachen, this was part of my oral exam on theoretical physics. I sat there in the office of Prof. H.A. Kastrup, and he asked me to derive a number of things, using pen and paper. One was the " Herleitung der Lagrange Gleichung zweiter Art aus der ersten Variation des Wirkungsintegrals".

7.2.2 Calculation of forces in molecules

As we will be calculating the forces in molecules using methods of quantum mechanics, we need to refer to a theorem that allows us to do so. In 1939, Feynman published a method for the calculation of forces in molecules based on wave mechanics (Feynman 1939). The method appears to be a relatively trivial application of variation calculus (in German: *Variationsrechnung*) and Lagrange formalism extended to the Schrödinger equation. A similar treatment was published in 1937 by Hellmann in his book *Einführung in die Quantenchemie* (Hellmann 1937)[6], and this relation has been hitherto known as the Hellmann–Feynman theorem:

$$\frac{dE_\lambda}{d\lambda} = \langle\psi_\lambda|\frac{d\hat{H}_\lambda}{d\lambda}|\psi_\lambda\rangle \tag{7.8}$$

The derivative of the total energy E to a parameter λ is equal to the expectation value of the Hamiltonian $\cap H$ to the same parameter; $|\psi\rangle$ is an eigenstate or eigenfunction of the Hamiltonian. The same relation was already shown in 1931 by Güttinger (1931)

$$\left[\left(\frac{\partial H}{\partial a}\right)_{p,q}\right]_{m,n} = \frac{\partial E_m}{\partial a} \tag{7.9}$$

(equation 11 in Güttinger 1931)

$$\frac{h}{2\pi i}\left[\left(\frac{\partial H}{\partial a}\right)_{p,q}\right]_{m,n} = k_{m,n}(E_n - E_m) \tag{7.10}$$

In quantum mechanical matrix form, this expression reads as

$$\frac{h}{2\pi i}\left[\left(\frac{\partial H}{\partial a}\right)\right] = kE - Ek = [k, E] \tag{7.11}$$

and $[k, E]$ is the commutator bracket (as compared to the Poisson bracket f, g in classical mechanics). Maybe it would, thus, be more appropriate to call it the Güttinger theorem, since Paul Güttinger used this relation in his thesis at ETH Zürich in 1931 (Güttinger 1931, 1932) – in a research topic suggested by Wolfgang Pauli, who was his thesis advisor (Pauli et al. 1997).

The potential energy V of a molecule is composed of the interactions of every nucleus with any other nucleus, the interaction of every electron with every nucleus, and the interaction of every electron with another electron (Feynman 1939). Feynman concludes from his calculation that the concentration of electric charges between the nuclei in a molecule causes the two nuclei to be attracted to these charges "in the middle" by strong forces. Then he explains further that in an H_2 molecule, the antisymmetri-

6 Vorwort von Hellmann: Verkuerzte und dabei teils umgearbeitete Ausgabe meiner in russischer Sprache erschienenen 'Kwantowaja chimija' (Moskau 1937). Sein Inhalt wurde 1935/1936 als Vorlesung am hiesigen Karpowinstitut fuer Physikalische Chemie vorgetragen.

cal wavefunction would be zero in the middle between the H atoms and, thus, cannot concentrate electric charge at this middle position. The symmetrical wavefunction, however, would provide for sufficient charge concentration in the middle between the nuclei, which attracts the nuclei and, thus, allows for the formation of a stable molecule. This justifies the term valence bonds.

7.2.3 Conical intersections

The potential energy surfaces can be approximated by parabolic curves – in one-dimensional systems we know it, for example, from Hooke's law with $V(x) = 1/2\,kx^2$ – and this is based on the linear force laws $F = kx$ of the harmonic oscillator, an idealization of reality in nature. As proteins have many atoms, the forces are, of course, manifold, but it is possible to focus on the more relevant and study them. This is when in the literature you see parabolic curves, such as the ones I show in Figure 4.21. When the potential energy surfaces of two or more subsystems intersect at some points, we call these points conical intersections. At conical intersections, the Born–Oppenheimer approximation for wavefunctions is no longer warranted, and nonadiabatic chemical reactions can take place.

In photosynthesis systems and in any biomolecular living systems, we have transport of energy and mass (heat and mass transfer, like in a thermodynamic machine, or machinery, factory, or industrial complex), and we can rationalize it first as transport in three-dimensional space. This can be an electron or a proton or a molecule. Next, we must consider that it moves along an energy landscape imposed by potential energies of oscillators and resonators, which are electronically excited by visible light and vibrationally excited by the warm environment, by microwaves. In this landscape, light absorption, emission, fluorescence resonance energy transfer, and chemical reactions take place. It is then worthwhile to look at the molecular chemical factory like a real industrial factory, where you have raw materials arriving at a shipping and receiving department, local carriers bringing precursors to the labs and picking up reaction products to the shipping and receiving department for further use elsewhere.

You can write this schematically in a map as reaction pathways and introduce a quantity of reaction coordinates. In the molecular picture, it would be the change of the spatial coordinate of the reacting species in an environment set by the energy landscape, modulated by thermal and electromagnetic excitation and catalytic lowering of activation barriers. You can consider raw materials or reactants with a quantum mechanical reactant state $|r\rangle$ and the expected chemical product with a product state $|p\rangle$, as originally proposed in Purchase and de Groot (2015)[7] and resketched for you in Figure 7.9.

7 Professor H.G.H. "Huub" de Groot was the *spiritus rector* of the SUNRISE flagship and coordinated it from Leiden University in The Netherlands, (Kupferschmidt 2019; Abbott 2019; Braun 2019b).

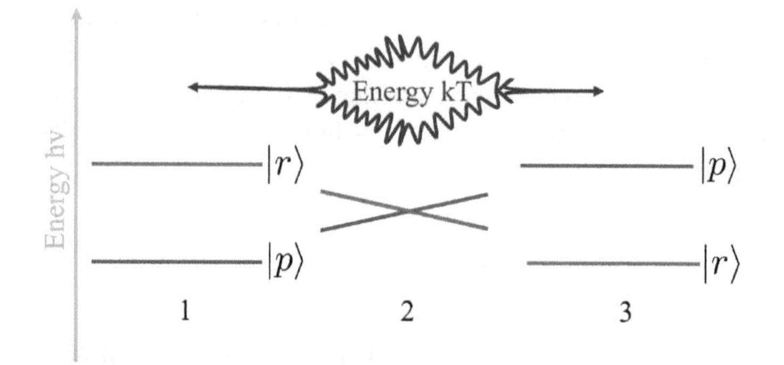

Fig. 7.9: Scheme for the quantum conversion of chemical reactants to chemical products by vibronic coupling of quantum states $|r\rangle$ and $|p\rangle$.

These quantum states are coupled due to the molecular framework and modulated by the molecular vibrations due to thermal excitation, which is indicated by the brown colored spring arrow with energy kT. The reactant states are energy levels and ordered along the energy axis with energies $h\nu$. The states a pure quantum states at timestamp (1). They may overlap at timestamp (2) induced by the molecular vibrations as shown by the brown spring. The two states may overlap and become unstable; chances are that there becomes a quantum coherent superposition of $|r\rangle$ and $|p\rangle$, where the energy levels of both are exchanged. Given the situation where the dynamics of the molecule nuclei is synchronous with the dynamics of the electronic states, resonant quantum chemical conversion is possible. Purchase and de Groot believe that after several oscillations, a chemically pure reactant state could be produced to a chemically pure product state, at timestamp (3) (Purchase and de Groot 2015).

It is clear that for the further mathematical description of this process, we could use creation and annihilation operators, where the chemical reactants are annihilated and the products are being created. I think we are also not necessarily bound to keep the spatial coordinate and energy as the natural coordinates for the mathematical description of the matter. There may be situations where it could be worthwhile to consider more abstract coordinates, not much different from the canonical conjugated variables that we know from the Hamilton–Jacobi formalism in theoretical mechanics. The action integral would then be included by the variables in the phase space.

It may depend on the proximity of the energy surfaces as to whether an energetic transition may take place between two oscillators[8].

---

8 We may expand the problem and think about what happens when there are more than two proximate energy surfaces for a potential transition to occur. This is similar to the problem where we have one seller and one buyer and the negotiation over which they may come to terms and close the transaction. How will the game change when there are several interested buyers or several interested sellers? To some extent, we may treat this like a perturbation problem. Maybe every buyer walks home with a share of the goods offered.

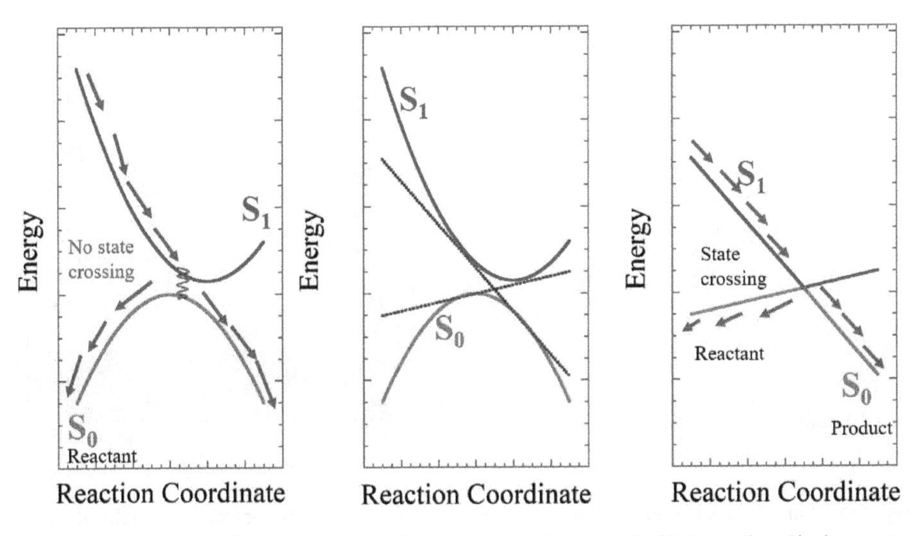

Fig. 7.10: Scheme for radiationless decay of the state towards the conical intersection. The image on the left-hand side shows an avoided crossing of levels of a, say, diatomic molecule with parabolic potential energy curves. In the middle image, the gray straight lines indicate the crossing curves. On the right-hand side, the crossing is successful by a conical intersection. Inspired by Teller (1969); Garavelli (2006).

Think of two parabolic curves, one pointing upwards and the other downwards, and their maximum and minimum are very close to each other in the spatial coordinate and energy coordinate. Edward Teller was one of the theoreticians who pioneered the idea of internal conversion in polyatomic molecules (Teller 1969). Internal conversion is the radiationless decay of an excited electronic state to a lower one. This is sketched in Figure 7.10. An example is shown of the decay of the S_1 state to the S_0 state in ketene via a conical intersection (Yarkony 1999). Today, the theory of the conical intersection is part of computational chemistry and, specifically, computational organic photochemistry (Garavelli 2006).

The β-carotene is a pigment molecule with a reddish-orange color (we know it naturally from carrots; one can also synthesize it in the laboratory or buy it in crystal form) which shows internal conversion. The structural formula of the molecule is shown in Figure 7.11. A detailed spectroscopic study was published in *Physical Review Letters* re-

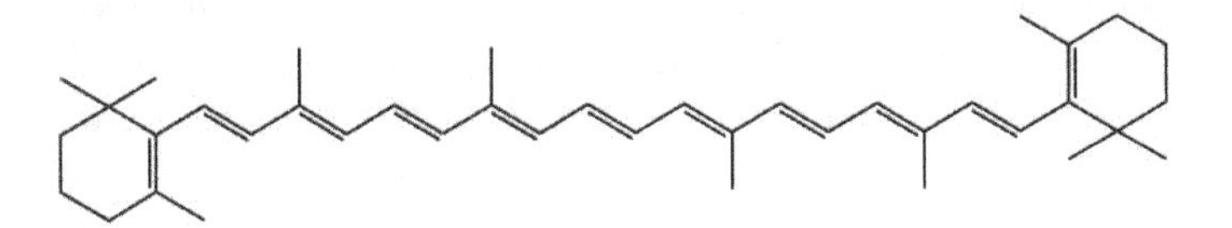

Fig. 7.11: Structural formula of β,β-carotene, also known as provitamin A. When sold as a chemical, it comes in orange crystals. CHEBI:17579

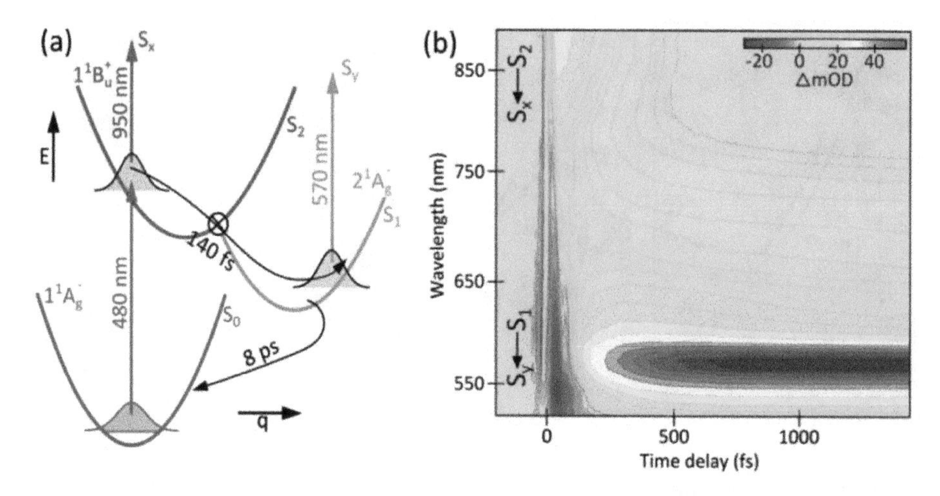

Fig. 7.12: Photophysics of beta-carotene. (a) Energy-level scheme for beta-carotene with the absorption maxima of the three lowest electronic singlet states. (b) Chirp corrected [19] differential absorbance map for beta-carotene in toluene. Reprinted with permission from Physical Review Letters: Liebel, M.; Schnedermann, C.; Kukura, P., Vibrationally Coherent Crossing and Coupling of Electronic States during Internal Conversion in beta-Carotene, 112/19, 83021–83025, 2014.] Copyright (2014) by the American Physical Society. 10.1103/PhysRevLett.112.198302.

cently by Liebel et al. (2014). Figure 7.12 shows a sketch of the three lowest electronic singlet states over the energy axis E and wave vector axis q. The energy surfaces are projected to two dimensions and sketched as parabolic profiles for the visualization of the concept, and, thus, should not be taken literally. For spectroscopic studies, the pigment is dissolved in a solvent, in this case toluene.

It is excited with a laser to the singlet state S_2, from where it decays to S_1 with a time constant of 140 fs. The molecule responds with an absorption peak at 950 nm, which is invisible infrared, and further decays with a time constant of 8,000 fs producing absorption peaks at 570 nm (lemon yellow) and 480 nm (turquoise-blue). The panel on the right shows a so-called differential absorbance map of the absorption wavelengths plotted versus the time delay in a pump-probe experiment. The authors showed that a transition through a conical intersection was vibrationally coherent, which led them to conclude on the breakdown of the Born–Oppenheimer approximation (Liebel et al. 2014).

The vibration structure of the molecules can be probed with vibration spectroscopy, for which we typically have infrared and Raman spectroscopy. A very advanced method is nuclear resonant vibration spectroscopy (NRVS, (Pelmenschikov et al. 2017)), which is not covered here. The combination of optical and vibrational spectroscopy helps to identify vibrational coherence in a molecular system, as demonstrated in their study. The vibrational properties of proteins also play an important role in the dissipation of primary energy in photosynthesis (Abramczyk 2012).

Woo et al. (2017) carried out an ultrafast spectroscopy study on the molecule thioanisole $C_6H_5SCH_3$, which shows a nonadiabatic bifurcation at the S–CH_3 bond. The molecules has several conical intersections. When excited to the S_1 level, and when only the vibrational state near the first conical intersection is accessed, they find two reaction pathways, an adiabatic one and a nonadibatic one. Reactive flux is funneled through two consecutive conical intersections faster through the nonadiabatic pathway than through an adiabatic pathway with minimum reaction energy. The bifurcation is the existence of two reaction pathways with product flux.

The conical intersections exist, of course, in photosynthesis functionalities. Warshel and Chu carried out a computational study on the "surface crossing" processes in bacteriorhodopsin (Warshel and Chu 2001), which is known for its proton pumping function in the thylakoid membrane. Bacteriorhodopsin undergoes a photoisomerization in a cycle that is shown in Figure 7.3. The proton pumping process is performed mechanically by the shape changes due to photoisomerization.

A semiclassical model for photoisomerization was developed by Benderskii et al. (2005) for the 1D case. The potential curves are shown in Figure 7.13 – with opposite curvature, as they write. The potential energy results from two harmonic potentials with parabolic profiles and can be written in a 2D matrix

$$V = \begin{bmatrix} -1 + (X + X_0)^2 & u_{12} \\ u_{12} & 1 - (X - X_0)^2 \end{bmatrix} .$$

The maximum and minimum of both parabola are erased because they overlap. Both extrema without overlap, as shown by the dashed curves, would constitute a diabatic potential. However, the overlap makes a convolution with the potential profile shown by the two solid lines, the adiabatic potential. There is no way for a direct transition along the extrema like tunneling; see Figure 7.10. Fast nonadiabatic transitions along the double-crossing points are an alternative to the tunneling process.

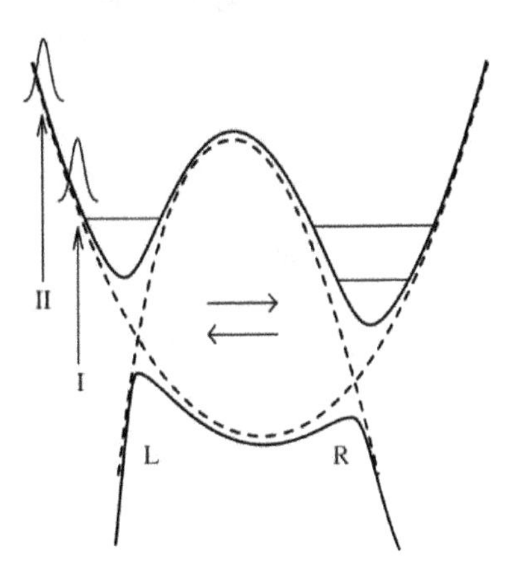

Fig. 7.13: Diabatic (dashed line) and adiabatic (solid line) potentials of the ground and excited states in the double-crossing region (the minima of the ground electronic state are not indicated, see text). The horizontal arrows indicate L to R isomerization. Positions I and II, at which wave packets are created, are indicated (see text). Reprinted from Chemical Physics Letters, 409, V.A. Benderskii, E.V. Vetoshkin, E.I. Kats, H.P. Trommsdorff, A semiclassical 1D model of ultrafast photoisomerization reactions, 240–244, Copyright (2005), with permission from Elsevier.

7.2.4 Quantum mechanical description of proton wire dynamics

Tubes and channels, roads and rivers, ways and paths are the topological elements that allow for the transport part in the logistics business. Electric transport is typically engineered by using wires, particularly electron transport. Then, the wires are from metals with a high electronic conductivity. But also ionic conductivity may sometime require channels, or proton wires. For fuel cell applications, Leem et al. have engineered proton wires by decorating pore channels in nonproton conducting polymer membranes with bio-organic molecules such as L-lysine and aspartic acid, for example Leem et al. (2008).

Proton transport is important in hydrogenases, which reversibly catalyze the production of molecular hydrogen H_2 from protons. Proton pathways are therefore important. It is possible that mutations arise in the cells which contain hydrogenases where such proton pathways are blocked or interrupted. There is, however, a second water channel in hydrogenases, which can be activated and can to some extent serve as a proton wire, as shown in computer simulations by Sode and Voth (2014).

The relevance of the theoretical concept of the proton wire originates from the idea of the water wires that exist in biological membranes. Imagine a one-dimensional arrangement of protons with not too much in distance from each other one by one, and the arrangement need be straight. Think of what is known as the bucket brigade, people pass buckets of water to each other in order to extinguish a fire or in order to empty a flooded area, for example. Bucket brigade devices are known in logic circuit engineering and also in protein biology. Or think of a group of mountain hikers, who walk on a narrow curvy path in the mountain range one after the other. Something like that.

As I am writing these lines in this book, I am involved in a manuscript on proton transport in metal oxide proton conductor membranes, where we discuss which pathways in the perovskite crystal structure the protons take for charge transport. A frequently heard model for the proton transport in liquids, and sometimes also in solids, is the one worked out by Grotthuss[9] over 200 years ago (Marx 2006). The Grotthuss process assumes the formation of intermediate hydronium species H_3O^+, where the proton is passed on to some other water molecule, which becomes H_3O^+, and so on, like in the aforementioned bucket brigade.

Hence, the H_3O^+ does not move; it only passes on the proton. An alternative transport model is the vehicle mechanism, where a molecule carries the proton, and the molecule moves and delivers the proton. Observe here the difference between the models in the bucket brigade picture. In the bucket brigade, the people do not walk, but they stand in their individual position and pass the bucket. In the vehicle mechanism, every human carries a bucket from front to end, and the human must walk the entire distance from front to end – carrying the bucket. Which of these only two alternatives

9 Grotthuss is credited for the first law of photochemistry, which says that only those radiations that are absorbed by a substance can make a chemical change in this substance (Bancroft and Clapp 1938).

applies and how they actually work at the molecular scale is experimentally not easy to resolve. Says Agmon in his paper from 1995 (Agmon 1995)

> Concluding, proton migration may be envisioned as a process propelled by hydrogen-bond cleavage, taking place in front of the moving proton, and hydrogen bond formation in its back. If an allegoric description is required, it may be found in Moses crossing the sea, with waves parting before him and reclosing behind his track.

He anticipated that progress in computational modeling could eliminate unrealistic models and refine which of the models could be realistic. Nagle and Morowitz, for example, explored the possibilities of electric charge transport by protons through biomembranes and resorted to the proton transport in ice and concluded that continuous chains of hydrogen bonds formed from protein side groups are the fundamental structural element. (Nagle and Morowitz 1978).

We consider the arrangement of water molecules with the respective spatial coordinates as laid out in the paper by Pomes and Roux (1996a, b). This exercise is based on the semiempirical PM6 quantum chemistry method (parameterized model number 6), which is introduced in an article of Stewart in Stewart (2009). The only structural elements used in this model are only the H^+ and the O^{2-}, which build up, for example, two water molecules plus one proton in between, which looks like $H_2O{-}H^+{-}OH_2$. This information is sufficient to write down the potential energy U of the water wire configuration:

$$U(\{r_O\},\{r_H\}) = \sum_{i<j=1}^{N_O} \phi_{OO}(|r_{O_i} - r_{O_j}|) + \sum_{i=1}^{N_O}\sum_{j=1}^{N_O} \phi_{OH}(|r_{O_i} - r_{H_j}|) + \sum_{i<j=1}^{N_H} \phi_{HH}(|r_{H_i} - r_{H_j}|) + \Phi_{pol}(\{r_O\},\{r_H\}) \tag{7.12}$$

You can notice that r_O and r_H are the vectors for the positions of the oxygen ions and the protons, respectively. The ϕ are the wavefunctions for the O–O, O–H, and H–H bonds and have radial symmetry. We also include a polarization energy that arises from the polarization of the nucleus of the oxygens, Φ_{pol}.

Figure 7.14 shows the potential energy barrier in kcal/mol for the proton (H nucleus) motion depending on the distance between two oxygen ions (O–O separation in Å) in the $H_2O{-}H^+{-}OH_2$ environment. Pomes and Roux compare[10] the PM6 method

10 I have been involved for many years in an activity with the meteorological institutes in Europe developing a technical platform for the quality assurance of batteries and their mass production with electroanalytical methods and x-ray spectroscopy (EURAMET EMPIR g10 program). The project included the computer simulation of NEXAFS L-edge multiplet spectra. One task is the comparison of the suitability and performance of two computer codes, which were the old and long-established Cowan code (Kramida 2019) and the new Ocean package, which now also includes the solving of the Bethe–Salpeter Equation (BSE) (Vinson et al. 2011).

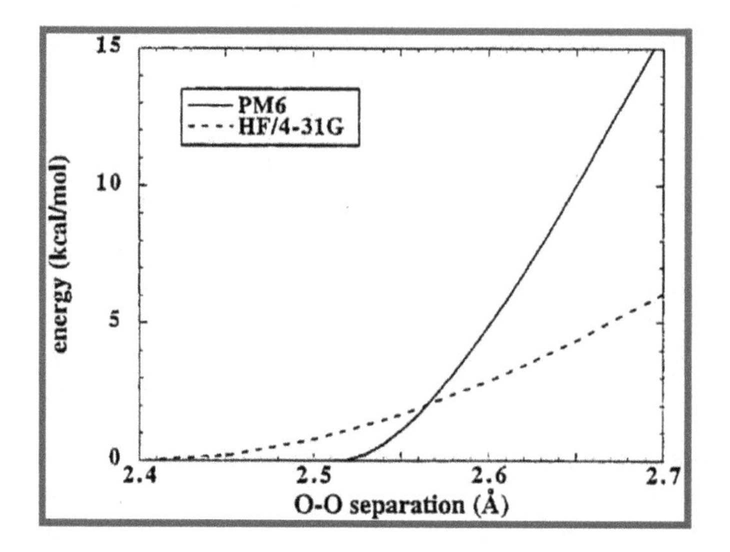

Fig. 7.14: Dependence of the potential energy barrier on H nucleus motion upon the donor–acceptor distance in H_2O–(H^+)–OH_2. The results from ab-initio calculations at the HF/4-31G level (dashed line) and from the PM6 model (solid line) are shown. Reprinted (adapted) with permission from Pomes and Roux, Theoretical Study of H^+ Translocation along a Model Proton Wire, J. Phys. Chem. 1996, 100 (7), 2519–2527. Copyright (1996) American Chemical Society.

with the HF/4-31G method and find, for example, 6 kcal/mol and 16 kcal/mol for an O–O separation of 2.7 Å. The calculated (!) potential energy barrier of this arrangement for H^+ can vary considerably, depending on which computational method you apply, as you can see from Figure 7.14.

The potential energy barrier calculated with the PM6 package comes out to zero kcal/mol when the separation is between 2.4 and 2.52 Å, and then sharply spikes up with increasing O–O separation, with a slope of around 75 kcal/mol Å. When the energy barrier is calculated with the HF/4-31G method, the barrier is already sensed by the proton slightly above Å and then increases with an approximated slope of around 24 kcal/mol Å. We are looking here at a difference of a factor of 3, depending on how large the separation is. Such differences are not uncommon among alternative computer codes, because they may be based on different physical models and on different mathematical algorithms.

Figure 7.15 shows the potential energy profiles for the movement of an excess proton in a linear water tetramer, which was calculated for a particular trajectory in molecular dynamics (MD) at ambient temperature. Four different cases are compared. On top is the case where there is no barrier for proton conductivity.

We recall that in this model, we are looking at two water molecules within which there is a proton that is supposed to move. The proton finds itself in a situation of acceptor–donor separation. When a proton participates in a strong hydrogen bond, then the probability of a transfer of that proton is larger than when the proton is in

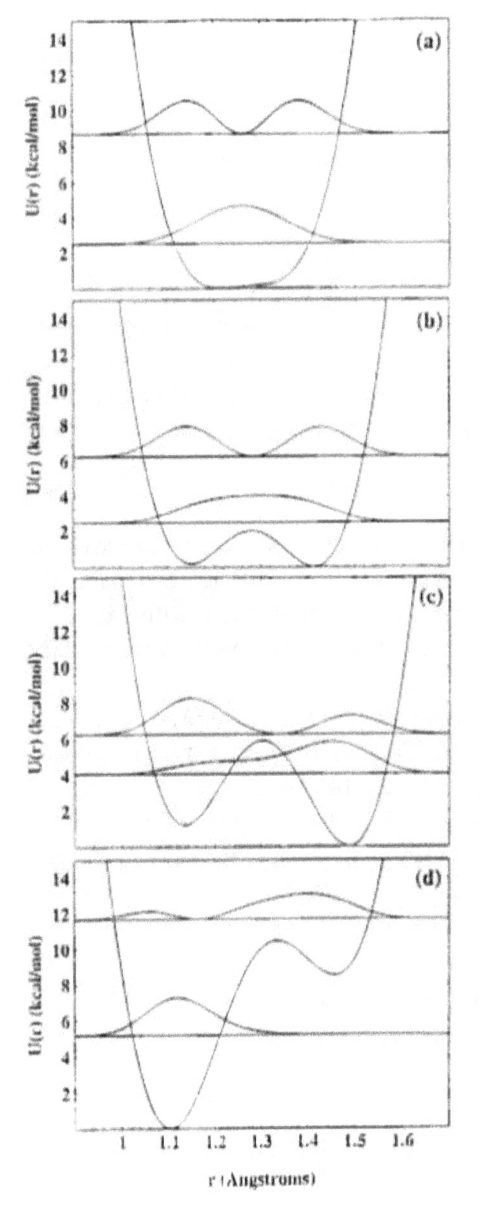

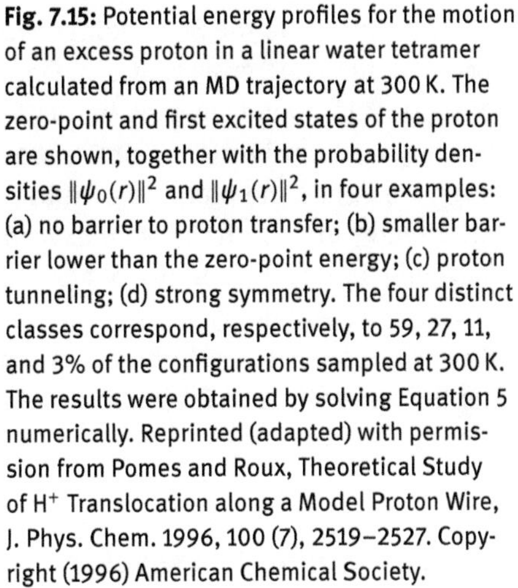

Fig. 7.15: Potential energy profiles for the motion of an excess proton in a linear water tetramer calculated from an MD trajectory at 300 K. The zero-point and first excited states of the proton are shown, together with the probability densities $\|\psi_0(r)\|^2$ and $\|\psi_1(r)\|^2$, in four examples: (a) no barrier to proton transfer; (b) smaller barrier lower than the zero-point energy; (c) proton tunneling; (d) strong symmetry. The four distinct classes correspond, respectively, to 59, 27, 11, and 3% of the configurations sampled at 300 K. The results were obtained by solving Equation 5 numerically. Reprinted (adapted) with permission from Pomes and Roux, Theoretical Study of H^+ Translocation along a Model Proton Wire, J. Phys. Chem. 1996, 100 (7), 2519–2527. Copyright (1996) American Chemical Society.

a weak hydrogen bond. In their simulation of the situation, Pomes and Roux focus on the case where the proton is in the center of the $O_4H_9^+$ unit. There is an energetic Coulomb barrier stopping the proton from moving.

Pomes and Roux build a discretized Feynman path integral by modification of the potential energy into an effective one U_{eff} [Equation (2) in Pomes and Roux 1996b] with the protons represented by an aromatic ring structure and formulate the equation of motion as a Langevin equation of the form

$$m_\alpha \ddot{x}_\alpha = -\partial_{x_\alpha} U_{\mathrm{eff}} - \gamma \dot{x}_\alpha + f(t) \tag{7.13}$$

with a friction term $\gamma\dot{x}_\alpha$ and with an extra force term $f(t)$, which originates from the statistics of the fluctuation dissipation theorem around the friction process.

$$\langle f(t)f(0)\rangle = 2k_\mathrm{B}T\gamma\delta(T) \tag{7.14}$$

Equation 7.14 introduces a stochastic component, which can be considered as thermal noise, into the proton motion (Equation 7.13). The authors considered protonated linear chains built from several water molecules, up to nine water molecules, which read ($O_9H_{19}^+$) and applied a similar dynamics simulation on them. Trajectories of the protons are exemplified in figure 2 in the original reference (Pomes and Roux 1996b).

The one-dimensional Schroedinger equation reads

$$-\frac{\hbar^2}{2M_\mathrm{H}}\Psi_n''(r) + U(r)\Psi_n(r) = E^n\Psi_n(r) \tag{7.15}$$

and was solved numerically in order to obtain the energy levels E^n and the wavefunctions $\Psi_n(r)$. You can play around with the water chain configuration by moving the central proton H^+ along the axis between two adjacent oxygen ions with a distance vector $\boldsymbol{r}$. The authors of (Pomes and Roux 1996b) used a square well potential of the width L and the function $\sin(n\pi r/L)$.

The solutions to the Schroedinger equation with this potential provides profiles for the potential energy as shown in Figure 7.15. Four different processes were considered. Type (a) is found when no barrier for the proton barrier is considered. When we allow for an energy barrier for proton transport that is even lower than the zero-point energy, we obtain the profile shown in (b) in Figure 7.15, where the peak for the potential energy around the oxygen–oxygen distance of 1.3 Å is broadened considerably when compared to the absence of a barrier.

The next scenario (c) is where proton tunneling is allowed to occur; note that tunneling is not a barrierless process. As a practical application, Pomes and Roux consider the water wire in a Gramidicin A peptide complex, which has a tubular cylinder structure.

7.3 Exercise: proton wires in an electric field

While chromophores are genuine subjects of quantum electrodynamics because they absorb photons that eventually cause electron–hole pair generation, protons do not participate directly in photon absorption processes[11]. In photosynthesis, it is the retinal that absorbs the photon and then changes its conformation. This conformation

11 There is a nuclear photoeffect and the photon energy required for it to happen is in the energy range of γ-rays, (Chadwick et al. 1937; Chadwick and Goldhaber 1934; Bethe and Critchfield 1938).

change acts on the conformation of the seven α-helices, which constitutes a mechanical action (not much different from the principle of a peristaltic pump). The proton movement that I am presenting here as an exercise was originally published by Natalie Pavlenko in the *Journal of Physics D: Condensed Matter* (Pavlenko 2003).

The following may help you in respect to the motivation of the exercise. In the last section, we encountered Gramicidin A structures. They can be extracted and isolated from the bacteria *Bacillus brevis*. Gramicidin are peptides that do not contain any ribosomes or DNA. They are made from amino acids, and these can form a tunnel structure that serves for the transport of ions between cytoplasma and the cell exterior. The exterior is hydrophobic, and the interior is hydrophilic. Gramicidin molecules form helices; two of them will build one such channel with cylinder structure, which preferentially allows for the transport of monovalent ions, such as K^+, or protons. See for example the diffusion constants in Table 7.1. In biochemistry and electrophysiology, Gramicidin is used in patch clamp experiments when it is given to the electrolyte in the pipette.

Tab. 7.1: The water diffusion constant inside the Gramidicin A channel was extracted from the experimental diffusional water permeability $P_w = 1.82 \cdot 10^{-5}$ cm^3/s (Finkelstein and Andersen, 1981), using $P_w = DS/L$, with $L = 23$ Å and $s = 7.84$ Å^2. The cation diffusion constant inside the GA channel was extracted from the experimental maximum conductance using $\lambda_{max} = De^2/k_B TL^2$, with $L = 23$ Å^2 (see Roux and Karplus, 1991). The bulk water data are from Hille (1992) (Pomes and Roux 1996c).

Species	**Diffusion constants (Å^2/ps)**	
	in Bulk water	**in GA channel ***
H_2O	$2.1 \cdot 10^{-1}$	$4.4 \cdot 10^{-3}$
K^+	$2.0 \cdot 10^{-1}$	$1.9 \cdot 10^{-3}$
H^+	$9.3 \cdot 10^{-1}$	$3.4 \cdot 10^{-2}$

The Gramicidin will form pore channels on the cell wall and, thus, allow for electrochemical measurements due to ion transport across the cell wall. An example of a study on thylakoids is shown in Cherkashin et al. (1999). The fact that monovalent but not higher valent ions are permitted through the channels causes an unregulated ion current in the presence of concentration gradients or electrochemical gradients. This can cause cell death, which is the reason why Gramicidin can be toxic and dangerous, and finally deadly. It is, therefore, of interest how, for example, the flow of protons through such an ion channel is modeled in the presence of an external electric field[12].

We consider the hopping of a proton across a "proton wire", which is built by water molecules that are arranged in a zig-zag profile, see Figure 7.16. We assume that

12 The parameterization method 6 (PM6) is one example of a computational code (Stewart 2007) that was used to model the proton transport in Gramicidin A, see, for example, Pomès (1999).

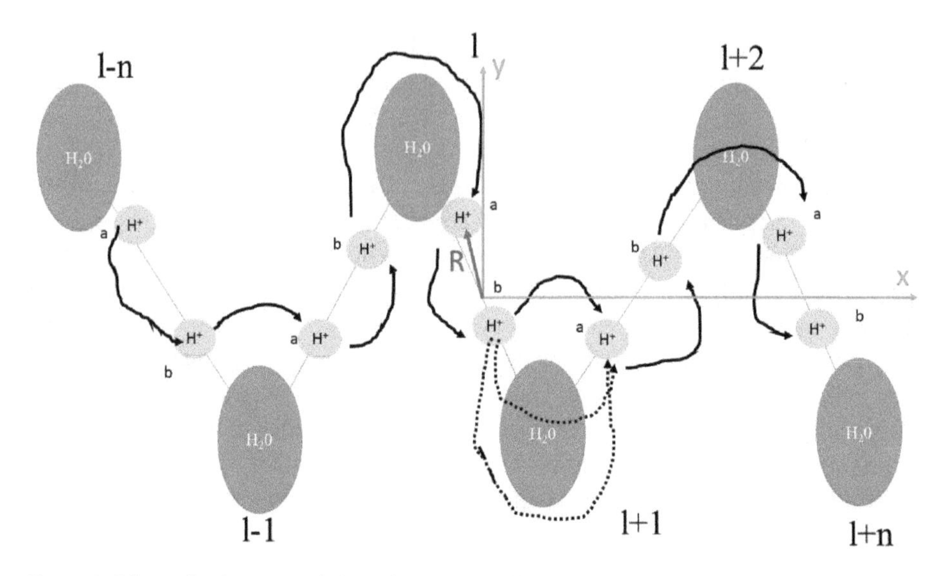

Fig. 7.16: Schematic of a proton chain with zig-zag profile set by water molecules with indexed positions $l-n, \ldots, l, \ldots, l+1$. The H^+ symbols denote potential proton positions and not protons. The arrows indicate potential proton paths along the chain. I have used a two-dimensional cartesian coordinate system (y, x) shown by green arrows, along with a red arrow R pointing to one proton in position R_{lv}.

the protons can make two different motions. One is a rotation of the proton around the oxygen ion of the water molecule. There is only very little activation energy necessary to overcome for this rotational mode. A higher activation barrier is posed when the proton is supposed to jump from one water molecule to an adjacent water molecule[13]. This motion is called the translational mode.

These two modes are known to occur, for example, in ceramic proton conductors and can be resolved with quasi elastic neutron scattering (QENS) (Matzke et al. 1996) (see also the application of QENS for bacteriorhodopsin, (Chan et al. 2016); neutrons are very good probes for protons and, thus, QENS is a very useful method for proton pump studies). The proton changes its position via the breaking and making of a hydrogen bond. The distance along the bond also represents the distance in a double-well potential, the activation energy of which is Ω_T, which the proton must overcome.

$$H_{\text{Trans}} = \Omega_T \sum_{l=1}^{N} \left(c_{la}^{+} c_{lb} + c_{lb}^{+} c_{la}\right) = \Omega_T \left(\left(c_{1a}^{+} c_{1b} + c_{1b}^{+} c_{1a}\right) + \ldots\right) \tag{7.16}$$

The operator c_{la}^{+} is the proton creation operator for the position la, and c_{lb}^{+} is the operator that creates a proton at position lb. Analogously, we have the proton annihilation

13 These two processes with different activation energies are also used to explain the proton conductivity in perovskite metal oxide proton conductors for intermediate temperature fuel cells.

operators c_{la} and c_{lb}. An existing proton at position la becomes annihilated, and at position lb, then, a new proton is created. This is the general quantum mechanical formalism that we must understand and follow: creation and annihilation of particles. In the previous chapters we learnt how to apply this to photons, i.e., Bosons with no electric charge and also no rest mass. Here, we apply the formalism to charged particles with a large mass. You can compare this to the "beaming" procedure for human beings in the Star Trek movie series from the 1970s.

The potential energy barrier to be overcome by a proton between two adjacent oxygen ions is lower than the zero-point vibration energy for protonated chains (Pomès 1999), thus there is no necessity for an intrabond transfer of protons, and hence the energy value for Ω_T is negligible and can be taken as zero (Pavlenko 2003). Therefore, the contribution H_Trans will not be taken into account in the exercise.

The Hamiltonian for the rotational motion reads:

$$H_\mathrm{Rotate} = \Omega_\mathrm{R} \sum_{l=1}^{N} \left(c_{l+1,a}^{+} c_{lb} + c_{lb}^{+} c_{l+1,a} \right) \tag{7.17}$$

Such a rotational motion is relevant when we consider a proton bound to a water molecule, and when we twist this arrangement, it may break at the proton bond, and the proton may bind to another water molecule in the chain shown in Figure 7.16. This proton, for example, would change its state from $|l+1, a\rangle$ to the state $|l, b\rangle$, as you can check in Figure 7.16. The energy necessary to carry out this chain dipole moment change is 5.5 kcal (Pomes and Roux 1996b) per mol in a chain of up to eight H_2O molecules. This contribution needs to be taken into account in the model.

When coupled the proton will experience a noticeable electrostatic repulsion toward energy U, as they are attached with the adjacent oxygen ions:

$$H_\mathrm{C} = U \sum_{l=1}^{N} n_{la} n_{lb} \,. \tag{7.18}$$

In addition to the Coulomb interaction, there may be also be an interaction of quantum mechanical nature, which is linked to the short-range configuration of the proton in proximity to an inner H_2O group or some outer molecular group. When there is a short bond, then this may be of covalent nature, or when it is a long bond then it may be of hydrogen bond nature. There can be various configurations of the proton with respect to its molecular neighborhood, which you should identify in Figure 7.16. I only give one example of one particular type of configuration; for further derivations of other configurations I refer the reader to the original paper by Pavlenko.

$$H_\mathrm{I} = \omega'(1 - n_{l+1,a})(1 - n_b) + \omega n_{l+1,a} n_{lb} + \epsilon(1 - n_{lb}) n_{l+1,a} + \epsilon n_{lb}(1 - n_{l+1,a}) \tag{7.19}$$

Finally, an external electric field E is applied (you can say "switched on") along the proton wire. The electric force qE by the electric field E is then

$$- e_\mathrm{p} E = U \sum_{l=1}^{N} R_{lv} n_{lv} \tag{7.20}$$

where l is the position of the protons and v the left-hand or right-hand side on the hydrogen bond. This convention basically allows for a coordinate system for the calculation with respect to the center of the proton chain as shown in Figure 7.16. The protons are either on site a or site b (compare this with the abscissa y), and the index l is the running number of the water molecule (compare this with the ordinate x). In terms of the coordinates l, v, the position of the proton is given by $R_{l,v}$.

The proton dynamics will by far be determined by the dipole moment P of the arrangement sketched in Figure 7.16, which is taken as the sum of the products of the proton position vectors with 0th statistical moment ($\langle \ldots \rangle$) of the proton occupancy n_{lv}:

$$\boldsymbol{P} = e_{\mathrm{p}} \sum_{l=1}^{N} \boldsymbol{R}_{l\boldsymbol{v}} \langle n_{lv} \rangle \tag{7.21}$$

We rationalize the protons in the coordinate system and must now find a way to express the system with an arbitrary number of hydrogen bonds on what Pavlenko calls multisite states. We must, therefore, set up a mathematical base system $|i\rangle$. When we consider only two hydrogen bonds, then $N = 2$, and the base will have discrete 2^{2N} states. For $N = 3$, we are looking at $2^{2\cdot 3} = 64$ states. With $N = 10$, we arrive at 1024 states, and so on. The matrix below indicates how the number of states grows with the number of hydrogen bonds that are being included in the calculation. The costs of computation will be higher, the larger the matrix is.

$$H_0 = \begin{bmatrix} |1\rangle = |00\ldots00\rangle & |2\rangle = |10\ldots00\rangle & \ldots & V_{1i} & \ldots & V_{1N} \\ V_{21} & \epsilon_2 & \ldots & V_{2i} & \ldots & V_{2N} \\ \vdots & \vdots & \ddots & \vdots & \vdots & \vdots \\ V_{i1} & V_{i2} & \ldots & \epsilon_i & \ldots & V_{iN} \\ \vdots & \vdots & \vdots & \vdots & \ddots & \vdots \\ V_{(N-1)1} & V_{(N-1)2} & \ldots & V_{(N-1)i} & \ldots & V_{(N-1)N} \\ V_{N1} & V_{N2} & \ldots & V_{Ni} & \ldots & |n\rangle = |11\ldots11\rangle \end{bmatrix}$$

It may be possible that the hydrogen bond, which usually has one proton, may have an additional proton, a defect, so to speak, and in this case, we speak of a Bjerrum D-defect. The other extreme case is that the hydrogen bond has no proton at all; then we speak of a Bjerrum L-defect (Bjerrum 1952).

$$|1\rangle = |0000\rangle \tag{7.22}$$

For the proton creation and annihilation operators c_{lv}, c^*_{lv}, we define the relation

$$c_{lv} = \sum_{i,j} \langle i|c_{lv}|j\rangle X^{ij} \tag{7.23}$$

Here, $\langle i|c_{lv}|j\rangle$ are the quantum mechanical expectation values, which we can determine by summing up as follows:

$$c_{0a} = X^{1,2} + X^{3,6} + X^{4,7} + X^{5,8} + X^{9,12} + X^{10,13} + X^{11,14} + X^{15,16} \tag{7.24}$$

It is an exercise for you to write down the expressions for $c_{1,a}$, $c_{0,b}$, and $c_{1,b}$. Watch out how the signs may change in the respective components X^{ij}.

The next task is to sum up the Hamiltonians for either degrees of freedoms, that is, the translational energy (H_P), the rotational energy (H_R), the Coulomb repulsion force between two protons (H_C), and the external electric field ($-e_\rho E$). Natalie Pavlenko also included a fifth energetic contribution (Pavlenko 2003), that is, the energy associated with different short-range configurations of the protons in their aqueous environment (H_l).

$$\frac{dE_\lambda}{d\lambda} = \langle\psi_\lambda|\frac{d\hat{H}_\lambda}{d\lambda}|\psi_\lambda\rangle \tag{7.25}$$

All the states $|i\rangle$ are orthogonal (it is an orthogonal base system with all axes perpendicular to each other) to each other, and thus we can abbreviate by using the Kronecker symbol $X^{ii'}X^{ll'} = \delta_{i'l}X^{il'}$.

A DFT study by Karahka and Kreuzer (2013) looks into the proton transfer in a water wire. They find that this is based on an atomic exchange mechanism, where the proton moves over a distance that is shorter than the oxygen–oxygen distance in a water molecule, where the water molecule becomes quasi neutral, and where a temporary hydronium ion H_3O^+ is formed. They compare their work with the results of Pavlenko (2003), who found that along the water–proton chain there are four proton adsorption sites between any two adjacent oxygen ions. Karahka and Kreuzer find support for this suggestion in their calculations.

7.4 Life beyond photosynthesis

7.4.1 Nonphotosynthetic life and bacterial chemosynthesis

In my scientific life, I have been to Washington DC several times. The first time I was there was in the year 2003 when I visited the neutron source at NIST, The National Institute of Standards and Technology in Gaithersburg, Maryland, and gave a talk on neutron small-angle scattering (SANS), and where I met a number of scientists from UC Irvine and Johns Hopkins University, who were working on lipid bilayers. To get to NIST, you use the Ronald Reagan Washington National Airport, which also serves Washington DC.

The second time I was in DC was in 2014, together with some of my group members, because we had been invited by Foreign Policy magazine for an award ceremony in Georgetown, right next to Washington DC (Rothkopf 2014). After that, I went to DC several times for various purposes. One day, I think it was in 2016, when I had time to look around, I decided to walk over to the Smithsonian Institute's Natural History Museum. I can recommend this museum for everyone who likes science and technology.

The Smithsonian museum also provides the eduroam service: students, teachers, researchers, and scientists from schools have free access to the Internet provided that their institutes and schools support it and are supported by eduroam with a corresponding subscription or membership. Eventually, I came across a video movie screen that presented a show on life in the deep oceans. The main story was that there exists life in the oceans that does not feed on photosynthesis, i.e., not on light from the sun. Instead, this form of life takes its energy from chemical compounds that are present in the hot environment near volcanos under water. We are thus not looking at photosynthesis, but at chemosynthesis – in deep-sea vents (Karl et al. 1980).

It was Karl from the University of Hawai'i, and Wirsen and Jannasch from the Woods Hole Oceanographic Institute (WHOI) in Massachusetts who reported that there were dense animal populations at over 2500 meter depth below sea level at the Galapagos Rift, and their food might be bacteria that would be the result of primary production, or even constitute primary production powered by reduced sulfur compounds emitted by deep-sea vents (I would call them volcanoes). The reduced sulfur compound eventually becomes oxidized in the water, and the resulting energy is used by bacteria to reduce CO_2 to organic matter (Karl et al. 1980). Currently, the WHOI is carrying out an expedition for further exploration of these deep-sea vents[14].

It is worthwhile mentioning that 50% of the oxygen produced by photosynthesis is produced in the oceans; 44% of the human population lives within 150 km of the seashores; 90% of all trade of material goods comes by ship; 95% of the oceans remains unexplored by humans. Further, 90% of the heat from global warming has been absorbed by the ocean (source: WHOI, 7 February 2019).

7.4.2 Artificial photosynthesis and how fossils are fossil fuels

The primary energy sources referred to in this book are either the well-established fossil fuels (coal, natural gas, mineral oil) or the so-called renewable energies, particularly solar power, wind power, and hydropower. When we include biomass as a still important energy source on the globe, we can trace this back immediately to solar power because it was produced by photosynthesis. Then, even fossil fuels can be traced back to photosynthesis products made millions of years ago.

Until recently I had no doubt that all life on Earth originated from photosynthetic life. The beginning of all life took place at the surface of the Earth. Then, when I visited the Smithsonian Institute in Washington DC in 2016, I saw an educational movie that showed how some form of life was created in darkness in the deep sea. The energy

14 https://web.whoi.edu/darklife/about-this-expedition/.

necessary for starting and maintaining life comes from chemical compounds in the water and deep-sea minerals or from the heat there. Hence, the paradigm "without light there can be no life, so let there be $h\nu$!" (Tien and Ottova-Leitmannova 2000) that light as energy source is necessary for the creation is not correct. There are at least a few examples that show that alternatives can work as well.

I did not mention yet the thermal energy that is produced and delivered from within the center of our own planet, mostly because of natural nuclear reactions in the Earth's core. There is an indication that "fossil" hydrocarbons can be formed under conditions that persist deep in our Earth between the crust and the core[15] (Kolesnikov et al. 2009; Kundt and Marggraf 2014), which implies that they are not necessarily of fossil origin and, thus, not of solar origin. Certainly, these findings can mount further to highly controversial theories, which are not necessarily supported by the established communities. However, whether or not a theory is right cannot depend on the consent in a community. Can it?

Photosynthesis is the process that drives the primary production that we use as food today and that in many parts of the world is still used as biomass fuel. Also, the oxygen produced during photosynthesis, which we need to breathe, is a product of primary production. Over 100 years ago, Ciamician suggested that we can produce chemicals with photochemistry by using light as the direct energy source (Ciamician 1912; Venturi et al. 2005; Braun et al. 2016). Still in 1979, Melvin Calvin believed we could build "petroleum plantations", factories that would produce fuel from sunlight (Calvin 1979). In 2016, Caltech Professor Nathan S. Lewis provided a Congressional Testimony (Lewis 2016) on the relevance and game changing power of artificial photosynthesis for the energy economy.

7.5 Photosynthesis as a thermodynamic machine

Since in this section we are dealing with quantum optics, we should pose the more general question as to what extent quantum mechanical principles would anyway apply to biology and biological systems. Can the phenotype of a cell be absolutely predicted by the DNA sequence? Strippoli and colleagues (Strippoli et al. 2005) conducted a number of Gedankenexperiments about genetics and concluded that the behavior of the cell cannot be known with absolute certainty, even if the DNA and nucleotide sequences are known exactly. There is an uncertainty, related with the terms of the base pairs, which is equal to or larger than the product of the cell size and the muta-

15 That study was conducted at the Geophysical Research Laboratory of the Carnegie Institution of Science in Washington DC, which I happened to visit in 2014, where an eminent high-pressure physics scientist was my host. They have a High Pressure Research Laboratory with facilities that can be interesting not only for geoscientists but also for materials scientists.

Tab. 7.2: Potential Quantum Dynamical Processes in Biology. Reproduced from Fleming et al. (2011). Reprinted from Procedia Chemistry, 3, Graham R. Fleming, Gregory D. Scholes, Yuan-Chung Cheng, Quantum effects in biology, 38–57, Copyright (2011), with permission from Elsevier.

	Excited states	Light particles	Radical pairs
Biological phenomena	Primary steps in photosynthesis Vision	Enzyme catalysis Photosynthesis	
Quantum Processes	Energy transfer Electron transfer Isomerization	Long-range e^- tunneling H atom transfer Proton-coupled electron transfer	Reactions producing radical pairs

tion rate. They, thus, identify an uncertainty principle in biology and conclude that the living cell is better described in a probabilistic way and not in a deterministic way.

Fleming et al. posed that question and summarized their answers in a table (Table 7.2), as shown below (Fleming et al. 2011). In the first column in the table they distinguish quantum processes, which are certainly well established as they are of quantum nature, and biological phenomena, where the quantum nature is assumed but not developed with all rigor.

Around 100 years ago, the founders of quantum mechanics interestingly also engaged in discussions on whether biological systems and life in general follow the fundamental principles of quantum mechanics, see this explained in the review in McFadden and Al-Khalili (2018). Around 100 years ago, when quantum physics became developed, or not long after, some of the founders of quantum physics wondered whether quantum concepts could be utilized to address questions about the origin and state of living systems, of life in general, which hitherto could not be answered. For example, Niels Bohr (Bohr 1933) and Erwin Schrödinger (Schrodinger and Penrose 2012) wrote pieces about life.

Many experiments in quantum physics are done at close to 0 K temperature in order to allow for coherence and to avoid thermal broadening of states. Sometimes such experiments are also done in the dark, so that no excitation of states can take place. It is, therefore, a common belief that quantum experiments can only be done in situations that are cold, dark, and without any noise, as Fleming et al. put it in their work in Fleming et al. (2011).

The interaction of light with matter, of photons with the photosynthesis machinery, is a genuine application of quantum physics and quantum electrodynamics. Consider, for example, an arrangement of N chromophores, which are excited by light into excited states $|m\rangle$ and $|m\rangle$ for the chromophores m and n by overcoming the transition energies ϵ_m and ϵ_n. The excitation creates exciton states. The transition dipoles of either chromophores are coupled by these excitons, which is quantified by a coupling constant J_{mn}. The exciton Hamiltonian of the chromophore system then reads

$$H_{\mathrm{e}} = \sum_{n=1}^{N} \epsilon_n |n\rangle\langle n| + \sum_{n<m} J_{nm}(|n\rangle\langle m| + |m\rangle\langle n|) . \tag{7.26}$$

Below, I give the LaTeXcode for Equation 7.26 by using the *verbatim* environment.

```
\begin{equation}
H_{\mathrm{e}} = \sum_{n=1}^N \epsilon_n \ket{n}\!\bra{n}
+ \sum_{n<m} J_{nm} (\ket{n}\!\bra{m} + \ket{m}\!\bra{n})
\end{equation}
```

7.6 Bioenergy and vectorial bioenergetics

There used to be the picture of the "bag of enzymes" that would perform the entire functions of living cells, including photosynthesis. "Bag of enzymes" is the same wild, or naive idea like when you take all parts of an automobile and pack them in a bag, maybe shake the bag, and then end up with a functional car. This certainly is not the case. All parts need to be properly assembled according to the blueprint of the engineers and designers who designed the car. Every part has its particular position in the car and also its particular function in the network with the other functional parts. The same concept certainly holds for the organisms of humans and animals, which are built up from the organs. Further, the same holds for the individual cells in humans, animals, and plants. The arrangement of the components has a vectorial character. Function arises from complexity, as was pointed out by Anderson (1972), and this is illustrated also in figure 1 in the paper on the structure of the living cell by Campbell (2008).

When we look into the images of the cells, for example in Section 6.3, we may have a first impression that the cells are isolated and closed systems. However, they are not closed systems. Plant cells like those sketched in Figure 6.8 are arranged in a compartmentalized system. However, they receive light, they receive CO_2 and H_2O, and they release O_2. In addition to light, they receive infrared radiation, the heat from the environment. Moreover, when it is very cold, they release heat to the environment – and then may freeze and die, unless they have antifreezing protein genes (see, for example, Makhalanyane et al. 2015 and Mishra et al. 2019). Biological cells, plant cells, chloroplasts, and thylakoids, thus, interact thermodynamically with their environment; they are open dynamic systems that exchange matter and energy with their environment. Topics known from hard-core mechanical engineering, such as multiphase flow and heat and mass transfer apply to biological cells and higher organized organisms, to vertebrates, primates, and human beings. The entire metabolism, including osmosis in cells, is vectorial bioenergetics (Mitchell 1991). The Gibbs free energy is the appropriate thermodynamic quantity to describe such a system, given by (see Vectorial bioenergetics (Melandri 1997)) the temperature T and the manifold of concentrations, or better, activities a of the chemical products and reactants

$$\Delta G = -RT \ln K + RT \ln \frac{\prod a_p^{v_p}}{\prod a_s^{v_s}} \tag{7.27}$$

Biological cells are, thus, molecular machines, factories, or even industrial complexes. However, they work only when minute details in the design at all levels are maintained. For example, Edman et al. identified the structure of an early intermediate in the photocycle of bacteriorhodopsin (Edman et al. 1999), specifically a vectorial translocation of a proton from one side of the purple membrane to the other. The research group also found that the deformation of the DNA helix is caused by a vectorial proton transport in the photocycle of bacteriorhodopsin (Royant et al. 2000).

Transport concepts of physics, such as ballistic charge carriers and soliton waves of fluids, can be applied to model processes in biological cells, or even proteins like in actin, the muscle protein. Tirosh (Tirosh 2006) applied this concept to ballistic protons and microwave-induced water solitons, for example.

Not all biological organisms take their necessary energy from the photons by solar light. Other organisms simply use chemical fuels to keep their molecular engines running. This can be easily sketched, as shown in Figure 7.17. In the center is the upright arrow for the Gibbs energy ΔG. On the left-hand side is the photoautotrophic system with initial energy state I, which is promoted to an excited state E by picking up photon energy $h\nu$. On the right-hand side is the energetic analog for the heterotrophic system in an initial state I', which is promoted to an excited state E' by consuming chemical energy in the form of adenosin triphosphate ATP, see Renger (2007).

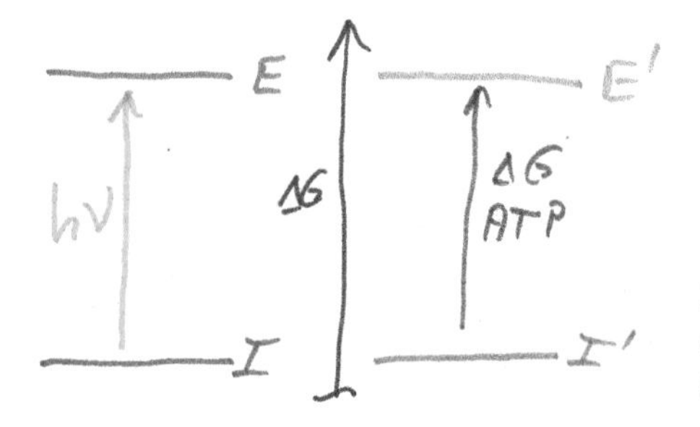

Fig. 7.17: Sketch of energy uptake by photoautotrophic and heterotrophic organisms. After (Renger 2007).

7.6.1 Coherent states

Fröhlich suggested that the dipolar properties of cell membranes cause them to have longitudinal electric modes; in the 100–1000 GHz range[16] (Fröhlich 1968). Biological

16 This happens to be the range where mobile phones would operate in the 5G frequency band. There is some controversial opinions on whether such millimeter waves could have adverse impacts on health, see Kostoff et al. (2020).

systems may have three types of coherent excitations (Fröhlich 1983). These can be the vibrations of biological membranes and proteins when the frequencies are above 10 GHz. This is a wavelength range of 30 mm. For comparison, the 4G mobile phone band operates at around 2 GHz. The new 5G band is expected to operate in the range of 30 to 100 GHz, which corresponds to 10 to 3 mm wavelengths.

Further coherent excitations may be the static excitation of highly polar metastable states. Periodic enzyme chemical reactions may have low-frequency coherent states. This is all based on experimental studies discussed in Fröhlich (1983). Two-dimensional spectroscopy shows the existence of quantum coherent states (Hildner et al. 2013), as Hildner et al. showed on the light harvesting LH2 complex of the purple bacterium *Rhodopseudomonas acidophila*. See Figures 7.1 and 7.2 for examples of how purple bacteria can make change the color of lakes and marshes.

Hildner et al. (2013) hypothesize that the long-lived coherence states are useful in plants for light absorption and energy transfer in order to overcome external disorder and perturbations exerted by the biological environment, such as physiological temperatures.

8 Agriculture and food supply

Humans and animals feed on plants, which are the products of photosynthesis. Figure 8.1 shows a pile of sugar beet that was harvested by the farmer from the acre behind the pile. It is right by Swiss Highway 1 and the railroad tracks near the city of Effretikon.

The brown field has a size of around 50 m × 250 m, around 3 acres. Note that the beets are *not the fruit of the plant but the root of the plant*, like carrots and potatoes. In the midst of the field with green grass, you can notice a wide brown strip of soil. This is where the roots, the beet, were harvested with heavy equipment.

Fig. 8.1: Pile of sugar beet in Switzerland. Photo taken 24 November 2019 by Artur Braun.

https://doi.org/10.1515/9783110629941-008

Fig. 8.2: Cattle on a meadow in a Swiss village. This is not a farm with extensive cattle industry. Photo taken 23 June 2019 by Artur Brown.

The sugar roots are used for the production of sugar or for food for cattle. When we eat meat instead, the animals that we eat may have eaten plants before. Figure 8.2 shows a meadow in a Swiss village. The grass is tall, and four young cattle are enjoying the fresh food. Here you can see how primary production turns into dairy products and potential meat. This looks very nostalgic and is certainly not the industry grade cattle farming that produces the enormous amounts of milk for butter, cheese, and yoghurt, and other dairy products. I know a young cattle farmer in Germany who owns 600 cows for industrial dairy production. Such large business cannot be run the way you see it in Figure 8.2. Instead, the cattle are crowded in factories.

However, not only food, but also the oxygen that we breathe is a result of photosynthesis. It is estimated that 50% of the oxygen on Earth is produced by splitting of the water molecules by photosystem II (PS II). Most of the plants grow on land, on soil. However, the oceans are full of plankton, and this produces the other half of the oxygen on the globe.

There are also plants that grow in water, such as seaweed. The photo in Figure 8.3 was taken on Jeju-do, an island in the Korea Strait, known for the tallest Korean mountain Hallasan and the women diving for shells. In the photo you can see an old gentleman collecting seaweed at the seashore near Choeyeong-ro. This is food. It is used

Fig. 8.3: A Korean gentleman collecting seaweed at the seashore near Choeyeong-ro on Jeju island in Korea. Photo taken 13 September 2013 by Artur Braun.

in Korean soups. Possibly you can get it in dried form in an Asian food store anywhere in the world. In Figure 8.4 you can see an old lady riding a scooter. Close inspection shows that she has a bag of seaweed on her back and a large loose pile of seaweed in front of her legs.

Agriculture (from the ancient Latin *agri colus*; care for the acre, the soil, the field) is the technology of preparing the soil and making it ready for farming plants on it so that you can grow plants, harvest them, and have the basis for food. When you grow algae, this is done in the water. You also can grow food "in the air", such as hanging tomato plants in a greenhouse, and you supply the nutrients in a matrix different from conventional soil.

According to data collected by the World Bank, around 10% of the area on the globe is used for agriculture – the production of food; this is one third of all land on the globe. Humans developed agriculture around 12,000 years ago when the climate became warmer and the ice age finished and conditions for systemic planting of crops made sense for human population. In the last 100 years, we have observed that the numbers of farms has drastically decreased and the size of farms have drastically increased. Agriculture is an industry that is based on many other industries, such as the chemical industry for the production of fertilizers, pesticides, and heavy equipment like trucks, trailers, harvesters, and silos. Agriculture is the foundation for the entire food processing industry and food waste management industries.

Fig. 8.4: A Korean lady transporting seaweed at the seashore near Choeyeong-ro on Jeju island in Korea on her scooter. Photo taken 13 September 2013 by Artur Braun.

Around 50% of land in Germany is currently used for agriculture and forestry. I grew up in a rural village in western Germany between the rivers Maas and Rhein, with small mountains up to 880 meters in height and many woods and fields for crops and cattle farming in between. My village with a population of 500 had maybe 25 families that engaged in cattle farming in the early 1970s. By 2019, the population was the same but the number of professional farmers was not larger than 5. Most of the land that they use and need for farming is rented by them from the other owners who stopped farming; 1 m^2 of farming and wood land costs 1 Euro (actually, the prize for one such m^2 is 1 Euro). The annual rent for farming land is around 160 Euros per $10{,}000\,m^2$. In return, farmers who operate (not necessarily their own) land for farming receive considerable subsidies from the government, which basically constitute their gross income. There are regions where farming gives a better yield, a larger primary production. When this land is in a region with a very good transportation infrastructure, the price for the land may be higher, when demand is higher. In Germany, rents then can go up to as high as 320 Euros per $10{,}000\,m^2$.

Because agriculture depends on climate and weather conditions (drought, floods, insect attacks, diseases that harm agriculture), crop growth is not fully predictable and, thus, bears fundamental risks, around which and against which the economy has taken measures and insurances. If you cannot buy food from the farmers in your neighborhood, you must purchase it elsewhere. Transport of food is, thus, a huge global business. All this is secured by a finance and insurance industry, which hires experts and specialists, including chemists, meteorologists, biologists, geographers, all disciplines, you name them, in order to assess the risks and the resulting consequences when the risks become reality.

Agriculture is very important for many countries in the world because of the food supply for its own population or because the country can trade the food against other goods with other nations. Countries that produce and sell a lot of agricultural products can come into a monopoly position that gives the country a tool for coercing other countries in need of food imports. Intelligence on a countries' agricultural industry can be important for the national policies of other countries (anonymous, cia 1969, 1967).

8.1 The strategic importance of photosynthesis

I explained already in a previous book of mine (chapter 3 on ammonia synthesis for fertilizer production in Braun 2019a), not every country on the globe provides to 100% for its population's food supply, although this would be technically possible for many countries from the perspective of land, fertilizers, farming, water and sunshine supply, and farming work force. Being independent from foreign food imports can give a country more freedom in various activities.

It is, therefore, not surprising that countries are interested in agriculture economy and its relevant components, such as advances in photosynthesis and artificial photosynthesis. In Figures 8.5 and 8.6 you can see a report, with a link to the University of (you guess) Radiation Laboratory Bio-Organic Chemistry Group (the Calvin Lab) for the Central Intelligence Agency (CIA) (CIA 1954) in 1954.

Specifically, the report assesses the quality of photosynthesis research in the Union of the Socialist Soviet Republics (USSR). You conclude from this that it is not necessarily the employees of an intelligence agency who provide the intelligence, the information on a topic, and the answer to a question. It can be external experts, and it can be also science attaches (anonymous, cia 1967) in embassies worldwide. The gathering of information can be overt and covert. When you after your studies work in a scientifically or technologically relevant field, it is throughout possible that some colleague from some other country approaches you friendlily over a beer, so that he can ask how, and particularly, what this well-known researcher or that eminent professor in an important field is doing right now. Particularly if the professor was nominated for a Nobel Prize.

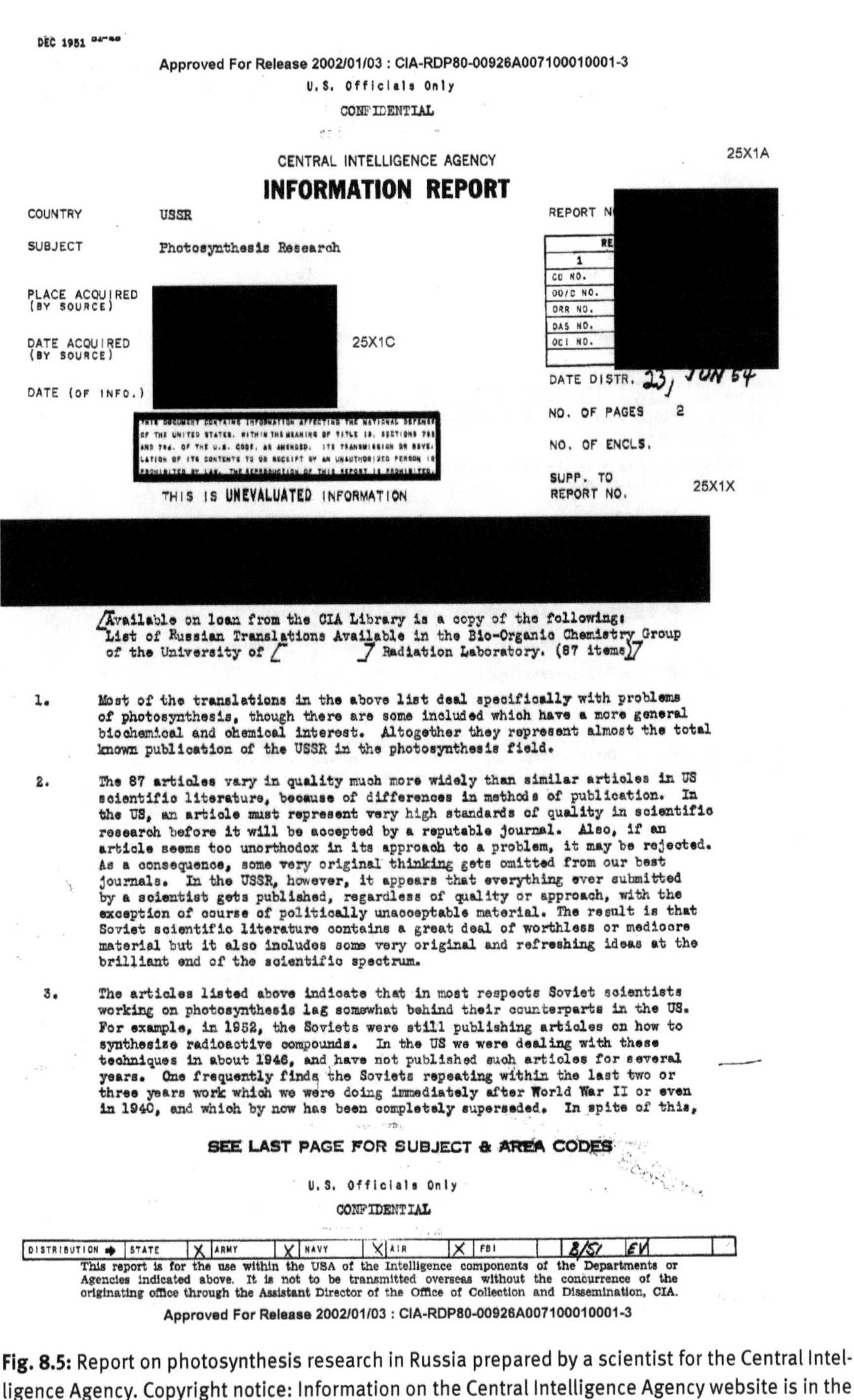
DEC 1951

Approved For Release 2002/01/03 : CIA-RDP80-00926A007100010001-3

U.S. Officials Only

CONFIDENTIAL

CENTRAL INTELLIGENCE AGENCY

25X1A

INFORMATION REPORT

COUNTRY USSR

SUBJECT Photosynthesis Research

PLACE ACQUIRED (BY SOURCE)

DATE ACQUIRED (BY SOURCE) 25X1C

DATE (OF INFO.)

THIS DOCUMENT CONTAINS INFORMATION AFFECTING THE NATIONAL DEFENSE OF THE UNITED STATES, WITHIN THE MEANING OF TITLE 18, SECTIONS 793 AND 794, OF THE U.S. CODE, AS AMENDED. ITS TRANSMISSION OR REVELATION OF ITS CONTENTS TO OR RECEIPT BY AN UNAUTHORIZED PERSON IS PROHIBITED BY LAW. THE REPRODUCTION OF THIS REPORT IS PROHIBITED.

THIS IS UNEVALUATED INFORMATION

REPORT N

RE

1

CD NO.

OO/C NO.

ORR NO.

DAS NO.

OCI NO.

DATE DISTR. 23 JUN 54

NO. OF PAGES 2

NO. OF ENCLS.

SUPP. TO REPORT NO. 25X1X

/Available on loan from the CIA Library is a copy of the following: List of Russian Translations Available in the Bio-Organic Chemistry Group of the University of / / Radiation Laboratory. (87 items)/

1. Most of the translations in the above list deal specifically with problems of photosynthesis, though there are some included which have a more general biochemical and chemical interest. Altogether they represent almost the total known publication of the USSR in the photosynthesis field.

2. The 87 articles vary in quality much more widely than similar articles in US scientific literature, because of differences in methods of publication. In the US, an article must represent very high standards of quality in scientific research before it will be accepted by a reputable journal. Also, if an article seems too unorthodox in its approach to a problem, it may be rejected. As a consequence, some very original thinking gets omitted from our best journals. In the USSR, however, it appears that everything ever submitted by a scientist gets published, regardless of quality or approach, with the exception of course of politically unacceptable material. The result is that Soviet scientific literature contains a great deal of worthless or mediocre material but it also includes some very original and refreshing ideas at the brilliant end of the scientific spectrum.

3. The articles listed above indicate that in most respects Soviet scientists working on photosynthesis lag somewhat behind their counterparts in the US. For example, in 1952, the Soviets were still publishing articles on how to synthesize radioactive compounds. In the US we were dealing with these techniques in about 1946, and have not published such articles for several years. One frequently finds the Soviets repeating within the last two or three years work which we were doing immediately after World War II or even in 1940, and which by now has been completely superseded. In spite of this,

SEE LAST PAGE FOR SUBJECT & AREA CODES

U.S. Officials Only

CONFIDENTIAL

DISTRIBUTION	STATE	X	ARMY	X	NAVY	X	AIR	X	FBI		8/51	EV	

This report is for the use within the USA of the Intelligence components of the Departments or Agencies indicated above. It is not to be transmitted overseas without the concurrence of the originating office through the Assistant Director of the Office of Collection and Dissemination, CIA.

Approved For Release 2002/01/03 : CIA-RDP80-00926A007100010001-3

Fig. 8.5: Report on photosynthesis research in Russia prepared by a scientist for the Central Intelligence Agency. Copyright notice: Information on the Central Intelligence Agency website is in the public domain and may be reproduced, published, or otherwise used without the Central Intelligence Agency's permission; (CIA 1954).

CONFIDENTIAL/US OFFICIALS ONLY

-2-

they refer to US work of rather recent date indicating that they receive and make use of the results of our research. Being as generous [redacted] with Soviet scientists, their published work indicates that they are at least two years behind us in isotope work.

4. In work on the photochemistry of chlorophyll, however, USSR research is about on a par with that of the US. In some aspects it appears to be a little ahead, in others somewhat behind. As an example, they claim to have accomplished photochemical reduction of a co-enzyme I or II using ascorbic acid as the reducing agent and chlorophyll for photo-sensitizing. This is an experiment which we in the US have tried and failed.

5. The most active workers in the photosynthesis field in the USSR appear to be the following:

PA Kolesnikov

AA Krasnovskii

AM Kuzin

EA Boichenko - This woman has been claiming for years accomplishments unknown in the US. I am inclined to think she is completely haywire.

NG Doman - Apparently quite young.

YB Evstigneef - Photochemical work.

6. Since 1952 Soviet work in photosynthesis has been increasing in both volume and quality. Their isotope work shows much improvement. Their photochemical work, however, was very good from the start.

-end-

CONFIDENTIAL/US OFFICIALS ONLY

Fig. 8.6: Report on photosynthesis research in Russia prepared by a scientist for the Central Intelligence Agency. Copyright notice: Information on the Central Intelligence Agency website is in the public domain and may be reproduced, published, or otherwise used without the Central Intelligence Agency's permission; (CIA 1954).

8.2 Food supply from the North American great plains

Saskatoon is a province in Canada that is known for its oil reserves, for its agricultural industry, particularly crops and grains, and for its potash mining, which is used and sold as fertilizer. During my visit to Saskatoon (August 2018) I cruised around in the north of Saskatoon. From what I saw, I can say that the area is at large wide and flat and underpopulated. The only natural barriers were frequent small lakes, and there were no mountains. A railway locomotive train announces itself with a bright light from far away, and it may take 10 to 15 minutes before it arrives at your place. The roads were, thus, straight and organized like in a square matrix in south to north and east to west directions. It appears these must be excellent conditions for large-scale farming.

Figure 8.7 gives you an impression of the landscape 15 km east of Saskatoon. We can see a long, straight highway (Highway 5) looking towards Saskatoon. The highway only has one lane because traffic only requires one in this sparsely populated area. On both sides we can see crop fields and ahead of us a small forest to both sides of the lane. On the left-hand side you can see an array of around 20 metal silos at a farm place.

Fig. 8.7: View at Highway 5 east of Saskatoon. The wide flat plains allow for easy industrial farming and growth of crops. Saskatchewan, Canada, August 2018. Photo taken by Artur Braun.

While only 7% of the land in Canada is used for farming, the plots are large enough to make it convenient for large-scale farming. Grain harvesters can spend an entire day or more on one wide grain field in Canada, whereas in small-grid Switzerland, a farmer is finished with the job in 2 hours, for example, because the patches are much smaller, and, thus, the track is much curvier. You get a glimpse of the differences between farming in wide Canada and narrow Switzerland when you look at Figure 8.8,

Fig. 8.8: Top: a corn harvester is finishing a very small corn field and loading a trailer with the grains while another truck brings a full trailer to the farm. Near the center of the Canton Zürich, 5. September 2018. Bottom: a small wheat field nearby, with roadside red corn poppy flowers. 23 June 2019. Photos taken by Artur Braun.

which shows a much smaller harvester and trailer on a small corn field right in the center of Canton Zürich. It takes two adults to carry out this operation. One drives the harvester, and the other drives the tractor. In March 2020, I saw that the brown soil was spread with blue fertilizer granulates, see the inset in Figure 8.14.

Switzerland, a country with a population of about 8 million, produces around half of its food by itself. The rest is imported. The relatively small volume of produce grown in this area in Switzerland is sold at local and regional markets, whereas the huge amount of produce grown in Canada is shipped all over the world.

On my trip through the great plains I came across a small farmhouse with a couple of silos, surrounded by a large wheat field on which several harvesters were busy with harvesting and loading wheat grains onto trucks. One truck would drive down the field road and then hit on the main road, drive across the street, and unload the grains at a range of silos, which you can see in Figure 8.9. The silos are located next to a railway track that runs parallel to the main road. You can see railway waggons on the track, waiting to be loaded during the harvesting season. The train is connected with over 200 such waggons and extends over several kilometers down the road.

Fig. 8.9: A truck delivers fresh harvested wheat from a nearby farm and fills the silo with grains at a railroad track. Hundreds of wagons are connected to the train, extending over miles of railway. Saskatchewan, Canada, August 2018. Photo taken by Artur Braun.

Fig. 8.10: Terminal for the handling and loading of grains, located right at the railway tracks. G3 is a grain handling company. https://www.g3.ca/ Saskatchewan, Canada, August 2018. Photo taken by Artur Braun.

At another road you can see a complex with an array of huge silos that apparently belong to the G3 company. G3 is a grain-handling company. The complex is in the middle of the great plains where the grains are grown, right next to a railroad, as you can see from Figure 8.10.

One multinational grain supplier is Cargill Inc., a family-owned business with global operations. One facility called "Cargill Inland Terminal" near the village of Clavet, not far away from Saskatoon, is shown in Figure 8.11. It is operated under the name of Cargill AG Horizons, which is a Minneapolis (Minnesota)-based company that provides grain contracts, a subsidiary of Cargill. The facility has a battery of grain elevators with direct connection to the railway and the highway.

Cargill also provides other farming services, including the development of new breeds of plants (Coonrod et al. 2008). Figure 8.12 shows a field of rape plants, which are derived from what farmers have known since the 1970s as Canola[1], an oil seed plant. Here, the new product VICTORY® Hybrid Canola grows on the field. The yellow "O" in the name symbolizes the yellow oil seed.

1 The name "Canola" was registered in the late 1970s by the Western Canadian Oilseed Crushers Association to describe the seed produced by a new market class of rape (Coonrod et al. 2008).

Fig. 8.11: Cargill Inland Terminal near Clavet in Saskatchewan, Canada, August 2018. Photo taken by Artur Braun.

Fig. 8.12: A field of rape near Saskatoon with advertising by Victory Hybrid Canola. Saskatchewan, Canada, August 2018. Photo taken by Artur Braun.

Nowadays, grain elevators are made from steel sheets. In earlier times, grain elevators were made from stone and concrete. At the back of the photo in Figure 8.13 you can see a concrete construction behind the field, which is one such grain elevator.

The railway does not only transport grain and oil seeds across the country to the ports, but also mineral oil, which Saskatchewan has plenty of. Figure 8.13 shows a

Fig. 8.13: Canadian Railway transporting petroleum in the small city of Young 1, Saskatchewan.

very long train in the city named Young 1, which is near Manitou Lake. The train was pulling around 200 waggons full of petroleum. It was actually not only one train, but a handful of trains that were distributed along the entire chain of waggons. Another major product that is shipped from Saskatchewan is potash, which is mined in the province. I will come to that in the section below.

So, in Saskatchewan, you can see a well-organized industrial complex, which is based on the availability of wide flat land with easy local transport by truck and efficient regional and continental transport by railway. Enough sunshine arrives on the ground, and the climate is warm enough to provide efficient growing periods. The soil is good enough, and still it also contains petroleum, which apparently does not cause big harm to agriculture. The soil actually contains large amounts of potassium phosphate, potash, which is good for growing plants and which is also strip mined and sold to other countries. On the global scale, primary production is limited by temperature, water, and phosphorus, when light and CO_2 are considered to be in sufficient quantity. In the photosynthesis apparatus, phosphorus is an important structural element and is also important in energy conversion processes. The transport of P is given naturally on the global scale by continental weathering and riverine transport (Hao et al. 2020).

8.2.1 Minerals, nitrogen, and fertilizers

In the first photo of this book, in Chapter 1, we saw the plants that do the photosynthesis and the blue sky that sends the necessary light energy, but we did not see the soil on which it grows. Figure 8.14 shows the farmland (800 meters away from the cornfield shown in Figure 8.8) after corn was harvested, and a plough reshaped the soil for further use (14 October 2018).

Fig. 8.14: An agricultural area near Zürich, Switzerland, with the Alps' panorama in the background and a small forest at the back. The brown strip is fresh soil after the corn harvesting was finished and a plough provided new surface for the fall season. 14. October 2018. The inset on the lower right-hand side shows the soil from a nearby farm on which blue NPK fertilizer was spread in March 2020. Photos by Artur Braun.

To have the necessary chemical elements to make photosynthesis, the plant cells need carbon dioxide (CO_2) and water (H_2O) from the atmosphere to grow grass, bushes, trees, and so on. However, we know from the past chapters that metaloproteins, such as hydrogenase, oxygenase, and chlorophyll, not only contain carbon, hydrogen, oxygen, sulfur, phosphorus, and nitrogen from the ambient. These proteins also have metal centers of iron, nickel, magnesium, and zinc; they contain sulfur and phosphorus and also potassium for the electrolyte.

These elements are added to soil as mineral fertilizers, which farmers buy from the agricultural chemical industry. The inset in Figure 8.14 at the right-hand bottom side shows the soil of a farming acre on which the farmer has spread blue fertilizer in the form of granulates. They have a size of 2 mm diameter and 2–3 mm height. The blue coloring is used so that farmers can distinguish the element composition of the fertilizers. The blue one is called 12/12/17 + 2, which means 12% nitrogen, 12% phosphate, 17% potassium, and 2% magnesium. This is a so-called NPK fertilizer, which was called Nitrophoska by the BASF company and registered as a trademark in Germany in 1926.

Plants pick up these elements from the ground, from the very soil that you can see as the brown region in Figure 8.14. These elements are all part of a cycle. There is a carbon cycle (see Melvin Calvin and Fridlyand and Scheibe 1999), a nitrogen cycle (Elishav et al. 2017; Norby 1998), a phosphorus cycle (Hao et al. 2020), an oxygen cycle (Falkowski and Godfrey 2008), an iron cycle (Stumm and Sulzberger 1992), and even a hydrological cycle (Michaelian 2012; Shvartsev 2018). Further, there is the CNO cycle in solar fusion by the proton–proton chain (Adelberger et al. 2011).

The CO_2 fixation is part of the carbon cycle and is essential for the primary production of biomass (Boyle and Morgan 2011). When the soil is not good enough for that purpose, because it lacks the minerals, we can add fertilizers that contain compounds that have these chemical elements, specifically potassium and phosphorus in the form of potassium phosphate and ammonium nitrate, provided that we have all those. If we lack the necessary fertilizers and if we cannot buy them, then we are in big trouble. Until 120 years ago, nitrogen fertilizer came to Germany as Chile saltpeter, a mineral containing $NaNO_3$. Before that, Peru had the monopoly, but in the late nineteenth century Chile became the main supplier of nitrates worldwide and basically had the monopoly on the nitrates when it won the War of the Pacific against Peru and Bolivia in 1891 (Sicotte et al. 2009; Watt 2003; Mora 2008).

In old ages and also still today, farmers bring the manure from their cattle and spread it over the soil; this is natural nitrogen fertilizer (read, for example, pages 563 and 575 in Chang and Halliday (2005). Figure 8.15 shows a farmer pulling a barrel of manure over a meadow and spraying it over the green grass. I was surprised when I took the photo that the liquid looks like water and not like brown manure. But it had the same smell, I can tell.

Not long ago, the pile of manure in front of a farm house was the direct measure and metric for the wealth of the farmer. Guano was the nitrate fertilizer produced from bird droppings on the coast in Chile and Peru. Table 8.1 shows the imports of nitrate minerals in the years 1912 to 1918. In the First World War, Germany was subject to a sea blockade by Great Britain and, thus, could no longer import nitrate minerals. The nitrates were used for gun powder and explosives, because nitrates are strong oxidizers. They were also needed also as fertilizers to feed the nation. The table lists the imports from Germany, Great Britain, and the United States, and other countries in Europe and the world. The Europeans were the major importers, certainly, because over 100 years

Fig. 8.15: Farmer spraying liquid manure over his meadow. Photo taken near the center of the Canton of Zürich 3 April 2020 by Artur Braun.

Tab. 8.1: Nitrate minerals imported from Chile in the years 1912–1918; data from Bastias Saavedra (2014); Monteón (1982).

Nitrate imports from Chile around World War I					
Year	**Germany**	**USA**	**Great Britain**	**Other Europe**	**Other world**
1912	37.9	23.6	5.7	31.6	1.1
1913	32.9	17.4	4.9	43.6	1.2
1914	23.0	23.0	13.9	37.5	2.6
1915	0	22.4	15.0	60.4	2.2
1916	0	40.4	23.7	34.3	1.6
1917	0	46.7	9.3	42.7	1.3
1918	0	57.1	19.4	23.3	0.2

ago, Europe was the center of world industrialization. The sea blockade against Germany worked, because imports immediately dropped to 0 from 1915 (Bastias Saavedra 2014; Monteón 1982).

The foreseeable need for more nitrate resources in modern society, the dependency on natural nitrate sources, which were limited, were the reasons why Fritz Haber worked persistently on ammonia synthesis from water and air – and succeeded along with his colleague Bosch (Haber 1914, 1920, 1922), and won the Noble Prize in chemistry for his invention in 1918.

Around Saskatoon, in between the crop fields, there are flat hills where mineral companies are located, such as Nutrien Corp. and Mosaic[2]. They take the potash minerals from underground and pile up the other soil of these hills and sell it as fertilizer worldwide. They process the potash on site and ship it with the railway, the tracks of which begin right at the facilities.

Figure 8.16 shows a sketch from the Annual Report 2017 of Potash Corp. with the potash mining and processing steps. Two holes are dug 1000 meter into the deep ground and constitute the service shaft and production shaft. Through the service shaft all equipment necessary for underground operation, including workers are brought downside to the layer of the underground potash ore body. Heavy equipment mines the ore and an assembly rail brings the ore debris to an underground ore storage silo near the production shaft.

Through the production shaft, the ore is lifted to the surface and sent to the ore milling facility. After the milling, the ore as a raw fertilizer product is put onto the

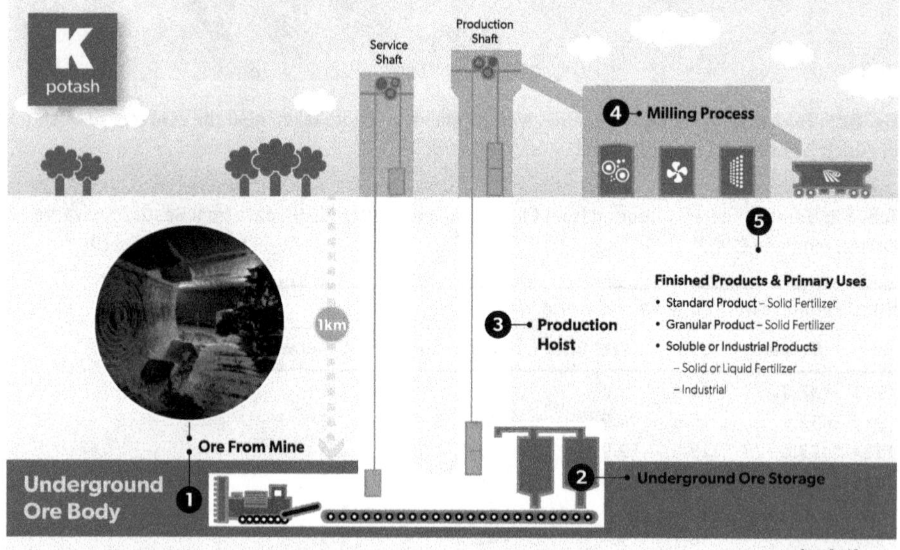

Fig. 8.16: Potash mining and production process. (1) Ore from mine: potash is mined using two and four-rotor continuous boring machines. (2) Underground ore storage: conveyor belts carry ore to underground bins, where it is stored until transportation to the loading pocket of the shaft hoist. (3) Production hoist: potash ore is hoisted to the surface through the production shaft. (4) Milling process: crushing, grinding, and desliming → flotation → drying and sizing → compaction and crystallization. (5) Finished products & Primary uses. Primary distribution methods: rail and vessel. Extracted from the Annual Report of Potash Corp from 2017, with kind permission from Nutrien Ltd. Image used under license from Nutrien Ltd.

2 Nutrien Corp. www.nutrien.com/what-we-do/about-nutrien and Mosaic www.mosaicco.com/.

railway waggons and transported away. The Potash Corporation of Saskatchewan (PotashCorp), with headquarters in Saskatoon, was founded in 1975 by the government of Canada. In 1989, it became a publicly traded company. The company produces potassium fertilizer in Saskatoon, and nitrogen and phosphate fertilizer in the United States. In 2012, they reported 8 billion US $ in revenue and 2.3 billion US $ in net profit. They had 5700 employees in 2011 and 4656 employees in 2017. Potash Corp. merged with Agrium Corp. in 2017 under the new name Nutrien Ltd.

Figure 8.17 shows a somewhat remote view of the mining facility of Potash Corp. at Patience Lake. The mining process there is somewhat different from the process shown in Figure 8.16. Instead, at Patience Lake, a hot liquid, the brine, is pumped into the underground, which leaches out the potash, and is then pumped up again to the surface where the brine cools down, and the potash crystallizes and settles. The residual brine is then separated, heated again, and again pumped into the mine for the next turn of the dissolution of the potash. This process is called solution mining.

Figure 8.18 shows the front gate of the mining plant, and Figure 8.19 an angle of view where you can see that the mining plant is connected to the railroad.

Behind the flat rock is the 5 km long and 1 km wide Patience Lake. The end of the lake close to the mining facility has four brine and crystallization basins with a length of 500 m and a width of 200 m. I could not access this part of the area, but you can find the lake on a map.

Fig. 8.17: The potash mining plant of Potash Corp. at Patience Lake near Saskatoon.

Fig. 8.18: The entrance gate to Potash Corp. mining plant at Patience Lake.

Fig. 8.19: Close-up view of the Potash. Corp. mine at Patience Lake.

8.3 Wildfires

8.3.1 Natural wildfires

In August 2018 I was invited to deliver a talk at the XRM2018 (X-ray Microscopy) conference in the city of Saskatoon in Saskatchewan, Canada. I had hoped that I could see the northern lights there. Somehow, lights are important in Saskatoon. When I arrived at the airport, there was a large advertising display at the luggage belt, which featured the Canadian Light Source in Saskatoon: the brightest light in Canada. See for yourself in Figure 8.20. I do not know whether the claim about the brightest light was in competition with the northern lights. During the conference, I had the impression that the organizers used a lot of green light to illuminate the stage for the presenters.

Fig. 8.20: The Synchrotron in Saskatoon advertises as the brightest light in Canada. Photo taken at Saskatoon Airport in August 2018, Artur Braun.

When I left the hotel in the morning to pick up my rental car from the airport, I noticed a smell from wood burning in Saskatoon. I asked a lady who was smoking a cigarette outside in front of the hotel, whether this was the smell from wildfires. She confirmed and said this was the smoke from the wildfires in British Columbia. I looked the place up on the Internet. British Columbia, on the Pacific coast, with the wildfires – that was 1500 km away from my place. With so much smoke in the sky, there was no chance I could see the northern lights, then. The weather forecast said it was a sunny day, but meteorologists apparently did not take into account the clouds from the wildfires. I did not see the sun.

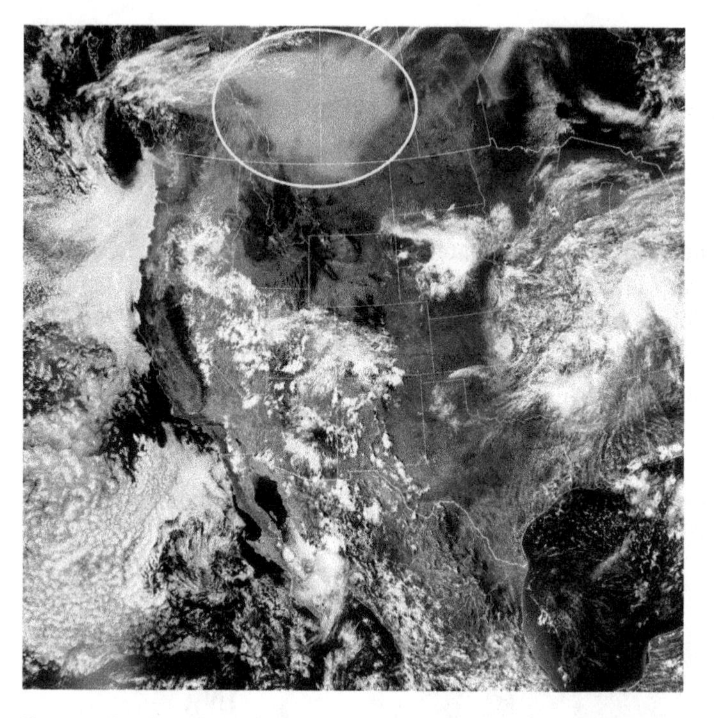

Fig. 8.21: Satellite photo of northern America showing a large cloud from wildfires over Canada. Credit: NASA Earth Observatory images by Lauren Dauphin, using VIIRS data from the Suomi National Polar-orbiting Partnership, GEOS-5 data from the Global Modeling and Assimilation Office at NASA GSFC, and data from DSCOVR EPIC. Story by Kathryn Hansen.

Figure 8.21 shows a photo taken from the DSCOVR (Deep Space Climate Observatory, (Frey and Davis 2016; Burt and Smith 2012), launched 11 February 2015) on August 15[3]. It shows the northern American continent, with the pacific on the left under many clouds, and the gulf of Mexico in the lower right. California is free of clouds. Oregon and Washington State show a light brownish fog, which looks different to the white clouds. There is a huge plume of slight brownish fog in the upper part of the photo showing the Canadian states. The yellow circle shows the huge smoke cloud that covers the state of Saskatoon and Manitoba.

Wildfires destroy the fruits of the primary production, the vegetation that does photosynthesis. Its clouds can also block the sunlight so that photosynthesis is affected for those plants who have not fallen victim to the flames. The latter holds also for volcano eruptions. We recall from Section 2.5 that satellite/based DSCOVR data can also be used to determine the portion of leaves that are exposed to sunlight (Yang et al. 2017).

3 https://earthobservatory.nasa.gov/images/92612/smoky-skies-in-north-america.

8.3.2 Manmade wildfires

In August 2019, I was invited to South Africa to various speaking events. This was the first time I went to Africa in the European summer, which in South Africa means winter. Typically, I go to South Africa in November to February. When our plane approached Johannesburg, I saw the city under a cover of whitish-brownish smog. "Damn, this must be a dirty city with air pollution now," were my thoughts. When I left the airport to walk to the rental car station, I smelled the air outside and immediately knew this was from the smoke of burning biomass, grass and bushes, and the like. On my way further to Pretoria, along the highway I saw that the strip to right and the left of the highway was black from burnt grass. I drove and I drove, and the whole way, I saw the black sides of the road, since the grass was burnt.

Eventually, due to the sheer magnitude of the black strips, I came to the conclusion that the Africans had been burning the grass along the highways by intention. So I told my wife that it appeared to me that the Africans were still doing what farmers did in my childhood in Germany: after every winter, when the snow melted, and the long yellow old grass from the previous year had grown too high, the farmers went along the fences and lit a fire. The farmers would cut a bundle of that old winter dry long grass, put a match to it to light it, and then walk along the fence and burn off the entire lines of high yellow old grass.

I know it because I saw it and I did it myself, almost 50 years ago in Germany. I burnt the old grass in March on my grandfathers meadows. The grass was still somewhat wet, and the smoke would be fat white from the humidity, and you could see the clouds from far away. This "flaming" (in local language: "vlämme") was part of the cultivation of the land to make it ready as a new meadow for the current year. However, in the 1980s, in Germany this practice was forbidden by law. I do not know the reasons for that, but overall it was for environmental purposes, I believe. When I drove through Italy in the Ligurain region in the 1990s, I saw fires in the hills along the ocean, which were of agricultural origin. People were obviously burning biomass for which they had no use. So, August 2019 in South Africa was obviously the phasing out of the South African winter, and people would simply burn off the old grass from the previous year.

I was wondering how and why it still seemed to be allowed, in South Africa, to burn biomass that way. To poison the environment. To kill the friendly good climate and foster adverse climate change. The black rims along the highways cannot have been the work of individual misguided local farmers; because this was not farmland. It was public land. So methinks it was the road workers that took care of the old grass and burnt it. Later, in the Pilanesberg Wildlife Reserve, I came across a group of workers preparing some concrete structure in the Savanna. It was about one dozen workers in uniforms, black people, who were building something there in the bush. I worried about the workers because 20 minutes earlier, we had seen a family of lions several kilometers away from that place. In that wildlife reserve, there are dangerous wild animals, not only lions, but rhinos and so on, that can be threatening to humans.

I was relieved when I saw a lady near the workers, a black woman in uniform with a rifle ready, apparently guarding the engineering operation in the bush with a weapon in order to be prepared for a potential attack by wild animals. Then I noticed that here too the bushes were burnt. There was no grass around. The large bushes, however, were still standing. I also saw animals (Figure 8.22), deer eating plants and leaves on the ground, on the black soil! In the above photos you can see the springbok and the zebra, eating from the black soil. I do not know whether they get food, maybe from the burnt biomass, which might contain ash with necessary minerals.

Anyway, the animal were eating there. A few hundred yards away, the nature was green, so there would not have been any need to feed on the black remains if they were really looking for green food. There was still plenty around.

Lately, we have heard about the gigantic fires in Brazil, and more recently about the wildfire catastrophe in Australia. Before that, we heard about the bad fires on the Californian West Coast, and I remember the huge wildfires in British Columbia in August 2018, which I smelled a 1000 miles away when I was in the midst of Canada in Saskatchewan. So, it seems we are now having now fire after fire, and in the news, this is sometimes blamed on climate change. I would blame it on arson, or maybe poor forest management.

When I was standing in the boarding queue in April, boarding my flight from Phoenix (I had attended the MRS Spring Meeting) to San Francisco, I got chatting with a lady from Oregon. She told me that many of the forest fires on the west coast are a result of poor forest management, based on ideology and environmentalism. There are frequently recurring forest fires in Yosemite Valley in California. Methinks the government and forestry management does not care too much about this. The fires are part of nature. They just make sure not too much of it is burnt away.

I was struck, however, this week, in January 2020, over a TV news report in Germany where an Australian voice said that many of the horrific fires could have been prevented if the old ways of forest management as practiced by the Aborigines (Solonec 2015) for tens of thousands of years, had been applied. Then they mentioned wet burning, the burning of the grass, the premature burning of the grass that would burn fast, like a *Strohfeuer* (straw fire), but one that would not ignite the trees. Right.

However, the fear of carbon dioxide, the *Klimagift*, from unnecessary biomass burning, made governments pass legislation that makes it illegal to light a fire in your garden today. Police will come even in very remote areas in Germany and fine you if you make a fire on your meadow from scrap wood or cut grass. Yet this is the way to get rid of combustibles, the fuel from the forests. At least, this is an easy and economic procedure that helps to "discharge" the fuel that naturally piles up in nature. Maybe we should reconsider some past policies in environmental protection and preservation of the last 40 years in Germany.

Fig. 8.22: Workers of the Pilanesberg Natural Park have laid fires in the wilderness. The photo shows a springbok and a zebra eating from the ground after the fire had destroyed the grass land. Photo taken 2 August 2019 by Artur Braun.

8.4 Oil from food. Food from oil

Around 10 years ago (maybe in 2008) when I had one of my numerous Synchrotron campaigns in Berkeley, at Berkeley National Laboratory, I took a couple of hours off to attend a podium discussion on the UCB campus. Speakers were nanoscience pioneer Paul Alivisatos and genetical biology pioneer Chris R. Somerville, both of UC Berkeley. I believe the topic was on solar fuels, and I remember it being said there that while today we are looking at making fuels from plants, the situation was reversed some 50 years ago. Back then, in the 1960s, the oil industry was looking into making food from oil.

At the time when this podium discussion was held, UC Berkeley was trying to enter a deal with British Petroleum BP and secure funding (very big funding, 500 Million USD) for what is now known as the Energy Biosciences Institute (EBI) (Sheridan 2007; Youngs and Somerville 2017; sugar 2011). At about the same time, the Helios Project on solar fuels was established at Berkeley Lab, and another large-scale national project on solar fuels was initiated at Berkeley and Caltech, the Joint Center for Artificial Photosynthesis.

8.5 Food supply in deserts and outer space

There are areas on the globe where you can grow food very well. However, food production in deserts, be they hot dry deserts in Australia, the Americas or Sahara Africa and Arabia, or cold ice deserts like in the Arctic and Antarctica, is a problem or largely impossible.

However, with specific precautions and preparations you can grow food in these desert places, including space and outer space. The Biosphere 2 project in the city of Oracle near Tucson, Arizona, is a large greenhouse that simulates an independent second biosphere, where eight researchers lived for 2 years and raised their own cattle and grew crops, with no external supply of water and air and food. Only the sunshine through the greenhouse windows and – this was cheating – the external electricity supply that powered the large climatization motor were resources from outside. (Torbert and Johnson 2001). Certainly, you have to build a new biosphere in such deserts.

8.5.1 Colonization of the Toshka region in Egypt

Egypt is a very large country (one million square kilometers) with a large population, much of which is concentrated along the Nile river region (El-Shabrawy and Dumont 2009). It can be looked at as five different land use/land cover (LULC) classes, which can be distinguished as agricultural land, barren land, urban areas, natural aquatic and terrestrial vegetation, and water bodies (Bakr and Bahnassy 2019).

Moreover, much of the country is desert land which is not populated because of the difficult living conditions there, such as high temperatures, no water, no food. But still, Egypt has a large population of 100 Million people, confined to a small valley along the Nile river. The colonization of the unpopulated regions is therefore a task also for architects (El Fiky 2002). Therefore, Egypt started a large development project in the desert.

> The fact remains that we cannot sustain ourselves in a narrow valley, with all the social and economic problems of overcrowding. We have to go to the desert and we have to build. Or the country will collapse.

says Mahmoud Abu-Zeid (Hope 2012)[4], Egypt's former Minister of Water and Irrigation when the Toshka project was launched in 1997, and who later headed the Arab Water Council in Cairo.

Access to water is very important for the agricultural industry. Who needs the water most? You and your family in the residential area? Or the large farm complexes out in the fields? Is the food produced on the farms for you and your family, or is it produced for export worldwide? Should the water in your communities' wells be used for residential purposes and nearby farming? Or should the water license be sold to a beverage company that uses so much water that in the long run, your community lacks access to water unless you pay the beverage company for water? To make natural water into potable water and residential water, it takes investments for which the community has to pay. How much of the burden is shared by residents, company owners, and farmers, is thus a good question. What is the fair water pricing for agriculture (Abu-Zeid 2010)?

Egypt started a mega project in its south eastern region at the Toshka lakes, the Toshka project. The purpose was to increase the inhabited land of Egypt from 5 to 25%, which is a considerable factor of 5. The Toshka is a desert region, but it has artificial lakes filled with water from the Nile river (known as Lake Nasser). It requires big pumping stations for that purpose (Malterre-Barthes 2016). The remote sensing with satellites provides useful data for the planning of such projects (Chipman 2019; Hereher 2015; Badreldin et al. 2014), but also for the impact of the drying out of the lakes and the changes in the regional climate (Hereher 2017). In 1997, the Saudi Kingdom Holding Company established the Kingdom Agricultural Development Company (KADCO) in order to buy 100,000 acres of desert land and prepare it for the farming of produce. This is an area of 20 km × 20 km. The overall land for the Toshka project

4 Mahmoud Abu Zeid received his PhD in Aquatic Engineering from the University of California at Davis in 1962. He was, for example, Chairman of the World Water Forum and of the INP Council of the World Meteorological Organization. He held offices in the Egyptian adminstration for various water affairs. He also was Head of Council of the African Water Facility (AWF). He joined the Egyptian governments under Atif Muhammad Nagib Sidqi, Ganzuri, and Abaid.

Fig. 8.23: Ramadan Nabil Abu Majed, 16, prunes grape vines with his classmates at the Kingdom Agricultural Development Company. Reproduction with permission from David Degner, Photographer.

would be an area of 45 km × 45 km. In Figure 8.23, you can see one of KADCO's plants, where a young man is pruning a grape vine tree[5].

Toshka was even once considered as a future solar hydrogen production region (Shaltout 1998)). The Egyptians want to reclaim land from the desert by distributing the water accordingly. They anticipate that the population in the new land will be low-income citizens. As the region is also prone to earthquakes, architects will have to take into account social geological and climate factors to build the settlements. The paper by Hassan and Lee is a theoretical concept for underground desert dwellings with passive cooling (Figure 8.24) that can accommodate the hot desert climate and be safe against seismic eruptions (Hassan and Lee 2014).

Today, the Toshka project is considered one of the various failed desert farming projects in Africa ("You can't look at this in a purely economic way," said Ahmed Sedky, a businessman who's involved in feasibility studies for parts of the "1.5 Million Feddans" project. "There is a lot of geopolitics in this." (Schwartzstein 2016). Warner believes there could also be inner political reasons for pushing the Toshka project

5 In his book on the Armagosa River, R.E. Lingenfelter explains how grapes were grown in Death Valley (Lingenfelter 1986).

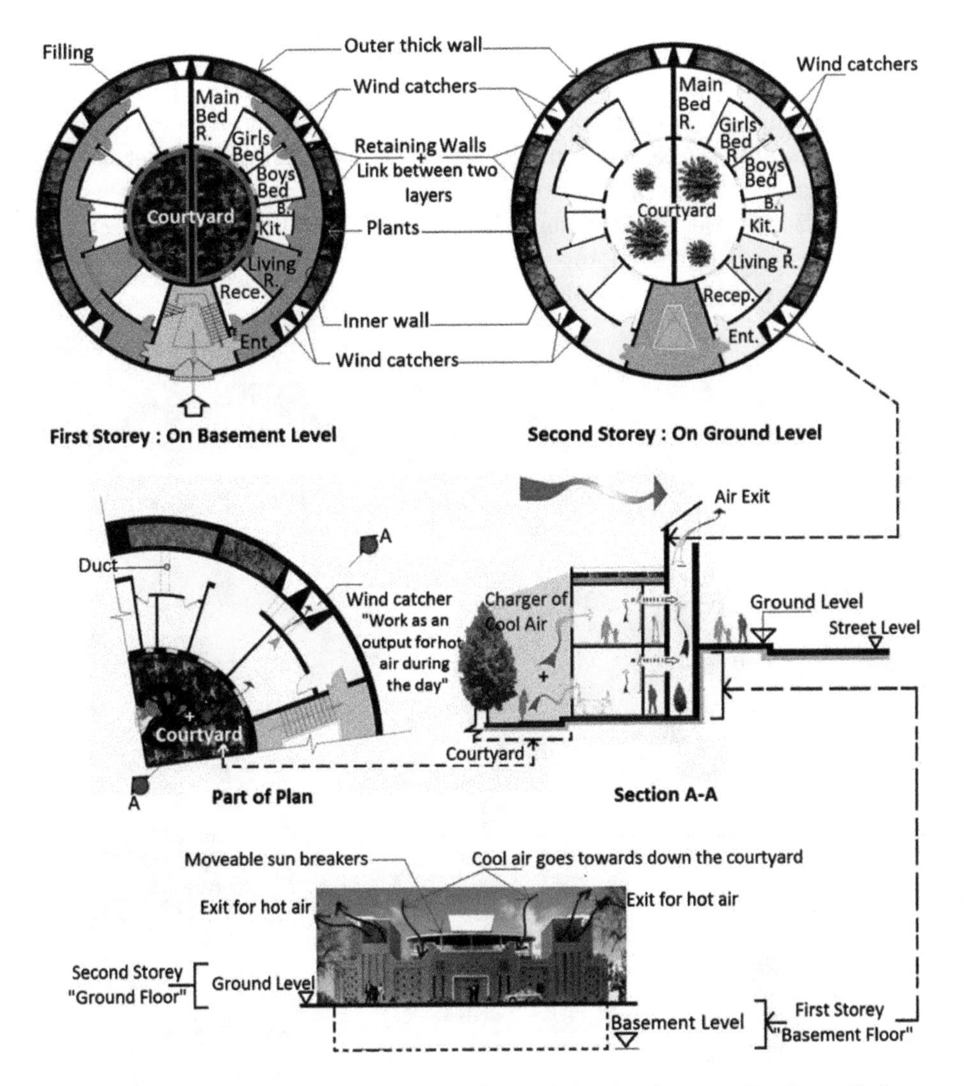

Fig. 8.24: First and second storey plan, partial section, and elevation demonstrating the ventilation system [(modified from Hassan (2009)]. Reprinted from Tunneling and Underground Space Technology, 40, Hassan, A.M. and H. Lee, A theoretical approach to the design of sustainable dwellings in hot dry zones: A Toshka case study, 251–262, Copyright (2014).

(Warner 2013). This is notwithstanding that farming is done at Toshka, as is shown by a fertilizer study (Awadalla and Morsy 2016). Moreover, there is still a belief that Egypt as a nation can be rebuilt "green and sustainable" (Sheweka 2012).

8.5.2 Lettuce and tomatoes grown in the Antarctica

A recent example of growing food in Antarctica is a project funded by the European Commission called EDEN-ISS c (Schubert 2017; Santos et al. 2016; Mauerer et al. 2016). Space engineer Paul Zabel has grown over 200 kg of lettuce, cucumbers, radishes, and other plants in the Neumayer Station, a laboratory of the Helmholtz-Gemeinschaft in Germany, which is located in Antarctica. The Neumayer Station is shown in the top panel in Figure 8.25

Fig. 8.25: Top: the Neumayer Station III of the Alfred Wegener Institute located at Atka Bay on the Ekstrom Ice Shelf in the Antarctica. Bottom: the Greenhouse of the EDEN ISS project at Neumayer Station III, 400 meters away from the main complex. Photos by DLR German Aerospace Center.

The food was grown in a greenhouse made of two sea containers, 400 meters distant from the main complex of the Neumayer Station. The bottom panel in Figure 8.25 shows the greenhouse container mounted on a large steel rig construction. On the container, you can read the words "Plant Cultivation Technologies for Space". The gardener, whose name is Paul Zabel, of the DLR Deutsche Versuchsanstalt für Luft- und Raumfahrt, needed to 400 meters from the main station to the greenhouse.

It appears that the greenhouse is made from two modules in the two containers. One module is the actual growing area with the plants, and the other is the control area, which hosts other necessary infrastructure. For the Biosphere 2 project, this module infrastructure was basically located down in the basement.

Figure 8.26 gives a glimpse of what it looks like in the grower's module of the greenhouse. There are shelves on the left and on the right, which have gray plastic trays that are filled with plants like lettuce and tomatoes – green and red food. Note that there is no window in the growth module that would allow for natural light to come in. The middle panel in Figure 8.26 shows beautiful red bell peppers hanging in the air.

The light for photosynthesis comes from arrays of light-emitting diodes (LED), which are commercially provided by the Swedish manufacturer Heliospectra AB. The necessary carbon dioxide for plant growth comes from pressurized gas bottles stationed outside in the cold. Nutrients are sprayed with water on the plant roots. There is no soil being used for such a kind of plant growth. So note that, here, the "farmer" does not use the very diluted concentration of carbon dioxide that is naturally in the atmosphere (currently around 400 ppm), but supplying CO_2 from a supplied gas bottle.

Industrial farmers elsewhere are doing the same. As the natural CO_2 concentration varies depending on the season, it actually in goes cycles, in addition to an overall increase, which is believed to force global warming, some greenhouse farmers supply extra CO_2 in order to increase the photosynthesis in their greenhouses (to increase the primary production, right? It is also called Greenhouse carbon dioxide supplementation). You can look up the webpage of the Ontario Ministry of Agriculture, Food, and Rural Affairs (OMAFRA) (Blom et al. 2002), which explains and shows how greenhouse farmers can use equipment that produces CO_2 for their greenhouses. The Factsheet of the OMAFRA begins with (Blom et al. 2002):

> The benefits of carbon dioxide supplementation on plant growth and production within the greenhouse environment have been well understood for many years.

A quantitative study on leaves from aspen trees was published by Jurik et al. (1984). The study shows the positive influence of the carbon dioxide concentration on the carbon-dioxide assimilation of aspen leaves, for example. The study also shows the effect of temperature and ageing of the leaves. When I visited the Biosphere 2 in Arizona in 2018, I took a guided tour through the greenhouse and learnt from the tour guide that the high concentration of CO_2 would eventually harm the plants.

Fig. 8.26: Inside the greenhouse of the EDEN ISS project at Neumayer Station III. The middle photo shows red bell peppers. The bottom photo shows roots without soil. Photo taken by DLR, the German Aerospace Center.

In Figure 8.27, I have schematically sketched the quantitative relationship between CO_2 concentration and its assimilation but with realistic values. There you can see that the relationship is sigmoidal. So over high concentrations, the photosynthesis will reach a plateau value. So, there is a maximum CO_2 concentration that plants can digest. When you look up another factsheet, a report by Poudel and Dunn at Oklahoma State University (Poudel and Dunn 2017), you will find a graph where the performance of the plants even goes down when a "toxic" concentration of CO_2 is reached, which appears to be 1200 ppm of CO_2.

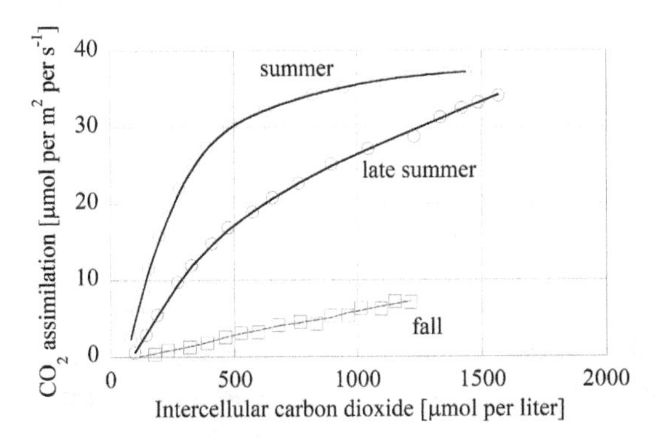

Fig. 8.27: Exemplary sketch of the sigmoidal variation of the CO_2 assimilation by a leaf depending on the concentration of CO_2. The ageing of the leaf will have an effect on the overall CO_2 assimilation and photosynthesis performance. It is highest in summer and will decline in late summer and become close to failing in fall. Inspired by Jurik et al. (1984).

As the current CO_2 concentration in the atmosphere is around 400 ppm, we still have some time to go before the CO_2 concentration story to photosynthesis turns toxic. Actually, with 400 ppm we are still at the growing tail for primary production. If we continue to plant more trees, the concentration may even decline, unless there are other ways to enormous carbon dioxide production, for example by burning of fossil fuels or biomass, or by volcanic eruptions, which may deliver carbon dioxide into the atmosphere. However, the CO_2 supplemented to the greenhouses for food production will again be turned into CO_2 by us when we eat and digest the food; this is part of our human metabolism. For the metabolism of plants, see the textbook (Smith et al. 2009).

8.6 Some thoughts about gas pipelines

It was shown almost 100 years ago that the transport of energy as hydrogen gas would be less costly than transport of electricity, at least when the distance is larger than 1000 km (Bockris 2002, 2013; Lawaczeck 1933). As a matter of fact, there are gas pipelines, including H_2 pipelines, and one should wonder whether it would make sense to also make CO_2 pipelines. One should also wonder how dangerous such pipelines can be (Doctor 2015); 25 years ago there was an incident in Africa where a volcano erupted softly and made a huge amount of CO_2, which caused the death of several thousand people.

8.6.1 Biosphere 2: a regenerative life-support system

When you live in outer space in a space station (or in a submarine or some other closed system) you have to make sure that you get enough food (and enough water to drink, oxygen from the air to breathe, and maybe some sunshine to produce vitamin D in your skin), see Mitchell (1994). For this, we need to have a life-support system[6]. Three potential systems are shown in Hoff et al. (1982). They are sketched as flow diagrams in Figures 8.28 and 8.29.

In the systems 1 and 2 find the human as the center of importance, and then the food, water, and O_2. Notice the arrows, which indicate that the human receives the three products necessary to live. The human also creates waste, and this is the CO_2 that we breathe out, and the feces as mostly dry waste and urine as liquid waste. The CO_2 is chemically converted to solid Li_2CO_3 (consider this as CO_2 trapping), and the water from feces and urine is electrochemically (electrolysis) split into H_2 and O_2. The somewhat more advanced system converts the CO_2 to methane CH_4 or solid carbon.

While the two life-support systems are quite sophisticated, they are considered nonregenerative because no food is produced; after some time, all food is used up, or for system 2, all water is used up and not regenerated, and then life is not supported further. These systems would work in a spaceship for some time, for example in Skylab (Johnston and Dietlein 1977). It is, thus, not surprising that the systems were proposed by NASA for their space missions (Hoff et al. 1982). Is it surprising that the book *Foundations of Space Biology and Medicine*, which includes lengthy details about life support systems, was coedited by Melvin Calvin (Calvin and Gazenko 1975)?

6 "A spacecraft's environmental control and life support (ECLS) system enables and maintains a habitable and sustaining environment for its crew. A typical ECLS system provides for atmosphere consumables and revitalization, environmental monitoring, pressure, temperature and humidity control, heat rejection (including equipment cooling), food and water supply and management, waste management, and fire detection and suppression. The following is a summary of ECLS systems used in United States (US) and Russian human spacecraft" writes Kathy Daues in her report (Daues 2006).

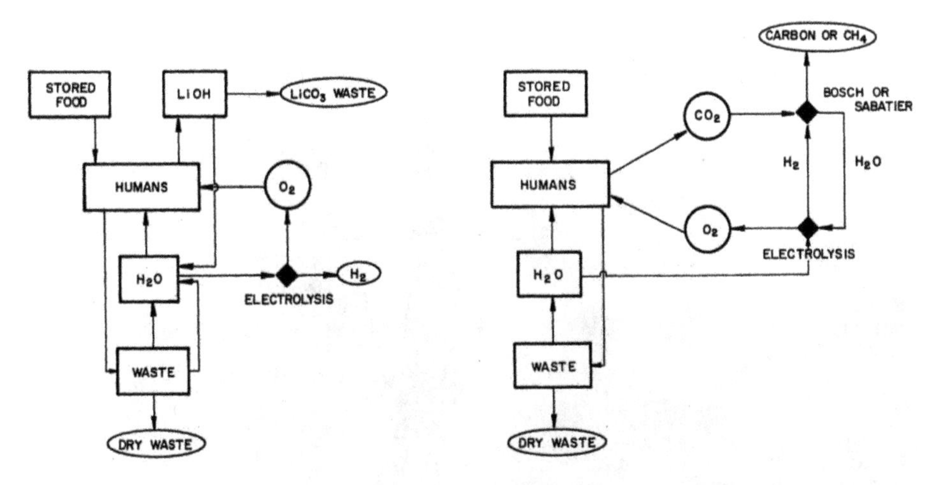

Fig. 8.28: Left: a nonregenerative life support system utilizing CO_2 trapping. Right: a nonregenerative life-support system utilizing catalytic O2 regeneration. Reproduced from Hoff, J.E., Howe, J.M., and Mitchell, C.A. Purdue University 1982, Nutritional and Cultural Aspects of Plant Species Selection for a Controlled Ecological Life Support System, NASA Report, National Aeronautics and Space Administration, Ames Research Center, (Hoff et al. 1982) Hoff 1982.

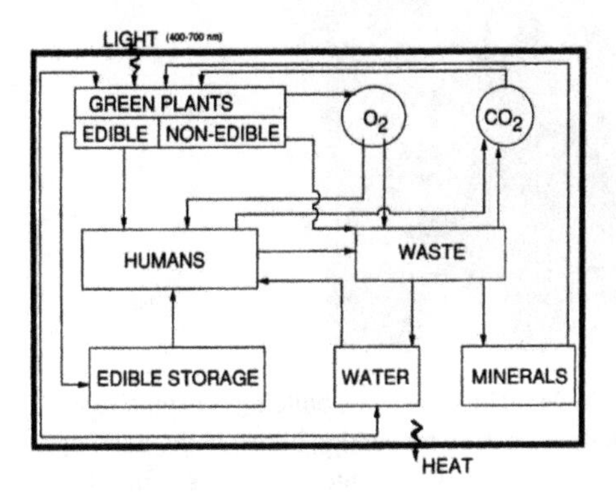

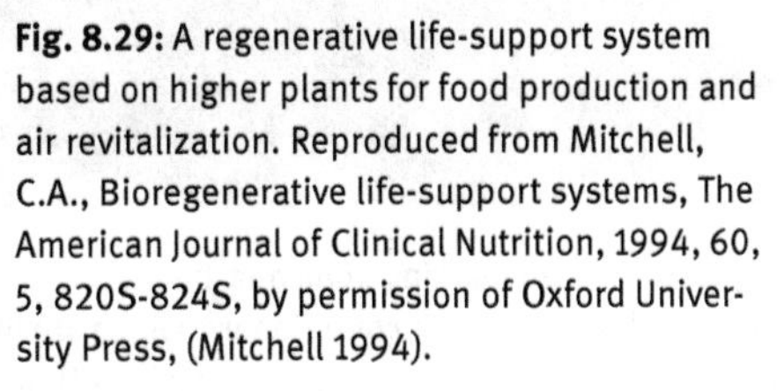

Fig. 8.29: A regenerative life-support system based on higher plants for food production and air revitalization. Reproduced from Mitchell, C.A., Bioregenerative life-support systems, The American Journal of Clinical Nutrition, 1994, 60, 5, 820S-824S, by permission of Oxford University Press, (Mitchell 1994).

A regenerative life-support system is explained in Figure 8.29. Observe that the authors drew a thick black line as a frame around the sketch, most likely to show that this is a closed system, with the exception that solar light radiation enters the frame, and dissipated heat leaves the frame. Figure 8.30 shows the front side of an apartment complex in Seoul, Korea. You can notice that almost 40% of the front side is covered with small air-conditioning systems, a really extreme situation which makes me wonder what the house is being used for.

In a regenerative life-support system, humans are locked up in the system, and they do not come out when they might run out of water or out of food, or out of oxygen, because all of this will be regenerated by the light as the primary and only energy source.

Fig. 8.30: You are looking at an apartment complex in Seoul in Korea. This building is extensively equipped with many air conditioners, which are likely to be used for venting hot air in the buildings to the outside. Compare this to the previous Figure 8.29, where the bioregenerative life-support system design shows or indicates an exit for heat.

The incoming light – probably through windows, otherwise the air might escape – shines on green plants inside the frame, and these produce biomass, which is edible food (fruit from trees, berries from bushes, grains, roots like carrots and potatoes, lettuce, cabbage, and so on) and nonedible biomass like wood and leaves. The edible food is eaten by the humans, maybe by cattle, and the other biomass can be converted to CO_2, which is used by the plants to continue the photosynthesis. Food that is not eaten immediately will be stored, as we know from farming communities. There must be a basic constant volume water supply to water the plants and for drinking water. The water may also be used for technical operation of the system, like cooling and processing water. There is a basic oxygen supply in the atmosphere inside the system, and further O_2 is produced as a byproduct of photosynthesis.

Feces and urine as human waste, along with biomass waste contains minerals that are necessary for plants to grow. The minerals are fed back to the plants. In theory, this regenerative system works. It also works in practice, because what is shown here is nothing less than our globe, observed from outer space. It has worked for millions of years. This is our biosphere.

In the 1980s, plans to build an artificial regenerative life-support system solidified, and it was eventually built in the Arizona desert, 20 miles north-west of Tucson, and called Biosphere 2. You can see part of the Biosphere 2 complex in Figure 8.31, with a large greenhouse that looks like a pyramid but without the top, and to the right of it some white sphere, which is part of the residential and laboratory part of the biosphere complex.

Fig. 8.31: The Biosphere 2 complex near Oracle, Arizona. Photo taken, 9 April 2018 by Artur Braun.

When Biosphere 2 was built and ready for operation, a team of eight scientists, four men and four women, were sent into the complex and left to themselves to live and feed on what was prepared for them for their mission. With one exception: the energy necessary to operate the complex came from outside and was mostly electric power. Biosphere 2 is a construction of concrete, steel, and glass. It must be glass so that enough light can enter the complex for the photosynthesis to work.

Fig. 8.32: The Biosphere 2 complex near Oracle, Arizona. Photo taken by Artur Braun, 9 April 2018.

The eight inhabitants were expected to do the farming, I think they also had some goats there, and also do scientific research. Figure 8.32 shows a photo I took on the large complex in April 2018. The photosynthesis of biomass is done in the large pyramid-like greenhouse, part of which you see on the left-hand side. In the center of the Figure, you can see the white buildings that contain the sleeping areas, the kitchen, work areas, laboratories, and so on. On the right, there is a large, flat white dome and this contains the power installations for Biosphere 2.

Today, Biosphere 2 is a museum and conference center. Only minor work is for original scientific research is carried out there. Most of the science work is done for the education of undergraduate students, to the best of my knowledge. It is a good thing that Biosphere 2 is a museum, because then this system is open to the general public, and we all can learn details about it on site. They offer guided tours, for example. I am not aware of any current experimental work at Biosphere 2, but there is a constant increase in publication output mentioning Biosphere 2, even in the year 2020. Further, Biosphere 2 as a research lab itself publishes papers; I found over 200.

The Biosphere 2 visitor will notice, for example, an experiment where tomatoes are grown in an artificial greenhouse, comparable to what you saw previously in the EDEN project in the Antarctica (Figures 8.26 and 8.33). You can see a cylinder structure, a large tube so to speak. It is illuminated by artificial lighting and keeps the tomato plants growing. I can assure you that I have seen tomatoes inside and that the experiment appears to be working properly. On my last visit there I happened to

Fig. 8.33: An experimental greenhouse with tomato plants in the housing and science module of Biosphere 2 complex near Oracle, Arizona. Photo taken by Artur Braun, 9 April 2018.

meet a young microbiologist, who was preparing experiments for soon-to-come high school students who had been selected for a project time-out at Biosphere 2.

The greenhouse in Biosphere 2 is divided into several sections, such as a savannah, a desert, and a rain forest. It contains water ponds and dry areas. The designers and architects of Biosphere 2 were very creative and wanted to test several kinds of habitats, like rainforests, oceans, upper savannahs, coastal fog, deserts, kitchen habitats, technospheres, in addition to the original idea of the project, which was the regenerative life-support system for the eight humans. Figure 8.34 shows the upper savannah. When I took the guided tour in 2018, I learnt that the concentration of CO_2 in Biosphere 2 had increased so much because of the eight people living in it, that this was harming the vegetation.

One problem of the project was that the oxygen was consumed by the steel structure (see the ceiling in Figure 8.33) of Biosphere 2. Steel contains iron, and iron likes to react with water and oxygen and then makes $Fe(OH)_2$, which is known as rust (consider also the pioneering paper (Stumm and Sulzberger 1992)). So, the infrastructure competed with the humans, and the goats, for oxygen, and there was no easy way to recover from that. Maybe you can understand that the sensors found that oxygen was lost from the system. However, the sensors did not say why this was so. Was there a leak in the structure? You would not know immediately, and it takes a lot of analyses to find out the true reasons for the observations.

Fig. 8.34: Inside the Biosphere 2 growing complex (greenhouse). Photo taken 9 April 2018 by Artur Braun.

The Biosphere 2 project had quite bad press. According to the press, it was a failure. The eight team members had to leave the project after 2 years because the regenerative system was not fully regenerative. They were not able to grow enough food for everyone, and there were personal conflicts among the eight, which made cooperation very problematic. This is what the press considered failure. I have a different opinion. I think Biosphere 2 was a great success because it shows the bottlenecks that need to be taken into account when planning for a long-term space mission where no help from outside, from the Earth, would be possible.

Even with support from the outside, the Biosphere 2 adventure, or expedition, did not last longer than 2 years. The support from the outside was the technical infrastructure that provided the electric power from the dome, and the heat and mass transfer, which was powered by machinery in the basement of Biosphere 2. It had a lung, a huge

Fig. 8.35: Walk through the Biosphere 2 basement with air-conditioning infrastructure. Photo taken 9 April 2018 by Artur Braun.

membrane that pumped the air and caused circulation of CO_2, which in real nature is done by the wind. When you book the guided tour, you will gain access to those parts of Biosphere 2 that are otherwise not open to the public. This also includes the basement with much technical infrastructure.

Figure 8.35 shows a photo I took in the basement during the guided tour. You can see pipes, and there are also pumps and vessels and reactors, which constitute the air conditioning for Biosphere 2. In so far, the whole Biosphere 2 seems to be a fake. However, this seems to be the first project of its kind. In the beginning, we often start from scratch, or we have to make concessions just in order to somehow get started. Mankind learns from Biosphere 2 that it is not so easy to make a self-sustaining small copy of our globe.

However, to climb the ladder to the top requires taking the steps in between; Biosphere 2 seems to me to be the first step on the ladder to a regenerative life-support system for the future colonization of space and outer space. While I am writing these lines in February 2020, I am preparing for a project[7], which is about the influence of γ-radiation exposure of cyanobacteria on its photosynthesis performance. This project is funded by a bilateral program between Switzerland and the province of Flanders in Belgium.

My collaborators on this project are from Hasselt University (Braun et al. 2014) and from the Belgian Nuclear Research Center in Mol, SCK·CEN (Badri et al. 2015; Janssen et al. 2014). The research group in Mol has expertise on cyanobacteria, particularly *arthrospira*. They are participating in the ARTEMIS project, which has sent some cyanobacteria experiments with the International Space Station not into outer space, not even into space, but into an orbit around the globe. γ-radiation can be harmful to the photosynthesis of cyanobacteria (Abomohra et al. 2016).

The world population has kept growing considerably since the invention of the Haber–Bosch process for ammonia synthesis (Haber 1920; Kyriakou et al. 2017), which produces synthetic fertilizer and helps produce food for mankind in an unprecedented way. In general, it has been the fossil fuels that have provided the power for this process, and for this growth of population; this is not withstanding that there are ways to produce ammonia without fossil fuels, for example by using hydropower plants to produce electricity (Egypt, Norway) and then using electrolyzers to produce the hydrogen H_2. At some point, if the population keeps growing at this rate, the world may become too small for dozens of billions of humans; we do not want to live in chicken cages with a mandatory limit on calories and living space. At least, I do not want people to have to live in such world.

On the other hand, a Malthusian limitation of population growth is not an ethical philosophy for human development either. As the radius of the Earth's globe is not expanding, we must consider the colonization of space and outer space. Mankind is not breaking with a tradition by leaving Earth and going elsewhere. When we did not have enough space, we went elsewhere and occupied far-away territories, in the Americas, in Africa, and in Australia. These endeavours were not peaceful. It was a colonization of other people's territory.

In America, the European settlers expelled the native Americans, who occupied way more territory per person than the settlers from Europe did. We (Europeans) had better farming technology, which allowed us to produce more food per area than the hunters and gatherers among the native Americans could. The same holds for Africa and for Australia. I think I wrote elsewhere in this book that 10,000 years ago people required an area of 10 km square per person for survival, as hunters and gatherers.

---

7 http://p3.snf.ch/Project-189455, Flanders/Swiss Lead Agency Process: Charge and energy transfer between cyanobacteria and semiconductor electrodes under gamma-irradiation.

Therefore, we depend on the progress of technology if we want to sustain the growth of the human population, and we will need way more space for that, which is not available on Earth. Therefore, it is logical for us to make the next settlement in space and in outer space. We must, therefore, develop the technology that will allow us to do so. This will also require that we move beyond mere technological boundaries and learn to work with each other in such a challenging mission. For example, at least one of the eight humans in Biosphere 2 was caught cheating, bringing in food from the outside, which would not work under real space conditions.

Another issue that I heard about during the guided tour at Biosphere 2 was that there was not enough food for the eight humans, given the circumstances during the project. So, they expelled one off the team of eight humans to continue living in normal civilization. Fine, but what if you have a colony of humans in outer space – millions of miles away from Earth, and not enough food for 1/8 of the population. Would it be possible to send all of them back to Earth to get the food they need, if Earth were already overpopulated? Certainly not.

A worst-case scenario of overpopulation and struggle for survival, at least by access to food, is shown in the scary 1973 movie Soylent Green (Fleischer 1973; Locher 2019). Based on a book from 1966 with the title *Make room! Make room!*, it is a science fiction with story about 20 million people living in New York, and the only food for the population is cookies, supposedly made from soya and lentils (SoyLent), and from krill, which comes from the oceans. However, the main characters in the movie find out that the oceans are dead, the green Soylent Green is a swindle, and the cookies are actually made from the corpses of the deceased population. Therefore, food security is an important topic (Uz et al. 2019). In the movie, the last resort to food security was food from the oceans, krill, and that failed for reasons not disclosed in the film.

According to eighteenth century scholar Malthus, population growth follows a geometric growth law, whereas food production follows an arithmetic growth law, with the consequence that there will be eventually a point of coincidence with an inevitable battle for food, known as the Malthusian catastrophe. The fear of such consequences has caused interest groups to influence governments in their policy making (Locher 2019; Fennell 2019).

8.7 Algal bloom

Cyanobacteria are the first organism in the evolution of life that have been able to produce oxygen from photosynthesis. This oxygen production was the source of all life that breathes air (oxygen), and is, thus, essential for the evolution of life in general. Blue-green algae are also called cyanobacteria; they are procaryotic algae. The blue color of the phycocyanin dye is the reason why they are called blue algae or blue-green algae. With a high concentration of nutrients, high light intensity, and warm water temperature, they have optimum growth conditions.

When there is an oversupply of nutrients in the water, the algae reproduce too fast, and this makes that the population of algae become too large and too concentrated, with the result that the water gets a greenish color. This is known as green tide (see, for example, Ye et al. 2011 and Wang et al. 2016). Because of light absorption by the dense and concentrated algal "soup", the transmission of light into lower regions in the water is decreased to the extent that life below the green algae layer is no longer sustained. When microbes do no longer do well because of reduced light intensity in the deeper regions, they may malfunction and release toxins into the water, which can be harmful to fish, for example (for a review on harmful algal blooms on aquatic organisms, see Landsberg 2010).

In Figure 8.36 you can see a collection of four photos that were taken at Bodega Bay in Northern California, 100 km north-west of Berkeley. The University of California Davis operates its Bodega Marine Reserve in Bodega Bay. While I was sitting in a cafe at the seashore in Bodega Bay, I noticed green patches in the water. The waitress told me these were algae, and I believe it could have been the sea lettuce, the *ulva spirulina*. The waitress told me that they were not harmful, at least not in the current situation. On the bluish water you can clearly identify the green patches that one may call bloom. In the lower right photo, which shows the bay from a far distance, you can see large colorized patches that make an optical contrast against the water color.

Fig. 8.36: Patches of green algae during an algal bloom on the Pacific Coast in Northern California. Photo taken at Bodega Bay CA, 16 August 2018 by Artur Braun.

The waitress told me that there had been a quite harmful red algal bloom several years ago. This is also called red tide. Back then, much of the seafood, the fish and abalone and shells, had died, and some fisherman lost their business because of this bloom. Such an algal bloom can begin when in shallow waters, the bottom of the sea is swirled up and nutrients in the soil are dispersed in the water. The algae then find fertile ground for reproduction and multiplication. The algae can form large further colonies, which grow and spread, in the extreme case over very large distances. This is, then, a large contribution to primary production. This, however, is at the cost of other species. This amounts to the population dynamics of a predator–prey system, where the predator is the algal bloom, and the prey may be other disadvantaged species, which, however, are not consumed by the algae, but the dynamics is the same.

You can grow such algae in a controlled environment and harvest and process them, and use them as a dietary supplement, such as the anabaena spirulina cyanobacteria. (Green et al. 2014; Olyarnik and Stachowicz 2012)

8.8 Food trade, stock exchange, and Warenterminhandel

When you are a gross sale trader of bulk commodities, for example wheat or other grains and cereals, you also depend on the weather conditions and the market conditions, pretty much like the farmer. The same holds, for example, for a grain mill that is in need of grain in order to produce and then sell the flour. Or the grain is needed as food for cattle farming elsewhere. The same holds for the buyer.

When you and the supply chain depend on the availability of such goods or the need of such goods, and when for whatever reason the goods cannot be delivered, you do at least want to limit your risk of financial loss. You may want to secure the availability of these commodities with an insurance policy.

This is institutionalized on the stock market with what is called a warrant. A common stock of a company may come with a warrant that constitutes an option that you can purchase that stock at a particular price until a particular deadline in the future, like 1 year or 5 years. When you observe the stock market as an investor or as a speculator, you may come to the conclusion that it may be worthwhile purchasing the common stock or the warrant, or both, under particular circumstances. In the development of the option practice, two different standards or practices have evolved, which are known as American options and European options, the differences of which I have listed in Table 8.2.

The evolution of markets and the conditions that shape them are a matter of fundamental uncertainty. It is, therefore, at large a matter of psychology how buyers and sellers decide on a purchase contract on a market, including the pricing[8].

---

8 There may certainly be technical and economic reasons why a purchase or sale must be made, irrespective of ny psychological bias. Yet the time when and the conditions how such transactions are made may throughout depend on psychological factors.

Tab. 8.2: The difference between the American option and the European option.

Difference between American option and European option	
American option	**European option**
The stock may be purchased and sold any time **before** the contract expires	The stock may be purchased and sold only on the day **when** the contract expires

Sheen Thomas Kassouf talks about the Never-Never Land of complete certainty when he believes there is a the need for some insurance on trade deals that come with an uncertainty that amounts in a potential risk, which needs to be controlled by a common stock purchase warrant.

This is why sellers and buyers, traders of such material commodities particularly from the volatile agricultural sector and farming business, agree on option pricing contracts, which basically insure the deals around the traded material. There exist theories and mathematical tools for the determination of fair pricing for option certificates. Today, even tools of quantum mechanics are applied for option pricing, see, for example, Inoue (2006); Contreras et al. (2010).

8.8.1 The option price model by Kassouf

When working as stock market analyst in New York, Kassouf made observations when he was evaluating stocks for his employer (Kassouf 1962), which raised his interest in common stock purchase warrants, for which little scientific basis and theory existed at the time. Kassouf discovered plenty of material for a qualitative and quantitative theory of the common stock purchase warrants.

Writes Kassouf in his doctoral thesis (Kassouf 1965):

> By examining the decision-making process of an individual, Chapter I revealed the futility of attempting to reconstruct the process from observed behavior; but Chapter III outlined a method by which inferences can be made from an existing option-common price relationship. This may provide economists and econometricians with a key to the house now occupied almost solely by psychologists and sociologists interested in motivations and "investor psychology", i.e., investor expectations and risk-attitudes.

Kassouf was able to show "that many different combinations of risk-attitudes and expectation distributions can give rise to the same observed result". This behavior is reminiscent to the path integrals and the global description of systems[9] by Hamilton formalism.

9 I write here systems with the distinct understanding that the reference is to physical systems and extensions to any systems, in as much they can be represented by the mathematical Hamilton formalism, certainly including the economy, as we will see later, more specifically in this chapter.

Kassouf then continues in the abstract of his doctoral thesis by stating "by describing the decision-making procedure quantitatively, the role of expectations and risk attitudes is made explicit and tractable". This description is reminiscent of Lagrange formalism, a local description of systems that can be derived (the second kind of Lagrange equations) by the first variation of the action integral.

The spirit of the language in Kassouf's thesis is, therefore, very helpful for paving the path of his classical treatment of the mathematics of stock options towards a quantum mechanical formalism of the same.[10] Shortly after writing his PhD thesis, which Kassouf finished at aged over 36, he coauthored a book about the stock market (Thorp and Kassouf 1967) with the title *Beat the Market: A Scientific Stock Market System*.

Kassouf argues that the decision-making process can be rationalized by probability distributions that can be mathematically expressed as utility functions $Q_t(x, S)$, with S being a set of parameters, such as the expectations E_i of the investor i, his risk attitude R_i, and also a quantity Q_i, which estimates the relation between common price and warrant. These are the exercise price a, the duration t of the option to exercise, the dividend yield r of the common plus the possible, or potential dilution d of the common stock. Finally, Kassouf takes a stochastic variable e, so that the price y of the warrant has the following structure:

$$y = Q(x, t, d, r, E, R, e) \tag{8.1}$$

Kassouf looked into actual stock data and found trends that suggested a linear relation between the logarithm of the warrant price y, after normalizing the warrant price y by dividing it by the exercise price a and then get a reduced quantity y^*, so that a does no longer shows up in the equations:

$$\log(y) = k_1 + k_2 \cdot \log(x) + f_3(t, d, r, e, R) + e \tag{8.2}$$

which he could parameterize into the following form:

$$\log(y) = k_1 + k_2 \cdot \log(x) + k_3 f_3(t) + k_4 f_4(t) + k_5 f_5(r) + k_6 f_6(E) + k_7 f_7(R) + e \tag{8.3}$$

and specify in more detail

$$\begin{aligned}\log(y) &= k_1 + k_2 \cdot \log(x) + k_3 \left(\frac{1}{t}\right) + k_4 \cdot d \\ &\quad + k_5 \cdot r + k_6 \cdot \log\left(\frac{X}{\bar{X}}\right) + k_7 \left(\frac{1}{t}\right) \cdot \log(x) \\ &\quad + k_8 \cdot d \cdot \log(x) + k_9 \cdot r \cdot \log(x) + k_{10} \cdot \log\left(\frac{X}{\bar{X}}\right) \cdot \log(x) + e \end{aligned} \tag{8.4}$$

10 Kassouf's works are available for free download at the institute website of the University of California at Irvine.

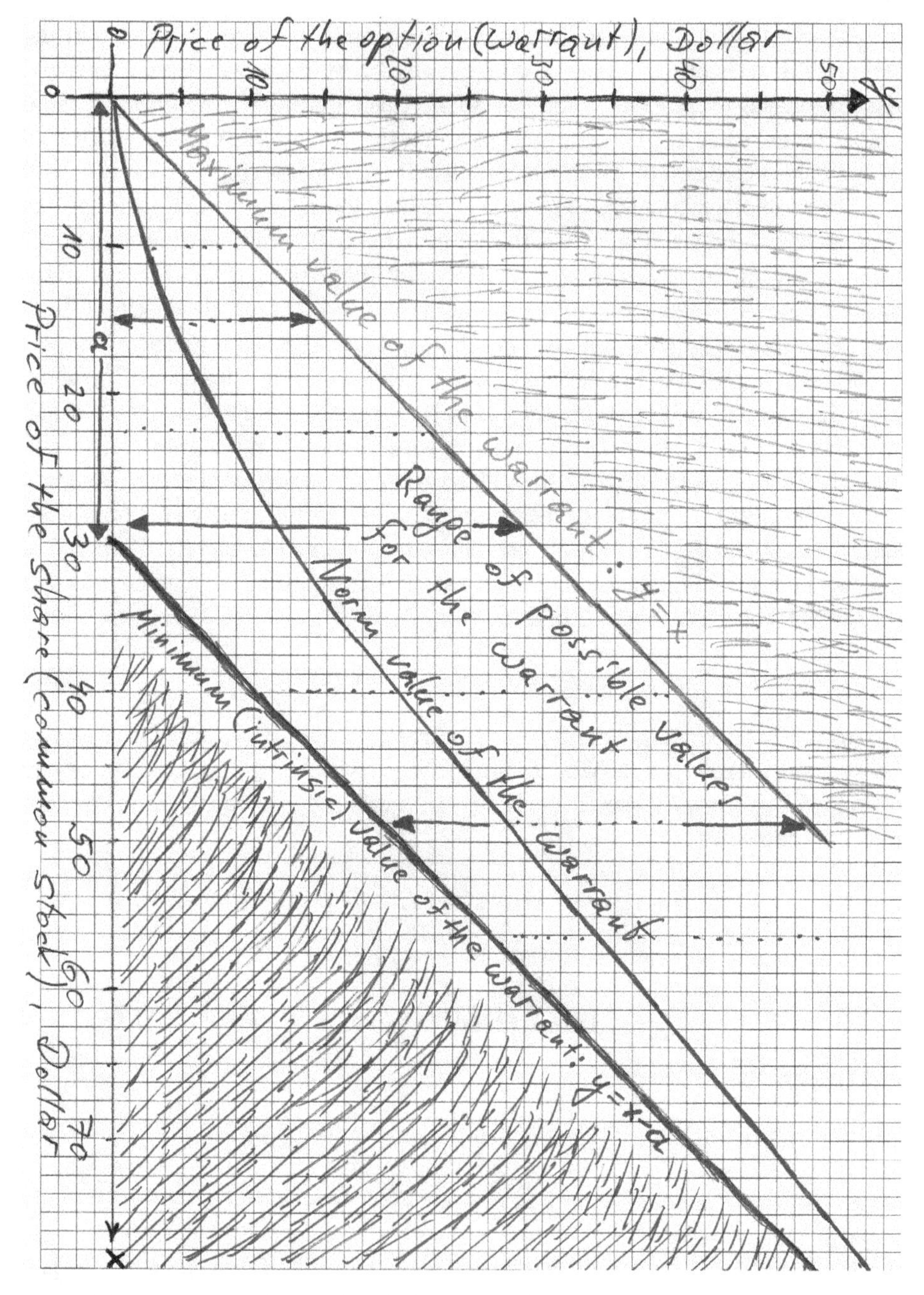

Fig. 8.37: A phase diagram showing the relation between the price of the warrant y and the price of the share x. The range of possible values (prices) for the warrants is enclosed by the straight lines $y = x$ (green), the maximum value for the warrant, and the straight line $y = (x-a)$ (red), the minimum value for the warrant. The exercise price is given by the constant a, which determines the minimum value by offset. The blue line within the range of possible values for the warrant is one potential of many possible ones.

The behavior of an investor with respect to the purchase of the warrant and the exercise of the warrant, i.e., the decision-making can be described by the trajectory of the points $P_t(x_t, y_t)$ over which Kassouf estimates the prize of the common stock and the warrant. Here, t is the time parameter, which can also be used to parameterize the natural coordinates y and x, which define the space of common stock and warrant prices.

Therefore, in Figure 8.37, I have plotted a coordinate system where the price of the common share is the ordinate and the price of the warrant the abscissa. How does the price of the warrant relate to the price of the common (the common stock)? It is a fundamental characteristic of warrants that they cannot have a higher value than the common share itself; they do not have any offsetting characteristics. This defines the upper limit for the warrant price. This value is trivially given as the linear $y = x$.

One relevant parameter that influences the price of the warrant is the option period, the length of time that the option may be executed, which we call a in Figure 8.37. This value will offset the value of the warrant and constitute its minimum value. We, thus, find $y = x - a$ for this lower limit of the price of the warrant. This is called the intrinsic value of the warrant and can be calculated by subtraction of the price of the execution from the selling price of the stock.

The range of the price of a warrants is, therefore, determined by the blue region, confined between the maximum value of the warrant (the lower limit of the green region) and the intrinsic value of the warrant (the upper limit of the green region). This blue range is a phase space within which the actual price of the warrant may evolve. However, the its actual curvature is not known.

Kassouf's model resorts in general to regression analysis (Webster 1997) as the principal mathematical tool, and he is aware of the limited analytical value of this method. This is not withstanding that this is a valid way to capture the human decision-making process in a more illustrative, yet mathematical way.

The increasing interest in such contracts led traders to find fair pricing for such insurance contracts. It turned out that the financial and economical systematics of these future options exhibited a similar diffusive behavior to the diffusion of heat in solid matter. The mathematical equation for this behavior is a partial differential equation derived by Black and Scholes (1974, 1973, 1972).

8.8.2 Perversion of warrants by futures

Let us imagine farmers who produce apples and bring them to town to sell at the market. I will present here a visual example how the tool of warrants has become a business needing business and more business, and which eventually has become perverted into a casino-style piece of crime. The example is illustrated by Ernst Wolff in a presentation that he delivered in Munich in 2018 beforeATTAC (I found no other WOS reference than (Achino 2014))[11], and which can be found on YouTube under (Wolff 2018). Here, I translate for you[12]:

> Imagine the world as a huge market place, a gross sale market where the farmers offer their produce to gross sale buyers. These gross sale buyers are the middle men who sell the produce to smaller buyers, the store owners, who sell the produce to the consumers. Next to this gross sale market is a bank, which runs the classical banking business of lending money, giving loans. The business of the bank runs well, until one day, all farmers own enough farming land, the gross dealers have enough cars and delivery trucks, and the store owners have paid off the loans for their businesses. Now, there is no more need for bank loans, and the business of the bank is ruined.
>
> On and off, farmers need new machinery, the middle men here and there one new truck, and the store owners sometimes need to refurnish or remodel their stores and shops, but the bank's big business of providing many large loans is finished. In this situation, one of the bankers has an idea. He opens a new booth and offers betting at the gross sale market; a gambling business. First, he lets people bet on whether on one day more apples or more potatoes are sold on the market. Then he lets them bet which sort of apple sells best. The gambling business starts running well, so that the banker offers ever new and more refined bets. Are red apples or green apples being sold, are domestic or foreign apples being sold, or apples with small or large cores? The plan of the banker works.
>
> More and more farmers and buyers at the market participate in the betting. This is not surprising, because betting is simple. It requires no work. It requires only the money for the bet. In addition, with gambling you can make high profits very fast. However, the betting and gambling also takes a negative turn. Over time, more and more farmers and dealers neglect their core business. The quality of produce at the gross sale market diminishes. Eventually, some of the farms, some of the trucking and dealing businesses, and some of the stores become run down. Also, the addiction to betting and gambling causes some farmers and dealers go bankrupt. A few, however, who focus on the betting, become professional gamblers, make huge profits, and quit their original profession.
>
> The biggest winner, however, is the bank that does the business in the back. The bank participates and profits from every single gambling transaction. The bank also has privileged knowledge about the financial and economical status of the players and can manipulate the game to their own advantage. This is exactly the situation that we are in today with the global finance system. The excessive trade with derivatives has caused the physical economy to degrade, and the gambling carousel is turning faster and faster. A smaller and smaller number of people is getting richer, and the standard of living of an increasing number of people is becoming worse and worse.

– End of my translation of Ernst Wolff's presentation

11 The 'Association pour la Taxation des Transactions financière et l'Aide aux Citoyens' (Association for the Taxation of Financial Transactions and Aid to Citizens) was founded in France in December 1998 after the publication in the Monde Diplomatique of an editorial entitled *Désarmer les marchés* (Disarm the markets) that launched the notion of creating an association to promote the Tobin tax.

12 Thanks to Ernst Wolff, who in 2019 gave me permission to translate his text for my book.

8.8.3 Is the second law of thermodynamics driving society into rich and poor?

The term primary production is used in this book because all our life, even technological life and economical wealth, depends in general on the primary production by photosynthesis. Mimkes and Aruka, thus, correctly begin their book (Mimkes and Aruka 2005) with the confirmation that economic growth is a result of production. They claim that production is a Carnot process that operates between two levels, i.e., poor and rich, in analogy to the mechanical levels cold and hot in a thermal bath. They even drive the analogy so far that they claim the efficiency of the economic Carnot process increases when the difference between rich and poor increases, and as a result, the gap between rich and poor would grow permanently.

For the mathematical description of their theory, they consider the production that leads to economic growth as a differential form where the variables are the labor L and capital K, expressed as $\delta q(L, K)$. The determination of the production function $q(L, K)$ is not possible, in general, unless the integration path x along the production process is exactly known. By introducing an integrating factor λ, it becomes possible to calculate the differential form exactly. In economy, λ can be defined as the gross domestic product GDP, for example. The second law of thermodynamics is identified by the relation $\delta q = \lambda\, df$, where f is considered the entropy (typically known as S).

Mimkes brings an agricultural example for the Carnot cycle in Mimkes (2006). I will phrase this as an exercise, so that the reader can sketch the Carnot cycle with pen and paper. The first phase goes from stage (1) to stage (2) and means that the farmer will collect apples from the trees and store them in the basement. Collecting apples is laborious and yields a reduction of the entropy of apple distribution, which would read $\Delta S < 0$. The *costs* of the production of apples is $C = T_f \Delta S$, where T_f is the price level of apples in fall. In the steps from (2) to (3), the apples are stored by the farmer in the basement. With respect to the entropy S, this step is neutral, because the distribution is not altered; therefore $S = \text{const}$. If you like, you can even “unstore” the apple in a specific process, which is also neutral with respect to the entropy.

In the new year, in spring, the farmer will distribute the apples from his farm to the produce market, $\Delta S > 0$. This is step (3) to (4) in the cycle we are building. He does this because the prices T_s are high for apples in spring. Now he is cashing in, and the *income* from the sale of the apples at the market is given as $I = T_s \Delta S$. When you construct the Carnot cycle, use the entropy S as the x-axis and the *price* T as the integrating factor on the y-axis. Give an interpretation for the area ΔQ enclosed by the cycle.

You can definitely compare classical physical quantities, let us call them mechanical quantities, and quantities from the economical sciences. I consider economy a part of the physical sciences; because it can successfully explain economics. Table 8.3, which was inspired by John Bryant (Bryant 2011; Dimitrijevic and Lovre 2015), lists a number of such comparable quantities.

Tab. 8.3: List of quantities from classical physics, i.e., mechanics, and the economical sciences which are comparable in treatment by thermodynamics. Inspired by John Bryant (Bryant 2011).

Thermodynamic similarity	
Statistical mechanics	**Economics**
Boltzmann constant	Productive content per unit
Energy value	flow rate
Entropy marginal utility	production growth
Number of molecules	Number of stock units
Pressure	Prices per units
Temperature	GDP per capita; average income
Temperature	Index of trading value
Time	Time
Volume physical	production volume (GDP)

What we learn from all this is that the economy is subject to similar statistical principles and the same statistical laws as the statistical ensembles in statistical physics, such as in kinetic gas theory. The temperature T is an intensive statistical variable. The temperature is intrinsic to the thermal bath. Whereas the volume of the bath is an extensive variable. The volume grows with the size of the bath, and so do the entropy and the energy. The two thermal baths with a difference in T are the drivers of the Carnot machine; let them be the high temperature T_s and the low temperature T_f, which we remember from the apple farmer. The efficiency η of a Carnot machine is defined by the ratio of the temperature difference $T_f - T_s$ versus the lowest temperature T_f, which then yields $\eta = 1 - T_s/T_f$.

What is the analog of the thermodynamic temperature in economy? It is the surplus, the profit, that results from the production, delivery, and sale of the product to the consumer. This economic Carnot cycle can occur between two individuals, two groups, and even between two countries, which in the extreme case can be considered an exchange between a rich and a poor trade partner (Dimitrijevic and Lovre 2015). The low temperature T_f is the situation of the system with the lower development, the poorer country. The economic situation of the well-developed or the richer portion of the population country is given as T_s. The surplus is being sourced from the transfer of the value of the resources from the less developed country, such as human labor, agricultural products, minerals, petroleum – to the better developed countries or regions; ΔT thus basically constitutes the difference in the standard of living in two economies. This can also be the difference in the standard of living in the same economy, one at time t_1 and the same economy at time t_2, for example, in spring and in fall.

Dimitrijevic and Lovre (2015) list a number of economically relevant quantities in which the “temperature” can be a process. Be aware that with temperature, we mean an abstract intensive thermodynamic quantity, which is, like in statistical mechanics, defined by the ratio of energy difference and entropy, via $F = U - T \cdot S$, so that $T \cdot S = (U - F)/S$. Thus, T determines the state of economic heat in an economic system in contact and exchange with another economic system. One such relevant quantity is the average amount of money in a system following a Gibbs–Boltzmann distribution P, similar to the type $P = \exp(-E/kT)$.

We can come back to the example of the apple farmer. Let us take, for example, an apple farmer at Lake Constance in Germany. The gross domestic product of Germany was $T_{DE} = 48{,}000$ USD per capita in 2018. The apple will be sold across the lake, in Kreuzlingen, Switzerland. The Swiss GDP was $T_{CH} = 83{,}000$ USD per capita in 2018[13]. The thermodynamic efficiency η for this transaction across the border is, thus, $\eta = (T_{CH} - T_{DE})/T_{DE} = (83{,}000 - 48{,}000)/48{,}000 = 35{,}000/48{,}000 \sim 73\%$. Theoretically, it should be possible to arrive at the same result by dividing the difference of *income* income I and *cost* cost C by the *cost*: $\eta = (I - C)/C$.

Let's say that the Braeburn apple costs 2.60 Euro per kilo at Lake Constance, a realistic retail price, and 3.80 Swiss Francs at the retail store in Switzerland, as my wife tells me. I have no better data now. Currently 1 CHF = 0.94 Euro. This makes $(3.80 - 2.60)/2.60 = 1.20/2.60 = 37\%$. This is only half of the above determined 73%. This is not surprising, because the GDP is a very integral quantity ranking the economy of an entire country, the national economy (in German: Volkswirtschaft). This is not reflected by the price and cost difference of a specific product class at the local market. Mimkes (Mimkes 2006) uses the example of furniture purchased by The Netherlands ($T_{NL} = 5300$ USD) from its former colony Java, Indonesia ($T_{IN} = 4000$ USD). This makes a 1200% efficiency. For further reading on this matter, I can recommend the introduction in the Thermodynamic-Economic Dictionary by Udriste et al. (2018) and maybe also the teachings on bioeconomics by Nicholas Georgescu-Roegen (Georgescu-Roegen 1975; Weber and Cabras 2019; Mazumi 2001).

8.8.4 The thermodynamics of molecular machines

The photosynthetic apparatus is also a system that follows a thermodynamic cycle. An early work on CO_2 assimilation by Bassham (1958) shows that the biological steady state can be compared to a chemical steady state with a partially cyclic metabolic pathway. The inclusion of light energy widens this kind of cycle in the thermodynamical sense. In the late 1950s, it became clear that the conversion of light with internal energy $U = h\nu$ into excited states of chlorophyll molecules, given by the free energy G, could be rationalized by a Carnot cycle of the type $G = Q(1 - T/T_r)$, where Q is U the lowest excited singlet state of the chlorophyll. The T and T_r are the ambient temperature of the chlorophyll and the radiation temperature, respectively (Jennings

13 Just for your reference, the corresponding GDP was 63,000 USD for USA. The GDP were not corrected for purchase power.

et al. 2005). As an exercise calculate the radiation temperature T_r for the absorption bands of chlorophyll. Also calculate the radiation energy Q for these wavelengths. You should be able to calculate the Carnot efficiency η and estimate the theoretically possible chemical work done by this photosynthesis. The temperature of the plant itself is sometimes considered in botanical and ecological modeling. Alexandrov and Yamagata developed a peaked function[14] that can be used for the modeling of the temperature dependence of primary production (Alexandrov and Yamagata 2007).

As the photon with internal energy U is absorbed by a photosystem, part of this energy is converted into the free energy ΔG of the excited state of the chlorophyll but some part is also lost into heat dissipation by excitation of the vibrational states in the structural infrastructure of the photosystem, which represents the entropy ΔS. This situation is comparable with heat engines, where the change of volume and pressure is largely negligible, nil. We can, thus, write (Jennings et al. 2005) $\Delta G = \Delta U - T\Delta S = h\nu_0 - T\Delta S$. However, the volume of the photosystem does, indeed, change, see, for example, Hou and Mauzerall (2011), where it states that light-driven volume changes – what the authors call electrostriction – are directly related to the photoreaction in PS II and can, thus, be a useful measurement of PS II activity and function. Also, the bacteriorhodopsin proton pump changes its conformation and, thus, changes the volume during operation. Similarly to nitrogenase when it "shakes" during catalysis (Xiao et al. 2006).

Kinesin is a protein that can be considered a molecular motor. It is very efficient because almost 100% of the chemical energy that it receives in the form of ATP is converted to mechanical energy. The operation of kinesin is based on the directional working cycle. The cyclicity is necessary, it can, for example, go over four states ABCD, and the direction is the kinetic result. The directionality D depends on the entropy changes ΔS along these states and reads, as worked out by Efremov and Wang (2011), as

$$D = \frac{\left(e^{\frac{\Delta S_{BC}}{k_B}} - 1\right)\left(e^{\frac{\Delta S_{DA}}{k_B}} - 1\right)}{e^{\frac{\Delta S_{BC}+\Delta S_{DA}}{k_B}} - 1}. \tag{8.5}$$

They also derived the mathematical proof for the perfect engine working conditions of the optimized directionality.

Before closing this section, let us come back to the role of temperature in thermodynamic efficiency. It was found that the thermodynamic efficiency of photosynthesis depends on what is known as air-dry bulb temperature T_{db} and air-wet bulb temperature T_{wb}. The former is the ambient temperature. The latter is the temperature when

14 As a scientist, on your crusade to pose the right questions or find the big answers, you sometimes come across problems for which you would need some special tools that have not been developed yet. These can also be mathematical tools and concepts. I did such on two occasions, see Braun (2003); Braun et al. (2002).

a damp cloth is laid over the sample. The studies of Nabil Swedan (Swedan 2019) suggest that the difference $T_{db} - T_{wb}$ is a limiting factor for the terrestrial photosynthesis, and the efficiency yields

$$\eta = 1 - \sqrt{\frac{T_{wb}}{T_{db}}} . \tag{8.6}$$

8.8.5 Instantaneous economic breakdown following crises and disasters

I wrote the previous Section 8.8.3 on 20–21 March 2020, when Europe has been busy fighting the Covid-19, the coronavirus. This year I again received many invitations to deliver talks and speeches at conferences, in Italy, France, USA, Australia – and they have all been canceled by now. We at Empa have received orders to do home office – a very easy task for me –; the entire country does home office now. However, many researchers in the lab, particularly Master's students and PhD candidates depend on experimental data, which they must produce in a laboratory. Entry is forbidden now to avoid people getting into contact and spreading the virus. Restaurants and bars are closed, many small and medium businesses and enterprises cannot work, are losing income.

While primary production is going on in nature, and also in agriculture for the time being, the government policies that many people can no longer work will make it problematic for technological production to proceed as usual. This will mean that not everybody, not every worker and not every business, can currently generate normal income as usual. Soon, people will not be able to pay their rent or their mortgages, businesses will not be able to pay their employees and will have to lay them off or send them – in Germany – into Kurzarbeit (short-time work, see Faia et al. 2013). This is a support measure by the government, which allows businesses to have employees work less than the usual 40 hours a week, while the government steps in with supplementary payments. Yet this does not create more production.

Figure 8.38 shows 13 out of the 90 airplanes of Swiss International Airlines that have been grounded by management because nobody wants to fly or is allowed to fly anymore. Many countries have banned international travel. Airlines must pay fees for every takeoff, which go by the maximum permissible load and not the actual weight of the plane[15]. Half-empty planes might, thus, create more costs for an airline than income. The measures by governments and corporations and organizations in

15 When I took the photo, I met a gentleman who was also taking photos of the grounded airplanes. He even brought a ladder to be able to reach over the fence to take his photos. I had an interesting chat with him; he told me he was a chief operations manager of the airline and that his professional life has always been around his airline, and this is why he also went to Dübendorf airport to just take these historical photos.

Fig. 8.38: 13 out of 90 airplanes of Swiss International Airlines, grounded at Dübendorf airport during the Covid-19 crisis. Photo taken 21 March 2020 by Artur Braun.

order to counter Covid-19 have turned out to be very harsh, including curfews and lockdowns of public areas.

In the Prologue in the beginning of the book, Chauncey Gardiner says that growth has its seasons. This is correct for the garden and the general economy. However, there are also catastrophes in nature and in civilization, which affect growth.

Even the New York Stock Exchange, NYSE, was locked down, because it has a lot of people, who could continue to spread the coronavirus. Therefore, it was decided to carry out trading at NYSE in full electronic mode, effective March 23, 2020, as was announced on their webpage:

> New York Stock Exchange to Move Temporarily to Fully Electronic Trading Monday, March 23. 'NYSE's trading floors provide unique value to issuers and investors, but our markets are fully capable of operating in an all-electronic fashion to serve all participants, and we will proceed in that manner until we can re-open our trading floors to our members...' Stacey Cunningham, President of the New York Stock Exchange

In Figure 8.39 I have plotted some generic stock exchange data, such as the Dow Jones index, a stock exchange index as US dollars versus time, covering the last 3 years. The index climbed steadily, with some intermediate dips, but around the 19 and 20 February 2020, there was a rapid decline of the index, which is indicated by the bold red arrow. This index is a rough and integral measure that shows how the world economy is doing. When you look up the index over the last 50 years, it is impressive how catastrophes and wars can leave their mark on the stock exchange index, either after

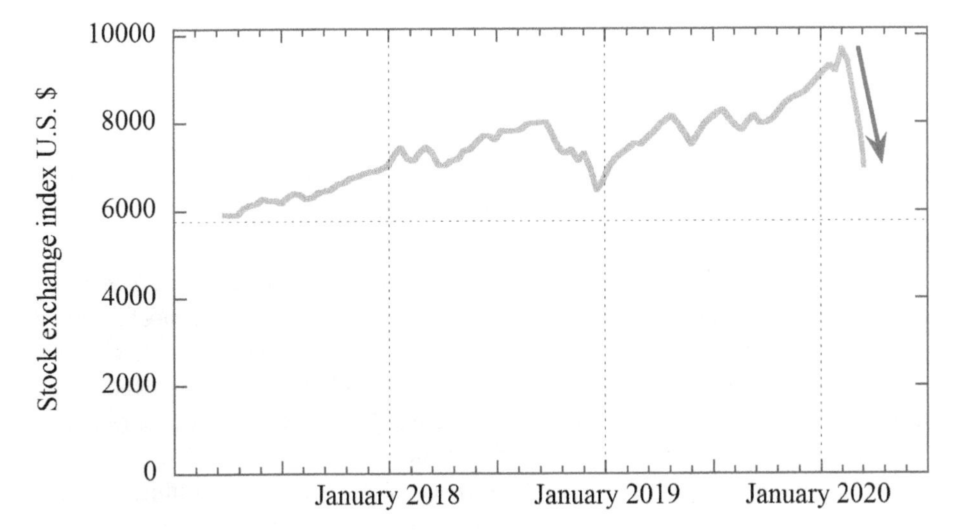

Fig. 8.39: Transient of the stock exchange index from summer 2017 until 21 March 2020. The index increases steadily, with minor dips here and there, the major decline setting in on 20 February 2020.

the catastrophe, or, very interestingly, prior to it. The financial crisis in 2008 was, says Joseph A. Tainter, a failure of complexity.

I also wanted to add here a section on the complexity of economy and society. It is important to know that the complexity of societies is an economical mechanism and depends on the availability of economy. With increasing complexity, more and more energy is necessary, also from the aspect of societal metabolism, see specifically the papers and book by Joseph A. Tainter (Taylor and Tainter 2016; Tainter and Taylor 2013; Tainter 1990). The economy of the Roman Empire was based entirely on solar energy, on nonmechanized agriculture, says Tainter (Wissner and Tainter 2012):

> When the Romans conquered a province, they would seize much of the accumulated surpluses of past solar energy, transformed into precious metals, works of art, and people.

A province (Province – in German: territorialer Zugewinn) is the territory that an empire gains by conquest. The gain of energy allows for an increase of complexity of the society. Competition for territory is part of life, in nature, in business, and in civilization. “Access to light is the big problem here,” says David Attenborough in one of his BBC documentaries (Attenborough 2007) on nature and photosynthesis, when a water lily opens its leaves and competes with other lilies over first access to space.

8.8.6 Random walk hypothesis and the diffusive nature of human behavior

For a better understanding of the physics of the theory of speculation, we have to go further back, more than half a century, and read the thesis of Louis Bachelier *Theorie de la speculation* (Bachelier 1900). It was Bachelier's aim to mathematically capture the speculation on a stock market with common warrants.[16]

For his thesis research, Bachelier had to consider the activity, the buying and selling on the stock market, as a game and the buyers and sellers as gamblers. The mathematical quantity that frames the possibilities of the outcomes (of games) is called probability. What is unknown is the probability of some future event on the stock price. This is what Bachelier calls "rayonnement de la probabilite" and this, indeed, is the diffusion of the probability, for which he introduces the concept of a random walk.

It turns out that Bachelier's theory on speculation is virtually identical to the well-known phenomenon of diffusion in the physical sciences, be it the diffusion of the heat or the diffusion of a species, be it an ion or atom or a representative of a species of a biological population (plants, animals, humans). Diffusion is a process where the species explores a range within a period, and, hence, a process that includes a spatial coordinate and a time coordinate.

Relevant for the idea of speculation is the derivation of the probability p around two different events A and B, which are complementary in such a way that when A happens with probability p, then the event B will happen with probability $1 - p = q$. If the events are tried on several occasions m, the outcome will be a times equal to event A and $m - a$ times equal to event B The probability of this happening is then (Bachelier identifies it as a mathematical expansion of $(p + q)^m$)

$$\frac{m!}{\alpha!(m - \alpha)!} p^{\alpha} q^{m-\alpha} \tag{8.7}$$

The maximum values for the sought probabilities are given as $\alpha = mp$ and $(m - \alpha) = mq$. Bachelier now introduces a quantity h, which he calls "spread", to express the exponent to p, which is $\alpha = mp + h$, and, thus, the probability reads

$$\frac{m!}{(mp + h)!(mq - h)!} p^{mp+h} q^{mq-h} . \tag{8.8}$$

16 Louis Bachelier was born in 1870 and wrote a doctoral thesis on mathematics with supervision from no lesser than the great Henri Poincaré. The thing is that back then, just before 1900 when classical physics turned into modern physics, including relativity theory, there was no field such as finance mathematics (see also the case of Hermann Grassmann (Schubring 1996)). Bachelier, therefore, worked in a field of mathematics where there was no suitable community to support or guide him. Now, with Bachelier's PhD thesis, the year 1900 is considered the year of the birth of finance mathematics (Courtault et al. 2000). Bachelier is one of the early pioneers of the field and certainly of the class of Poincaré.

Bachelier introduces the *spread* h for the expectation that a gambler will receive an amount of money that is equal to h, when the condition $h > 0$ holds. We can now write the probability that h is included in an interval $[h, h + \mathrm{d}h]$, which is given by

$$\frac{\mathrm{d}h}{\sqrt{2\pi mpq}} \mathrm{e}^{-\frac{h^2}{2mpq}} \tag{8.9}$$

Let us then consider the time interval Δt during which (the sum of the products of) spreads h exist and formulate their corresponding probabilities $\int_0^\infty p \cdot h \, \mathrm{d}h$, where we eventually arrive at the probability

$$p = \frac{1}{2\pi k \sqrt{t}} \mathrm{e}^{-\frac{x^2}{4\pi k^2 t}} \tag{8.10}$$

It is an exercise for the reader to show that by taking the derivative $\mathrm{d}p/\mathrm{d}t = 0$, we obtain a relation $t \sim h^2$. This is basically a representation known as Fick's law (Fick 1855, 2009). The proportionality factor between time and space is known as the diffusion constant (usually termed D).

Bachelier consequently derives the second-order partial differential equation, originally derived by Jean Baptiste Fourier, which for the heat conduction reads as

$$c^2 \frac{\partial P}{\partial t} - \frac{\partial^2 P}{\partial h^2} = 0 \tag{8.11}$$

The differential equation has the same mathematical structure as Fick's second law of the diffusion of chemical species. There is a fundamental shortcoming of Bachelier's theory. It predicts that with increasing time t, the value of the option will increase with square-root behavior. However, this is not true, as experience shows. Otherwise, if you were speculating on the stock market, you would purchase the option and simply wait and make a large profit.

Writes Samuelson (Samuelson 2015):

> Under this formula, the value of a warrant grows proportionally with the square-root of the time to go before elapsing; this is a good approximation to actual pricing of short-lived warrants, but it leads to the anomalous result that a long-lived warrant will increase in price indefinitely, coming even to exceed the price of the common stock itself—even though ownership of the stock is equivalent to a perpetual warrant exercisable at zero price!

Samuelson is right. The square-root function does not converge, it diverges to ∞, although relatively slowly. In my doctoral thesis[17] I unwillingly faced a similar problem.

17 I call it a doctoral thesis and not a PhD thesis because I do not hold a PhD degree but a Doctor of Science degree. There is confusion around academic degrees, and even those who hold a doctoral degree get the meaning wrong surprisingly often. I refuse to call my degree a PhD, because it is not a PhD. I only have the degrees that I earned as certified by the schools who grant them. From RWTH

I was working on the activation of glassy carbon, and the goal was to produce porous carbon films on a nonporous carbon substrate. The film evolved in a hot-oxidizing gas atmosphere, and the thickness of the film depended on the temperature of the furnace, the time of exposure, and also on the gas concentration. The thickness versus time profile was a square-root-like curve, but it was a scientist in our research group who pointed me to the fact that the thickness reached a constant plateau after some time and not a divergence into infinity. He was right about that, but this was not in line with Fick's diffusion law or Tammann's Zunder law. Eventually, I realized that I was wrong, and I had to develop a new model for the film growth. My supervisor insisted I make a model for the film growth. For me, this meant that I had to make a mathematical model. I found that the combustion of the carbon and, thus, volatization of solid matter competed with the growth of the film, and my system was a predator–prey system comparable to a Lotka–Volterra model. Eventually, I was able to write down the differential equation for the combined and competing processes, and after a tough time and some struggles, I found an analytical solution to the equation, which predicted the actual experimental film growth data very well (Braun et al. 2000, 2003; Braun 1999). Therefore, there is a process that counteracts the square-root trend for infinite growth, and this must also hold for the value of the warrant. This one needs to be identified, and this is an exercise for the reader.

Aachen in Germany, I have a Diploma in Physics and not a Master's degree. There were no Master's degrees for technical disciplines when I enrolled. A Bachelor's degree did not exist at all. The *Diplom* was established in Germany in 1899 by a decree of Kaiser Wilhelm. Back in my time, you could get a Magister degree (*Magister artium*) only in Arts and Letters and Humanities. Scientists and engineers would become *Diplom-Physiker, Diplom-Chemiker, Diplom-Mineraloge, Diplom-Ingenieur*, and so on. If you continue your studies for a Doctoral degree, some schools offered a Doctor of Philosophy degree, but in my time you would, particulary in Germany, go for a *Dr.rer.nat.* or *Dr.Ing.* degree. These were the doctoral degrees you could get. There were no medical schools or law schools in Germany either. Instead, there was the Faculty of Medicine and the Faculty of Law. Unlike in the USA or the UK, you do not need a college degree to study medicine or law. You need the 13-year education system in Germany behind you and then you can enroll at university. For some studies, like medicine, law, architecture, and few more, there is an extra threshold that limits the number of students. Further, when you were finished with the tough studies in medicine, you did not graduate with a degree! You had to take the government exams and, after you passed, you were allowed to bear the title *Arzt* (doctor). If you call yourself *Arzt* without having passed the exam, you could get up to 1 year in prison. If, as an *Arzt*, you want to get a doctor's degree, you have to write a dissertation, which is typically way shorter than a dissertation in any other field. Then, you may be called *Dr.med.*, which corresponds to MD in USA. In Switzerland, as a physicist or chemist you get a *Dr.sc.nat.* (Doctor of Science) at ETH Zürich and a PhD (Doctor of Philosophy) at the University of Basel. Some of the doctoral students in my research group got a PhD from Basel. Others got a *Dr.sc.* from ETH Zürich and EPFL in Lausanne. Stay with the degree that you have on your diploma, on your certificate from the school. Don't say something that is not true just to make it easier for other people to understand. They will get the wrong understanding if you do.

For a general case, the probability of a cash purchase or a forward contract purchase, we need to know the spread nb between the quoted price and the actual, true price, which is then given by

$$\int_{-nb}^{\infty} p(x)\,\mathrm{d}x = \frac{1}{2\pi k\sqrt{t}} \int_{-nb}^{\infty} \mathrm{e}^{-\frac{x^2}{4\pi k^2 t}}\,\mathrm{d}x \tag{8.12}$$

with $p(x)$ known from Equation 8.10. Let us think of a cash purchase that we want to resell in $t = 30$ days. You notice that we do not integrate over t; n is the number of days; b is the quantity by which a "Rente" (Bachelier uses in his French thesis the term "rente", which means the financial gain, the "return". In Germany the Rente is for example the money which you receive from your retirement funds.) increases within one single day. The lower integration border $-nb$ is to be set at 25. You can try to carry out the integration as an exercise and will find that p becomes 0.64, which is almost $2/3$. The interpretation for this result is that with two out of three chances there will a net profit from the transaction of purchase and resale (Bachelier 1900).

8.8.7 The known and unknown unknowns

When we make observations, be they in the natural world or in the technical world, like in physics or chemistry, or on the stock exchange and finance market, we can try to see some rational pattern and logic in it and then we may try to model it mathematically. This is the natural activity of a scientist, in the physical sciences and also in the economy or social sciences. However, our observations are not necessarily complete. We observe with our senses, optically – visually, acoustically, with our skin. Yet even with full function of our senses, there are things that we do not sense, although these are parameters that may affect what we can sense.

The other case is that the existence of some event may be proven. But to prove that someone or something not exists is very difficult, maybe even impossible. Which does not rule out the possibility that it exists, or not.

In the year 2002, the Defense Secretary Donald Rumsfeld made a remarkable statement at a press conference, where he gave a briefing on the situation in Afghanistan (Rumsfeld 2002):

> Reports that say that something hasn't happened are always interesting to me, because as we know, there are known knowns; there are things we know we know. We also know there are known unknowns; that is to say we know there are some things we do not know. But there are also unknown unknowns – the ones we don't know we don't know. And if one looks throughout the history of our country and other free countries, it is the latter category that tend to be the difficult ones.

In 2003, Rumsfeld was awarded the Foot for Mouth Prize because of this statement. Chang (Chang 2012) says that the prize committee probably was not fully aware of the implications of Rumsfeld's statements for the rationalism of science and cognitive science. This is why I brought it up here in my book, for you to consider it thoroughly.

8.8.8 The search for extraterrestrial life and intelligence: SETI

At some point in the development of human intelligence, man started wondering whether there could be life out there in the starry night. At some level of higher intelligence he may have wondered, with an educated background, whether in some other solar system there would be the air to breathe, water to drink, and food to eat. A yet higher level of intelligence asks whether there could be life with some intelligence out there, not only as primitive as a bacteria or virus, but animals that can make decisions, or transhumans that might be way more intelligent than humans on our Earth. The search for extraterrestrial intelligence is abbreviated to SETI. My former colleagues from Berkeley National Laboratory (Stephen P. Cramer and Hongxin Wang) are now researching SETI.

The Drake equation is a simple arithmetic tool to estimate the number N_C of communicative technological civilizations in our galaxy; N_C is the product of a number of factors, which are listed in the Table 8.4. Basically, we look at a galaxy that is divided into subsets of stars, planets, planets with life, and planets with intelligent life, while the galaxy also has a star production rate. So it is a dynamic system as well. The goal of this arithmetic exercise was to help radio astronomers with some numbers and give them a hint as to what extent it is theoretically possible to detect radio signals from intelligent populations somewhere in the universe. In early times, N_C was estimated to be 1 million. Later, reconsiderations yielded a much smaller number, like 10^0 to 10^2 (Wallenhorst 1981; Walters et al. 1980; Kreifeldt 1971).

Tab. 8.4: Parameters for the Drake equation. Data from Wallenhorst (1981).

Definitions of Drake equation parameters	
Symbol	**Parameter**
R	Average star production rate
f_g	Fraction of stars
f_s	Number of civilizations per star
f_p	Fraction of stars with planets
n_e	Number of planets per star
f_L	Fraction of planets which evolve life
f_i	Fraction of planets with life that develop intelligent life
f_c	Fraction of planets with intelligent life that develop a technological civilization
H_T	Characteristic time for the evolution of a civilization
H_*	Characteristic decay time for the galactic star formation rate

8.8.9 Central limit theorem of statistics

Maccone (2012) looked into the parameter f_{L} in Table 8.4 and applied a statistical generalization to the Drake equation by assuming a logarithmic normal distribution for the evolution of life by Darwin's theory. He assumes that the number of different species evolves exponentially during the 3.5 billions of years of the evolution of life on Earth and puts this into the SETI context of the last 10 billions of years (Figure 6.1). Using the central limit theorem of statistics, Maccone concludes that also the entire Drake equation should have a logarithmic normal characteristic. Maccone shows that the Darwinian exponential is the envelope of a number of log normal distributions, the maxima of which follow the exponential.

The central limit theorem is a general tool in statistics that says when independent variables are added, their normalized sum will eventually assume a normal distribution irrespective of whether their original variables were normally distributed. For example, this can be used for the calculation of the distribution of spins with $\uparrow$ and $\downarrow$ with 50% probabilities. In the case of n particles with $n \to \infty$, the average value for the "spin" has a normal distribution. This no longer holds for quantum spins, which follow the Bose–Einstein statistics. This is because they are described by noncommuting operators. Dorlas (1996) developed a theorem for the noncommutative central limit and used the Feynman–Kac formula for the proof of his theorem.

The Feynman–Kac formula (Kac 1949) gives a link between parabolic partial differential equations and stochastic processes, for example the Brownian motion. The Brownian motion is the irregular zig-zag trajectory of an object in some medium. The formula could prove the case for Feynman's path integrals. However, when spins are included in the system, the equations have complex variables, and for such a case there has not yet been any proof.

We can define the Feynman–Kac formula as done in Faris (2004). We consider a Hamiltonian

$$H = H_0 + V \tag{8.13}$$

with the potential V a real function in the set $\mathbf{R}^d$ being an operator defined by separability as $(Vf)(x) = V(x)f(x)$. The Hamiltonian H is, then, a self-adjoint operator. Then, for $t > 0$, the operator $\exp(-tH)$ is also a self-adjoint operator. When V is not negative, the interpretation for V is that of a diffusion process.

This could be, for example, an electron–hole pair generated in a semiconductor upon illumination, which diffuses some distance and time until it finds arbitrary recombination centers and disappears again; $V(y)$ could then be the rate of "killing" the pair, or one of its constituents (hole, electron) when it reaches the position y. We can define a quantity $g_t^V(x, y)$ as the defective probability density (defective, like your team member who disappears overnight or a citizen from a totalitarian country), for which the following relation holds

$$(\mathrm{e}^{-tH}f)(x) = \int g_t^V(x, y)f(y)\,\mathrm{d}y\,. \tag{8.14}$$

For the above Hamiltonian, the Feynman–Kac formula holds accordingly

$$(\mathrm{e}^{-tH}f)(x) = E_x \mathrm{e}^{-\int_0^t V(\omega(s))\,\mathrm{d}s)f(\omega(t))} . \tag{8.15}$$

In quantitative finance, the Feynman–Kac formula is used to calculate solutions to the Black–Scholes equation to price options on stocks.

A jump risk is an investment risk that comes when there is a sudden jump in the volatility of the prices of stocks. The Lehman crisis in 2008 was such a sudden jump. Basically, these are spikes in the stock exchange prices. Merton developed a stock price model for discontinuous stock data (Merton 1976).

8.8.10 Option model by Fischer Black and Myron Scholes

In the 1960s, finance mathematics began to develop further and also addressed the valuation of warrants and option contracts, see, for example, the essay by Case M. Sprenkle (Sprenkle 1961), who regarded warrant prices as indicators for expectations and preferences of the investors (Lo 2008).

The Black–Scholes equation describes the fair value of stock options pretty well. Both enter trade deals, and they want to make sure that they can deliver and receive, respectively, the goods at some agreed time and price in the near future. The quantity S is here the price of the asset (S: the stock) at some time t. At the same time t, the option for the asset S will have the price V. Both quantities are related to the interest rate r and the standard deviation σ of the returns from the stock. This is all contained in the Black–Scholes equation, which we can consider (Equation (8.16)) as a Hamiltonian:[18]

$$\frac{\partial V}{\partial t} + \frac{1}{2}\sigma^2 S^2 \frac{\partial^2 V}{\partial S^2} + rS\frac{\partial V}{\partial S} - rV = 0 \tag{8.16}$$

In Equation (8.16) we read that the potential V comes with the first derivative to time. This is certainly useful, when we wish to have at some point a Schrödinger equation with the first derivative of a wavefunction to the time; V is the option price, which depends on the variables S, the asset price, and time t. We do have a second-order and a first-order derivative with respect to the asset price S, and it will require a coordinate transformation $S \to \mathrm{e}^x$ in order to arrive at a simplified equation that reads

$$\frac{\partial V}{\partial t} + \frac{1}{2}\sigma^2 \frac{\partial^2 V}{\partial x^2} = 0 . \tag{8.17}$$

---

18 In 1997, Robert Merton and Myron Scholes were awarded with the Nobel Prize in Economy in acknowledgement for their method for the determination of the value of stock derivatives. Both Nobel Laureates were on the Advisory Board of Long-Term Capital Management (LTCM), a hedge fonds company that nearly went bankrupt in 1998 and went bankrupt in 2000 as a result of the Russian finance crisis in 1998. Scholes founded his own company, Platinum Grove Asset Management (PGAM) in 1999. That company evaporated in 2008. In 2009, the Trinsom Group, with Merton as Scientific Head, went bankrupt. Apparently, the two Nobel Laureates do not know their field so well that it can be considered safe (Chang 2012).

We know this equation (8.17) from the heat conduction for one dimension. So, there is some idea of the propagation of some observable included, such as the heat energy or the temperature. We can apply a simple mathematical trick, another coordinate transformation $t \to \tau/\mathrm{i}$, which is known as the Wick rotation. Some of you may know this transformation from your undergraduate studies in mathematics, which is typically mentioned in the conformal mappings curriculum. The Wick rotation is an operation known in quantum electrodynamics (Schilcher 2019), but it is also used here in finance in the Black–Scholes model. Largely, I am following here the exercise by Contreras et al. (2010).

$$\mathrm{i}\frac{\partial V}{\partial t} = -\frac{1}{2}\sigma^2\frac{\partial^2 V}{\partial x^2} = 0\,. \tag{8.18}$$

Equation 8.18 does not look much different to the heat Equation 8.17, but visually we are now dealing with a type of Schrödinger equation, which we know from the motion of the free particle, except that Planck's constant here now reads $\hbar = 1$, and the mass is $m = 1/\sigma^2$. With the Wick rotation, we can carry out the transformation of an equation for a problem in Euclidean space (this is the space with the three spatial coordinates x, y, and z, which is the relevant space for the architect and carpenter) into the Minkowski space, and vice versa. The Minkowski space extends in four dimensions, as it has the additional dimension of the time t included by forming the coordinate $\mathrm{i}ct$, with c being the speed of light. Wick exercises this rotation in the complex plane in his paper on the properties of the Bethe–Salpeter wavefunctions (Wick 1954).

It was Henri Poincaré who figured that this fourth dimension would mean that the Lorentz transformation was basically a simple rotation (Poincaré 1906). We recall that Poincaré was thesis advisor to Bachelier, who developed the theory on speculation. The change of the metric from Euclidean space to Minkowski space, along with the Wick rotation, considerably removes obstacles in the mathematical treatment, specifically in connection with an integral equation and its kernel (O'Brien 1975).

When the path integral is subject to the Wick rotation, it will look like a partition function, like in statistical mechanics with a number of oscillators. This transformation will point us to the formal change from the reciprocal temperature $1/k_{\mathrm{B}}T$ to the reciprocal quantum $\mathrm{i}t/\hbar$ (just think of the simple relation $h\nu = k_{\mathrm{B}}T$). The Wick rotation, thus, combines statistical mechanics with quantum mechanics. Therefore, Equation 8.18 shows that the Black–Scholes model for the option price is a dynamic quantum mechanical model without any interactions; the equation stands for the "free particle".

Getting back to the Black and Scholes model, let us calculate the value of a European call option. It is for a stock that does not pay any dividends. Let $T - t$ be the time in years for the maturity of the option, K be the strike price, and S be the stock price. We need a cumulative distribution function from the standard normal distribution, which we call N. The quantity r is the risk-free rate, and σ is the volatility of the returns of the stock.

The call option value is then

$$W(s,t) = N(x_a) \cdot S - N(x_b) \cdot Q(K) \tag{8.19}$$

with x_a

$$x_a = \frac{1}{\sigma\sqrt{T-t}} \left(\ln\left(\frac{S}{K}\right) + \left(r + \frac{\sigma^2}{2}\right)(T-t) \right) \tag{8.20}$$

and x_b

$$x_b = x_a - \sigma\sqrt{T-t} \tag{8.21}$$

and $Q(K)$ is an exponential, with the rate and time in the argument,

$$Q(K) = Ke^{r(t-T)} . \tag{8.22}$$

As this is a European call option, in Equation 8.19 we are counting with Euro cents.

8.8.11 Quantum mechanical interpretation of the Black–Scholes model

In the introduction of his doctoral thesis (Bachelier 1900), Bachelier writes "[...] la Bourse agit sur elle-meme [...]"; this means the Stock Exchange acts on itself. He means that the Stock Exchange has an interaction with its external environment. Segal and Segal (1998) obviously adopt this interpretation, as they write that the attempt to gain information on stock exchange effects act back on the very effects, and this could be one reason for the extreme irregularities observed, which are reminiscent of what we know today as the Wiener process and which represents the generalized Brownian motion, (Brown 2009) in the evolution of prices on the financial markets. Do we know of such a retroaction from an accelerated electric charge, which produces an electric field that acts back on the charge?

Over 20 years ago Segal and Segal showed how the Black–Scholes pricing formula can be interpreted within the framework of quantum theory. With a simple *analogon* in physics, they resort to the ensemble of photons (light particles) and the observables of the electromagnetic field strength of the light. For this, let us simply consider the light vector $\boldsymbol{E}_i$[19]. While the ensembles of photons are made from mutually commutative observables, the two ensembles of photons and field strengths do not commute with each other.

Segal and Segal incorporate the quantum effect in the Black–Scholes model by resorting to the Wiener process $W(t)$, which represents the evolution of publicly available information, which can affect the market, about some specific process $X(t)$, while $X(t)$ represents only those factors that are not observable (the known unknowns) at the same time as those that affect $W(t)$.

19 The electric vector in electromagnetic waves is historically considered as the light vector because it already engages with electric charges that are not in motion, unlike the magnetic vector.

For the mathematical formulation of the uncertainty, they resort to an infinite number of degrees of freedom and, thus, base it on infinite dimensional vector space as is usual in quantum mechanics, i.e., using the Hilbert space. After clarifying a number of algebraic requirements, Segal and Segal need to establish a theorem about an arbitrary function F for which it holds that:

$$F(W(t)) = F(0) + \int_0^t [F'(W(s)) \circ w(s) + k(s)F''(W(s))]\,\mathrm{d}s \tag{8.23}$$

Note here that the Jordan[20] product operator $\circ$ is applied to $F'(W) \circ w$ in the integrand. They can now integrate the quantum process $P(t)$, which is the solution of an integral equation of the kind

$$E(t) = 1 + \int_0^t P'(s) \circ E(s)\,\mathrm{d}s \tag{8.24}$$

and with the help of the aforementioned theorem, it is shown that with the validity of $P(t) = r + w(t)$, the relation

$$E(t) = \exp\left[rt + W(t) - \frac{1}{2}kt\right] \tag{8.25}$$

becomes valid. This is an example that the conventional equation for the time evolution of the price $P(t)$, which is considered a quantum process for a stock item, needs only Jordan's product.

The Jordan product operator $\circ$ is defined as follows:

$$a \circ b = \frac{1}{2}(ab + ba) \tag{8.26}$$

The Jordan product $a \circ b = 1/2\,(ab + ba)$ can transform the associative algebra $\mathcal{A}$, which has an associative product xy, into a so-called Jordan algebra $\mathcal{A}^*$. Segal and Segal use Jordan's product in Equation 8.23, with the understanding that a product Fw of observables F and w can be defined in terms of observability only when they can be observed simultaneously, which finds its mathematical, group theoretical equivalent in commutativity (Segal and Segal 1998). When the commutator $ab - ba \neq 0$, then the observables a and b cannot be observed at the same time.

The zig-zag structure of the stock prices such as shown in Figure 8.39 is reminiscent of a Wiener process, the transient of which Segal and Segal call $W(t)$. In addition to this, they introduce a pseudo Wiener process $V(t)$; $P(t)$ is the prize of the stock at time t and behaves like a quantum stochastic process; P is its initial price, and K is the strike price. (The German translation for strike is "Zuschlag", which corresponds

20 Ernst Pascual Jordan (Ehlers 2007; Schroer 2011).

to the beat of the hammer in an auction when the highest bid is accepted and the transaction executed.) T is the time until the option matures.

$$V = Pe^{\frac{1}{2}T(y^2s^2-k)}N\frac{\log\frac{P}{K} + T\left(r + y^2s^2 - \frac{k}{2}\right)}{ys\sqrt{T}} - Ke^{rT}N\frac{\log\frac{P}{K} + T\left(r - \frac{k}{2}\right)}{ys\sqrt{T}} \tag{8.27}$$

here, N stands for the stochastic integral

$$N(x) = \frac{1}{\sqrt{2\pi}}\int_0^x e^{\frac{y^2}{2}}\,dy\,. \tag{8.28}$$

It is an exercise for the reader to identify the similarities in the classical Black–Scholes equation and the quantum Black–Scholes equation. Consider, therefore, the parameter $s = \sigma$, $k = \sigma^2$, and what happens when you set $y = 1$. According to Segal and Segal, the quantum model displays greater extreme deviations than one would expect from the classical model. An alternative to the Wiener process for describing the pricing of the option is the application of quantum game theory (Piotrowski et al. 2006).

8.8.12 Arbitrage

One central quantity in finance and market is the arbitrage. This is a French word and is not easy for me to translate into German, and thus not easy to understand. The literal translation means what you know in English as arbitrarily. Without any of your reasonable input. Technically, it means the yield for you from a financial, stock, or trade transaction, which is based only on the increase of the value. Originally it would be the added value, the value added in a production chain where work, labor, is involved, which puts the materials that are used into the production and finally make a higher-value product.

Take, for example, the huge sequoia tree in Figure 6.6 and the wood in Figures 6.3 and 6.4. It may have a monetary value as it stands there in the wood. If you cut it, load it onto a truck, and bring it to a sawmill company that makes bars and boards from it, you can sell them at a higher price than the stem alone. You pay the wood cutter, the truck, and the sawmill for their work and services and, hopefully, after subtracting all the investments, a profit from sale is left for you. The components that raised the value are due to the work; in German this is called *Arbeit*. The Polish and Slovak word for work or worker, or slave even, is *Robotnik*. The robot is the worker.

If you have a good sense for words and language, you notice the linguistic similarity between *Arbeiter* and *Roboter*, which has a similar meaning to the French *arbitrage*. What is neglected in this value-adding process is that of the market value, which depends on supply and demand. If nobody wants your wooden bars and boards because they are not necessary, since the market is saturated, then there is no current value in them, and you better lay them aside and trade something else, if you can. If you are the only supplier, and there is high demand for wooden boards and bars, you may be able to sell them at a price that is obscenely higher than justified by the efforts you put into making them.

So, the value and price of your goods may fluctuate and depend on the supply and demand situation (see Piotrowski et al. 2010 for a model on trader strategy when there is little information available on supply and demand). A friend of yours may have bought up all your useless bars and boards when nobody needed them. Further, your friend may have given you some friendly money for it, more than you would have gotten elsewhere. 1 year later, your friend may be able to sell the boards and bars on the market because, for some strange reason, there is now high demand. He sells them for three times the price that he bought them from you the year before.

Hence, your friend made a profit on the trade because the price rose, although he did not do anything on his part raise the price. The only *work* he had put in was waiting for the price to increase. This is arbitrage – the change in price. The same holds for stocks and foreign currency. You try to buy when they are cheap and you sell them when they are expensive. The proper German word for this is *Gewinnmitnahme*, and in English, profit taking. So, the arbitrage is the amount of "work" in the deal, the "action". The deal has "worked", *worked out* for you.

8.8.13 Quantum-mechanical features in decision making

Currently, while finishing writing this book, I am trying to buy a home. Real estate prices have increased in the past couple of years, and homes are now considered overpriced. It is not wise to buy a home now, but by ways of life, people must sell and buy homes. Owners who bought 10 years ago for 400,000 Swiss Francs may find that their home now has a higher price, such as 750,000 Swiss Francs.

However, this does not mean that the first interested buyer actually wants to spend so much money on that home. He may make an offer of 650,000 Swiss Francs. The current owner is not inclined to accept this offer because he feels his home has a net worth of what the current real estate market statistics say. Yet, for some reason, maybe because he wants to retire in a foreign country, he needs to sell soon and, thus, makes a counteroffer at 700,000 Swiss Francs. The interested buyer does not have that much money and cannot afford more than 680,000 Swiss Francs. We do not know how the prospective seller and buyer will decide.

Emmanuel E. Haven explains in his paper (Haven 2014) how the theory of Brownian motion found its justification in the theory of option pricing, whereas real economists may argue that this theory completely ignores the use of preferences, which would be the "mainstay tool of applied and theoretical economics". Haven further argues that the econophysics community is not recognized by the economics and finance community. We remember Bachelier's difficulty to get promoted in the mathematics community (Courtault et al. 2000). However, the decision-making process does not only originate from the differences in supply and demand, but also from persona and, individual appreciation of the seller and buyer and is, thus an issue of sociology and psychology, two fields, where economists have no dominance and no monopoly in saying and ruling.

Haven does not claim that phenomena in the social sciences are macroscopic manifestations of quantum mechanics, but considers the mathematical tools in quantum mechanics and fluid dynamics[21] that could be borrowed to tackle social science questions that have a relevance in economy and finance. One example that Haven brings forward makes use of the Fisher information, which is the measurable amount of information that an observable random variable V carries about an unknown parameter p of a distribution that models V. This is exemplified in the field of finance by Hawkins and Frieden (Frieden 1998; Hawkins and Frieden 2012), with the price p at time t_0 of the kind of asset x_0, while it has fluctuations of size x; the information I is written as a statistical expectation value with angle brackets:

$$I = \left\langle \frac{\partial \ln(p(x_{\mathrm{obs}}|x_0, t))}{\partial x_0} \right\rangle^2 = \int \left| \frac{\mathrm{d}\psi(x)}{\mathrm{d}x} \right|^2 \mathrm{d}x \tag{8.29}$$

They model the fluctuations around the real prize with a probability amplitude $\psi(x)$, which in my opinion has a wave-mechanical interpretation (whereas Hawkins and Frieden emphasized it is independent of any quantum mechanical interpretation.). This metric contains a high level of information provided that the density function is sharp, or condensed (Courtault et al. 2000).

8.8.14 Inclusion of information in the decision-making process

Haven (2002) showed how the option price can be interpreted as a state function that can satisfy the Schrödinger equation. Bagarello and Haven introduce the aforementioned wavefunction $\psi(x)$, which, as they call it, can produce a *mental force* and with it along with the real or *hard forces* of the economic system, a Hamiltonian that can follow to a conventional physical differential equation (Bagarello and Haven 2014). Bagarello and Haven now use an additional potential, which constitutes the *reservoir of information*. This makes them develop the problem in terms of open and closed systems, where the closed system is represented by a two-modes bosonic operator, which obeys commutation rules. What follows is that information, cash, and shares are operators of the same mathematical kind.

The behavior of two traders will also be guided by the information available in the reservoir, which condenses to the decisions in buying and selling. Bagarello and

21 The term from finance known as "contract for difference", abbreviated to CFD, shall not be mistaken with "computational fluid dynamics", also abbreviated with CFD.

Haven consider three different cases for the information reservoir. The simple case is where the reservoir of information is empty or not present at all. The second case is where there is information in the reservoir. The third case is where the reservoir generates information. This latter case is an example where the trader acts back on the market.

The creation and annihilation operators for shares, cash, and information read s, c, and I, respectively. The Hamiltonian then has the following structure:

$$\begin{cases} H = H_0 + H_{\text{inf}} \\ H_0 = \sum_{j=1}^{2} \left(\omega_j^s \hat{S}_j + \omega_j^c \hat{K}_j + \Omega_j \hat{I}_j + \int_R \Omega_j^{(r)}(k) \hat{R}_j(k)\, \mathrm{d}k \right) \\ H_{\text{inf}} = \sum_{j=1}^{2} \left[\lambda_{\text{inf}} \left(i_j (s_j^\dagger + c_j^\dagger) + i_j^\dagger (s_j + c_j) \right) + \gamma_j \int_R (i_j^\dagger r_j(k) + i_j r_j^\dagger(k))\, \mathrm{d}k \right] . \end{cases}$$

Using this kind of quantum mechanics formalism helps us to formalize and formulate information dynamics in the macroscopic setting of economy. Bagarello and Haven find that the *economic intuition* would remain robust in the sense that the loss of information would influence the incremental value of asset portfolios.

8.8.15 Ronald Reagan's imperial circle

We know since Bachelier that the stock market acts on itself. We can call this reflexivity, or reciprocity, or internal feedback. This is known from cybernetics. Star investor George Soros made similar observations when looking at developments and patterns in the currency markets and called it reflexivity, and developed a theory from it. He published this theory and further ideas in his book on finance (Soros 1994). It is, thus, not surprising that his book was reviewed in the journal *Kybernetika* and that Soros' theory has been compared to fundamental ideas in physics and mathematics.

Soros explains the economy and currency policy in the Ronald Reagan administration (Aberbach 2008). For reasons not to be detailed here, the interest rates in the United States became very high, which attracted foreign money into the US. Also imports to the US were cheap because of the strong dollar. A self-reinforcing period emerged that saw a strong American economy and currency, a large budget deficit, and a large trade deficit, all of which mutually reinforce and produce a noninflationary economic growth. George Soros calls this situation Reagan's imperial circle, which could finance a strong military force by attracting material goods and money from foreign countries. Inside, this circle was benign, and on the outside it was vicious.

"A benign circle for the United States is a vicious circle for the debtor nations," says Soros. In order to put the economy status into a graphical scheme, Soros uses a notation as follows: strong economy is $\uparrow v$; strong currency $\uparrow e$, growing budget deficit ↓ B, and a growing trade deficit is ↓ T. Speculative capital goes by S, and growing debt service is ↑ N. A growing interest rate is denoted by ↑ i. Many of these parameters may be coupled to each other, and there may be more parameters, the naming and plotting of which would make the mechanism impenetrable like a jungle – says Soros (1994).

Figure 8.40 shows Soros' integrated model, which combines these partially counterintuitive relationships of the aforementioned economic parameters into the imperial circle (Soros 1994). The model is based on a simple *if* – *then* logics and shall visualize what will happen to the currency e when the other parameters increase or decrease. The numerals (2)–(7) denote the relationships shown and discussed in the book (Soros 1994), which are implemented in the shown model here. Soros provides no further quantitative formulation for the parameters, which, I believe, it would be possible to develop.

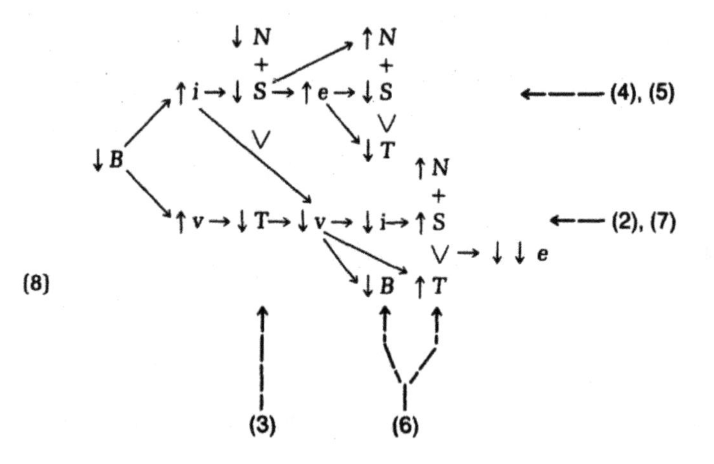

Fig. 8.40: George Soros' integrated model for Ronald Reagan's imperial circle linking economic parameters. ↑ v signifies a strong economy, ↑ e signifies a strong currency, ↓ B signifies a growing budget deficiency, and ↓ T signifies a growing trade deficit. Speculative capital goes by S, and growing debt service is ↑ N. A growing interest rate is denoted ↑ i. Reprinted from The Alchemy of Finance, (Soros 1994), with permission from John Wiley & Sons.

8.8.16 Quantum econodynamics

The randomness of the Brownian motion is reflected in the theory of speculation on the stock market (Bachelier 1900). The economy of the market follows to some extent the physical principles of statistical mechanics. Schinckus has called for quantum econophysics (Schinckus 2014) in order to better understand where randomness yields to particular results in economy. Important in this aspect is the term "emergence". What does emergence mean? Schinckus cites C. Craver (Craver 2006), who thinks emergence is a filler term, like some other words frequently used when it becomes difficult to answer questions of "why" something happens.

Some years ago, I was at one of the MRS meetings, I do not recollect whether it was in Phoenix or in Boston, where an eminent nanoscientist from Washington State gave an award talk (De Yoreo 2016) on the crystallization of materials from liquid solutions. He showed a movie of microscopy images, where the audience could see how nanocrystals would suddenly "emerge" in the liquid here and there and then disappear again by redissolution. Only very few of them would survive the initial crystallization phase and make it into manifest crystals, which would then further grow to larger crystals.

When the session was open for questions, I took the microphone, stood up, and said this process of emergent crystallization, or solidification from the liquid, and subsequent dissolution reminded me of the many startup companies, for example, in Silicon Valley, who would pop up and then vanish; and only very few of them would really make it into solid business. Do you think your model applies also to this economic problem of company foundation, dissolution, or solidification? What is it, then, that makes winning companies win?

Now, such a question is important. I remember a TEDX talk in which a researcher explained how the companies that rule the world have become such companies (Glattfelder 2015). His and his colleagues' study showed that there is a network of organizations that rule our world, and that this is not the result of a so-called conspiracy but a result of emergence. Now we do not know yet where emergence comes from. However, there are paths in business development that are successful and some that are not, and it is hard to predict who will win, and who will not.

Hiroshi Inoue (Inoue 2006) defines a Hamiltonian after the parameter transformation $S \to \mathrm{e}^x$ in Black–Scholes Equations 8.16 and 8.17, which yields

$$\frac{\partial f}{\partial t} = H_{\mathrm{BS}} f \tag{8.30}$$

so that he obtains a generalized Black–Scholes Hamiltonian that satisfies also the statistical Martingale condition of fair players

$$H_V = -\frac{1}{2}\sigma^2 \frac{\partial^2}{\partial x^2} + \left(\frac{1}{2}\sigma^2 - V(x)\right)\frac{\partial}{\partial x} + V(x) \tag{8.31}$$

with some, as yet, arbitrary potential $V(x)$. The next step is the application of the path integral including the Lagrangian L

$$K(x, t|y, s) = \int_{\Omega} \mathrm{e}^{\frac{i}{\hbar}\int_s^t L(u,r,\dot{r})\,\mathrm{d}u}\,\mathrm{d}r \tag{8.32}$$

where for K holds the relation $|K^2| = P$, and P is the probability that a particular trajectory in the time-space domain is taken. The integration boundaries in the Lagrangian is the time $s \to t$, which goes over the time parameter u, in this case. Then, r is the spatial coordinate; K is, then, the integral kernel and the Feynman path integral, and the basic solution for the Schrödinger equation.

Green's function

$$K_V(x_T, T|x, t) = \int_{x}^{x_T} e^{-\int_t^T (L_0+V)\,dt}\, dx(t) \tag{8.33}$$

satisfies the Black–Scholes equation; L_0 is a Lagrangian for a process with zero drift. So, what we are doing here now is not quantum electrodynamics but quantum econodynamics.

8.8.17 Planck's constant and the Feynman integral in economics

By formulating the Black–Scholes model into a quantum mechanical model, specifically formulating the Black–Scholes equation as having the shape of a Schrödinger equation, the question for Planck's constant h may eventually come up. In the Black–Scholes model, there is no arbitrage potential included. It is arbitrage free. Haven included an arbitrage potential in the Schrödinger formulation of Black–Scholes model and worked out that in the case where $h = 0$, there would be no arbitrage, and the corresponding Feynman path integral would simplify to the classical path, whereas $h \neq 0$ would allow for a nonzero arbitrage (Haven 2002).

The question is what the role and interpretation of Planck's constant h in quantum econodynamics is. The numeric value for h in this context is not the value we know from Physics. The adoption of the Black–Scholes equation into the Schrödinger formalism yields – in a particular example in Haven (2002) – a relation of the kind

$$N(d_2) = N(d_1)e^{\left(\frac{\sigma}{2}\right)^2 + \frac{\sigma}{\sqrt{2}} + \left(7.8203 - 0.39852\left(\frac{1}{h}\right)\right)} = 0.73495\,, \tag{8.34}$$

which yields $h^* = 0.044933$.

8.8.18 Feynman diagram on the stock market

I want to finish the formal part of this book with an example where the Feynman diagrams are used for the description and illustration of a problem from the stock market. The example was published by Bagarello in a physics journal around 10 years ago (Bagarello 2009) and, therefore, stands in the more than 100-year old tradition of how physics concepts are bound and found in economy and social sciences. We recall the example of Bachelier's development of the theory of Brownian motion (Bachelier 1900; Courtault et al. 2000), which was reproduced 5 years later by Albert Einstein (Samuelson 2015).

You can consider the stock market as a statistical ensemble where a number of N different stock brokers $b_1, b_2, \ldots, b_N$ trade a number of L different shares $S_1, S_2, \ldots, S_L$. This model was suggested by Bagarello (2009). As the shares change owners, the shares vanish from one portfolio by an annihilation operator and show up in a portfolio elsewhere by a creation operator. At the same time, the money will vanish by the amount of the purchase prize from the wallet of the buyer of the stocks, and the money of seller will increase accordingly. The change of money in the wallets will also be described by the monetary annihilation and creation operators. As we are dealing here with the volatility of the price of shares, we also need operators that can lower and increase the value of a share, as listed in Table 8.5

Tab. 8.5: The Boson type operators accounting for the economical transactions at the stock market after the model of Bagarello. Reprinted from Physica A, 388, F. Bagarello, A quantum statistical approach to simplified stock markets, 4397–4406, Copyright (2009), with permission from Elsevier.

Quantum mechanical operators and their economical interpretation	
QM operator	**Economical interpretation**
$a_{j,\alpha}$	annihilates a share σ_α in the portfolio of τ_j
$a^\dagger_{j,\alpha}$	annihilates a share σ_α in the portfolio of τ_j
$\hat{n}_{j,\alpha} = a^\dagger_{j,\alpha} a_{j,\alpha}$	counts the number of shares σ_α in the portfolio of τ_j
c_j	annihilates a monetary unit in the portfolio of τ_j
$c^\dagger_j$	creates a monetary unit in the portfolio of τ_j
$\hat{k}_j = a^\dagger_j a_j$	counts the number of monetary units in the portfolio of τ_j
p_α	lowers the price of the share σ_α of one unit of cash
$p^\dagger_\alpha$	increases the price of the share σ_α of one unit of cash
$\hat{P}_\alpha = p^\dagger_\alpha p_\alpha$	gives the value of the share σ_α

These operators are all of Bosonic nature and satisfy particular commutator rules such as

$$[c_j, c^\dagger_k] = \mathbf{1}\delta_{j,k} \tag{8.35}$$

$$[p_\alpha, p^\dagger_\beta] = \mathbf{1}\delta_{\alpha,\beta} \tag{8.36}$$

$$[a_{j,\alpha}, a^\dagger_{k,\beta}] = \mathbf{1}\delta_{j,k}\delta_{\alpha,\beta} \tag{8.37}$$

and the other commutators are nihil. Furthermore, *selling* operators $\chi_{j,\alpha} = a_{j,\alpha} c_j^{\dagger \hat{P}_\alpha}$ and *buying* operators like $\chi^\dagger_{j,\alpha} = a^\dagger_{j,\alpha} c_j^{\hat{P}_\alpha}$, are introduced. With the appropriate Hamiltonians and wavefunctions as developed in the paper by Bagarello (2009), and the inclusion of time dependency, transactions with constant prices or weakly varying prices are exercised. Two representative Feynman diagrams are shown in Figures 8.41 and 8.42:

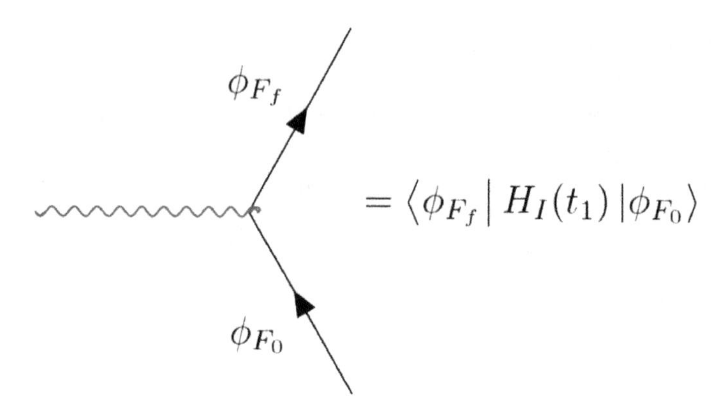

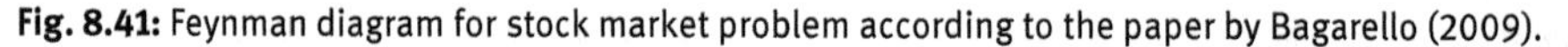

Fig. 8.41: Feynman diagram for stock market problem according to the paper by Bagarello (2009).

```
\tikzfeynmanset{
every edge = {green},
every boson = {red},
every anti fermion = {blue},
every fermion = {black},
every photon = {green},
}

\begin{equation}
\feynmandiagram [baseline=(d.base), horizontal=d to b] {
a -- [fermion, edge label = \(\phi_{F_0}\)]
b -- [fermion, edge label = \(\phi_{F_f}\)] c,
b -- [boson, out=0, in=180] d,
};
=   \bra{\phi_{F_f}} H_I(t_1)\ket{\phi_{F_0}}
\end{equation}
```

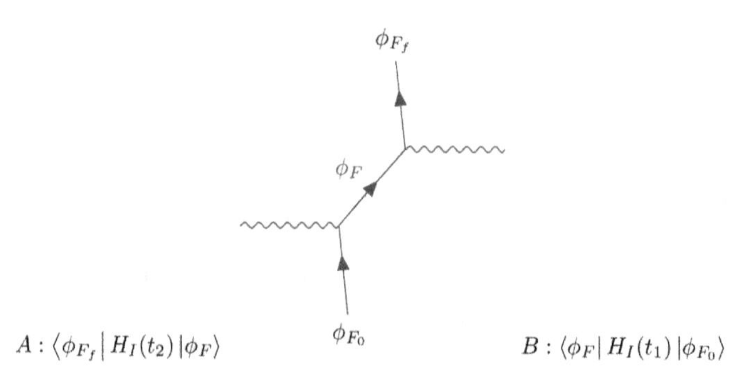

Fig. 8.42: Feynman diagram for stock market problem according to the paper by Bagarello (2009).

```
\tikzfeynmanset{
every edge = {green},
every boson = {red},
every anti fermion = {blue},
every fermion = {black},
every photon = {green},
}

\begin{equation}
A: \bra{\phi_{F_f}} H_I(t_2)\ket{\phi_F}
\feynmandiagram [horizontal=a to b] {
f2 [particle=\(\phi_{F_f}\)] -- [anti fermion] c -- [boson] f3 [particle],
a [particle] -- [boson] b -- [anti fermion] f1 [particle=\(\phi_{F_0}\)],
b -- [fermion, edge label=\(\phi_F\)] c,
};
B: \bra{\phi_{F}} H_I(t_1)\ket{\phi_{F_0}}
\end{equation}
```

We have seen in this chapter on *agriculture and food supply* how important the ingredients of the Earth, the minerals and nutrients found in the soil, are for the proteins in the photosynthetic apparatus. They are necessary for the growth of organisms and carbon-dioxide assimilation, oxygen production, and water splitting. Mankind is bound to the soil, but we are thinking about building human life-support systems for habitats in hostile regions, including outer space.

The energy for driving the biomolecular and chemical processes comes primarily from the sun, and quantum electrodynamics describes the interaction of the photons from the sun with all matter hit by the photons on their way to Earth and the biosphere, and photosynthesis organisms. Also, the production of photons by the nuclear reactions in the sun and other stars are an effect of quantum electrodynamics. The fruit of primary production are traded on the market and stock market. On the latter, no fruit are physically visible. Only numbers, the numerals of supply and demand, meet on the stock market, with a commercial interest in the mind of the traders and their clients.

Farmers' fear of losses due to crop failure, flood, drought, and beetle infestation causes them to stock more food than necessary at a certain time. This is the need for surplus as insurance against future losses. The inherent uncertainty about losses mandates the need for profit to warrant surplus. This mindset is not different at the stock exchange and translates into negotiation, like on any other market. The uncertainty when looking in the future is mathematically expressed by diffuse (see Piotrowski and Sladkowski 2005 for quantum diffusion of prices and profits) operations and probability.

The mathematical description of the processes in economy is the same like the processes that take place in classical mechanics and in quantum mechanics (or projective geometry (Piotrowski and Sladkowski 2007), although this is not treated here), to the extent that even quantum electrodynamics tools can be employed with the stock market transactions. We have seen this in the last examples shown here by Bagarello (2009). Further progress in describing and solving problems in economy and social sciences with the mathematical tools from quantum electrodynamics remains to be seen.

8.9 When the sun sets

I am closing this book with a section that was meant to be in Chapter 5. However, I felt it would be better suited here at the end of the book. As I explained in Section 2.3, the anthropologic relevance of the sun is deeply rooted in its practical importance for mankind, its simple utility in daily human life. The sun bears strong symbolism in ancient and contemporary and popular cultures. Daylight has become the center of attention for some of my colleagues, and this is why we founded, as a spinoff of the VELUX Foundation, the Daylight Academy (Norton et al. 2017). Daylight is very important for the well-being of humans, and daylight is given by the sun. There are countless songs and music featuring the sun, literally.

In my book *Electrochemical Energy Systems* (Braun 2019a), I began the chapter on nuclear energy with a photo of the sunrise over Death Valley and ended it with a photo of the sunset I took in Berkeley. It shows the sunset over Mount Tamalpais in Mill Valley, California, the native home of the Coast Miwok people, an Indian tribe. To the Miwok people, Mount Tamalpais was the bride of the sun god. The rays of the sun setting over the Pacific Ocean at Point Reyes were the paths for Miwok souls into eternity. Meanwhile, the sun reliably maintains its eternal cycle of sunrise and sunset for us. As night dawns over the San Francisco Bay area, people in Australia and Eastern Asia are preparing for the new day. Night and winter are the dark phases of the sun's cycle. Since the creation of life billions of years ago, nature has coped with night and with winter by energy storage. Scientists, engineers, and technologists are working worldwide to make sure that in future you have all the energy you need – every time you need it – with solar energy storage by artificial photosynthesis.

You can consider fossil fuels like an energy source, like a primary battery that is used up more and more until nothing is left (Schramski et al. 2015). Many people believe that we will run out of fossil fuels in the next few hundreds of years. After that, we will have to get our energy from the sun as renewable energy or from nuclear power plants. Maybe geothermal energy will help. Yet can solar energy and its derivatives, hydropower and wind power, be considered renewable energy, when the cycles of day and night, and summer and winter will fade away as the sun, our hydrogen star, will have entirely turned into helium in the next 4–5 billion years?

If for any reason we can no longer get our energy from the sun, think, for example, about a huge volcano explosion that would lay an ash cloud over the skies, we should be prepared for this by having preserved enough fossil fuels and by having developed a viable nuclear energy technology, preferably nuclear fusion, as featured in Chapter 5. However, irrespective of all the bad scenarios, it is worthwhile tapping the sun as long it is there. Did you know that in only 1 hour we receive the same amount of energy from our sun as we use in one entire year?

Epilogue

1. In the beginning God created the heaven and the Earth.
2. And the Earth was without form, and void; and darkness was upon the face of the deep. And the Spirit of God moved upon the face of the waters.
3. And God said, Let there be **light**: and there was **light**.
4. And God saw the **light**, that it was good: and God divided the **light** from the darkness.
5. And God called the **light** Day, and the darkness he called Night. And the evening and the morning were the first day.
6. And God said, Let there be a firmament in the midst of the waters, and let it divide the waters from the waters.
7. And God made the firmament, and divided the waters which were under the firmament from the waters which were above the firmament: and it was so.
8. And God called the firmament Heaven. And the evening and the morning were the second day.
9. And God said, Let the waters under the heaven be gathered together unto one place, and let the dry land appear: and it was so.
10. And God called the dry land Earth; and the gathering together of the waters called he Seas: and God saw that it was good.
11. And God said, Let the Earth bring forth grass, the herb yielding seed, and the fruit tree yielding fruit after his kind, whose seed is in itself, upon the Earth: and it was so.
12. And the Earth brought forth grass, and herb yielding seed after his kind, and the tree yielding fruit, whose seed was in itself, after his kind: and God saw that it was good.
13. And the evening and the morning were the third day.
14. And God said, Let there be **lights** in the firmament of the heaven to divide the day from the night; and let them be for signs, and for seasons, and for days, and years:
15. And let them be for **lights** in the firmament of the heaven to give **light** upon the Earth: and it was so.
16. And God made two great **lights**; the greater **light** to rule the day, and the lesser **light** to rule the night: he made the stars also.
17. And God set them in the firmament of the heaven to give **light** upon the Earth,
18. And to rule over the day and over the night, and to divide the **light** from the darkness: and God saw that it was good.
19. And the evening and the morning were the fourth day.
20. And God said, Let the waters bring forth abundantly the moving creature that hath life, and fowl that may fly above the Earth in the open firmament of heaven.

https://doi.org/10.1515/9783110629941-009

21. And God created great whales, and every living creature that moveth, which the waters brought forth abundantly, after their kind, and every winged fowl after his kind: and God saw that it was good.
22. And God blessed them, saying, Be fruitful, and multiply, and fill the waters in the seas, and let fowl multiply in the Earth.
23. And the evening and the morning were the fifth day.
24. And God said, Let the Earth bring forth the living creature after his kind, cattle, and creeping thing, and beast of the Earth after his kind: and it was so.
25. And God made the beast of the Earth after his kind, and cattle after their kind, and every thing that creepeth upon the Earth after his kind: and God saw that it was good.
26. And God said, Let us make man in our image, after our likeness: and let them have dominion over the fish of the sea, and over the fowl of the air, and over the cattle, and over all the Earth, and over every creeping thing that creepeth upon the Earth.
27. So God created man in his own image, in the image of God created he him; male and female created he them.
28. And God blessed them, and God said unto them, Be fruitful, and multiply, and replenish the Earth, and subdue it: and have dominion over the fish of the sea, and over the fowl of the air, and over every living thing that moveth upon the Earth.
29. And God said, Behold, I have given you every herb bearing seed, which is upon the face of all the Earth, and every tree, in the which is the fruit of a tree yielding seed; to you it shall be for meat.
30. And to every beast of the Earth, and to every fowl of the air, and to every thing that creepeth upon the Earth, wherein there is life, I have given every green herb for meat: and it was so.
31. And God saw every thing that he had made, and, behold, it was very good. And the evening and the morning were the sixth day.

From the Book of Genesis, the first book of the Hebrew Bible and the Old Testament (Didot 1839).

Appendix 1 – Electromagnetic spectrum

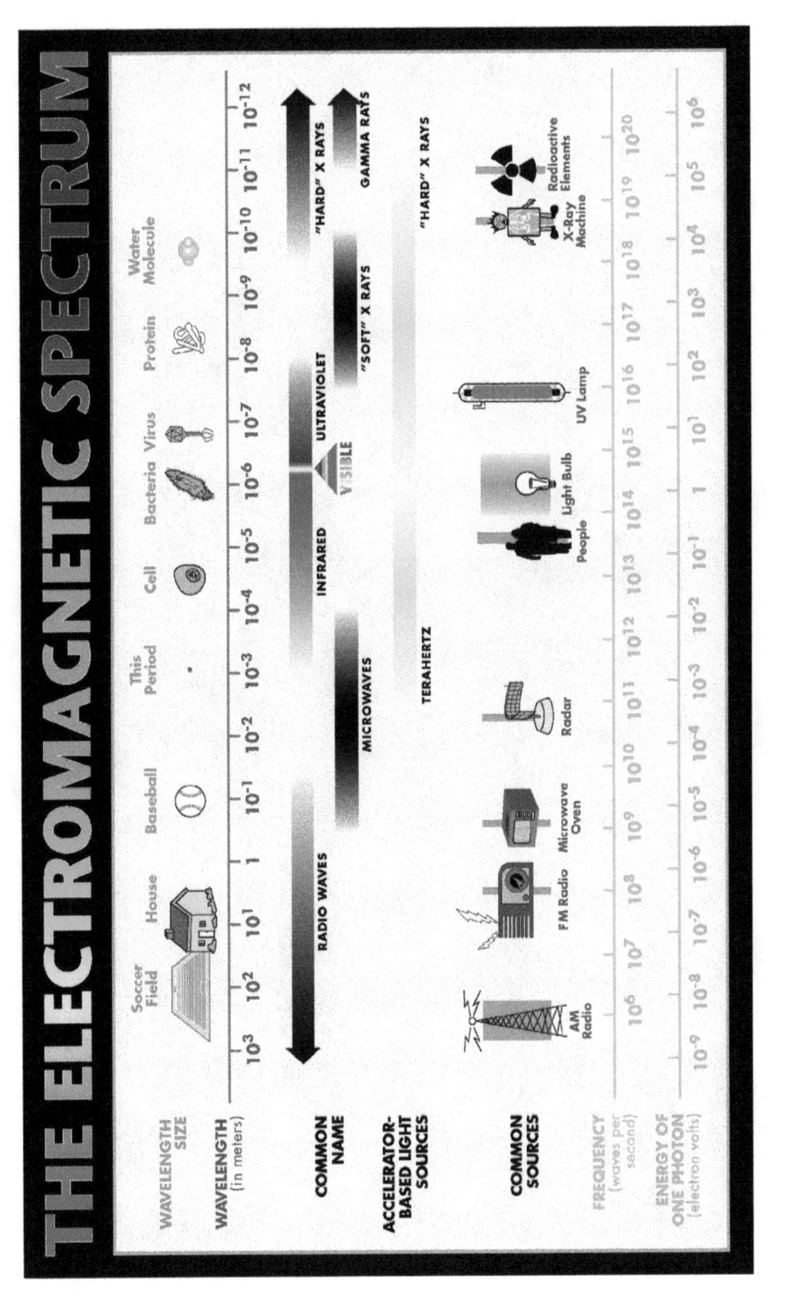

Fig. A.1: The electromagnetic spectrum. Courtesy of the Advanced Light Source, Lawrence Berkeley National Laboratory. ©2010–2020 The Regents of the University of California, Lawrence Berkeley National Laboratory.

https://doi.org/10.1515/9783110629941-010

Appendix 2 – The Gospel of John

1. In the beginning was the Word, and the Word was with God, and the Word was God.
2. The same was in the beginning with God.
3. All things were made by him; and without him was not any thing made that was made.
4. In him was life; and the life was the **light** of men.
5. And the **light** shineth in darkness; and the **darkness** comprehended it not.
6. There was a man sent from God, whose name was John.
7. The same came for a witness, to bear witness of the **Light**, that all men through him might believe.
8. He was not that **Light**, but was sent to bear witness of that **Light**.
9. That was the true **Light**, which **lighteth** every man that cometh into the world.
10. He was in the world, and the world was made by him, and the world knew him not.
11. He came unto his own, and his own received him not.

From John 1, the first chapter of the Gospel of John in the New Testament (testamentum 1994).

https://doi.org/10.1515/9783110629941-011

Bibliography

[Abbott 2019] ABBOTT, A.: Europe's next euro 1-billion science projects: six teams make it to final round. In: *Nature* 566 (2019), Nr. 7743, pp. 164–165. – URL https://www.nature.com/articles/d41586-019-00541-y. – ISSN 1476-4687 (Electronic) 0028-0836 (Linking)

[Aberbach 2008] ABERBACH, J. D.: *Transforming the Presidency: The Administration of Ronald Reagan* . pp. 191–207. In: *Ronald Reagan and the 1980s: Perceptions, Policies, Legacies*, Palgrave Macmillan US, 2008 (Studies of the Americas). – URL https://doi.org/10.1057/9780230616196. – ISBN 978-0-230-60302-8

[Abitan et al. 2008] ABITAN, H. ; BOHR, H. ; BUCHHAVE, P.: Correction to the Beer-Lambert-Bouguer law for optical absorption. In: *Appl Opt* 47 (2008), Nr. 29, pp. 5354–7. – URL https://www.ncbi.nlm.nih.gov/pubmed/18846176. – ISSN 1539-4522 (Electronic) 1559-128X (Linking)

[Abomohra et al. 2016] ABOMOHRA, A. E. ; EL-SHOUNY, W. ; SHARAF, M. ; ABP-ELENEEN, M.: Effect of Gamma Radiation on Growth and Metabolic Activities of Arthrospira Platensis. In: *Brazilian Archives of Biology and Technology* 59 (2016), p. 11. – URL http://www.scielo.br/pdf/babt/v59/1516-8913-babt-16150476.pdf. – ISSN 1516-8913

[Abramczyk 2012] ABRAMCZYK, H.: Mechanisms of energy dissipation and ultrafast primary events in photostable systems: H-bond, excess electron, biological photoreceptors. In: *Vibrational Spectroscopy* 58 (2012), pp. 1–11. – ISSN 0924-2031

[Abu-Zeid 2010] ABU-ZEID, M.: Water Pricing in Irrigated Agriculture. In: *International Journal of Water Resources Development* 17 (2010), Nr. 4, pp. 527–538. – URL https://www.tandfonline.com/doi/pdf/10.1080/07900620120094109. – ISSN 0790-0627 1360-0648

[Acharya et al. 2016] ACHARYA, B. ; CARLSSON, B. D. ; EKSTROM, A. ; FORSSEN, C. ; PLATTER, L.: Uncertainty quantification for proton-proton fusion in chiral effective field theory. In: *Physics Letters B* 760 (2016), pp. 584–589. – URL https://www.sciencedirect.com/science/article/pii/S0370269316303732. – ISSN 0370-2693

[Acheson 1910] ACHESON, E. G.: *A Pathfinder. Discovers, Invention, and Industry*. New York: The Press Scrap Book, 1910

[Achino 2014] ACHINO, E.: What is your gender? Gender and participation in Attac-Italy. In: *Journal of Gender Studies* 25 (2014), Nr. 2, pp. 155–168. – ISSN 0958-9236 1465-3869

[Adelberger et al. 1998] ADELBERGER, E. G. ; AUSTIN, S. M. ; BAHCALL, J. N. ; BALANTEKIN, A. B. ; BOGAERT, G. ; BROWN, L. S. ; BUCHMANN, L. ; CECIL, F. E. ; CHAMPAGNE, A. E. ; BRAECKELEER, L. de ; DUBA, C. A. ; ELLIOTT, S. R. ; FREEDMAN, S. J. ; GAI, M. ; GOLDRING, G. ; GOULD, C. R. ; GRUZINOV, A. ; HAXTON, W. C. ; HEEGER, K. M. ; HENLEY, E. ; JOHNSON, C. W. ; KAMIONKOWSKI, M. ; KAVANAGH, R. W. ; KOONIN, S. E. ; KUBODERA, K. ; LANGANKE, K. ; MOTOBAYASHI, T. ; PANDHARIPANDE, V. ; PARKER, P. ; ROBERTSON, R. G. H. ; ROLFS, C. ; SAWYER, R. F. ; SHAVIV, N. ; SHOPPA, T. D. ; SNOVER, K. A. ; SWANSON, E. ; TRIBBLE, R. E. ; TURCK-CHIEZE, S. ; WILKERSON, J. F.: Solar fusion cross sections. In: *Reviews of Modern Physics* 70 (1998), Nr. 4, pp. 1265–1291. – URL https://journals.aps.org/rmp/pdf/10.1103/RevModPhys.70.1265. – ISSN 0034-6861

[Adelberger et al. 2011] ADELBERGER, E. G. ; GARCIA, A. ; ROBERTSON, R. G. H. ; SNOVER, K. A. ; BALANTEKIN, A. B. ; HEEGER, K. ; RAMSEY-MUSOLF, M. J. ; BEMMERER, D. ; JUNGHANS, A. ; BERTULANI, C. A. ; CHEN, J. W. ; COSTANTINI, H. ; PRATI, P. ; COUDER, M. ; UBERSEDER, E. ; WIESCHER, M. ; CYBURT, R. ; DAVIDS, B. ; FREEDMAN, S. J. ; GAI, M. ; GAZIT, D. ; GIALANELLA, L. ; IMBRIANI, G. ; GREIFE, U. ; HASS, M. ; HAXTON, W. C. ; ITAHASHI, T. ; KUBODERA, K. ; LANGANKE, K. ; LEITNER, D. ; LEITNER, M. ; VETTER, P. ; WINSLOW, L. ; MARCUCCI, L. E. ; MOTOBAYASHI, T. ; MUKHAMEDZHANOV, A. ; TRIBBLE, R. E. ; NOLLETT, K. M. ; NUNES, F. M. ; PARK, T. S. ; PARKER, P. D. ; SCHIAVILLA, R. ; SIMPSON, E. C. ; SPITALERI, C. ; STRIEDER, F. ; TRAUTVETTER, H. P. ; SUEMMERER, K. ; TYPEL, S.: Solar fusion cross sections. II. The p p chain and CNO cycles. In: *Reviews of Modern Physics* 83

https://doi.org/10.1515/9783110629941-012

(2011), Nr. 1, pp. 195–245. – URL https://journals.aps.org/rmp/pdf/10.1103/RevModPhys.83.195. – ISSN 0034-6861 1539-0756

[Agmon 1995] AGMON, N.: The Grotthuss mechanism. In: *Chemical Physics Letters* 244 (1995), Nr. 5-6, pp. 456–462. – URL 10.1016/0009-2614(95)00905-J. – ISSN 00092614

[Agostiano et al. 1992] AGOSTIANO, A. ; GOETZE, D. C. ; CARPENTIER, R.: Cyclic Voltammetry Measurements of the Photoelectrogenic Reactions of Thylakoid Membranes. In: *Photochemistry and Photobiology* 55 (1992), Nr. 3, pp. 449–455. – URL https://onlinelibrary.wiley.com/doi/abs/10.1111/j.1751-1097.1992.tb04260.x. – ISSN 0031-8655

[Agostiano et al. 1993] AGOSTIANO, A. ; GOETZE, D. C. ; CARPENTIER, R.: Photoelectrochemistry of Thylakoid and Sub-Thylakoid Membrane Preparations– Cyclic Voltammetry and Action Spectra. In: *Electrochimica Acta* 38 (1993), Nr. 6, pp. 757–762. – URL https://www.sciencedirect.com/science/article/pii/0013468693850303. – ISSN 0013-4686

[Ahmadi et al. 2012] AHMADI, M. ; OSEI, E. K. ; YEOW, J. T. W.: Bacteriorhodopsin for superficial X-ray sensing. In: *Sensors and Actuators B: Chemical* 166-167 (2012), pp. 177–183. – URL https://www.sciencedirect.com/science/article/pii/S0925400512001694. – ISSN 09254005

[Alakavuk and Yağmuroğlu 2010] ALAKAVUK, E. ; YAĞMUROĞLU, Z.: *Damlatas Cave, Alanya*. Book Section Chapter 35, pp. 311–314. In: *Natural Heritage from East to West*, 2010. – ISBN 978-3-642-01576-2 978-3-642-01577-9

[Albers and Knorr 1936] ALBERS, V. M. ; KNORR, H. V.: Spectroscopic Studies of the Simpler Porphyrins I. The Absorption Spectra of Porphin, ms-Methyl Porphin, ms-Ethyl Porphin, ms-Propyl Porphin and ms-Phenyl Porphin. In: *The Journal of Chemical Physics* 4 (1936), Nr. 7, pp. 422–425. – URL https://aip.scitation.org/doi/pdf/10.1063/1.1749873. – ISSN 0021-9606 1089-7690

[Albertsson and Andreasson 2004] ALBERTSSON, P. A. ; ANDREASSON, E.: The constant proportion of grana and stroma lamellae in plant chloroplasts. In: *Physiologia Plantarum* 121 (2004), Nr. 2, pp. 334–342. – URL https://onlinelibrary.wiley.com/doi/full/10.1111/j.0031-9317.2004.00315.x. – ISSN 0031-9317

[Albertsson 2001] ALBERTSSON, P.-Å.: A quantitative model of the domain structure of the photosynthetic membrane. In: *Trends in Plant Science* 6 (2001), Nr. 8, pp. 349–354. – URL https://doi.org/10.1016/S1360-1385(01)02021-0. – ISSN 13601385

[Alexandrov and Yamagata 2007] ALEXANDROV, G. A. ; YAMAGATA, Y.: A peaked function for modeling temperature dependence of plant productivity. In: *Ecological Modelling* 200 (2007), Nr. 1-2, pp. 189–192. – URL doi:10.1016/j.ecolmodel.2006.07.012. – ISSN 03043800

[Alivisatos et al. 2014] ALIVISATOS, A. P. ; PERLMUTTER, S. ; BISSELL, M.: *Paul Alivisatos Accepts ALS Ice Bucket Challenge*. 22 August 2014 2014. – URL https://www.youtube.com/watch?v=e2aoxqSXPpc

[Allakhverdiev 2008] ALLAKHVERDIEV, S. I.: Recent perspectives of photosystem II: structure, function and dynamics. In: *Photosynth Res* 98 (2008), Nr. 1-3, pp. 1–5. – URL https://link.springer.com/content/pdf/10.1007%2Fs11120-008-9390-2.pdf. – ISSN 0166-8595 (Print) 0166-8595 (Linking)

[Allam et al. 2011] ALLAM, N. K. ; YEN, C.-W. ; NEAR, R. D. ; EL-SAYED, M. A.: Bacteriorhodopsin/TiO2 nanotube arrays hybrid system for enhanced photoelectrochemical water splitting. In: *Energy & Environmental Science* 4 (2011), Nr. 8. – URL https://pubs.rsc.org/en/content/articlepdf/2011/ee/c1ee01447a. – ISSN 1754-5692 1754-5706

[Anderson 1972] ANDERSON, P. W.: More is different. In: *Science* 177 (1972), Nr. 4047, pp. 393–6. – URL http://www.ncbi.nlm.nih.gov/pubmed/17796623. – ISSN 0036-8075 (Print) 0036-8075 (Linking)

[anonymous, cia 1967] ANONYMOUS, CIA: *Intelligence for National Policy*. Virginia, United States: Central Intelligence Agency – Office of Training, 1967 [anonymous, cia 1969] ANONYMOUS, CIA: *Intelligence School – Training Manual*. Vol. 4: *Intelligence School – Training Manual Number 4*. Washington, D.C.: The Office of Economic Research, 1969. – URL https://www.cia.gov/library/readingroom/document/cia-rdp80-00317a000100040001-3

[Arnold and Oppenheimer 1950] ARNOLD, W. ; OPPENHEIMER, J. R.: Internal conversion in the photosynthetic mechanism of blue-green algae. In: *J Gen Physiol* 33 (1950), Nr. 4, pp. 423–35. – URL http://jgp.rupress.org/content/jgp/33/4/423.full.pdf. – ISSN 0022-1295 (Print) 0022-1295 (Linking)

[Aro et al. 2017] ARO, E.-M. ; VINCENT, A. ; AZZOLINI, R. ; BARBIERI, A. ; BAUMANN, S. ; BERCEGOL, H. ; BRAUN, A. ; CAMPUS, P. ; CHANDEZON, F. ; DE GROOT, H. ; PENA, V. de la ; DURRANT, J. ; FABER, C. ; FLEISCHER, M. ; FRESNO, F. ; GAERTNER, T. ; HAMMARSTROM, L. ; JOANNA, K. ; ANTONI, L. ; LOPEZ, L. ; LORENA, T. ; LORETO, F. ; MEIER, M. ; NOBLE, A. ; TONDELLI, L. ; VELDEN, E. Van der: *SUNRISE – Solar energy for a circular economy*. 2017. – URL www.sunriseflagship.eu

[Artsimov et al. 1969] ARTSIMOV, L. ; BOBROVSK, G. ; GORBUNOV, E. P. ; IVANOV, D. P. ; KIRILLOV, V. D. ; KUZNETSOV, E. ; MIRNOV, S. V. ; PETROV, M. P. ; RAZUMOVA, K. A. ; STRELKOV, V. S. ; SHCHEGLO, D.: Experiments in Tokamak Devices. In: *Nuclear Fusion* S (1969), p. 17. – ISSN 0029-5515

[Ashby 1979] ASHBY, H.: *Being there*. 1979. – URL https://www.imdb.com/title/tt0078841/

[Atkinson and Houtermans 1929] ATKINSON, R. d. E. ; HOUTERMANS, F. G.: Zur Frage der Aufbaumoeglichkeit der Elemente in Sternen. In: *Zeitschrift fuer Physik* 54 (1929), Nr. 9-10, pp. 656–665. – URL https://link.springer.com/article/10.1007%2FBF01341595. – ISSN 1434-6001 1434-601X

[Attenborough 2007] ATTENBOROUGH, D.: *Giant waterlillies in the Amazon*. 9 Feb 2007 2007. – URL https://www.youtube.com/watch?v=igkjcuw_n_U

[Awadalla and Morsy 2016] AWADALLA, A. ; MORSY, A. S. M.: Response of Some Yellow Maize Crosses to N-fertilizer Rates and Plant Densities at Toshka Region. In: *Egyptian Journal of Agronomy* 38 (2016), Nr. 3, pp. 337–354. – URL https://agro.journals.ekb.eg/article_1273_be9c0531d0f15b285047aa9e29d0853e.pdf. – ISSN 0379-3575

[Babusiaux et al. 2018] BABUSIAUX, C. ; LEEUWEN, F. van ; BARSTOW, M. A. ; JORDI, C. ; VALLENARI, A. ; BOSSINI, D. ; BRESSAN, A. ; CANTAT-GAUDIN, T. ; LEEUWEN, M. van ; BROWN, A. G. A. ; PRUSTI, T. ; BRUIJNE, J. H. J. de ; BAILER-JONES, C. A. L. ; BIERMANN, M. ; EVANS, D. W. ; EYER, L. ; JANSEN, F. ; KLIONER, S. A. ; LAMMERS, U. ; LINDEGREN, L. ; LURI, X. ; MIGNARD, F. ; PANEM, C. ; POURBAIX, D. ; RANDICH, S. ; SARTORETTI, P. ; SIDDIQUI, H. I. ; SOUBIRAN, C. ; WALTON, N. A. ; ARENOU, F. ; BASTIAN, U. ; CROPPER, M. ; DRIMMEL, R. ; KATZ, D. ; LATTANZI, M. G. ; BAKKER, J. ; CACCIARI, C. ; CASTAÑEDA, J. ; CHAOUL, L. ; CHEEK, N. ; DE ANGELI, F. ; FABRICIUS, C. ; GUERRA, R. ; HOLL, B. ; MASANA, E. ; MESSINEO, R. ; MOWLAVI, N. ; NIENARTOWICZ, K. ; PANUZZO, P. ; PORTELL, J. ; RIELLO, M. ; SEABROKE, G. M. ; TANGA, P. ; THÉVENIN, F. ; GRACIA-ABRIL, G. ; COMORETTO, G. ; GARCIA-REINALDOS, M. ; TEYSSIER, D. ; ALTMANN, M. ; ANDRAE, R. ; AUDARD, M. ; BELLAS-VELIDIS, I. ; BENSON, K. ; BERTHIER, J. ; BLOMME, R. ; BURGESS, P. ; BUSSO, G. ; CARRY, B. ; CELLINO, A. ; CLEMENTINI, G. ; CLOTET, M. ; CREEVEY, O. ; DAVIDSON, M. ; DE RIDDER, J. ; DELCHAMBRE, L. ; DELL'ORO, A. ; DUCOURANT, C. ; FERNÁNDEZ-HERNÁNDEZ, J. ; FOUESNEAU, M. ; FRÉMAT, Y. ; GALLUCCIO, L. ; GARCÍA-TORRES, M. ; GONZÁLEZ-NÚÑEZ, J. ; GONZÁLEZ-VIDAL, J. J. ; GOSSET, E. ; GUY, L. P. ; HALBWACHS, J. L. ; HAMBLY, N. C. ; HARRISON, D. L. ; HERNÁNDEZ, J. ; HESTROFFER, D. ; HODGKIN, S. T. ; HUTTON, A. ; JASNIEWICZ, G. ; JEAN-ANTOINE-PICCOLO, A. ; JORDAN, S. ; KORN, A. J. ; KRONE-MARTINS, A. ; LANZAFAME, A. C. ; LEBZELTER, T. et al.: Gaia Data Release 2. In: *Astronomy and Astrophysics* 616 (2018). – URL https://www.aanda.org/articles/aa/abs/2018/08/aa32843-18/aa32843-18.html. – ISSN 0004-6361 1432-0746

[Bachelier 1900] BACHELIER, L.: *Theorie de la Speculation*, Sorbonne, Thesis, 1900. – URL archive.numdam.org/article/ASENS_1900_3_17__21_0.pdf

[Bader 2011] BADER, R. F.: Worlds apart in chemistry: a personal tribute to J. C. Slater. In: *J Phys Chem A* 115 (2011), Nr. 45, pp. 12667–76. – URL https://pubs.acs.org/doi/pdfplus/10.1021/jp203531x. – ISSN 1520-5215 (Electronic) 1089-5639 (Linking)

[Badreldin et al. 2014] BADRELDIN, N. ; FRANKL, A. ; GOOSSENS, R.: Assessing the spatiotemporal dynamics of vegetation cover as an indicator of desertification in Egypt using multi-temporal MODIS satellite images. In: *Arabian Journal of Geosciences* 7 (2014), Nr. 11, pp. 4461–4475. – URL https://link.springer.com/content/pdf/10.1007/s12517-013-1142-8.pdf. – ISSN 1866-7511

[Badri et al. 2015] BADRI, H. ; MONSIEURS, P. ; CONINX, I. ; NAUTS, R. ; WATTIEZ, R. ; LEYS, N.: Temporal Gene Expression of the Cyanobacterium Arthrospira in Response to Gamma Rays. In: *Plos One* 10 (2015), Nr. 8, p. 29. – URL https://journals.plos.org/plosone/article?id=10.1371/journal.pone.0135565. – ISSN 1932-6203

[Bagarello 2009] BAGARELLO, F.: A quantum statistical approach to simplified stock markets. In: *Physica A: Statistical Mechanics and its Applications* 388 (2009), Nr. 20, pp. 4397–4406. – URL https://www.sciencedirect.com/science/article/pii/S0378437109005470. – ISSN 03784371

[Bagarello and Haven 2014] BAGARELLO, F. ; HAVEN, E.: The role of information in a two-traders market. In: *Physica a-Statistical Mechanics and Its Applications* 404 (2014), pp. 224–233. – URL https://www.sciencedirect.com/science/article/pii/S0378437114001617. – ISSN 0378-4371

[Bakr and Bahnassy 2019] BAKR, N. ; BAHNASSY, M. H.: *Land Use/Land Cover and Vegetation Status*. pp. 51–67. In: *Soils of Egypt*. Cham: Springer International Publishing Ag, 2019 (World Soils Book Series). – URL https://link.springer.com/content/pdf/10.1007%2F978-3-319-95516-2_4.pdf. – ISBN 978-3-319-95516-2; 978-3-319-95515-5

[Balasubramanian et al. 2013] BALASUBRAMANIAN, S. ; WANG, P. ; SCHALLER, R. D. ; RAJH, T. ; ROZHKOVA, E. A.: High-performance bioassisted nanophotocatalyst for hydrogen production. In: *Nano Lett* 13 (2013), Nr. 7, pp. 3365–71. – URL https://www.ncbi.nlm.nih.gov/pubmed/23808953. – ISSN 1530-6992 (Electronic) 1530-6984 (Linking)

[Baldwin 2017] BALDWIN, M.: A look inside Feynman's calculus notebook. In: *Physics Today* (2017). – URL https://physicstoday.scitation.org/do/10.1063/pt.5.9099/full/. – ISSN 19450699

[Bancroft and Clapp 1938] BANCROFT, W. D. ; CLAPP, R. C.: Grotthuss and Einstein. In: *Journal of the Franklin Institute* 225 (1938), Nr. 1, pp. 23–43. – ISSN 00160032

[Barsoukov and Macdonald 2005] BARSOUKOV, E. ; MACDONALD, J. R.: *Impedance Spectroscopy: Theory, Experiment, and Applications*. 2nd edition. Hoboken NJ: John Wiley & Sons, Inc., 2005. – p. 616. – URL https://onlinelibrary.wiley.com/doi/book/10.1002/0471716243. – ISBN 9780471716242 9780471647492

[Bashkin and Stoner 1975] BASHKIN, S. ; STONER, J. O.: *Atomic Energy Levels and Grotrian Diagrams*. Elsevier, North-Holland, 1975. – ISBN 9780720403220

[Bassham 1958] BASSHAM, J. A.: National Academy of Sciences: Abstracts of Papers Presented at the Autumn Meeting, 6-8 November 1958, University of California, Berkeley. In: *Science* 128 (1958), Nr. 3332, pp. 1142–9. – URL https://www.ncbi.nlm.nih.gov/pubmed/17835852. – ISSN 0036-8075 (Print) 0036-8075 (Linking)

[Bastias Saavedra 2014] BASTIAS SAAVEDRA, M.: *Nitrate*. 8 October 2014 2014. – URL https://encyclopedia.1914-1918-online.net/article/nitrate

[Baum 2016] BAUM, C.: *Opinion: The economic wisdom of Chauncey Gardiner: There will be growth in the spring*. 19 MAy 2016 2016. – URL https://www.marketwatch.com/story/the-economic-wisdom-of-chauncey-gardiner-there-will-be-growth-in-the-spring-2016-05-19

[Bean and Dutton 1966] BEAN, B. R. ; DUTTON, E. J.: *Radio Meteorology*. U.S. Department of Commerce, 1966 (National Bureau of Standards Monograph 92). – URL https://archive.org/details/radiometeorology92bean/mode/2up

[Bederson and Stroke 2011] BEDERSON, B. ; STROKE, H. H.: History of the New York University Physics Department. In: *Physics in Perspective* 13 (2011), Nr. 3, pp. 260–328. – URL https://link.springer.com/article/10.1007/s00016-011-0056-7. – ISSN 1422-6944

[Belyaeva et al. 2016] BELYAEVA, N. E. ; BULYCHEV, A. A. ; RIZNICHENKO, G. Y. ; RUBIN, A. B.: Thylakoid membrane model of the Chl a fluorescence transient and P700 induction kinetics in plant leaves. In: *Photosynthesis Research* 130 (2016), Nr. 1-3, pp. 491–515. – URL https://link.springer.com/content/pdf/10.1007%2Fs11120-016-0289-z.pdf. – ISSN 0166-8595

[Benderskii et al. 2005] BENDERSKII, V. A. ; VETOSHKIN, E. V. ; KATS, E. I. ; TROMMSDORFF, H. P.: A semiclassical 1D model of ultrafast photoisomerization reactions. In: *Chemical Physics Letters* 409 (2005), Nr. 4-6, pp. 240–244. – ISSN 0009-2614

[Berberan-Santos 2001] BERBERAN-SANTOS, M. N.: *Pioneering Contributions of Jean and Francis Perrin to Molecular Luminescence*. Book Section Chapter 2, pp. 7–33. In: *New Trends in Fluorescence Spectroscopy*, Springer, 2001 (Springer Series on Fluorescence). – URL https://doi.org/10.1007/978-3-642-56853-4_2. – ISBN 978-3-642-63214-3 978-3-642-56853-4

[Bergdolt 1922] BERGDOLT, E.: Ibn Hazms Abhandlung uber die Farben. In: *Zeitschrift fuer Semitistik und verwandte Gebiete / hrsg. im Auftr. der Deutschen Morgenlaendischen Gesellschaft* 9 (1922), pp. 139–146. – URL http://menadoc.bibliothek.uni-halle.de/dmg/periodical/titleinfo/117087

[Bernstein 2010] BERNSTEIN, J.: John von Neumann and Klaus Fuchs: an Unlikely Collaboration. In: *Physics in Perspective* 12 (2010), Nr. 1, pp. 36–50. – URL https://link.springer.com/content/pdf/10.1007%2Fs00016-009-0001-1.pdf. – ISSN 1422-6944

[Bethe 1938] BETHE, H. A.: The Oppenheimer-Phillips process. In: *Physical Review* 53 (1938), Nr. 1, pp. 39–50. – URL https://journals.aps.org/pr/pdf/10.1103/PhysRev.53.39. – ISSN 0031-899x

[Bethe 1939] BETHE, H. A.: Energy Production in Stars. In: *Physical Review* 55 (1939), Nr. 5, pp. 434–456. – URL https://journals.aps.org/pr/pdf/10.1103/PhysRev.55.434. – ISSN 0031-899X

[Bethe 1997] BETHE, H. A.: J. Robert Oppenheimer, 1904-1967. In: *Biographical Memoirs of Fellows of the Royal Society* 14 (1997), pp. 390–416. – ISSN 0080-4606 1748-8494

[Bethe and Critchfield 1938] BETHE, H. A. ; CRITCHFIELD, C. L.: The Formation of Deuterons by Proton Combination. In: *Physical Review* 54 (1938), Nr. 4, pp. 248–254. – URL https://journals.aps.org/pr/pdf/10.1103/PhysRev.54.248. – ISSN 0031-899X

[Birge 1990] BIRGE, R. R.: Photophysics and molecular electronic applications of the rhodopsins. In: *Annu Rev Phys Chem* 41 (1990), pp. 683–733. – URL https://www.annualreviews.org/doi/10.1146/annurev.pc.41.100190.003343. – ISSN 0066-426X (Print) 0066-426X (Linking)

[Bjerrum 1952] BJERRUM, N.: Structure and Properties of Ice. In: *Science* 115 (1952), Nr. 2989, pp. 385–90. – URL https://www.ncbi.nlm.nih.gov/pubmed/17741864. – ISSN 0036-8075 (Print) 0036-8075 (Linking)

[Bjorkman et al. 1972] BJORKMAN, O. ; PEARCY, R. W. ; HARRISON, A. T. ; MOONEY, H.: Photosynthetic adaptation to high temperatures: a field study in death valley, california. In: *Science* 175 (1972), Nr. 4023, pp. 786–9. – URL https://science.sciencemag.org/content/sci/175/4023/786.full.pdf. – ISSN 0036-8075 (Print) 0036-8075 (Linking)

[Black and Scholes 1972] BLACK, F. ; SCHOLES, M.: Valuation of Option Contracts and a Test of Market Efficiency. In: *Journal of Finance* 27 (1972), Nr. 2, pp. 399–417. – URL https://www.jstor.org/stable/2978484. – ISSN 0022-1082

[Black and Scholes 1973] BLACK, F. ; SCHOLES, M.: Pricing of Options and Corporate Liabilities. In: *Journal of Political Economy* 81 (1973), Nr. 3, pp. 637–654. – URL https://www.journals.uchicago.edu/doi/10.1086/260062. – ISSN 0022-3808

[Black and Scholes 1974] BLACK, F. ; SCHOLES, M.: Individual Investors and Mutual Funds – from Theory to a New Financial Product. In: *Journal of Finance* 29 (1974), Nr. 2, pp. 399–412. – URL https://onlinelibrary.wiley.com/doi/abs/10.1111/j.1540-6261.1974.tb03054.x. – ISSN 0022-1082

[Blom et al. 2002] BLOM, T. J. ; STRAVER, W. A. ; INGRATTA, F. J. ; KHOSLA, S. ; BROWN, W.: Carbon Dioxide In Greenhouses. Ontario Ministry of Agriculture, Food, and Rural Affairs, December 2002 (1198-712X). – Report. – URL http://www.omafra.gov.on.ca/english/crops/facts/00-077.htm

[Bockris 2002] BOCKRIS, J.: The origin of ideas on a Hydrogen Economy and its solution to the decay of the environment. In: *International Journal of Hydrogen Energy* 27 (2002), Nr. 7-8, pp. 731–740. – ISSN 03603199

[Bockris 2013] BOCKRIS, J. O. M.: The hydrogen economy: Its history. In: *International Journal of Hydrogen Energy* 38 (2013), Nr. 6, pp. 2579–2588. – ISSN 03603199

[Boeker et al. 1997] BOEKER, E. ; GRONDELLE, R. van ; ERNST, R. D. ; KURRE, K.: *Physik und Umwelt*. Wiesbaden: Vieweg+Teubner Verlag, 1997. – p. 443. – URL https://www.amazon.de/Physik-Umwelt-Egbert-Boeker/dp/3528067802/. – ISBN 978-3528067809

[Bohr 1933] BOHR, N.: Light and Life*. In: *Nature* 131 (1933), Nr. 3308, pp. 421–423. – ISSN 0028-0836 1476-4687

[Boltzmann 1884] BOLTZMANN, L.: Ableitung des Stefan'schen Gesetzes, betreffend die Abhaengigkeit der Waermestrahlung von der Temperatur aus der electromagnetischen Lichttheorie. In: *Annalen der Physik* 258 (1884), Nr. 6, pp. 291–294. – URL https://onlinelibrary.wiley.com/doi/pdf/10.1002/andp.18842580616. – ISSN 00033804 15213889

[Bondi 1950] BONDI, H.: On the Interpretation of the Hertzsprung-Russell Diagram. In: *Monthly Notices of the Royal Astronomical Society* 110 (1950), Nr. 6, pp. 595–606. – URL https://academic.oup.com/mnras/article/110/6/595/2603633. – ISSN 0035-8711

[Born 1926] BORN, M.: Quantenmechanik der Stossvorgange. In: *Zeitschrift fur Physik* 38 (1926), Nr. 11-12, pp. 803–827. – URL https://link.springer.com/content/pdf/10.1007%2FBF01397184.pdf. – ISSN 1434-6001 1434-601X

[Born 1933] BORN, M.: *Optik*. 3rd edition. Berlin, Heidelberg, New York, Tokyo: Springer-Verlag, 1933. – ISBN 3-540-05954-7

[Born 1949] BORN, M.: *Natural Philosophy of Cause and Chance*. Oxford at the Clarendon Press, 1949 (The Waynflete Lectures)

[Born and Oppenheimer 1927] BORN, M. ; OPPENHEIMER, R.: Zur Quantentheorie der Molekeln. In: *Annalen der Physik* 389 (1927), Nr. 20, pp. 457–484. – URL https://onlinelibrary.wiley.com/doi/abs/10.1002/andp.19273892002. – ISSN 00033804 15213889

[Boudoire 2015] BOUDOIRE, F.: *Self-assembled photonic mesostructures for water splitting photoanodes*, Universität Basel, Doctoral Thesis, 2015. – URL http://edoc.unibas.ch/36693/

[Boudoire et al. 2014] BOUDOIRE, F. ; TOTH, R. ; HEIER, J. ; BRAUN, A. ; CONSTABLE, E. C.: Photonic light trapping in self-organized all-oxide microspheroids impacts photoelectrochemical water splitting. In: *Energy and Environmental Science* 7 (2014), Nr. 8, pp. 2680–2688. – URL https://pubs.rsc.org/en/content/articlepdf/2014/ee/c4ee00380b. – ISSN 1754-5692

[Bouguer 1729] BOUGUER, P.: *Essai d'optique, sur la gradation de la lumière*. Paris: Gauthier-Villars et Cie, 1729

[Boyle and Morgan 2011] BOYLE, N. R. ; MORGAN, J. A.: Computation of metabolic fluxes and efficiencies for biological carbon dioxide fixation. In: *Metab Eng* 13 (2011), Nr. 2, pp. 150–8. – URL https://www.sciencedirect.com/science/article/pii/S1096717611000061. – ISSN 1096-7184 (Electronic) 1096-7176 (Linking)

[Bradley 2008] BRADLEY, R.: Midsummer and Midwinter in the Rock Carvings of South Scandinavia. In: *TEMENOS* 44 (2008), Nr. 2, pp. 223–232. – URL https://journal.fi/temenos/article/download/4587/6789

[Bradonjić et al. 2009] BRADONJIĆ, K. ; SWAIN, J. D. ; WIDOM, A. ; SRIVASTAVA, Y. N.: The Casimir Effect in Biology: The Role of Molecular Quantum Electrodynamics in Linear Aggregations of Red Blood Cells. In: *Journal of Physics: Conference Series* 161 (2009). – URL https://iopscience.iop.org/article/10.1088/1742-6596/161/1/012035/pdf. – ISSN 1742-6596

[Brauer 2003] BRAUER, R.: Fair? In: *Geier. autonomes Info-Flugi fuer die Fachschaft Mathe/Physik/Info* (2003), 20. Oktober 2003, Nr. 119. – URL https://www.fsmpi.rwth-aachen.de/news/geier/geier-119.html

[Braun 1999] BRAUN, A.: *Development and characterization of glassy carbon electrodes for a bipolar electrochemical double layer capacitor*, ETH Zürich, Doctoral Thesis, 1999. – URL https://e-collection.library.ethz.ch/view/eth:23232?lang=de

[Braun 2003] BRAUN, A.: Conversion of Thickness Data of Thin Films with Variable Lattice Parameter from Monolayers to Angstroms: An Application of the Epitaxial Bain Path. In: *Surface Review and Letters* 10 (2003), Nr. 06, pp. 889–894. – URL http://www.worldscientific.com/doi/abs/10.1142/S0218625X03005761http://www.worldscientific.com/doi/pdfplus/10.1142/S0218625X03005761. – ISSN 0218-625X 1793-6667

[Braun 2006] BRAUN, A.: Some comments on "Soot surface area evolution during air oxidation as evaluated by small angle X-ray scattering and CO2 adsorption". In: *Carbon* 44 (2006), Nr. 7, pp. 1313–1315. – ISSN 00086223

[Braun 2017] BRAUN, A.: *X-ray Studies on Electrochemical Systems – Synchrotron Methods for Energy Materials*. Berlin/Boston: Walter De Gruyter GmbH, 2017 (De Gruyter Textbook). – p. 454. – URL http://www.degruyter.com/books/978-3-11-043750-8. – ISBN 978-3-11-043750-8

[Braun 2019a] BRAUN, A.: *Electrochemical Energy Systems – Foundations, Energy Storage and Conversion*. Boston, Berlin: Walter de Gruyter GmbH, 2019. – p. 600. – URL https://www.degruyter.com/view/product/495747. – ISBN 978-3-11-056183-8

[Braun 2019b] BRAUN, A.: *Von der Nordsee bis Venedig: Mit Wasserstoff und Brennstoffzelle Europa "erfahren". Eine "Auto"-Biographie*. Erste Auflage. Zürich: Kindle Direct Publishing, 2019. – p. 373. – URL https://www.amazon.de/dp/1790981980. – ISBN 978-1790981984

[Braun 2020] BRAUN, A.: *Quantum Electrodynamics of Photosynthesis – Mathematical Description of Light, Life and Matter*. Berlin: De Gruyter, 2020. – URL https://www.degruyter.com/view/product/512407. – ISBN 9783110626926

[Braun et al. 2000] BRAUN, A. ; BÄRTSCH, M. ; SCHNYDER, B. ; KÖTZ, R.: A model for the film growth in samples with two moving reaction frontiers — an application and extension of the unreacted-core model. In: *Chemical Engineering Science* 55 (2000), Nr. 22, pp. 5273–5282. – URL http://www.sciencedirect.com/science/article/pii/S0009250900001433. – ISSN 00092509

[Braun et al. 2015a] BRAUN, A. ; BOUDOIRE, F. ; BORA, D. K. ; FACCIO, G. ; HU, Y. ; KROLL, A. ; MUN, B. S. ; WILSON, S. T.: Biological components and bioelectronic interfaces of water splitting photoelectrodes for solar hydrogen production. In: *Chemistry a European Journal* 21 (2015), Nr. 11, pp. 4188–99. – URL https://www.ncbi.nlm.nih.gov/pubmed/25504590https://onlinelibrary.wiley.com/doi/full/10.1002/chem.201405123. – ISSN 1521-3765 (Electronic) 0947-6539 (Linking)

[Braun et al. 2002] BRAUN, A. ; BRIGGS, K. M. ; BÖNI, P.: Analytical solution to Matthews' and Blakeslee's critical dislocation formation thickness of epitaxially grown thin films. In: *Journal of Crystal Growth* 241 (2002), Nr. 1-2, pp. 231–234. – URL http://www.sciencedirect.com/science/article/pii/S0022024802009417. – ISSN 00220248

[Braun and Chen 2017] BRAUN, A. ; CHEN, Q.: Experimental neutron scattering evidence for proton polaron in hydrated metal oxide proton conductors. In: *Nat Commun* 8 (2017), p. 15830. – URL https://www.ncbi.nlm.nih.gov/pmc/articles/PMC5474746/pdf/ncomms15830.pdf. – ISSN 2041-1723 (Electronic) 2041-1723 (Linking)

[Braun et al. 2019] BRAUN, A. ; CHEN, Q. ; YELON, A.: Hole and protonic polarons in perovskites. In: *Chimia* 73 (2019), Nr. 11, pp. 936–942

[Braun et al. 2014] BRAUN, A. ; FAN, H. Y. ; HAENEN, K. ; STANCIU, L. ; THEIL, J. A.: Braun, Fan, Haenen, Stanciu, and Theil to chair 2015 MRS Spring Meeting. In: *MRS Bulletin* 39 (2014), Nr. 08, pp. 740–741. – URL https://doi.org/10.1557/mrs.2014.183. – ISSN 0883-7694 1938-1425

[Braun et al. 2016] BRAUN, A. ; GAILLARD, N. ; MILLER, E. L. ; WANG, H.: Introduction: Advanced Materials and Structures for Solar Fuels. In: *Journal of Materials Research* 31 (2016), Nr. 11, pp. 1545–1546. – URL https://www.cambridge.org/core/services/aop-cambridge-core/content/view/E1A6B1908B54DBBBC76F4A6880CE8DEF/S0884291416002223a.pdf/introduction.pdf. – ISSN 0884-2914 2044-5326

[Braun et al. 2008] BRAUN, A. ; HUGGINS, F. E. ; KUBATOVA, A. ; WIRICK, S. ; MARICQ, M. M. ; MUN, B. S. ; MCDONALD, J. D. ; KELLY, K. E. ; SHAH, N. ; HUFFMAN, G. P.: Toward distinguishing woodsmoke and diesel exhaust in ambient particulate matter. In: *Environ Sci Technol* 42 (2008), Nr. 2, pp. 374–80. – URL https://pubs.acs.org/doi/pdf/10.1021/es071260k. – ISSN 0013-936X (Print) 0013-936X (Linking)

[Braun et al. 2015b] BRAUN, A. ; NORDLUND, D. ; SONG, S. W. ; HUANG, T. W. ; SOKARAS, D. ; LIU, X. S. ; YANG, W. L. ; WENG, T. C. ; LIU, Z.: Hard X-rays in-soft X-rays out: An operando piggyback view deep into a charging lithium ion battery with X-ray Raman spectroscopy. In: *Journal of Electron Spectroscopy and Related Phenomena* 200 (2015), pp. 257–263. – URL https://www.sciencedirect.com/science/article/pii/S0368204815000626. – ISSN 0368-2048

[Braun et al. 2009] BRAUN, A. ; OVALLE, A. ; POMJAKUSHIN, V. ; CERVELLINO, A. ; ERAT, S. ; STOLTE, W. C. ; GRAULE, T.: Yttrium and hydrogen superstructure and correlation of lattice expansion and proton conductivity in the BaZr0.9Y0.1O2.95 proton conductor. In: *Applied Physics Letters* 95 (2009), Nr. 22. – URL https://aip.scitation.org/doi/pdf/10.1063/1.3268454. – ISSN 0003-6951;1077-3118

[Braun et al. 2012] BRAUN, A. ; SIVULA, K. ; BORA, D. K. ; ZHU, J. ; ZHANG, L. ; GRAETZEL, M. ; GUO, J. ; CONSTABLE, E. C.: Direct Observation of Two Electron Holes in a Hematite Photoanode during Photoelectrochemical Water Splitting. In: *Journal of Physical Chemistry C* 116 (2012), Nr. 32, pp. 16870–16875. – URL https://pubs.acs.org/doi/pdfplus/10.1021/jp304254k. – ISSN 1932-7447

[Braun et al. 2007] BRAUN, A. ; WANG, H. ; SHIM, J. ; LEE, S. S. ; CAIRNS, E. J.: Lithium K(1s) synchrotron NEXAFS spectra of lithium-ion battery cathode, anode and electrolyte materials. In: *Journal of Power Sources* 170 (2007), Nr. 1, pp. 173–178. – URL http://www.sciencedirect.com/science/article/pii/S0378775307007148. – ISSN 03787753

[Braun et al. 2003] BRAUN, A. ; WOKAUN, A. ; HERMANNS, H. G.: Analytical solution to a growth problem with two moving boundaries. In: *Applied Mathematical Modelling* 27 (2003), Nr. 1, pp. 47–52. – ISSN 0307904X

[Braun 1909] BRAUN, K. F.: Nobel Lecture – Electrical Oscillations and Wireless Telegraphy. In: *Nobel Lecture* (1909)

[Breit and Teller 1940] BREIT, G. ; TELLER, E.: Metastability of Hydrogen and Helium Levels. In: *The Astrophysical Journal* 91 (1940). – URL https://ui.adsabs.harvard.edu/abs/1940ApJ....91..215B/abstract. – ISSN 0004-637X 1538-4357

[Brooksbank et al. 2005] BROOKSBANK, C. ; CAMERON, G. ; THORNTON, J.: The European Bioinformatics Institute's data resources: towards systems biology. In: *Nucleic Acids Res* 33 (2005), Nr. Database issue, pp. D46–53. – URL https://www.ncbi.nlm.nih.gov/pubmed/

15608238https://www.ncbi.nlm.nih.gov/pmc/articles/PMC539980/pdf/gki026.pdf. – ISSN 1362-4962 (Electronic) 0305-1048 (Linking)

[Brown 2002] Brown, L. M.: The Compton effect as one path to QED. In: *Studies in History and Philosophy of Science Part B: Studies in History and Philosophy of Modern Physics* 33 (2002), Nr. 2, pp. 211–249. – URL https://www.sciencedirect.com/science/article/pii/S1355219802000059. – ISSN 13552198

[Brown and Feynman 1951] Brown, L. M. ; Feynman, R. P.: Radiative Corrections to the Klein-Nishina Formula. In: *Physical Review* 82 (1951), Nr. 2, pp. 321–321. – URL https://doi.org/10.1103/PhysRev.82.291. – ISSN 0031-899x

[Brown and Feynman 1952] Brown, L. M. ; Feynman, R. P.: Radiative Corrections to Compton Scattering. In: *Physical Review* 85 (1952), Nr. 2, pp. 231–244. – URL https://journals.aps.org/pr/pdf/10.1103/PhysRev.85.231. – ISSN 0031-899x

[Brown 2009] Brown, R.: XXVII. A brief account of microscopical observations made in the months of June, July and August 1827, on the particles contained in the pollen of plants; and on the general existence of active molecules in organic and inorganic bodies. In: *The Philosophical Magazine* 4 (2009), Nr. 21, pp. 161–173. – URL https://www.tandfonline.com/doi/pdf/10.1080/14786442808674769. – ISSN 1941-5850 1941-5869

[Brown and Kumar 2011] Brown, R. A. ; Kumar, A.: A New Perspective on Eratosthenes' Measurement of the Earth. In: *The Physics Teacher* 49 (2011), Nr. 7, pp. 445–447. – URL https://aapt.scitation.org/doi/pdf/10.1119/1.3639158. – ISSN 0031-921X

[Bryant 2011] Bryant, J.: *Thermoeconomics – A Thermodynamic Approach to Economics*. 2nd edition. Harpenden, Hertfordshire, United Kingdom : VOCAT International Ltd, 2011 (Entropy & Economics). – p. 232. – ISBN 978-0956297525

[Bukartė et al. 2020] Bukartė, E. ; Haufe, A. ; Paleček, D. ; Büchel, C. ; Zigmantas, D.: Revealing vibronic coupling in chlorophyll c1 by polarization-controlled 2D electronic spectroscopy. In: *Chemical Physics* 530 (2020). – URL https://www.sciencedirect.com/science/article/pii/S0301010419306111. – ISSN 03010104

[Burt and Smith 2012] Burt, J. ; Smith, B.: *Deep Space Climate Observatory: The DSCOVR Mission*. In: *2012 IEEE Aerospace Conference*. New York: IEEE, 2012 (IEEE Aerospace Conference Proceedings). – URL https://ieeexplore.ieee.org/document/6187025. – ISBN 978-1-4577-0557-1

[Busbey 1999] Busbey, A. B.: IGOR Pro 3.1. In: *Geotimes* 44 (1999), Nr. 4, pp. 43–43. – URL www.geotimes.org/apr99/. – ISSN 0016-8556

[Butkus et al. 2012] Butkus, V. ; Zigmantas, D. ; Valkunas, L. ; Abramavicius, D.: Vibrational vs. electronic coherences in 2D spectrum of molecular systems. In: *Chemical Physics Letters* 545 (2012), pp. 40–43. – URL https://www.sciencedirect.com/science/article/pii/S0009261412008044. – ISSN 00092614

[Calvin 1961a] Calvin, M.: *Nobel Lecture – The Path of Carbon in Photosynthesis*. Amsterdam: Elsevier Publishing Company, 1961 (Nobel Lectures, Chemistry 1942-1962). – URL https://www.nobelprize.org/prizes/chemistry/1961/calvin/biographical/

[Calvin 1961b] Calvin, M.: Quantum conversion in photosynthesis. In: *Journal of Theoretical Biology* 1 (1961), Nr. 2, pp. 258–287. – ISSN 00225193

[Calvin 1979] Calvin, M.: *Petroleum Plantations*. Book Section Chapter 1, pp. 1–30. In: *Solar Energy*. Nagoya, Japan: Humana Press, 1979. – URL http://escholarship.org/uc/item/6xp5p246https://link.springer.com/chapter/10.1007%2F978-1-4612-6245-9_1. – ISBN 978-1-4612-6247-3 978-1-4612-6245-9

[Calvin and Benson 1948] Calvin, M. ; Benson, A. A.: The Path of Carbon in Photosynthesis. In: *Science* 107 (1948), Nr. 2784, pp. 476–80. – URL https://science.sciencemag.org/content/sci/107/2784/476.full.pdf. – ISSN 0036-8075 (Print) 0036-8075 (Linking)

[Calvin and Gazenko 1975] CALVIN, M. ; GAZENKO, O. G.: *Volume 3: Space Medicine and Biotechnology – Joint USA/USSR Publication in Three Volumes*. NASA, 1975 (Foundations of Space Biology and Medicine). – URL https://archive.org/details/NASA_NTRS_Archive_19760019741/mode/2up

[Camargo et al. 2015] CAMARGO, F. V. ; ANDERSON, H. L. ; MEECH, S. R. ; HEISLER, I. A.: Full characterization of vibrational coherence in a porphyrin chromophore by two-dimensional electronic spectroscopy. In: *J Phys Chem A* 119 (2015), Nr. 1, pp. 95–101. – URL https://pubs.acs.org/doi/pdf/10.1021/jp511881a. – ISSN 1520-5215 (Electronic) 1089-5639 (Linking)

[Campbell 2008] CAMPBELL, I. D.: The Croonian lecture 2006. Structure of the living cell. In: *Philos Trans R Soc Lond B Biol Sci* 363 (2008), Nr. 1502, pp. 2379–91. – URL https://www.ncbi.nlm.nih.gov/pmc/articles/PMC1955230/pdf/rstb20061960.pdf. – ISSN 0962-8436 (Print) 0962-8436 (Linking)

[Canfora 1999] CANFORA, L.: The vanished library – The fate of the Great Library of Alexandria. In: *Index on Censorship* 28 (1999), Nr. 2, pp. 46–53. – ISSN 0306-4220

[Canfora and Coleman 2016] CANFORA, L. ; COLEMAN, A.: The vanished library. In: *Index on Censorship* 28 (2016), Nr. 2, pp. 46–53. – URL https://journals.sagepub.com/doi/pdf/10.1080/03064229908536542. – ISSN 0306-4220 1746-6067

[Carroll et al. 2018] CARROLL, A. L. ; SILLETT, S. C. ; PALLADINI, M. ; CAMPBELL-SPICKLER, J.: Dendrochronological analysis of Sequoia sempervirens in an interior old-growth forest. In: *Dendrochronologia* 52 (2018), pp. 29–39. – URL https://www.sciencedirect.com/science/article/pii/S1125786518300961. – ISSN 11257865

[Carvalhal and Marques 2015] CARVALHAL, M. J. ; MARQUES, M. B.: Adam Hilger revisited: a museum instrument as a modern teaching tool. In: *Education and Training in Optics and Photonics: Etop 2015* 9793 (2015). – URL https://www.spiedigitallibrary.org/conference-proceedings-of-spie/9793/979328/Adam-Hilger-revisited--a-museum-instrument-as-a-modern/10.1117/12.2223202.full?SSO=1. – ISSN 0277-786x

[Casimir and Polder 1948] CASIMIR, H. B. G. ; POLDER, D.: The Influence of Retardation on the London-van der Waals Forces. In: *Physical Review* 73 (1948), Nr. 4, pp. 360–372. – ISSN 0031-899X

[Chadwick et al. 1937] CHADWICK, J. ; FEATHER, N. ; BRETSCHER, E.: Measurements of range and angle of projection for the protons produced in the Photo-disintegration of deuterium. In: *Proceedings of the Royal Society of London Series a-Mathematical and Physical Sciences* 163 (1937), Nr. A914, pp. 0366–0375. – URL http://rspa.royalsocietypublishing.org/content/royprsa/163/914/366.full.pdf. – ISSN 0080-4630

[Chadwick and Goldhaber 1934] CHADWICK, J. ; GOLDHABER, M.: A 'Nuclear photo-effect' – Disintegration of the diplon by gamma-rays. In: *Nature* 134 (1934), pp. 237–238. – URL https://www.nature.com/articles/134237a0.pdf. – ISSN 0028-0836

[Chadwick and Goldhaber 1935] CHADWICK, J. ; GOLDHABER, M.: The nuclear photoelectric effect. In: *Proceedings of the Royal Society of London Series a-Mathematical and Physical Sciences* 151 (1935), Nr. A873, pp. 0479–0493. – URL http://rspa.royalsocietypublishing.org/content/royprsa/151/873/479.full.pdf. – ISSN 0080-4630

[Chan et al. 2016] CHAN, C. K. ; TUYSUZ, H. ; BRAUN, A. ; RANJAN, C. ; LA MANTIA, F. ; MILLER, B. K. ; ZHANG, L. X. ; CROZIER, P. A. ; HABER, J. A. ; GREGOIRE, J. M. ; PARK, H. S. ; BATCHELLOR, A. S. ; TROTOCHAUD, L. ; BOETTCHER, S. W.: *Topics in Current Chemistry-Series*. Vol. 371: *Advanced and In Situ Analytical Methods for Solar Fuel Materials*. pp. 253–324. In: *Solar Energy for Fuels* Vol. 371. Cham: Springer Int Publishing Ag, 2016. – URL https://link.springer.com/content/pdf/10.1007%2F128_2015_650.pdf. – ISBN 978-3-319-23099-3; 978-3-319-23098-6

[Chang 2012] CHANG, H.-J.: *23 Luegen, die sie uns ueber den Kapitalismus erzaehlen*. 1. Auflage. Muenchen: Goldmann, 2012. – ISBN 978-3-442-15728-0

[Chang and Halliday 2005] CHANG, J. ; HALLIDAY, J.: *Mao. Das Leben eines Mannes. Das Schicksal eines Volkes*. 4. Auflage. Muenchen: Karl Blessing Verlag, 2005. – ISBN 978-3-89667-200-1

[Chen et al. 2013] CHEN, Q. ; BANYTE, J. ; ZHANG, X. ; EMBS, J. P. ; BRAUN, A.: Proton diffusivity in spark plasma sintered BaCe0.8Y0.2O3: In-situ combination of quasi-elastic neutron scattering and impedance spectroscopy. In: *Solid State Ionics* 252 (2013), pp. 2–6. – ISSN 01672738

[Chen et al. 2012] CHEN, Q. ; HOLDSWORTH, S. ; EMBS, J. ; POMJAKUSHIN, V. ; FRICK, B. ; BRAUN, A.: High-temperature high pressure cell for neutron-scattering studies. In: *High Pressure Research* 32 (2012), Nr. 4, pp. 471–481. – URL https://www.tandfonline.com/doi/pdf/10.1080/08957959.2012.725729. – ISSN 0895-7959;1477-2299

[Cherkashin et al. 1999] CHERKASHIN, A. A. ; BULYCHEV, A. A. ; VREDENBERG, W. J.: Outward photocurrent component in chloroplasts of Peperomia metallica and its assignment to the 'closed thylakoid' recording configuration. In: *Bioelectrochemistry and Bioenergetics* 48 (1999), Nr. 1, pp. 141–148. – URL https://www.sciencedirect.com/science/article/pii/S0302459898002268. – ISSN 03024598

[Chipman 2019] CHIPMAN, J.: A Multisensor Approach to Satellite Monitoring of Trends in Lake Area, Water Level, and Volume. In: *Remote Sensing* 11 (2019), Nr. 2, p. 22. – URL https://res.mdpi.com/remotesensing/remotesensing-11-00158/article_deploy/remotesensing-11-00158.pdf. – ISSN 2072-4292

[Chluba and Sunyaev 2008] CHLUBA, J. ; SUNYAEV, R. A.: Two-photon transitions in hydrogen and cosmological recombination. In: *Astronomy & Astrophysics* 480 (2008), Nr. 3, pp. 629–645. – URL https://www.aanda.org/articles/aa/pdf/2008/12/aa7921-07.pdf. – ISSN 0004-6361 1432-0746

[Chung 2015] CHUNG, D.: Cold Fusion: A Study in Scientific Controversy. In: *Submitted as coursework for PH241, Stanford University, Winter 2015* (2015)

[CIA 1954] CIA: CIA Information Report on Photosynthesis Research in the USSR. Central Intelligence Agency, 23 June 1954 1954. – Report. – URL https://archive.org/details/CIA-RDP80-00926A007100010001-3/page/n0

[CIA 1980] CIA: West Europe Report – Science and Technology / Central Intelligence Agency. 1980. – Report

[Ciamician 1912] CIAMICIAN, G.: The Photochemistry of the Future. In: *Science* 36 (1912), Nr. 926, pp. 385–94. – URL https://www.ncbi.nlm.nih.gov/pubmed/17836492. – ISSN 0036-8075 (Print) 0036-8075 (Linking)

[Clegg 2004] CLEGG, R. M.: *Nuts and Bolts of Excitation energy Migration and Energy Transfer*. Vol. 19. Book Section Chapter 4. In: *Chlorophyll a Fluorescence. A Signature of Photosynthesis* Vol. 19, Springer, 2004

[Cohen and Kupferschmidt 2020] COHEN, J. ; KUPFERSCHMIDT, K.: Strategies shift as coronavirus pandemic looms. In: *Science* 367 (2020), Nr. 6481, pp. 962–963. – URL https://science.sciencemag.org/content/sci/367/6481/962.full.pdf. – ISSN 1095-9203 (Electronic) 0036-8075 (Linking)

[Cole and Cole 1941] COLE, K. S. ; COLE, R. H.: Dispersion and Absorption in Dielectrics I. Alternating Current Characteristics. In: *The Journal of Chemical Physics* 9 (1941), Nr. 4, pp. 341–351. – URL https://doi.org/10.1063/1.1750906. – ISSN 0021-9606 1089-7690

[Combet et al. 2008] COMBET, S. ; PIEPER, J. ; CONEGGO, F. ; AMBROISE, J. P. ; BELLISSENT-FUNEL, M. C. ; ZANOTTI, J. M.: Coupling of laser excitation and inelastic neutron scattering: attempt to probe the dynamics of light-induced C-phycocyanin dynamics. In: *European Biophysics Journal with Biophysics Letters* 37 (2008), Nr. 5, pp. 693–700. – URL https://link.springer.com/content/pdf/10.1007%2Fs00249-008-0320-1.pdf. – ISSN 0175-7571

[Constantin 2016] CONSTANTIN, A.: *Fourier Analysis*. Cambridge University Press, 2016. – ISBN 9781107358508

[Constantin et al. 2015] CONSTANTIN, D. ; REHAK, M. ; AKHTMAN, Y. ; LIEBISCH, F.: Bestimmung von Kulturpflanzeneigenschaften mittels hyperspektraler Fernerkundung von einem Mikro-UAV – Detection of crop properties by means of hyperspectral remote sensing from a micro UA V . In: *Bornimer Agrartechnische Berichte* (2015), Nr. 88. – URL https://opus4.kobv.de/opus4-slbp/frontdoor/index/index/docId/7595. – ISSN ISSN 0947-7314

[Contreras et al. 2010] CONTRERAS, M. ; PELLICER, R. ; VILLENA, M. ; RUIZ, A.: A quantum model of option pricing: When Black–Scholes meets Schroedinger and its semi-classical limit. In: *Physica A: Statistical Mechanics and its Applications* 389 (2010), Nr. 23, pp. 5447–5459. – URL https://doi.org/10.1016/j.physa.2010.08.018. – ISSN 03784371

[Coonrod et al. 2008] COONROD, D. ; BRICK, M. A. ; BYRNE, P. F. ; DEBONTE, L. ; CHEN, Z. Z.: Inheritance of long chain fatty acid content in rapeseed (Brassica napus L.). In: *Euphytica* 164 (2008), Nr. 2, pp. 583–592. – URL https://link.springer.com/article/10.1007%2Fs10681-008-9781-7. – ISSN 0014-2336

[Cornish 1998] CORNISH, N. J.: The Lagrange Points. In: *WMAP Education and Outreach* (1998). – URL http://map.gsfc.nasa.gov/ContentMedia/lagrange.pdf

[Cottet et al. 2015] COTTET, A. ; KONTOS, T. ; DOUCOT, B.: Electron-photon coupling in mesoscopic quantum electrodynamics. In: *Physical Review B* 91 (2015), Nr. 20, p. 16. – URL https://journals.aps.org/prb/pdf/10.1103/PhysRevB.91.205417. – ISSN 2469-9950

[Courtault et al. 2000] COURTAULT, J.-M. ; KABANOV, Y. ; BRU, B. ; CREPEL, P. ; LEBON, I. ; LE MARCHAND, A.: Louis Bachelier on the Centenary of Theorie de la Speculation. In: *Mathematical Finance* 10 (2000), Nr. 3, pp. 339–353. – URL https://onlinelibrary.wiley.com/doi/pdf/10.1111/1467-9965.00098. – ISSN 0960-1627 1467-9965

[Craver 2006] CRAVER, C. F.: When mechanistic models explain. In: *Synthese* 153 (2006), Nr. 3, pp. 355–376. – URL https://link.springer.com/article/10.1007/s11229-006-9097-x. – ISSN 0039-7857

[Crepeau and Asme 2009] CREPEAU, J. ; ASME: *A BRIEF HISTORY OF THE T(4) RADIATION LAW*. New York: Amer Soc Mechanical Engineers, 2009 (Ht2009: Proceedings of the Asme Summer Heat Transfer Conference 2009, Vol 1). – pp. 59–65. – URL https://www.researchgate.net/publication/267650295_A_Brief_History_of_the_T4_Radiation_Law. – ISBN 978-0-7918-4356-7

[Crowley and Gregori 2014] CROWLEY, B. J. B. ; GREGORI, G.: Quantum theory of Thomson scattering. In: *High Energy Density Physics* 13 (2014), pp. 55–83. – URL https://www.sciencedirect.com/science/article/pii/S1574181814000548. – ISSN 15741818

[D'Amici et al. 2009] D'AMICI, G. M. ; HUBER, C. G. ; ZOLLA, L.: Separation of thylakoid membrane proteins by sucrose gradient ultracentrifugation or blue native-SDS-PAGE two-dimensional electrophoresis. In: *Methods Mol Biol* 528 (2009), pp. 61–70. – URL https://link.springer.com/protocol/10.1007%2F978-1-60327-310-7_4. – ISSN 1064-3745 (Print) 1064-3745 (Linking)

[Daues 2006] DAUES, K.: A History of Spacecraft Environmental Control and Life Support Systems. NASA Johnson Space Center, 2006. – Report. – URL https://ntrs.nasa.gov/archive/nasa/casi.ntrs.nasa.gov/20080031131.pdf

[De Broglie 1929] DE BROGLIE, L.: The wave nature of the electron. In: *Nobel Lecture, December 12, 1929* (1929), pp. 244–256

[De Yoreo 2016] DE YOREO, J.: *David Turnbull Lectureship Award Winner Jim De Yoreo*. 1 Dec 2016 2016. – URL https://www.youtube.com/watch?v=afPHYXoDNUw

[Dean and Miskiewicz 2003] DEAN, R. L. ; MISKIEWICZ, E.: Rates of electron transport in the thylakoid membranes of isolated, illuminated chloroplasts are enhanced in the presence of ammonium chloride. In: *Biochemistry and Molecular Biology Education* 31 (2003), Nr. 6, pp. 410–417. – URL https://iubmb.onlinelibrary.wiley.com/doi/full/10.1002/bmb.2003.494031060265. – ISSN 1470-8175

[Dencher et al. 1989] DENCHER, N. A. ; DRESSELHAUS, D. ; ZACCAI, G. ; BULDT, G.: Structural changes in bacteriorhodopsin during proton translocation revealed by neutron diffraction. In: *Proc Natl Acad Sci U S A* 86 (1989), Nr. 20, pp. 7876–9. – URL https://www.pnas.org/content/pnas/86/20/7876.full.pdf. – ISSN 0027-8424 (Print) 0027-8424 (Linking)

[Didot 1839] DIDOT, F.: *Vetus Testamentum Graece et Latine*. EX AUTORITATE SIXTI QUINTI PONTIFICIS MAXIMI EDITUM JUXTA EXEMPLAR ORIGINALE VATICANUM, TOMUS ALTERUS, FIRMIN DIDOT, 1839. 1839

[Dimitrijevic and Lovre 2015] DIMITRIJEVIC, B. ; LOVRE, I.: The Role of Temperature in Economic Exchange – An Empirical Analysis. In: *Journal of Central Banking Theory and Practice* 4 (2015), Nr. 3, pp. 65–89. – URL https://doi.org/10.1515/jcbtp-2015-0012. – ISSN 1800-9581

[Dirac 1927] DIRAC, P. A. M.: The quantum theory of the emission and absorption of radiation. In: *Proceedings of the Royal Society of London Series a-Containing Papers of a Mathematical and Physical Character* 114 (1927), Nr. 767, pp. 243–265. – URL https://doi.org/10.1098/rspa.1927.0039. – ISSN 0950-1207

[Dirac 1928] DIRAC, P. A. M.: The Quantum Theory of the Electron. In: *Proceedings of the Royal Society A: Mathematical, Physical and Engineering Sciences* 117 (1928), Nr. 778, pp. 610–624. – URL http://rspa.royalsocietypublishing.org/content/royprsa/117/778/610.full.pdf. – ISSN 1364-5021 1471-2946

[Doctor 2015] DOCTOR, R.: Carbon Dioxide on the Move. In: *Bulletin of the Atomic Scientists* 63 (2015), Nr. 4, pp. 19–20. – ISSN 0096-3402 1938-3282

[Dorlas 1996] DORLAS, T. C.: A non-commutative central limit theorem. In: *Journal of Mathematical Physics* 37 (1996), Nr. 9, pp. 4662–4682. – URL https://doi.org/10.1063/1.531646. – ISSN 0022-2488 1089-7658

[Dutka 1993] DUTKA, J.: Eratosthenes Measurement of the Earth Reconsidered. In: *Archive for History of Exact Sciences* 46 (1993), Nr. 1, pp. 55–66. – URL https://link.springer.com/content/pdf/10.1007%2FBF00387726.pdf. – ISSN 0003-9519

[Eastman 1935] EASTMAN, K.: *Photographic plates for use in spectroscopy and astronomy*. 2nd edition. Rochester, New York: Eastman Kodak Company, 1935

[Ebel 2014] EBEL, J.: Kommentierter Neusatz von „Ueber die Beziehung zwischen der Waermestrahlung und der Temperatur, J. Stefan". In: *Kommentierung und Bearbeitung* (2014). – URL http://www.ing-buero-ebel.de/Treib/Stefan.pdf

[Economist 2014] ECONOMIST: Tiny balls of fire – How to gather more light for solar power. In: *The Economist* (2014), 1st July 2014. – URL https://www.economist.com/science-and-technology/2014/07/01/tiny-balls-of-fire

[Edman et al. 1999] EDMAN, K. ; NOLLERT, P. ; ROYANT, A. ; BELRHALI, H. ; PEBAY-PEYROULA, E. ; HAJDU, J. ; NEUTZE, R. ; LANDAU, E. M.: High-resolution X-ray structure of an early intermediate in the bacteriorhodopsin photocycle. In: *Nature* 401 (1999), Nr. 6755, pp. 822–6. – URL https://www.nature.com/articles/44623.pdf. – ISSN 0028-0836 (Print) 0028-0836 (Linking)

[Efremov and Wang 2011] EFREMOV, A. ; WANG, Z.: Universal optimal working cycles of molecular motors. In: *Phys Chem Chem Phys* 13 (2011), Nr. 13, pp. 6223–33. – URL https://pubs.rsc.org/en/content/articlepdf/2011/cp/c0cp02118k. – ISSN 1463-9084 (Electronic) 1463-9076 (Linking)

[Ehlers 2007] EHLERS, J.: Pascual Jordan (1902–1980) Mainzer Symposium zum 100. Geburtstag. In: EHLERS, J. (editor): *Mainzer Symposium zum 100. Geburtstag von Pascual Jordan (1902-1980)* Vol. Preprint 329, Max-Planck-Institut für Wissenschaftsgeschichte, 2007, p. 206. – URL https://www.mpiwg-berlin.mpg.de/sites/default/files/Preprints/P329.pdf

[El Fiky 2002] El Fiky, U.: Arid desert colonization – Toshka Region, Egypt, as a case study. In: *Xxx Iahs World Congress on Housing, Housing Construction: an Interdisciplinary Task, Vols 1-3* (2002), pp. 303–311. – URL https://www.academia.edu/434879/Arid_Desert_Colonization_Toshka_region-Egypt_as_a_case_study

[El-Shabrawy and Dumont 2009] El-Shabrawy, G. M. ; Dumont, H. J.: *Monographiae Biologicae*. Vol. 89: *The Toshka Lakes*. pp. 157–162. In: *Nile: Origin, Environments, Limnology and Human Use* Vol. 89. Dordrecht: Springer, 2009. – URL https://doi.org/10.1007/978-1-4020-9726-3_8. – ISBN 978-1-4020-9725-6

[Elbers et al. 1957] Elbers, P. F. ; Minnaert, K. ; Thomas, J. B.: Submicroscopic Structure of Some Chloroplasts. In: *Acta Botanica Neerlandica* 6 (1957), Nr. 3, pp. 345–350. – URL https://onlinelibrary.wiley.com/doi/abs/10.1111/j.1438-8677.1957.tb00581.x. – ISSN 00445983

[Elishav et al. 2017] Elishav, O. ; Lewin, D. R. ; Shter, G. E. ; Grader, G. S.: The nitrogen economy: Economic feasibility analysis of nitrogen-based fuels as energy carriers. In: *Applied Energy* 185 (2017), pp. 183–188. – URL https://www.sciencedirect.com/science/article/pii/S030626191631532X?via%3Dihub. – ISSN 03062619

[Ellis 2017] Ellis, J. P.: Ti k Z-Feynman: Feynman diagrams with Ti k Z. In: *Computer Physics Communications* 210 (2017), pp. 103–123. – URL https://www.sciencedirect.com/science/article/pii/S0010465516302521. – ISSN 00104655

[Engel et al. 2007] Engel, G. S. ; Calhoun, T. R. ; Read, E. L. ; Ahn, T. K. ; Mancal, T. ; Cheng, Y. C. ; Blankenship, R. E. ; Fleming, G. R.: Evidence for wavelike energy transfer through quantum coherence in photosynthetic systems. In: *Nature* 446 (2007), Nr. 7137, pp. 782–6. – URL https://www.nature.com/articles/nature05678.pdf. – ISSN 1476-4687 (Electronic) 0028-0836 (Linking)

[Euler 1767] Euler, L.: De motu rectilineo trium corporum se mutuo attrahentium. In: *Novi Comm. Acad. Sci. Imp. Petrop.* 11 (1767), Nr. 144

[Faia et al. 2013] Faia, E. ; Lechthaler, W. ; Merkl, C.: Fiscal stimulus and labor market policies in Europe. In: *Journal of Economic Dynamics and Control* 37 (2013), pp. 483–499

[Falkowski and Godfrey 2008] Falkowski, P. G. ; Godfrey, L. V.: Electrons, life and the evolution of Earth's oxygen cycle. In: *Philos Trans R Soc Lond B Biol Sci* 363 (2008), Nr. 1504, pp. 2705–16. – URL https://www.ncbi.nlm.nih.gov/pubmed/18487127. – ISSN 0962-8436 (Print) 0962-8436 (Linking)

[Faris 2004] Faris, W. G.: The Feynman-Kac formula. University of Arizona, 11 Feb 2004 2004. – Report. – URL https://www.math.arizona.edu/~faris/talks/FKac.pdf

[Fenna and Matthews 1975] Fenna, R. E. ; Matthews, B. W.: Chlorophyll arrangement in a bacteriochlorophyll protein from Chlorobium limicola. In: *Nature* 258 (1975), Nr. 5536, pp. 573–577. – URL https://www.nature.com/articles/258573a0.pdf. – ISSN 0028-0836 1476-4687

[Fennell 2019] Fennell, S.: Malthus, Statistics, and the State of Indian Agriculture. In: *The Historical Journal* 63 (2019), Nr. 1, pp. 159–185. – ISSN 0018-246X 1469-5103

[Fermi 1949] Fermi, E.: *Nuclear Physics*. Chicago: The University of Chicago Press, 1949 (A Course Given by Enrico Fermi at the University of Chicago). – p. 258. – ISBN 978-0226243658

[Fernandez-Alonso and Zare 2002] Fernandez-Alonso, F. ; Zare, R. N.: Scattering resonances in the simplest chemical reaction. In: *Annu Rev Phys Chem* 53 (2002), pp. 67–99. – URL https://www.annualreviews.org/doi/10.1146/annurev.physchem.53.091001.094554. – ISSN 0066-426X (Print) 0066-426X (Linking)

[Feyerabend 1975] Feyerabend, P.: Lakatos,I. In: *British Journal for the Philosophy of Science* 26 (1975), Nr. 1, pp. 1–18. – URL https://doi.org/10.1093/bjps/26.1.1. – ISSN 0007-0882

[Feyerabend 1994] Feyerabend, P.: *Quantum theory and our view of the world*. Cambridge: Cambridge Univ Press, 1994 (Physics and Our View of the World). – pp. 149–168. – URL https://www.cambridge.org/core/books/physics-and-our-view-of-the-world/quantum-theory-and-our-view-of-the-world/C13E24A0AF17F04DC7B03286354E6C70. – ISBN 0-521-45372-0

[Feynman 1939] FEYNMAN, R. P.: Forces in molecules. In: *Physical Review* 56 (1939), Nr. 4, pp. 340–343. – URL https://journals.aps.org/pr/abstract/10.1103/PhysRev.56.340. – ISSN 0031-899x

[Feynman 1949] FEYNMAN, R. P.: Space-Time Approach to Quantum Electrodynamics. In: *Physical Review* 76 (1949), Nr. 6, pp. 769–789. – URL https://journals.aps.org/pr/pdf/10.1103/PhysRev.76.769. – ISSN 0031-899x

[Feynman 1955] FEYNMAN, R. P.: *Public address: The Value of Science*. 2-4 November 1955 1955. – URL http://calteches.library.caltech.edu/40/2/Science.htm

[Feynman 1979] FEYNMAN, R. P.: *Richard Feynman Video – The Douglas Robb Memorial Lectures – Part 1: Photons – Corpuscles of Light*. 1979. – URL http://vega.org.uk/video/programme/45

[Fick 1855] FICK, A.: Ueber Diffusion. In: *Annalen der Physik und Chemie* 170 (1855), Nr. 1, pp. 59–86. – URL https://doi.org/10.1002/andp.18551700105. – ISSN 00033804 15213889

[Fick 2009] FICK, A.: V. On liquid diffusion. In: *The London, Edinburgh, and Dublin Philosophical Magazine and Journal of Science* 10 (2009), Nr. 63, pp. 30–39. – URL https://doi.org/10.1080/14786445508641925. – ISSN 1941-5982 1941-5990

[Fieber 1960] FIEBER, H.: Zu den Lagrange'schen Lösungen des n-Körperproblems. In: *Monatshefte für Mathematik* 64 (1960), Nr. 1, pp. 1–14

[Flagg 2019] FLAGG, D. A.: *NSF press conference on first result from the Event Horizon Telescope project*. 10 April 2019 2019

[Fleischer 1973] FLEISCHER, R.: *Soylent Green*. 1973. – URL https://www.imdb.com/title/tt0070723/

[Fleischmann 2006a] FLEISCHMANN, M.: *Background to Cold Fusion: The genesis of a concept*. Singapore: World Scientific Publ Co Pte Ltd, 2006 (Condensed Matter Nuclear Science). – pp. 1–11. – URL https://pubs.acs.org/doi/pdfplus/10.1021/bk-2008-0998.ch002. – ISBN 981-256-564-7

[Fleischmann 2006b] FLEISCHMANN, M.: *Background to Cold Fusion: the Genesis of a Concept*. 2006

[Fleischmann and Pons 1989] FLEISCHMANN, M. ; PONS, S.: Electrochemically Induced Nuclear-Fusion of Deuterium. In: *Journal of Electroanalytical Chemistry* 261 (1989), Nr. 2a, pp. 301–308. – URL https://www.sciencedirect.com/science/article/pii/0022072889800063. – ISSN 0022-0728

[Fleischmann and Pons 1993] FLEISCHMANN, M. ; PONS, S.: Calorimetry of the Pd-D2o System – from Simplicity Via Complications to Simplicity. In: *Physics Letters A* 176 (1993), Nr. 1-2, pp. 118–129. – URL https://doi.org/10.1016/0375-9601(93)90327-V. – ISSN 0375-9601

[Fleischmann et al. 1989] FLEISCHMANN, M. ; PONS, S. ; HAWKINS, M. ; HOFFMAN, R. J.: Measurement of Gamma-Rays from Cold Fusion. In: *Nature* 339 (1989), Nr. 6227, pp. 667–667. – URL https://www.nature.com/articles/339667a0. – ISSN 0028-0836

[Fleischmann et al. 1994] FLEISCHMANN, M. ; PONS, S. ; PREPARATA, G.: Possible Theories of Cold-Fusion. In: *Nuovo Cimento Della Societa Italiana Di Fisica a-Nuclei Particles and Fields* 107 (1994), Nr. 1, pp. 143–156. – URL https://link.springer.com/article/10.1007%2FBF02813078. – ISSN 1124-1861

[Fleming et al. 2011] FLEMING, G. R. ; SCHOLES, G. D. ; CHENG, Y.-C.: Quantum effects in biology. In: *Procedia Chemistry* 3 (2011), Nr. 1, pp. 38–57. – URL https://www.sciencedirect.com/science/article/pii/S1876619611000507. – ISSN 18766196

[Flohr et al. 2004] FLOHR, S. ; PROTSCH, Z. R. von ; BERG, A. von: Morphological analysis of the neanderthal calotte from Ochtendung, Germany. In: *Human Evolution* 19 (2004), Nr. 1, pp. 1–18. – URL https://doi.org/10.1007/BF02438906. – ISSN 0393-9375 1824-310X

[Forster 1938a] FORSTER, T.: Die Lichtabsorption aromatischer Kohlenwasserstoffe // The light absorption of hydrocarbon according to quantum mechanics. In: *Physikalische Zeitschrift* 39 (1938), pp. 925–928

[Forster 1938b] FORSTER, T.: The light absorption of aromatic hydrocarbons. In: *Zeitschrift Fur Physikalische Chemie-Abteilung B-Chemie Der Elementarprozesse Aufbau Der Materie* 41 (1938), Nr. 4, pp. 287–306. – ISSN 0372-9664

[Forster 1946] FORSTER, T.: Energiewanderung und Fluoreszenz. In: *Naturwissenschaften* 33 (1946), Nr. 6, pp. 166–175. – URL https://link.springer.com/article/10.1007/BF00585226. – ISSN 0028-1042

[Förster 1950] FÖRSTER, T.: Elektrolytische Dissoziation angeregter Moleküle. In: *Zeitschrift für Elektrochemie und angewandte physikalische Chemie* 54 (1950), Nr. 1, pp. 42–46. – URL https://onlinelibrary.wiley.com/doi/epdf/10.1002/bbpc.19500540111

[Frackowiak 1988] FRACKOWIAK, D.: The Jablonski diagram. In: *Journal of Photochemistry and Photobiology B: Biology* 2 (1988), Nr. 3. – URL https://doi.org/10.1016/1011-1344(88)85060-7. – ISSN 10111344

[Frey and Davis 2016] FREY, N. P. ; DAVIS, E. P.: *Advances in the Astronautical Sciences*. Vol. 157: *Launch and Commissioning the Deep Space Climate Observatory*. pp. 999–1010. In: *Guidance, Navigation, and Control 2016* Vol. 157. San Diego: Univelt Inc, 2016. – ISBN 978-0-87703-631-9

[Fridlyand and Scheibe 1999] FRIDLYAND, L. E. ; SCHEIBE, R.: Regulation of the Calvin cycle for CO2 fixation as an example for general control mechanisms in metabolic cycles. In: *Biosystems* 51 (1999), Nr. 2, pp. 79–93. – URL https://doi.org/10.1016/S0303-2647(99)00017-9. – ISSN 03032647

[Frieden 1998] FRIEDEN, B. R.: *Physics from Fisher Information*. Cambridge University Press, 1998. – URL https://doi.org/10.1017/CBO9780511622670. – ISBN 9780511622670

[Fröhlich 1968] FRÖHLICH, H.: Long-range coherence and energy storage in biological systems. In: *International Journal of Quantum Chemistry* 2 (1968), Nr. 5, pp. 641–649. – URL https://doi.org/10.1002/qua.560020505. – ISSN 00207608

[Fröhlich 1983] FRÖHLICH, H.: Evidence for coherent excitation in biological systems. In: *International Journal of Quantum Chemistry* 23 (1983), Nr. 4, pp. 1589–1595. – URL https://doi.org/10.1002/qua.560230440. – ISSN 00207608

[Fuerth 1933] FUERTH, R.: Ueber einige Beziehungen zwischen klassischer Statistik und Quantenmechanik. In: *Zeitschrift fuer Physik* 81 (1933), Nr. 3-4, pp. 143–162. – URL https://link.springer.com/article/10.1007%2FBF01338361. – ISSN 1434-6001 1434-601X

[Funk et al. 1966a] FUNK, J. P. ; DEACON, E. L. ; COLLINS, B. G.: A radiosonde radiometer. In: *Pure and Applied Geophysics PAGEOPH* 64 (1966), Nr. 1, pp. 212–219. – URL https://doi.org/10.1007/BF00875548. – ISSN 0033-4553 1420-9136

[Funk et al. 1966b] FUNK, J. P. ; DEACON, E. L. ; COLLINS, B. G.: A radiosonde radiometer. In: *Pure and Applied Geophysics PAGEOPH* 64 (1966), Nr. 1, pp. 212–219. – URL https://doi.org/10.1007/BF00875548. – ISSN 0033-4553 1420-9136

[Gaffron and Wohl 1936a] GAFFRON, H. ; WOHL, K.: Zur Theorie der Assimilation. In: *Die Naturwissenschaften* 24 (1936), Nr. 6, pp. 81–90. – URL https://doi.org/10.1007/BF01473561https://link.springer.com/content/pdf/10.1007/BF01473561.pdf. – ISSN 0028-1042 1432-1904

[Gaffron and Wohl 1936b] GAFFRON, H. ; WOHL, K.: Zur Theorie der Assimilation (I-III). In: *Die Naturwissenschaften* 24 (1936), Nr. 7, pp. 103–107. – URL https://link.springer.com/content/pdf/10.1007/BF01474887.pdf. – ISSN 0028-1042 1432-1904

[Gamow 1928] GAMOW, G.: The quantum theory of nuclear disintegration. In: *Nature* 122 (1928), pp. 805–806. – URL https://www.nature.com/articles/122805b0.pdf. – ISSN 0028-0836

[Gamow 1938] GAMOW, G.: Nuclear Energy Sources and Stellar Evolution. In: *Physical Review* 53 (1938), Nr. 7, pp. 595–604. – ISSN 0031-899X

[Garavelli 2006] GARAVELLI, M.: Computational organic photochemistry: strategy, achievements and perspectives. In: *Theoretical Chemistry Accounts* 116 (2006), Nr. 1-3, pp. 87–105. – URL https://link.springer.com/content/pdf/10.1007/s00214-005-0030-z.pdf. – ISSN 1432-881x

[Gashev et al. 1965] GASHEV, M. A. ; GUSTOV, G. K. ; DYACHENK.KK ; KOMAR, E. G. ; MALYSHEV, I. F. ; MONOSZON, N. A. ; POPKOVIC.AV ; RATNIKOV, B. K. ; ROZHDEST.BV ; RUMYANTS.NN ; SAKSAGAN.GL ; SPEVAKOV.RM ; STOLOV, A. M. ; STRELTSO.NS ; YAVNO, A. K.: Basic Technical Characteristics of Experimental Thermonuclear Device Tokamak-3. In: *Journal of Nuclear Energy Part C-Plasma Physics Accelerators Thermonuclear Research* 7 (1965), Nr. 5pc, pp. 491–&

[Gassner 2018] GASSNER, J. M.: *Pfadintegralformalismus Doppelspalt reloaded Aristoteles Stringtheorie (38)*. 25 Dez 2018 2018. – URL https://www.youtube.com/watch?v=j-eUhdfyPoU

[Gauthier 2009] GAUTHIER, Y.: Orientation and Distribution of Various Dry Stone Monuments of the Sahara. In: RUBINO MARTIN, J. A. (editor) ; BELMONTE, J. A. (editor) ; PRADA, F. (editor) ; ALBERDI, A. (editor): *Workshop on Cosmology Across Cultures* Vol. 409, 2009, pp. 317–330

[Gauthier and Gauthier 1996] GAUTHIER, Y. ; GAUTHIER, C.: Monuments originaux du Messak et du Fezzân occidental (Libye). In: *Bul. Soc. Études et de Rech. des Eyzies* 45 (1996), pp. 46–65

[Georgescu-Roegen 1975] GEORGESCU-ROEGEN, N.: Energy and Economic Myths. In: *Southern Economic Journal* 41 (1975), Nr. 3. – URL http://www.jstor.org/stable/1056148

[Gerlach 1936] GERLACH, W.: Theorie und Experiment in der exakten Wissenschaft. In: *Die Naturwissenschaften* 24 (1936), Nr. 46-47, pp. 721–741. – URL https://link.springer.com/content/pdf/10.1007%2FBF01504074.pdf. – ISSN 0028-1042 1432-1904

[Geru 2018] GERU, I. I.: *Time-Reversal Symmetry*. Springer, 2018 (Springer Tracts in Modern Physics). – URL https://doi.org/10.1007/978-3-030-01210-6. – ISBN 978-3-030-01209-0 978-3-030-01210-6

[Gestrich 1997] GESTRICH, H.: Nikolaus von Kues speech. In: *Nuclear Physics A* 626 (1997), Nr. 1-2, pp. Xxvii–Xxxi. – ISSN 0375-9474

[Glattfelder 2015] GLATTFELDER, J. B.: *James B. Glattfelder: Who controls the world?* 13 Feb 2013 2015. – URL https://www.youtube.com/watch?v=NgbqXsA62Qs

[Glazer 1994] GLAZER, A. N.: Phycobiliproteins — a family of valuable, widely used fluorophores. In: *Journal of Applied Phycology* 6 (1994), Nr. 2, pp. 105–112. – URL https://doi.org/10.1007/BF02186064. – ISSN 0921-8971 1573-5176

[Goldstein 1984] GOLDSTEIN, B. R.: Eratosthenes on the Measurement of the Earth. In: *Historia Mathematica* 11 (1984), Nr. 4, pp. 411–416. – URL https://www.sciencedirect.com/science/article/pii/0315086084900259. – ISSN 0315-0860

[Gorelik 2009] GORELIK, G.: The Paternity of the H-Bombs: Soviet-American Perspectives. In: *Physics in Perspective* 11 (2009), Nr. 2, pp. 169–197. – URL https://link.springer.com/content/pdf/10.1007%2Fs00016-007-0377-8.pdf. – ISSN 1422-6944

[Gorelik 2013] GORELIK, G.: *Tamms Doktorand Andrej Sacharow*. Book Section Chapter 6, pp. 83–96. In: *Andreĭ Sakharov. A life for science and freedom*, Birkhäuser, 2013. – URL https://link.springer.com/content/pdf/10.1007%2F978-3-0348-0474-5_6.pdf. – ISBN 978-3-0348-0473-8 978-3-0348-0474-5

[Gorman et al. 2019] GORMAN, E. T. ; KUBALAK, D. A. ; PATEL, D. ; DRESS, A. ; MOTT, D. B. ; MEISTER, G. ; WERDELL, P. J.: The NASA Plankton, Aerosol, Cloud, ocean Ecosystem (PACE) mission: An emerging era of global, hyperspectral Earth system remote sensing. In: *Sensors, Systems, and Nextgeneration Satellites Xxiii* 11151 (2019). – URL https://www.spiedigitallibrary.org/conference-proceedings-of-spie/11151/111510G/The-NASA-Plankton-Aerosol-Cloud-ocean-Ecosystem-PACE-mission/10.1117/12.2537146.pdf. – ISSN 0277-786x

[Gouterman 1961] GOUTERMAN, M.: Spectra of porphyrins. In: *Journal of Molecular Spectroscopy* 6 (1961), pp. 138–163. – ISSN 00222852

[Govorov et al. 2016] GOVOROV, A. O. ; HERNÁNDEZ MARTÍNEZ, P. L. ; DEMIR, H. V.: *Understanding and Modeling Foerster-type Resonance Energy Transfer (FRET)*. 2016 (Springer Briefs in Applied Sciences and Technology). – ISBN 978-981-287-377-4 978-981-287-378-1

[Graneau and Graneau 2006] GRANEAU, P. ; GRANEAU, N.: *Johannes Kepler – The Astronomer who Coined the Word Inertia*. pp. 22–40. In: *In the Grip of the Distant Universe*, 2006. – ISBN 978-981-256-754-3 978-981-277-380-7

[Grassmann 1854] GRASSMANN, H. G.: XXXVII. On the theory of compound colours. In: *The London, Edinburgh, and Dublin Philosophical Magazine and Journal of Science* 7 (1854), Nr. 45, pp. 254–264. – URL https://www.tandfonline.com/doi/pdf/10.1080/14786445408647464?needAccess=true. – ISSN 1941-5982 1941-5990

[Green et al. 2018a] GREEN, D. ; F, V. A. C. ; HEISLER, I. A. ; DIJKSTRA, A. G. ; JONES, G. A.: Spectral Filtering as a Tool for Two-Dimensional Spectroscopy: A Theoretical Model. In: *J Phys Chem A* 122 (2018), Nr. 30, pp. 6206–6213. – URL https://www.ncbi.nlm.nih.gov/pubmed/29985004; https://pubs.acs.org/doi/pdf/10.1021/acs.jpca.8b03339. – ISSN 1520-5215 (Electronic) 1089-5639 (Linking)

[Green et al. 2018b] GREEN, D. ; F, V. A. C. ; HEISLER, I. A. ; DIJKSTRA, A. G. ; JONES, G. A.: Spectral Filtering as a Tool for Two-Dimensional Spectroscopy: A Theoretical Model (Supporting Information). In: *J Phys Chem A* 122 (2018), Nr. 30, pp. 6206–6213. – URL https://pubs.acs.org/doi/abs/10.1021/acs.jpca.8b03339;http://eprints.whiterose.ac.uk/133173/7/Green_JPCA_r3_SI_for_publication.pdf. – ISSN 1520-5215 (Electronic) 1089-5639 (Linking)

[Green et al. 2014] GREEN, L. ; SUTULA, M. ; FONG, P.: How much is too much? Identifying benchmarks of adverse effects of macroalgae on the macrofauna in intertidal flats. In: *Ecol Appl* 24 (2014), Nr. 2, pp. 300–14. – URL https://esajournals.onlinelibrary.wiley.com/doi/full/10.1890/13-0524.1. – ISSN 1051-0761 (Print) 1051-0761 (Linking)

[Greenemeier 2014] GREENEMEIER, L.: Moth Eyes Inspire Different Solar Cell. In: *Scientific American* (2014), 25 July 2014. – URL https://www.scientificamerican.com/podcast/episode/moth-eye-solar-cell/

[Greiner 1989] GREINER, W.: *Theoretische Physik*. Vol. 4A: *Quantentheorie – Spezielle Kapitel*. Frankfurt: Verlag Harri Deutsch, 1989. – ISBN 3-8171-1073-1

[Grotrian 1921] GROTRIAN, W.: The L-duplicate in neon. In: *Zeitschrift Fur Physik* 8 (1921), pp. 116–125. – ISSN 0044-3328

[Gu 2003] GU, W.: *Study of Active Sites in Ni Enzymes Using X-ray Absorption Spectroscopy*, University of California at Davis, Thesis, 2003

[Güttinger 1931] GÜTTINGER, P.: *Diplomarbeit in Physik*, ETH Zürich, Thesis, 1931

[Güttinger 1932] GÜTTINGER, P.: Das Verhalten von Atomen im magnetischen Drehfeld. In: *Zeitschrift fuer Physik* 73 (1932), Nr. 3-4, pp. 169–184. – URL https://link.springer.com/content/pdf/10.1007%2FBF01351211.pdf. – ISSN 1434-6001 1434-601X

[Haber 1914] HABER, F.: Untersuchungen ueber Ammoniak. Sieben Mitteilungen. In: *Zeitschrift für Elektrochemie und angewandte physikalische Chemie* 20 (1914), Nr. 22/23. – URL https://onlinelibrary.wiley.com/doi/10.1002/bbpc.191400016

[Haber 1920] HABER, F.: *Fritz Haber – Nobel Lecture: The synthesis of ammonia from its elements*. 1920. – URL https://www.nobelprize.org/prizes/chemistry/1918/haber/lecture/

[Haber 1922] HABER, F.: Über die Darstellung des Ammoniaks aus Stickstoff und Wasserstoff. In: *Die Naturwissenschaften* 10 (1922), Nr. 49, pp. 1041–1049. – URL https://link.springer.com/content/pdf/10.1007/BF01565394.pdf. – ISSN 0028-1042 1432-1904

[Hale 1900] HALE, G. E.: On Some Attempts to Detect the Solar Corona in Full Sunlight with a Bolometer. In: *The Astrophysical Journal* 12 (1900). – ISSN 0004-637X 1538-4357

[Hamidi et al. 2015] HAMIDI, H. ; HASAN, K. ; EMEK, S. C. ; DILGIN, Y. ; AKERLUND, H. E. ; ALBERTSSON, P. A. ; LEECH, D. ; GORTON, L.: Photocurrent generation from thylakoid membranes on osmium-redox-polymer-modified electrodes. In: *ChemSusChem* 8 (2015), Nr. 6, pp. 990–3. – URL https://onlinelibrary.wiley.com/doi/full/10.1002/cssc.201403200. – ISSN 1864-564X (Electronic) 1864-5631 (Linking)

[Hamm and Zanni 2011] HAMM, P. ; ZANNI, M.: *Concepts and Methods of 2D Infrared Spectroscopy*. Cambridge University Press, 2011. – URL https://doi.org/10.1017/CBO9780511675935. – ISBN 9780511675935

[Hantos et al. 1997] HANTOS, Z. ; PETAK, F. ; ADAMICZA, A. ; ASZTALOS, T. ; TOLNAI, J. ; FREDBERG, J. J.: Mechanical impedance of the lung periphery. In: *J Appl Physiol (1985)* 83 (1997), Nr. 5, pp. 1595–601. – URL https://www.physiology.org/doi/pdf/10.1152/jappl.1997.83.5.1595. – ISSN 8750-7587 (Print) 0161-7567 (Linking)

[Hao et al. 2020] HAO, J. ; KNOLL, A. H. ; HUANG, F. ; HAZEN, R. M. ; DANIEL, I.: Cycling phosphorus on the Archean Earth: Part I. Continental weathering and riverine transport of phosphorus. In: *Geochimica et Cosmochimica Acta* 273 (2020), pp. 70–84. – URL https://www.sciencedirect.com/science/article/pii/S001670372030048X. – ISSN 00167037

[Haraux and de Kouchkovsky 1998] HARAUX, F. ; KOUCHKOVSKY, Y. de: Energy coupling and ATP synthase. In: *Photosynthesis Research* 57 (1998), Nr. 3, pp. 231–251. – URL https://link.springer.com/content/pdf/10.1023%2FA%3A1006083802715.pdf. – ISSN 0166-8595

[Harley et al. 1989] HARLEY, D. ; GAJDA, M. ; RAFELSKI, J.: *Review of the Current Status of Cold Fusion*. Book Section Chapter 42, pp. 541–556. In: *The Nuclear Equation of State*, 1989 (NATO ASI Series). – ISBN 978-1-4612-7877-1 978-1-4613-0583-5

[Harrop et al. 2013] HARROP, S. J. ; WILK, K. E. ; CURMI, P. M. G.: *Light harvesting complex PC645 from the cryptophyte Chroomonas sp. CCMP270*. 2013-07-11 2013. – URL https://www.rcsb.org/structure/4LMS

[Hassan and Lee 2014] HASSAN, A. M. ; LEE, H.: A theoretical approach to the design of sustainable dwellings in hot dry zones: A Toshka case study. In: *Tunnelling and Underground Space Technology* 40 (2014), pp. 251–262. – URL https://www.sciencedirect.com/science/article/pii/S0886779813001739?via%3Dihub. – ISSN 0886-7798

[Hasselmann 1966] HASSELMANN, K.: Feynman diagrams and interaction rules of wave-wave scattering processes. In: *Reviews of Geophysics* 4 (1966), Nr. 1. – ISSN 8755-1209

[Haven 2014] HAVEN, E.: *Hydrodynamic Equations and Finance*. Book Section Chapter 29, pp. 317–323. In: *Quantum Interaction*, Springer, 2014 (Lecture Notes in Computer Science). – URL https://link.springer.com/chapter/10.1007%2F978-3-642-54943-4_29. – ISBN 978-3-642-54942-7 978-3-642-54943-4

[Haven 2002] HAVEN, E. E.: A discussion on embedding the Black-Scholes option pricing model in a quantum physics setting. In: *Physica A Statistical Mechanics and its Applications* 304 (2002), Nr. 3-4, pp. 507–524. – URL https://www.sciencedirect.com/science/article/pii/S0378437101005684. – ISSN 03784371

[Hawkins and Frieden 2012] HAWKINS, R. J. ; FRIEDEN, B. R.: Asymmetric Information and Quantization in Financial Economics. In: *International Journal of Mathematics and Mathematical Sciences* 2012 (2012), pp. 1–11. – URL https://doi.org/10.1155/2012/470293. – ISSN 0161-1712 1687-0425

[Hecht 2000] HECHT, K. T.: *Quantum Mechanics*. New York: Springer, 2000 (Graduate Texts in Contemporary Physics). – URL https://link.springer.com/content/pdf/10.1007%2F978-1-4612-1272-0.pdf. – ISBN 978-1-4612-7072-0 978-1-4612-1272-0

[Heisenberg 1925] HEISENBERG, W.: Uber quantentheoretische Umdeutung kinematischer und mechanischer Beziehungen. In: *Zeitschrift fur Physik* 33 (1925), Nr. 1, pp. 879–893. – URL https://link.springer.com/content/pdf/10.1007%2FBF01328377.pdf. – ISSN 1434-6001 1434-601X

[Heisenberg 1943] HEISENBERG, W.: Die beobachtbaren Groessen in der Theorie der Elementarteilchen. In: *Zeitschrift fuer Physik* 120 (1943), Nr. 7-10, pp. 513–538. – URL https://link.springer.com/content/pdf/10.1007%2FBF01329800.pdf. – ISSN 1434-6001 1434-601X

[Heisenberg and Pauli 1929] HEISENBERG, W. ; PAULI, W.: Zur Quantendynamik der Wellenfelder. In: *Zeitschrift fur Physik* 56 (1929), Nr. 1-2, pp. 1–61. – URL https://link.springer.com/content/pdf/10.1007%2FBF01340129.pdf. – ISSN 1434-6001 1434-601X

[Heisenberg and Pauli 1930] Heisenberg, W. ; Pauli, W.: Zur Quantentheorie der Wellenfelder. II. In: *Zeitschrift fur Physik* 59 (1930), Nr. 3-4, pp. 168–190. – URL https://link.springer.com/content/pdf/10.1007%2FBF01341423.pdf. – ISSN 1434-6001 1434-601X

[Heisler et al. 2014] Heisler, I. A. ; Moca, R. ; Camargo, F. V. ; Meech, S. R.: Two-dimensional electronic spectroscopy based on conventional optics and fast dual chopper data acquisition. In: *Rev Sci Instrum* 85 (2014), Nr. 6, p. 063103. – URL https://aip.scitation.org/doi/pdf/10.1063/1.4879822. – ISSN 1089-7623 (Electronic) 0034-6748 (Linking)

[Heitler 1947] Heitler: *The Quantum Theory of Radiation*. 2nd edition. Oxford: Oxford University Press, 1947 (The International Series on Monographs in Physics)

[Hellmann 1937] Hellmann, H.: *Einführung in die Quantenchemie*. Leipzig: F. Deuticke, 1937. – URL http://www.worldcat.org/title/einfuhrung-in-die-quantenchemie/oclc/8233915

[Hempelmann 2000] Hempelmann, R.: *Quasielastic Neutron Scattering and Solid State Diffusion*. Oxford Series on Neutron Scatt, 2000 (Oxford Series on Neutron Scattering in Condensed Matter). – p. 320. – ISBN 978-0198517436

[Hereher 2015] Hereher, M. E.: Environmental monitoring and change assessment of Toshka lakes in southern Egypt using remote sensing. In: *Environmental Earth Sciences* 73 (2015), Nr. 7, pp. 3623–3632. – URL https://link.springer.com/content/pdf/10.1007/s12665-014-3651-5.pdf. – ISSN 1866-6280

[Hereher 2017] Hereher, M. E.: Effects of land use/cover change on regional land surface temperatures: severe warming from drying Toshka lakes, the Western Desert of Egypt. In: *Natural Hazards* 88 (2017), Nr. 3, pp. 1789–1803. – URL https://link.springer.com/content/pdf/10.1007/s11069-017-2946-8.pdf. – ISSN 0921-030X

[Hermanns 2004] Hermanns, H.-G.: *Der Einfluß retardierter hydrodynamischer Wechselwirkungen auf die Bewegung von Kugeln in einer Suspension*, RWTH Aachen, Thesis, 2004. – URL http://publications.rwth-aachen.de/record/61865/files/Hermanns_Heinz-Guenter.pdf

[Hewitt and Hewitt 1979] Hewitt, E. ; Hewitt, R. E.: The Gibbs-Wilbraham phenomenon: An episode in fourier analysis. In: *Archive for History of Exact Sciences* 21 (1979), Nr. 2, pp. 129–160. – URL https://link.springer.com/content/pdf/10.1007%2FBF00330404.pdf. – ISSN 0003-9519 1432-0657

[Higgins 2012] Higgins, D. J.: *Positron lifetime modulation by electric field induced positronium formation on a gold surface*, Air Force Institute of Technology, Thesis, 2012. – URL https://apps.dtic.mil/dtic/tr/fulltext/u2/a558538.pdf

[Hildner et al. 2013] Hildner, R. ; Brinks, D. ; Nieder, J. B. ; Cogdell, R. J. ; Hulst, N. F. van: Quantum coherent energy transfer over varying pathways in single light-harvesting complexes. In: *Science* 340 (2013), Nr. 6139, pp. 1448–51. – URL https://science.sciencemag.org/content/340/6139/1448. – ISSN 1095-9203 (Electronic) 0036-8075 (Linking)

[Hindman 1967] Hindman, J. V.: A High Resolution Study of the Distribution and Motions of Neutral Hydrogen in the Small Cloud of Magellan. In: *Australian Journal of Physics* 20 (1967), Nr. 2. – ISSN 0004-9506

[Hoff et al. 1982] Hoff, J. E. ; Howe, J. ; Mitchell, C.: *Nutritional and Cultural Aspects of Plant Species Selection for a Controlled Ecological Life Support System*. March 1982 1982

[Holcomb 1969] Holcomb, R. W.: Fusion power: optimism and a tokamak gap at dubna. In: *Science* 166 (1969), Nr. 3903, pp. 363–4. – URL https://science.sciencemag.org/content/sci/166/3903/363.full.pdf. – ISSN 0036-8075 (Print) 0036-8075 (Linking)

[Holder 2011] Holder, M.: *Kepler's differential equations*. arXiv:11053964. [physics.hist-ph]. 2011

[Holder 2015] Holder, M.: *Die Kepler-Ellipse. Eine alte Geschichte neu erzählt*. Siegen: Universitätsverlag Siegen, 2015

[Hope 2012] HOPE, B.: *Egypt's new Nile Valley: grand plan gone bad.* 22 April 2012 2012. – URL https://www.thenational.ae/world/mena/egypt-s-new-nile-valley-grand-plan-gone-bad-1.402214

[Horn and Steinem 2005] HORN, C. ; STEINEM, C.: Photocurrents generated by bacteriorhodopsin adsorbed on nano-black lipid membranes. In: *Biophys J* 89 (2005), Nr. 2, pp. 1046–54. – URL https://www.ncbi.nlm.nih.gov/pubmed/15908580. – ISSN 0006-3495 (Print) 0006-3495 (Linking)

[Hou and Mauzerall 2011] HOU, H. J. ; MAUZERALL, D.: Listening to PS II: enthalpy, entropy, and volume changes. In: *J Photochem Photobiol B* 104 (2011), Nr. 1-2, pp. 357–65. – URL https://www.ncbi.nlm.nih.gov/pubmed/21530300. – ISSN 1873-2682 (Electronic) 1011-1344 (Linking)

[Huang et al. 1995] HUANG, T. Y. ; HAN, C. H. ; YI, Z. H. ; XU, B. X.: What Is the Astronomical Unit of Length? In: *Astronomy and Astrophysics* 298 (1995), Nr. 2, pp. 629–633. – ISSN 0004-6361

[Hubert 1969] HUBERT, P.: Thermonuclear Future of Tokamak Magnetic Confinement Device. In: *Nuclear Fusion* 9 (1969), Nr. 3, pp. 209–&. – URL https://iopscience.iop.org/article/10.1088/0029-5515/9/3/003/meta. – ISSN 0029-5515

[Inoue 2006] INOUE, H.: On Option Pricing by Quantum Mechanics Approach. In: *Information Processing and Management of Uncertainty in Knowledge-Based Systems*, IPMU, 2006. – URL http://ipmu2006.lip6.fr/program.php

[Ito et al. 1991] ITO, T. ; HIRAMATSU, M. ; HOSODA, M. ; TSUCHIYA, Y.: Picosecond time-resolved absorption spectrometer using a streak camera. In: *Review of Scientific Instruments* 62 (1991), Nr. 6, pp. 1415–1419. – URL https://aip.scitation.org/doi/pdf/10.1063/1.1142460. – ISSN 0034-6748 1089-7623

[Jablonski 1933] JABLONSKI, A.: Efficiency of Anti-Stokes Fluorescence in Dyes. In: *Nature* 131 (1933), Nr. 3319, pp. 839–840. – URL https://www.nature.com/articles/131839b0.pdf. – ISSN 0028-0836 1476-4687

[Jablonski 2013] JABLONSKI, A.: Ueber das Entstehen der breiten Absorptions- und Fluoreszenzbanden in Farbstoffloesungen. In: *Zeitschrift fuer Physik* 73 (2013), Nr. 7-8, pp. 460–469. – URL https://link.springer.com/content/pdf/10.1007%2FBF01349853.pdf. – ISSN 1434-6001 1434-601X

[Janssen 1983] JANSSEN, H.: *Horst Janssen: Die Kopie.* 2. Auflage. München: Deutscher Taschenbuch Verlag GmbH Co. KG dtv Kunst, 1983. – ISBN 3-423-02877-7

[Janssen et al. 2014] JANSSEN, P. J. ; LAMBREVA, M. D. ; PLUMERE, N. ; BARTOLUCCI, C. ; ANTONACCI, A. ; BUONASERA, K. ; FRESE, R. N. ; SCOGNAMIGLIO, V. ; REA, G.: Photosynthesis at the forefront of a sustainable life. In: *Front Chem* 2 (2014), p. 36. – URL https://www.ncbi.nlm.nih.gov/pubmed/24971306. – ISSN 2296-2646 (Linking)

[Jennings et al. 2005] JENNINGS, R. C. ; ENGELMANN, E. ; GARLASCHI, F. ; CASAZZA, A. P. ; ZUCCHELLI, G.: Photosynthesis and negative entropy production. In: *Biochim Biophys Acta* 1709 (2005), Nr. 3, pp. 251–5. – URL https://www.ncbi.nlm.nih.gov/pubmed/16139784. – ISSN 0006-3002 (Print) 0006-3002 (Linking)

[Johnson 1954] JOHNSON, F. S.: The Solar Constant. In: *Journal of Meteorology* 11 (1954), Nr. 6, pp. 431–439. – ISSN 0095-9634 0095-9634

[Johnston and Dietlein 1977] JOHNSTON, R. S. ; DIETLEIN, L. F.: *Biomedical Results from Skylab.* Washington, D.C. : Scientific and Technical Information Office, National Aeronautics and Space Administration, 1977 (NASA SP (Series), 377). – URL https://archive.org/download/NASA_NTRS_Archive_19770026836/NASA_NTRS_Archive_19770026836.pdf

[Jones et al. 1989] JONES, S. E. ; PALMER, E. P. ; CZIRR, J. B. ; DECKER, D. L. ; JENSEN, G. L. ; THORNE, J. M. ; TAYLOR, S. F. ; RAFELSKI, J.: Observation of Cold Nuclear-Fusion in Condensed Matter. In: *Nature* 338 (1989), Nr. 6218, pp. 737–740. – URL https://www.nature.com/articles/338737a0.pdf. – ISSN 0028-0836

[Jurik et al. 1984] JURIK, T. W. ; WEBER, J. A. ; GATES, D. M.: Short-Term Effects of CO(2) on Gas Exchange of Leaves of Bigtooth Aspen (Populus grandidentata) in the Field. In: *Plant Physiol* 75 (1984), Nr. 4, pp. 1022–6. – URL https://www.ncbi.nlm.nih.gov/pubmed/16663727http://www.plantphysiol.org/content/plantphysiol/75/4/1022.full.pdf. – ISSN 0032-0889 (Print) 0032-0889 (Linking)

[Kac 1949] KAC, M.: On distributions of certain Wiener functionals. In: *Transactions of the American Mathematical Society* 65 (1949), Nr. 1, pp. 1–1. – URL https://doi.org/10.1090/S0002-9947-1949-0027960-X. – ISSN 0002-9947

[Kahl 1999] KAHL, O.: A Note on Sabur Ibn Sahl. In: *Journal of Semitic Studies* XLIV (1999), Nr. 2, pp. 245–249. – URL https://doi.org/10.1093/jss/XLIV.2.245. – ISSN 0022-4480 1477-8556

[Kahle 1987] KAHLE, A. B.: Surface emittance, temperature, and thermal inertia derived from Thermal Infrared Multispectral Scanner (TIMS) data for Death Valley, California. In: *Geophysics* 52 (1987), Nr. 7, pp. 858–874. – ISSN 0016-8033 1942-2156

[Kananenka et al. 2018] KANANENKA, A. A. ; SUN, X. ; SCHUBERT, A. ; DUNIETZ, B. D. ; GEVA, E.: A comparative study of different methods for calculating electronic transition rates. In: *J Chem Phys* 148 (2018), Nr. 10, p. 102304. – URL https://aip.scitation.org/doi/pdf/10.1063/1.4989509. – ISSN 1089-7690 (Electronic) 0021-9606 (Linking)

[Kansy et al. 2017] KANSY, M. ; GUROWIETZ, A. ; WILHELM, C. ; GOSS, R.: An optimized protocol for the preparation of oxygen-evolving thylakoid membranes from Cyclotella meneghiniana provides a tool for the investigation of diatom plastidic electron transport. In: *BMC Plant Biol* 17 (2017), Nr. 1, p. 221. – URL https://www.ncbi.nlm.nih.gov/pubmed/29178846. – ISSN 1471-2229 (Electronic) 1471-2229 (Linking)

[Karahka and Kreuzer 2013] KARAHKA, M. L. ; KREUZER, H. J.: Charge transport along proton wires. In: *Biointerphases* 8 (2013), Nr. 1, p. 13. – URL https://avs.scitation.org/doi/pdf/10.1186/1559-4106-8-13. – ISSN 1559-4106 (Electronic) 1559-4106 (Linking)

[Karakostas and Zafiris 2015] KARAKOSTAS, V. ; ZAFIRIS, E.: Contextual semantics in quantum mechanics from a categorical point of view. In: *Synthese* 194 (2015), Nr. 3, pp. 847–886. – ISSN 0039-7857 1573-0964

[Karl et al. 1980] KARL, D. M. ; WIRSEN, C. O. ; JANNASCH, H. W.: Deep-Sea Primary Production at the Galapagos Hydrothermal Vents. In: *Science* 207 (1980), Nr. 4437, pp. 1345–1347. – URL https://science.sciencemag.org/content/sci/207/4437/1345.full.pdf. – ISSN 0036-8075 1095-9203

[Kasha 1999] KASHA, M.: From Jabłoński To Femtoseconds. Evolution of Molecular Photophysics. In: *Acta Physica Polonica A* 95 (1999), Nr. 1, pp. 15–36. – URL www.doi.org/10.12693/APhysPolA.95.15. – ISSN 0587-4246 1898-794X

[Kassouf 1962] KASSOUF, S. T.: *Evaluation of Convertible Securities*. Brooklyn, New York: Analytic Investors, Inc., 1962. – URL https://www.economics.uci.edu/files/kassouf/pdfs/evaluation.pdf

[Kassouf 1965] KASSOUF, S. T.: *A Theory and an Econometric Model for common Stock Purchase Warrants*, Columbia University, Dissertation, 1965. – URL http://www.economics.uci.edu/files/kassouf/pdfs/a_theory.pdf

[Katsapov and Braun 2010] KATSAPOV, G. Y. ; BRAUN, A.: Deuterium Tracer Experiments Prove the Thiophenic Hydrogen Involvement During the Initial Step of Thiophene Hydrodesulfurization. In: *Catalysis Letters* 138 (2010), Nr. 3-4, pp. 224–230. – URL https://link.springer.com/content/pdf/10.1007%2Fs10562-010-0400-6.pdf. – ISSN 1011-372X 1572-879X

[Katzir 2019] KATZIR, S.: Employment Before Formulation. In: *Historical Studies in the Natural Sciences* 49 (2019), Nr. 1, pp. 1–40. – URL https://doi.org/10.1525/HSNS.2019.49.1.1. – ISSN 1939-1811 1939-182X

[Kauffmann and Mayo 1996] KAUFFMANN, G. B. ; MAYO, I.: Multidisciplinary scientist – Melvin Calvin – His life and work . In: *Journal of Chemical Education* 73 (1996), Nr. 5, pp. 412–416

[Kaur 2019] KAUR, G.: *Intermediate Temperature Solid Oxide Fuel Cells – Electrolytes, Electrodes and Interconnects*. 1st edition. Springer, 2019. – p. 516. – ISBN 9780128174456

[Kayes et al. 2005] KAYES, B. M. ; ATWATER, H. A. ; LEWIS, N. S.: Comparison of the device physics principles of planar and radial p-n junction nanorod solar cells. In: *Journal of Applied Physics* 97 (2005), Nr. 11, p. 114302. – URL http://scitation.aip.org/docserver/fulltext/aip/journal/jap/97/11/1.1901835.pdf?expires=1481012692&id=id&accname=2085902&checksum=95636B2E74755C5178D7CC1EB1D9CCF0https://aip.scitation.org/doi/pdf/10.1063/1.1901835. – ISSN 0021-8979 1089-7550

[Kellermann et al. 2016] KELLERMANN, M. Y. ; YOSHINAGA, M. Y. ; VALENTINE, R. C. ; WORMER, L. ; VALENTINE, D. L.: Important roles for membrane lipids in haloarchaeal bioenergetics. In: *Biochim Biophys Acta* 1858 (2016), Nr. 11, pp. 2940–2956. – URL https://www.ncbi.nlm.nih.gov/pubmed/27565574. – ISSN 0006-3002 (Print) 0006-3002 (Linking)

[King 2000] KING, D. J.: Airborne remote sensing in forestry: Sensors, analysis and applications. In: *The Forestry Chronicle* 76 (2000), Nr. 6, pp. 859–876. – URL https://pubs.cif-ifc.org/doi/10.5558/tfc76859-6. – ISSN 0015-7546 1499-9315

[Klein and Nishina 1928] KLEIN, O. ; NISHINA, Y.: Über die Streuung von Strahlung durch freie Elektronen nach der neuen relativistischen Quantendynamik von Dirac. In: *Zeitschrift für Physik* 52 (1928), Nr. 11-12, pp. 853–868. – URL https://link.springer.com/content/pdf/10.1007%2FBF01366453.pdf. – ISSN 0044-3328

[Klemas 2012] KLEMAS, V.: Remote Sensing of Algal Blooms: An Overview with Case Studies. In: *Journal of Coastal Research* 278 (2012), pp. 34–43. – URL https://doi.org/10.2112/JCOASTRES-D-11-00051.1. – ISSN 0749-0208 1551-5036

[Kobayashi et al. 2016] KOBAYASHI, K. ; SAKKA, Y. ; SUZUKI, T. S.: Development of an electrochemical impedance analysis program based on the expanded measurement model. In: *Journal of the Ceramic Society of Japan* 124 (2016), Nr. 9, pp. 943–949. – URL https://www.jstage.jst.go.jp/article/jcersj2/124/9/124_16120/_pdf. – ISSN 1348-6535 1882-0743

[Kocks et al. 1995] KOCKS, P. ; ROSS, J. ; BJOERKMAN, O.: Thermodynamic Efficiency and Resonance of Photosynthesis in a C3 Plant. In: *The Journal of Physical Chemistry* 99 (1995), Nr. 44, pp. 16483–16489. – URL https://doi.org/10.1021/j100044a043. – ISSN 0022-3654 1541-5740

[Kolesnikov et al. 2009] KOLESNIKOV, A. ; KUTCHEROV, V. G. ; GONCHAROV, A. F.: Methane-derived hydrocarbons produced under upper-mantle conditions. In: *Nature Geoscience* 2 (2009), Nr. 8, pp. 566–570. – URL http://www.nature.com/articles/ngeo591https://www.nature.com/articles/ngeo591.pdf. – ISSN 1752-0894 1752-0908

[Komura and Itoh 2009] KOMURA, M. ; ITOH, S.: Fluorescence measurement by a streak camera in a single-photon-counting mode. In: *Photosynth Res* 101 (2009), Nr. 2-3, pp. 119–33. – URL https://link.springer.com/content/pdf/10.1007%2Fs11120-009-9463-x.pdf. – ISSN 1573-5079 (Electronic) 0166-8595 (Linking)

[Koningsberger and Prins 1988] KONINGSBERGER, D. C. ; PRINS, R.: *X-Ray Absorption: Principles, Applications, Techniques of EXAFS, SEXAFS and XANES*. 1st edition. Wiley-Interscience, 1988. – ISBN 978-0471875475

[Koon et al. 2011] KOON, W. S. ; LO, M. W. ; MARSDEN, J. E. ; ROSS, S. D.: *Dynamical Systems, the Three-Body Problem and Space Mission Design*. Marsden Books, 2011. – ISBN 978-0-615-24095-4

[Kosinski 1970] KOSINSKI, J. N.: *Being there: the startling novel of a new american hero*. New York: Bantam Books, 1970. – ISBN 9780553279306

[Kostoff et al. 2020] KOSTOFF, R. N. ; HEROUX, P. ; ASCHNER, M. ; TSATSAKIS, A.: Adverse health effects of 5G mobile networking technology under real-life conditions. In: *Toxicol Lett* 323 (2020), pp. 35–40. – URL https://www.ncbi.nlm.nih.gov/pubmed/31991167. – ISSN 1879-3169 (Electronic) 0378-4274 (Linking)

[Kowalski 2010] KOWALSKI, L.: Commentary Letter in "Hot topics in cold fusion". In: *Physics Today* 63 (2010), Nr. 6, p. 1. – URL http://physicstoday.scitation.org/doi/pdf/10.1063/1.3455240https://physicstoday.scitation.org/doi/pdf/10.1063/1.3455240. – ISSN 0031-9228 1945-0699

[Kramers 1930] KRAMERS, H. A.: General theory of the paramagnetic rotation in crystals. In: *Proceedings of the Koninklijke Akademie Van Wetenschappen Te Amsterdam* 33 (1930), Nr. 6/10, pp. 959–972. – URL https://www.dwc.knaw.nl/DL/publications/PU00014621.pdf

[Kramida 2019] KRAMIDA, A.: Cowan Code: 50 Years of Growing Impact on Atomic Physics. In: *Atoms* 7 (2019), Nr. 3. – ISSN 2218-2004

[Kreifeldt 1971] KREIFELDT, J. G.: A formulation for the number of communicative civilizations in the galaxy. In: *Icarus* 14 (1971), Nr. 3, pp. 419–430. – ISSN 00191035

[Kundt and Marggraf 2014] KUNDT, W. ; MARGGRAF, O.: *Physikalische Mythen auf dem Pruefstand*. Berlin, Heidelberg: Springer Spektrum, 2014. – ISBN 978-3-642-37705-1

[Kupferschmidt 2019] KUPFERSCHMIDT, K.: Europe abandons plans for 'flagship' billion-euro research projects. In: *Science* (2019). – ISSN 0036-8075 1095-9203

[Kurashige et al. 2013] KURASHIGE, Y. ; CHAN, G. K. ; YANAI, T.: Entangled quantum electronic wavefunctions of the Mn(4)CaO(5) cluster in photosystem II. In: *Nat Chem* 5 (2013), Nr. 8, pp. 660–6. – URL https://www.nature.com/articles/nchem.1677.pdf. – ISSN 1755-4349 (Electronic) 1755-4330 (Linking)

[Kuzmenko 1969] KUZMENKO, G. I.: Quantum-mechanical explanation of Liesegang rings and periodic deposition at electrodes. In: *Russian Journal of Physical Chemistry, USSR* 43 (1969), Nr. 9, pp. 1349–&. – ISSN 0036-0244

[Kyriakou et al. 2017] KYRIAKOU, V. ; GARAGOUNIS, I. ; VASILEIOU, E. ; VOURROS, A. ; STOUKIDES, M.: Progress in the Electrochemical Synthesis of Ammonia. In: *Catalysis Today* 286 (2017), pp. 2–13. – URL https://doi.org/10.1016/j.cattod.2016.06.014. – ISSN 0920-5861

[Laasonen et al. 1993] LAASONEN, K. ; SPRIK, M. ; PARRINELLO, M. ; CAR, R.: "Ab initio" liquid water. In: *The Journal of Chemical Physics* 99 (1993), Nr. 11, pp. 9080–9089. – URL https://aip.scitation.org/doi/pdf/10.1063/1.465574. – ISSN 0021-9606 1089-7690

[Lampert 2000] LAMPERT, T.: *Zur Wissenschaftstheorie der Farbenlehre: Aufgaben, Texte, Loesungen*. Bern: Libri Books on Demand, 2000. – p. 390. – URL http://www.worldcat.org/title/zur-wissenschaftstheorie-der-farbenlehre-aufgaben-texte-losungen/oclc/76289447. – ISBN 9783898118934

[Landsberg 2010] LANDSBERG, J. H.: The Effects of Harmful Algal Blooms on Aquatic Organisms. In: *Reviews in Fisheries Science* 10 (2010), Nr. 2, pp. 113–390. – URL https://doi.org/10.1080/20026491051695. – ISSN 1064-1262 1547-6553

[Lanyi 1993] LANYI, J. K.: Pathways of proton transfer in the light-driven pump bacteriorhodopsin. In: *Experientia* 49 (1993), Nr. 6-7, pp. 514–7. – URL https://www.ncbi.nlm.nih.gov/pubmed/11536537. – ISSN 0014-4754 (Print) 0014-4754 (Linking)

[Lawaczeck 1933] LAWACZECK, F.: *Nationalsozialistische Bibliothek*. Vol. 38: *Technik und Wirtschaft im Dritten Reich – Ein Arbeitsbeschaffungsprogramm*. 3. Auflage. München: Verlag Franz Eher Nachfahren, 1933

[Lazár 1999] LAZÁR, D.: Chlorophyll a fluorescence induction. In: *Biochimica et Biophysica Acta (BBA) – Bioenergetics* 1412 (1999), Nr. 1, pp. 1–28. – URL https://doi.org/10.1016/S0005-2728(99)00047-X. – ISSN 00052728

[Lee et al. 2015] LEE, H. ; MOON, Y. J. ; NAKARIAKOV, V. M.: Radial and Azimuthal Oscillations of Halo Coronal Mass Ejections in the Sun. In: *The Astrophysical Journal* 803 (2015), Nr. 1. – URL http://iopscience.iop.org/article/10.1088/2041-8205/803/1/L7/pdf. – ISSN 2041-8213

[Lee et al. 2017] LEE, M. K. ; BRAVAYA, K. B. ; COKER, D. F.: First-Principles Models for Biological Light-Harvesting: Phycobiliprotein Complexes from Cryptophyte Algae. In: *J Am Chem Soc* 139

(2017), Nr. 23, pp. 7803–7814. – URL https://pubs.acs.org/doi/abs/10.1021/jacs.7b01780. – ISSN 1520-5126 (Electronic) 0002-7863 (Linking)

[Leem et al. 2008] LEEM, H. J. ; DORBANDT, I. ; ROJAS-CHAPANA, J. ; FIECHTER, S. ; TRIBUTSCH, H.: Bio-analogue amino acid-based proton-conduction wires for fuel cell membranes. In: *Journal of Physical Chemistry C* 112 (2008), Nr. 7, pp. 2756–2763. – URL https://pubs.acs.org/doi/abs/10.1021/jp077547l. – ISSN 1932-7447

[Lehmann et al. 1955] LEHMANN, H. ; SYMANZIK, K. ; ZIMMERMANN, W.: Zur Formulierung quantisierter Feldtheorien. In: *Il Nuovo Cimento* 1 (1955), Nr. 1, pp. 205–225. – URL https://doi.org/10.1007/BF02731765. – ISSN 0029-6341 1827-6121

[Lemonde 2014] LEMONDE: *De l'hydrogène produit avec un oeil de mite artificiel.* 23 June 2014 2014

[Lewis 2016] LEWIS, N. S.: *Innovation in solar fuels, electricity storage, and advanced materials.* Hearing before the subcommittee on energy committee on science, space, and technology house of representatives. Washington, DC 20402–0001: U.S. Government Publishing Office, 2016. – URL https://docs.house.gov/meetings/SY/SY20/20160615/105071/HHRG-114-SY20-20160615-SD003.pdf

[Leyon 1954] LEYON, H.: The structure of chloroplasts IV. The development and structure of the Aspidistra chloroplast. In: *Experimental Cell Research* 7 (1954), Nr. 1, pp. 265–273. – URL https://doi.org/10.1016/0014-4827(54)90061-0. – ISSN 00144827

[Li et al. 1984] LI, J. ; PEAT, R. ; PETER, L. M.: Surface recombination at semiconductor electrodes. In: *Journal of Electroanalytical Chemistry and Interfacial Electrochemistry* 165 (1984), Nr. 1-2, pp. 41–59. – URL https://doi.org/10.1016/S0022-0728(84)80085-6. – ISSN 00220728

[Li et al. 2014] LI, L. ; GIOKAS, P. G. ; KANAI, Y. ; MORAN, A. M.: Modeling time-coincident ultrafast electron transfer and solvation processes at molecule-semiconductor interfaces. In: *J Chem Phys* 140 (2014), Nr. 23, p. 234109. – URL https://aip.scitation.org/doi/pdf/10.1063/1.4882664. – ISSN 1089-7690 (Electronic) 0021-9606 (Linking)

[Liebel et al. 2014] LIEBEL, M. ; SCHNEDERMANN, C. ; KUKURA, P.: Vibrationally Coherent Crossing and Coupling of Electronic States during Internal Conversion in beta-Carotene. In: *Physical Review Letters* 112 (2014), Nr. 19, p. 5. – URL https://journals.aps.org/prl/abstract/10.1103/PhysRevLett.112.198302. – ISSN 0031-9007

[Lin et al. 1990] LIN, G. H. ; KAINTHLA, R. C. ; PACKHAM, N. J. C. ; BOCKRIS, J. O.: Electrochemical fusion: a mechanism speculation. In: *Journal of Electroanalytical Chemistry and Interfacial Electrochemistry* 280 (1990), Nr. 1, pp. 207–211. – ISSN 00220728

[Lingenfelter 1986] LINGENFELTER, R. E.: *Death Valley & The Amargosa.* Berkeley and Los Angeles, California: University of California Press, 1986. – p. 664. – ISBN 978-0-520-06356-3

[Liptay 1965] LIPTAY, W.: Die Lösungsmittelabhängigkeit der Wellenzahl von Elektronenbanden und die chemisch-physikalischen Grundlagen. In: *Zeitschrift für Naturforschung A* 20 (1965), Nr. 11. – URL https://www.degruyter.com/downloadpdf/j/zna.1965.20.issue-11/zna-1965-1109/zna-1965-1109.pdf. – ISSN 1865-7109 0932-0784

[Lo 2008] LO, A.: *15.401 Finance Theory I, Options.* Boston MA: MIT Sloan MBA Program, 2008 (MIT Lecture Fall 2008). – URL https://ocw.mit.edu/courses/sloan-school-of-management/15-401-finance-theory-i-fall-2008/video-lectures-and-slides/MIT15_401F08_lec01.pdf

[Locher 2019] LOCHER, F.: Neo-Malthusianism, World Fisheries Crisis, and The Global Commons, 1950s–1970s. In: *The Historical Journal* 63 (2019), Nr. 1, pp. 187–207. – ISSN 0018-246X 1469-5103

[Maccone 2012] MACCONE, C.: *A mathematical model for evolution and SETI.* Book Section Chapter 8, pp. 215–237. In: *Mathematical SETI*, Springer, 2012. – URL https://link.springer.com/content/pdf/10.1007%2F978-3-642-27437-4_8.pdf. – ISBN 978-3-642-27436-7 978-3-642-27437-4

[Madiba 2017] Madiba, I. G.: *Space radiations like hardness of VO2 based active smart radiator device for nano-satellites applications*, University of South Africa, Doctoral Thesis, 2017

[Madiba et al. 2018] Madiba, I. G. ; Braun, A. ; Emond, N. ; Chaker, M. ; Tadadjeu, S. I. ; Khanyile, B. S. ; Maaza, M.: Resonant photoemission spectroscopy of gamma irradiated VO2 films. In: *MRS Advances* (2018), pp. 1–5. – ISSN 2059-8521

[Madiba et al. 2017] Madiba, I. G. ; Emond, N. ; Chaker, M. ; Thema, F. T. ; Tadadjeu, S. I. ; Muller, U. ; Zolliker, P. ; Braun, A. ; Kotsedi, L. ; Maaza, M.: Effects Of gamma irradiations on reactive pulsed laser deposited vanadium dioxide thin films. In: *Applied Surface Science* 411 (2017), pp. 271–278. – ISSN 0169-4332

[Makhalanyane et al. 2015] Makhalanyane, T. P. ; Valverde, A. ; Velázquez, D. ; Gunnigle, E. ; Van Goethem, M. W. ; Quesada, A. ; Cowan, D. A.: Ecology and biogeochemistry of cyanobacteria in soils, permafrost, aquatic and cryptic polar habitats. In: *Biodiversity and Conservation* 24 (2015), Nr. 4, pp. 819–840. – URL https://link.springer.com/content/pdf/10.1007%2Fs10531-015-0902-z.pdf. – ISSN 0960-3115 1572-9710

[Malterre-Barthes 2016] Malterre-Barthes, C.: The Toshka Project Colossal Water Infrastructures, Biopolitics and Territory in Egypt. In: *Architectural Design* 86 (2016), Nr. 4, pp. 98–105. – URL https://onlinelibrary.wiley.com/doi/abs/10.1002/ad.2074. – ISSN 0003-8504

[Manna and Dunietz 2014] Manna, A. K. ; Dunietz, B. D.: Communication: Charge-transfer rate constants in zinc-porphyrin-porphyrin-derived dyads: a Fermi golden rule first-principles-based study. In: *J Chem Phys* 141 (2014), Nr. 12, p. 121102. – URL https://aip.scitation.org/doi/pdf/10.1063/1.4896826. – ISSN 1089-7690 (Electronic) 0021-9606 (Linking)

[Marx 2006] Marx, D.: Proton transfer 200 years after von Grotthuss: insights from ab initio simulations. In: *Chemphyschem* 7 (2006), Nr. 9, pp. 1848–70. – URL https://www.ncbi.nlm.nih.gov/pubmed/16929553https://onlinelibrary.wiley.com/doi/pdf/10.1002/cphc.200600128https://onlinelibrary.wiley.com/doi/full/10.1002/cphc.200600128. – ISSN 1439-4235 (Print) 1439-4235 (Linking)

[Matveev and Sokolov 1961] Matveev, V. V. ; Sokolov, A. D.: Hard X-Ray Radiation from Tokamak-2, a Toroidal System. In: *Soviet Physics-Technical Physics* 5 (1961), Nr. 10, pp. 1084–1088. – ISSN 0038-5662

[Matzke et al. 1996] Matzke, T. ; Stimming, U. ; Karmonik, C. ; Soetratmo, M. ; Hempelmann, R. ; Guthoff, F.: Quasielastic thermal neutron scattering experiment on the proton conductor SrCe0.95Yb0.05H0.02O2.985. In: *Solid State Ionics* 86-8 (1996), pp. 621–628. – ISSN 0167-2738

[Mauerer et al. 2016] Mauerer, M. ; Schubert, D. ; Zabel, P. ; Bamsey, M. ; Kohlberg, E. ; Mengedoht, D.: Initial survey on fresh fruit and vegetable preferences of Neumayer Station crew members: Input to crop selection and psychological benefits of space-based plant production systems. In: *Open Agriculture* 1 (2016), Nr. 1, pp. 179–188. – URL https://www.degruyter.com/downloadpdf/j/opag.2016.1.issue-1/opag-2016-0023/opag-2016-0023.pdf. – ISSN 2391-9531

[Maxwell and Zaidi 1993] Maxwell, J. C. ; Zaidi, Q.: Commentary: On the theory of compound colours, and the relations of the colours of the spectrum. In: *Color Research and Application* 18 (1993), Nr. 4, pp. 270–287. – ISSN 03612317 15206378

[Mazumi 2001] Mazumi, K.: *The Origins of Ecological Economics: The Bioeconomics of Georgescu-Roegen*. London, New York: Routledge, 2001 (Routledge Explorations in Environmental Economics). – p. 176. – ISBN 978-0415235235

[McFadden and Al-Khalili 2018] McFadden, J. ; Al-Khalili, J.: The origins of quantum biology. In: *Proc Math Phys Eng Sci* 474 (2018), Nr. 2220, p. 20180674. – URL https://www.ncbi.nlm.nih.gov/pubmed/30602940. – ISSN 1364-5021 (Print) 1364-5021 (Linking)

[Meier et al. 2009] Meier, R. R. ; Englert, C. ; Chua, D. ; Socker, D. ; Picone, J. M. ; Carter, T. ; Huba, J. ; Slinker, S. ; Krall, J. ; Vincent, W.: Geospace imaging using Thomson scattering.

In: *Journal of Atmospheric and Solar-Terrestrial Physics* 71 (2009), Nr. 1, pp. 132–142. – URL https://www.sciencedirect.com/science/article/pii/S1364682608002794. – ISSN 13646826

[Melandri 1997] MELANDRI, B. A.: *Vectorial bioenergetics*. Book Section Chapter 3, pp. 95–138. In: *Bioenergetics*, Birkhäuser, 1997. – URL https://link.springer.com/chapter/10.1007%2F978-3-0348-8994-0_3. – ISBN 978-3-0348-9860-7 978-3-0348-8994-0

[Merton 1976] MERTON, R. C.: Option pricing when underlying stock returns are discontinuous. In: *Journal of Financial Economics* 3 (1976), Nr. 1-2, pp. 125–144. – URL https://doi.org/10.1016/0304-405X(76)90022-2. – ISSN 0304405X

[Meza et al. 2019] MEZA, E. ; SICARDY, B. ; ASSAFIN, M. ; ORTIZ, J. L. ; BERTRAND, T. ; LELLOUCH, E. ; DESMARS, J. ; FORGET, F. ; BÉRARD, D. ; DORESSOUNDIRAM, A. ; LECACHEUX, J. ; OLIVEIRA, J. M. ; ROQUES, F. ; WIDEMANN, T. ; COLAS, F. ; VACHIER, F. ; RENNER, S. ; LEIVA, R. ; BRAGA-RIBAS, F. ; BENEDETTI-ROSSI, G. ; CAMARGO, J. I. B. ; DIAS-OLIVEIRA, A. ; MORGADO, B. ; GOMES-JÚNIOR, A. R. ; VIEIRA-MARTINS, R. ; BEHREND, R. ; TIRADO, A. C. ; DUFFARD, R. ; MORALES, N. ; SANTOS-SANZ, P. ; JELÍNEK, M. ; CUNNIFFE, R. ; QUEREL, R. ; HARNISCH, M. ; JANSEN, R. ; PENNELL, A. ; TODD, S. ; IVANOV, V. D. ; OPITOM, C. ; GILLON, M. ; JEHIN, E. ; MANFROID, J. ; POLLOCK, J. ; REICHART, D. E. ; HAISLIP, J. B. ; IVARSEN, K. M. ; LACLUYZE, A. P. ; MAURY, A. ; GIL-HUTTON, R. ; DHILLON, V. ; LITTLEFAIR, S. ; MARSH, T. ; VEILLET, C. ; BATH, K. L. ; BEISKER, W. ; BODE, H. J. ; KRETLOW, M. ; HERALD, D. ; GAULT, D. ; KERR, S. ; PAVLOV, H. ; FARAGÓ, O. ; KLÖS, O. ; FRAPPA, E. ; LAVAYSSIÈRE, M. ; COLE, A. A. ; GILES, A. B. ; GREENHILL, J. G. ; HILL, K. M. ; BUIE, M. W. ; OLKIN, C. B. ; YOUNG, E. F. ; YOUNG, L. A. ; WASSERMAN, L. H. ; DEVOGÈLE, M. ; FRENCH, R. G. ; BIANCO, F. B. ; MARCHIS, F. ; BROSCH, N. ; KASPI, S. ; POLISHOOK, D. ; MANULIS, I. ; AIT MOULAY LARBI, M. ; BENKHALDOUN, Z. ; DAASSOU, A. ; EL AZHARI, Y. ; MOULANE, Y. ; BROUGHTON, J. ; MILNER, J. ; DOBOSZ, T. ; BOLT, G. ; LADE, B. ; GILMORE, A. ; KILMARTIN, P. ; ALLEN, W. H. ; GRAHAM, P. B. ; LOADER, B. ; MCKAY, G. ; TALBOT, J. ; PARKER, S. et al.: Lower atmosphere and pressure evolution on Pluto from ground-based stellar occultations, 1988–2016. In: *Astronomy & Astrophysics* 625 (2019). – ISSN 0004-6361 1432-0746

[Michaelian 2012] MICHAELIAN, K.: HESS Opinions "Biological catalysis of the hydrological cycle: life's thermodynamic function". In: *Hydrology and Earth System Sciences* 16 (2012), Nr. 8, pp. 2629–2645. – URL https://www.hydrol-earth-syst-sci.net/16/2629/2012/hess-16-2629-2012.pdf. – ISSN 1607-7938

[Michelson and Morley 1887] MICHELSON, A. A. ; MORLEY, E. W.: On the relative motion of the Earth and the luminiferous ether. In: *American Journal of Science* s3-34 (1887), Nr. 203, pp. 333–345. – URL http://www.ajsonline.org/content/s3-34/203/333.full.pdf. – ISSN 0002-9599

[Migulin and Menger 2001] MIGULIN, V. A. ; MENGER, F. M.: Adamantane-based crystals with rhythmic morphologies. In: *Langmuir* 17 (2001), Nr. 5, pp. 1324–1327. – URL https://pubs.acs.org/doi/10.1021/la001311q. – ISSN 0743-7463

[Miles 2000] MILES, M. H.: Calorimetric studies of Pd/D2O+LiOD electrolysis cells. In: *Journal of Electroanalytical Chemistry* 482 (2000), Nr. 1, pp. 56–65. – ISSN 0022-0728

[Miles et al. 1994] MILES, M. H. ; BUSH, B. F. ; STILWELL, D. E.: Calorimetric Principles and Problems in Measurements of Excess Power during Pd-D2o Electrolysis. In: *Journal of Physical Chemistry* 98 (1994), Nr. 7, pp. 1948–1952. – ISSN 0022-3654

[Miles et al. 2001] MILES, M. H. ; IMAM, M. A. ; FLEISCHMANN, M.: *Energy and Electrochemical Processes for a Cleaner Environment, Proceedings*. Vol. 2001: *Calorimetric analysis of a heavy water electrolysis experiment using a Pd-B alloy cathode*. Pennington: Electrochemical Society Inc, 2001. – pp. 194–205. – ISBN 1-56677-356-3

[Miles et al. 1990a] MILES, M. H. ; PARK, K. H. ; STILWELL, D. E.: Electrochemical Calorimetric Evidence for Cold Fusion in the Palladium Deuterium System. In: *Journal of Electroanalytical Chemistry* 296 (1990), Nr. 1, pp. 241–254. – ISSN 0022-0728

[Miles et al. 1990b] MILES, M. H. ; PARK, K. H. ; STILWELL, D. E. ; UNIV UTAH, NATL COLD FUSION INST: *Electrochemical calorimetric studies of the cold diffusion effect*. Salt Lake City: Natl Cold Fusion Inst, 1990 (First Annual Conference on Cold Fusion: Conference Proceedings). – pp. 328–334

[Milonni 2009] MILONNI, P. W.: *Zero-Point Energy*. Book Section Chapter 242, pp. 864–866. In: *Compendium of Quantum Physics*. Berlin, Heidelberg: Springer, 2009. – URL https://link.springer.com/content/pdf/10.1007%2F978-3-540-70626-7_242.pdf. – ISBN 978-3-540-70622-9 978-3-540-70626-7

[Milton 2009] MILTON, K. A.: *Quantum Electrodynamics (QED)*. Book Section Chapter 162, pp. 539–543. In: *Compendium of Quantum Physics*, Springer, 2009. – URL https://link.springer.com/content/pdf/10.1007%2F978-3-540-70626-7_162.pdf. – ISBN 978-3-540-70622-9 978-3-540-70626-7

[Mimkes 2006] MIMKES, J.: *A Thermodynamic Formulation of Economics*. pp. 1–33. In: *Econophysics and Sociophysics*. Weinheim: WILEY-VCH Verlag GmbH & Co. KGaA, 2006. – URL https://doi.org/10.1002/9783527610006.ch1. – ISBN 9783527610006 9783527406708

[Mimkes and Aruka 2005] MIMKES, J. ; ARUKA, Y.: *Carnot Process of Wealth Distribution*. Book Section Chapter 8, pp. 70–78. In: *Econophysics of Wealth Distributions*. Milano: Springer, 2005 (New Economic Windows). – URL https://doi.org/10.1007/88-470-0389-X_8. – ISBN 978-88-470-0329-3 978-88-470-0389-7

[Mishra et al. 2019] MISHRA, A. K. ; TIWARI, D. N. ; RAI, A. N.: *Cyanobacteria – From Basic Science to Applications*. London UK, San Diego CA, Cambridge MA, Oxford UK: Academic Press, 2019. – p. 513. – URL https://www.elsevier.com/books/cyanobacteria/mishra/978-0-12-814667-5. – ISBN 978-0-12-814667-5

[Mitchell 1994] MITCHELL, C. A.: Bioregenerative life-support systems. In: *Am J Clin Nutr* 60 (1994), Nr. 5, pp. 820S–824S. – URL https://www.ncbi.nlm.nih.gov/pubmed/7942592. – ISSN 0002-9165 (Print) 0002-9165 (Linking)

[Mitchell 1991] MITCHELL, P.: Foundations of vectorial metabolism and osmochemistry. In: *Biosci Rep* 11 (1991), Nr. 6, pp. 297–344; discussion 345–6. – URL https://www.ncbi.nlm.nih.gov/pubmed/1823594. – ISSN 0144-8463 (Print) 0144-8463 (Linking)

[Montechiaro and Giordano 2019] MONTECHIARO, F. ; GIORDANO, M.: Effect of prolonged dark incubation on pigments and photosynthesis of the cave-dwelling cyanobacterium Phormidium autumnale (Oscillatoriales, Cyanobacteria). In: *Phycologia* 45 (2019), Nr. 6, pp. 704–710. – ISSN 0031-8884 2330-2968

[Monteón 1982] MONTEÓN, M.: *Chile in the Nitrate Era. The Evolution of Economic Dependence, 1880-1930*. Madison WI: University of Wisconsin Press, 1982. – ISBN 978-0299088200

[Mora 2008] MORA, M. R.: The Gibbs' firm and the Peruvian nitrate monopoly / LA CASA GIBBS Y EL MONOPOLIO SALITRERO PERUANO: 1876-1878. In: *HISTORIA-SANTIAGO* 41 (2008), Nr. 1, pp. 63–77. – ISSN 0073-2435

[Mott 2008] MOTT, N. F.: On the Theory of Excitation by Collision with Heavy Particles. In: *Mathematical Proceedings of the Cambridge Philosophical Society* 27 (2008), Nr. 4, pp. 553–560. – URL https://doi.org/10.1017/S0305004100009816. – ISSN 0305-0041 1469-8064

[Mugasha et al. 2013] MUGASHA, W. A. ; BOLLANDSÅS, O. M. ; EID, T.: Relationships between diameter and height of trees in natural tropical forest in Tanzania. In: *Southern Forests: a Journal of Forest Science* 75 (2013), Nr. 4, pp. 221–237. – URL https://www.tandfonline.com/doi/pdf/10.2989/20702620.2013.824672. – ISSN 2070-2620 2070-2639

[Mulec et al. 2008] MULEC, J. ; KOSI, G. ; VRHOVSEK, D.: Characterization of cave aerophytic algal communities and effects of irradiance levels on production of pigments . In: *Journal of Cave and Karst Studies* 70 (2008), Nr. 1, pp. 3–12

[Myneni and Ross 1991] MYNENI, R. B. ; ROSS, J.: *Photon-Vegetation Interactions*. 1991. – ISBN 978-3-642-75391-6 978-3-642-75389-3

[Nagel 1998] NAGEL, D. J.: The status of 'cold fusion'. In: *Radiation Physics and Chemistry* 51 (1998), Nr. 4-6, pp. 653–668. – URL https://www.sciencedirect.com/science/article/pii/S0969806X97002302

[Nagle and Morowitz 1978] NAGLE, J. F. ; MOROWITZ, H. J.: Molecular Mechanisms for Proton Transport in Membranes. In: *Proceedings of the National Academy of Sciences of the United States of America* 75 (1978), Nr. 1, pp. 298–302. – URL https://www.ncbi.nlm.nih.gov/pmc/articles/PMC411234/pdf/pnas00013-0303.pdf. – ISSN 0027-8424

[Nelms and Oppenheim 1955] NELMS, A. T. ; OPPENHEIM, I.: Data on the atomic form factor: Computation and survey. In: *Journal of Research of the National Bureau of Standards* 55 (1955), Nr. 1, p. 53. – URL http://dx.doi.org/10.6028/jres.055.006. – ISSN 0091-0635

[Niklitschek 1949] NIKLITSCHEK, A.: *Ausflug ins Sonnensystem*. Wien, Oesterreich: Bruecken-Verlag, 1949 (Wunder in und um uns). – p. 180

[Norby 1998] NORBY, R. J.: Nitrogen deposition: a component of global change analyses. In: *New Phytologist* 139 (1998), Nr. 1, pp. 189–200. – URL https://nph.onlinelibrary.wiley.com/doi/pdf/10.1046/j.1469-8137.1998.00183.x. – ISSN 0028-646x

[Norton et al. 2017] NORTON, B. ; BALICK, M. ; HOBDAY, R. ; FOURNIER, C. ; SCARTEZZINI, J. L. ; SOLT, J. ; BRAUN, A.: Sponsored Collection | Changing perspectives on daylight: Science, technology, and culture. In: *Science* 358 (2017), Nr. 6363, pp. 680.2–680. – ISSN 0036-8075 1095-9203

[O'Brien 1975] O'BRIEN, D. M.: The Wick Rotation. In: *Australian Journal of Physics* 28 (1975), Nr. 1. – ISSN 0004-9506

[Oh et al. 2019] OH, S. A. ; COKER, D. F. ; HUTCHINSON, D. A. W.: Variety, the spice of life and essential for robustness in excitation energy transfer in light-harvesting complexes. In: *Faraday Discuss* 221 (2019), Nr. 0, pp. 59–76. – URL https://pubs.rsc.org/en/content/articlepdf/2020/fd/c9fd00081j. – ISSN 1364-5498 (Electronic) 1359-6640 (Linking)

[Oliver and Rawlinson 1955] OLIVER, I. T. ; RAWLINSON, W. A.: The absorption spectra of porphyrin alpha and derivatives. In: *Biochem J* 61 (1955), Nr. 4, pp. 641–6. – URL https://www.ncbi.nlm.nih.gov/pubmed/13276349. – ISSN 0264-6021 (Print) 0264-6021 (Linking)

[Olyarnik and Stachowicz 2012] OLYARNIK, S. V. ; STACHOWICZ, J. J.: Multi-year study of the effects of Ulva sp. blooms on eelgrass Zostera marina. In: *Marine Ecology Progress Series* 468 (2012), pp. 107–117. – URL https://www.int-res.com/articles/meps2012/468/m468p107.pdf. – ISSN 0171-8630 1616-1599

[Omar et al. 2018] OMAR, A. H. ; TZORTZIOU, M. ; CODDINGTON, O. ; REMERD, L. A.: Plankton Aerosol, Cloud, Ocean Ecosystem mission: atmosphere measurements for air quality applications. In: *Journal of Applied Remote Sensing* 12 (2018), Nr. 4. – URL https://www.spiedigitallibrary.org/journals/Journal-of-Applied-Remote-Sensing/volume-12/issue-4/042608/Plankton-Aerosol-Cloud-ocean-Ecosystem-mission--atmosphere-measurements-for/10.1117/1.JRS.12.042608.pdf. – ISSN 1931-3195

[Onsager 1931] ONSAGER, L.: Reciprocal Relations in Irreversible Processes. I. In: *Physical Review* 37 (1931), Nr. 4, pp. 405–426. – URL https://doi.org/10.1103/PhysRev.37.405. – ISSN 0031-899X

[Oppenheimer 1941] OPPENHEIMER, J. R.: Internal Conversion in Photosynthesis. In: *Minutes of the American Physical Society Meeting, June 18-20, 1941* 60 (1941), Nr. 2, pp. 158–165. – ISSN 0031-899X

[Pal 1960] PAL, G.: *The Time Machine*. 1960 1960. – URL https://www.imdb.com/title/tt0054387/?ref_=fn_al_tt_2

[Papageorgiou and Govindjee 2004] PAPAGEORGIOU, G. C. ; GOVINDJEE: *Chlorophyll a Fluorescence. A Signature of Photosynthesis*. Springer Netherlands, 2004 (Advances in Photosynthesis and Respiration). – p. 820. – URL https://www.springer.com/gp/book/9781402032172. – ISBN 978-1-4020-3217-2 978-1-4020-3218-9

[Park et al. 2019] PARK, M.-G. ; AHMAD, H. A. ; KAHDAR, K.: Combining Korean Traditional Patterns and Batik Cirebon Banji Pattern in Daily Hanbok. In: *Journal of Visual Art and Design* 11 (2019), Nr. 1, pp. 59–70. – URL http://journals.itb.ac.id/index.php/jvad/article/download/8746/4235. – ISSN 23375795 23385480

[Paschen 1897] PASCHEN, F.: Ueber Gesetzmaessigkeiten in den Spectren fester Koerper. In: *Annalen der Physik und Chemie* 296 (1897), Nr. 4, pp. 662–723. – URL https://onlinelibrary.wiley.com/doi/pdf/10.1002/andp.18972960408. – ISSN 00033804 15213889

[Patzelt et al. 2002] PATZELT, H. ; SIMON, B. ; TERLAAK, A. ; KESSLER, B. ; KUHNE, R. ; SCHMIEDER, P. ; OESTERHELT, D. ; OSCHKINAT, H.: The structures of the active center in dark-adapted bacteriorhodopsin by solution-state NMR spectroscopy. In: *Proceedings of the National Academy of Sciences of the United States of America* 99 (2002), Nr. 15, pp. 9765–9770. – URL https://www.ncbi.nlm.nih.gov/pmc/articles/PMC125008/pdf/pq1502009765.pdf. – ISSN 0027-8424

[Pauli et al. 1997] PAULI, W. ; ENZ, C. P. ; GLAUS, B.: *Wolfgang Pauli und sein Wirken an der ETH Zuerich: aus den Dienstakten der Eidgenoessischen Technischen Hochschule*. Zürich: vdf Hochschulverlag an der ETH, 1997. – ISBN 3-7281-2317-X

[Pauling 1992] PAULING, L.: The Nature of the Chemical-Bond – 1992. In: *Journal of Chemical Education* 69 (1992), Nr. 7, pp. 519–521. – URL https://pubs.acs.org/doi/abs/10.1021/ed069p519. – ISSN 0021-9584

[Pauling and Wheland 1933] PAULING, L. ; WHELAND, G. W.: The nature of the chemical bond. V. The quantum-mechanical calculation of the resonance energy of benzene and naphthalene and the hydrocarbon free radicals'. In: *Journal of Chemical Physics* 1 (1933), Nr. 6, pp. 362–374. – URL https://aip.scitation.org/doi/pdf/10.1063/1.1749304. – ISSN 0021-9606

[Pauling and Wheland 1934] PAULING, L. ; WHELAND, G. W.: The nature of the chemical bond. V The quantum-mechanical calculation of the resonance energy of benzene and naphthalene and the hydrocarbon free radicals (vol 1, pg 362, 1933). In: *Journal of Chemical Physics* 2 (1934), Nr. 8, p. 1. – URL https://aip.scitation.org/doi/pdf/10.1063/1.1749514. – ISSN 0021-9606

[Pavlenko 2003] PAVLENKO, N.: Proton wires in an electric field: the impact of the Grotthuss mechanism on charge translocation. In: *Journal of Physics-Condensed Matter* 15 (2003), Nr. 2, pp. 291–307. – URL https://iopscience.iop.org/article/10.1088/0953-8984/15/2/329/pdf. – ISSN 0953-8984

[Pawlowski and Zielenkiewicz 2013] PAWLOWSKI, P. H. ; ZIELENKIEWICZ, P.: The quantum Casimir effect may be a universal force organizing the bilayer structure of the cell membrane. In: *J Membr Biol* 246 (2013), Nr. 5, pp. 383–9. – URL https://www.ncbi.nlm.nih.gov/pubmed/23612889. – ISSN 1432-1424 (Electronic) 0022-2631 (Linking)

[Pelmenschikov et al. 2017] PELMENSCHIKOV, V. ; BIRRELL, J. A. ; PHAM, C. C. ; MISHRA, N. ; WANG, H. ; SOMMER, C. ; REIJERSE, E. ; RICHERS, C. P. ; TAMASAKU, K. ; YODA, Y. ; RAUCHFUSS, T. B. ; LUBITZ, W. ; CRAMER, S. P.: Reaction Coordinate Leading to H2 Production in [FeFe]-Hydrogenase Identified by Nuclear Resonance Vibrational Spectroscopy and Density Functional Theory. In: *J Am Chem Soc* 139 (2017), Nr. 46, pp. 16894–16902. – URL https://www.ncbi.nlm.nih.gov/pubmed/29054130https://www.ncbi.nlm.nih.gov/pmc/articles/PMC5699932/pdf/nihms914692.pdf. – ISSN 1520-5126 (Electronic) 0002-7863 (Linking)

[Perlmutter et al. 1999] PERLMUTTER, S. ; ALDERING, G. ; GOLDHABER, G. ; KNOP, R. A. ; NUGENT, P. ; CASTRO, P. G. ; DEUSTUA, S. ; FABBRO, S. ; GOOBAR, A. ; GROOM, D. E. ; HOOK, I. M. ; KIM, A. G. ; KIM, M. Y. ; LEE, J. C. ; NUNES, N. J. ; PAIN, R. ; PENNYPACKER, C. R. ; QUIMBY, R. ; LIDMAN, C. ; ELLIS, R. S. ; IRWIN, M. ; MCMAHON, R. G. ; RUIZ-LAPUENTE, P. ; WALTON, N. ; SCHAEFER, B. ; BOYLE, B. J. ; FILIPPENKO, A. V. ; MATHESON, T. ; FRUCHTER, A. S. ; PANAGIA, N. ; NEWBERG, H. J. M. ; COUCH, W. J. ; PROJECT, T. S. C.: Measurements of Omega and Lambda from 42 High-Redshift Supernovae. In: *The Astrophysical Journal* 517 (1999), Nr. 2, pp. 565–586. – URL https://iopscience.iop.org/article/10.1086/307221/pdf. – ISSN 0004-637X 1538-4357

[Perrin 1948] PERRIN, F. H.: Whose absorption law? In: *J Opt Soc Am* 38 (1948), Nr. 1, pp. 72–4. – URL https://www.osapublishing.org/josa/abstract.cfm?uri=josa-38-1-72. – ISSN 0030-3941 (Print) 0030-3941 (Linking)

[Peter 2019] PETER, L.: Photoelectrochemical Kinetics: Hydrogen Evolution on p-Type Semiconductors. In: *Journal of The Electrochemical Society* 166 (2019), Nr. 5, pp. H3125–H3132. – URL https://iopscience.iop.org/article/10.1149/2.0231905jes/pdf. – ISSN 0013-4651 1945-7111

[phoenix 2019] PHOENIX: *Pressekonferenz zu Wasserstoff und Energiewende mit Altmaier, Scheuer und Müller am 05.11.19.* 2019. – URL https://www.youtube.com/watch?time_continue=3&v=75H1CndhlAo

[Pieper 2010] PIEPER, J.: Time-resolved quasielastic neutron scattering studies of native photosystems. In: *Biochimica Et Biophysica Acta-Proteins and Proteomics* 1804 (2010), Nr. 1, pp. 83–88. – ISSN 1570-9639

[Pieper et al. 2009] PIEPER, J. ; BUCHSTEINER, A. ; DENCHER, N. A. ; LECHNER, R. E. ; HAUSS, T.: Light-induced Modulation of Protein Dynamics During the Photocycle of Bacteriorhodopsin. In: *Photochemistry and Photobiology* 85 (2009), Nr. 2, pp. 590–597. – URL https://onlinelibrary.wiley.com/doi/full/10.1111/j.1751-1097.2008.00501.x. – ISSN 0031-8655

[Pieper and Renger 2009] PIEPER, J. ; RENGER, G.: Protein dynamics investigated by neutron scattering. In: *Photosynth Res* 102 (2009), Nr. 2-3, pp. 281–93. – URL https://link.springer.com/content/pdf/10.1007%2Fs11120-009-9480-9.pdf. – ISSN 1573-5079 (Electronic) 0166-8595 (Linking)

[Pikuta et al. 2003] PIKUTA, E. V. ; ROZANOV, A. Y. ; DETKOVA, E. N. ; BEJ, A. K. ; PAEPE, R. R. ; MARSIC, D. ; HOOVER, R. B.: *Anaerobic halo- alkaliphilic bacterial community of athalassic, hypersaline Mono Lake and Owens Lake in California*. 2003. – URL https://doi.org/10.1117/12.463322

[Piotrowski et al. 2006] PIOTROWSKI, E. W. ; SCHROEDER, M. ; ZAMBRZYCKA, A.: Quantum extension of European option pricing based on the Ornstein-Uhlenbeck process. In: *Physica a-Statistical Mechanics and Its Applications* 368 (2006), Nr. 1, pp. 176–182. – URL https://www.sciencedirect.com/science/article/pii/S0378437105012756. – ISSN 0378-4371

[Piotrowski and Sladkowski 2005] PIOTROWSKI, E. W. ; SLADKOWSKI, J.: Quantum diffusion of prices and profits. In: *Physica a-Statistical Mechanics and Its Applications* 345 (2005), Nr. 1-2, pp. 185–195. – URL https://www.sciencedirect.com/science/article/pii/S0378437104009793. – ISSN 0378-4371

[Piotrowski and Sladkowski 2007] PIOTROWSKI, E. W. ; SLADKOWSKI, J.: Geometry of financial markets – Towards information theory model of markets. In: *Physica a-Statistical Mechanics and Its Applications* 382 (2007), Nr. 1, pp. 228–234. – URL https://www.sciencedirect.com/science/article/pii/S0378437107001549. – ISSN 0378-4371

[Piotrowski et al. 2010] PIOTROWSKI, E. W. ; SLADKOWSKI, J. ; SYSKA, J.: Subjective modelling of supply and demand-the minimum of Fisher information solution. In: *Physica a-Statistical Mechanics and Its Applications* 389 (2010), Nr. 21, pp. 4904–4912. – URL https://www.sciencedirect.com/science/article/pii/S0378437110006163. – ISSN 0378-4371

[Planck 1900a] PLANCK, M.: Kritik zweier Sätze des Hrn. W. Wien. In: *Annalen Der Physik* 3 (1900), Nr. 12, pp. 764–766. – URL https://onlinelibrary.wiley.com/doi/abs/10.1002/andp.19003081215. – ISSN 0003-3804

[Planck 1900b] PLANCK, M.: Ueber eine Verbesserung der Wienschen Spectralgleichung. In: *Verhandlungen der Deutschen physikalischen Gesellschaft* 2 (1900), Nr. 13

[Planck 1901a] PLANCK, M.: Law of energy distribution in normal spectra. In: *Annalen Der Physik* 4 (1901), Nr. 3, pp. 553–563. – ISSN 0003-3804

[Planck 1901b] PLANCK, M.: Ueber das Gesetz der Energieverteilung im Normalspectrum. In: *Annalen der Physik* 309 (1901), Nr. 3, pp. 553–563. – ISSN 00033804 15213889

[Poincaré 1906] Poincaré, M. H.: Sur la dynamique de l'électron. In: *Rendiconti del Circolo matematico di Palermo* 21 (1906), Nr. 1, pp. 129–175. – URL https://link.springer.com/content/pdf/10.1007/BF03013466.pdf. – ISSN 0009-725X 1973-4409

[Pollard 1927] Pollard, A. F. C.: Notes upon the mechanical design of some instruments shown at the exhibition of the physical and optical societies, 1927. In: *Journal of Scientific Instruments* 4 (1927), Nr. 6, pp. 184–190. – URL https://doi.org/10.1088/0950-7671/4/6/302. – ISSN 0950-7671

[Pomes and Roux 1996a] Pomes, R. ; Roux, B.: Structure and dynamics of a proton wire: a theoretical study of H+ translocation along the single-file water chain in the gramicidin A channel. In: *Biophys J* 71 (1996), Nr. 1, pp. 19–39. – URL https://www.ncbi.nlm.nih.gov/pubmed/8804586. – ISSN 0006-3495 (Print) 0006-3495 (Linking)

[Pomes and Roux 1996b] Pomes, R. ; Roux, B.: Theoretical study of H+ translocation along a model proton wire. In: *Journal of Physical Chemistry* 100 (1996), Nr. 7, pp. 2519–2527. – URL https://pubs.acs.org/doi/pdfplus/10.1021/jp9525752. – ISSN 0022-3654

[Pomes and Roux 1996c] Pomes, R. ; Roux, B.: Theoretical study of the structure and dynamics of biological proton wires. In: *Biophysical Journal* 70 (1996), Nr. 2, pp. TUPM4–TUPM4. – ISSN 0006-3495

[Pomès 1999] Pomès, R.: Theoretical Studies of the Grotthuss Mechanism in Biological Proton Wires. In: *Israel Journal of Chemistry* 39 (1999), Nr. 3-4, pp. 387–395. – URL https://onlinelibrary.wiley.com/doi/abs/10.1002/ijch.199900044. – ISSN 00212148

[Popper 1956] Popper, K. R.: The Arrow of Time. In: *Nature* 177 (1956), Nr. 4507, pp. 538–538. – ISSN 0028-0836 1476-4687

[Porschke 1996] Porschke, D.: Electrostatics and electrodynamics of bacteriorhodopsin. In: *Biophys J* 71 (1996), Nr. 6, pp. 3381–91. – URL https://www.ncbi.nlm.nih.gov/pubmed/8968607. – ISSN 0006-3495 (Print) 0006-3495 (Linking)

[Poudel and Dunn 2017] Poudel, M. ; Dunn, B.: Greenhouse Carbon Dioxide Supplementation. The Oklahoma Cooperative Extension Service, 2017. – Report. – URL https://extension.okstate.edu/fact-sheets/greenhouse-carbon-dioxide-supplementation.html

[Pozar 2011] Pozar, D. M.: *Microwave Engineering*. 2011. – ISBN 978-0-470-63155-3

[Pradhan and Khare 1976] Pradhan, T. ; Khare, A.: Feynman Diagram Method for Atomic-Collisions. In: *Pramana* 6 (1976), Nr. 5, pp. 312–322. – URL https://link.springer.com/content/pdf/10.1007/BF02872210.pdf. – ISSN 0304-4289

[Prochaska et al. 2015] Prochaska, J. X. ; O'Meara, J. M. ; Fumagalli, M. ; Bernstein, R. A. ; Burles, S. M.: The Keck + Magellan Survey for Lyman Limit Absorption. Iii. Sample Definition and Column Density Measurements. In: *The Astrophysical Journal Supplement Series* 221 (2015), Nr. 1. – ISSN 1538-4365

[PubChem 2019] PubChem: *Porphyrin;zinc*. 2019. – URL https://pubchem.ncbi.nlm.nih.gov/compound/66593577#section=2D-Structure

[Purchase and de Groot 2015] Purchase, R. L. ; Groot, H. J. de: Biosolar cells: global artificial photosynthesis needs responsive matrices with quantum coherent kinetic control for high yield. In: *Interface Focus* 5 (2015), Nr. 3, p. 20150014. – URL https://www.ncbi.nlm.nih.gov/pmc/articles/PMC4410567/pdf/rsfs20150014.pdf. – ISSN 2042-8898 (Print) 2042-8898 (Linking)

[Rabinowitch 1944] Rabinowitch, E.: Spectra of porphyrins and chlorophyll. In: *Reviews of Modern Physics* 16 (1944), Nr. 3/4, pp. 0226–0235. – URL https://journals.aps.org/rmp/pdf/10.1103/RevModPhys.16.226. – ISSN 0034-6861

[Rabinowitch 1961] Rabinowitch, E.: Photochemical Utilization of Light Energy. In: *Proceedings of the National Academy of Sciences* 47 (1961), Nr. 8, pp. 1296–1303. – URL https://www.pnas.org/content/pnas/47/8/1296.full.pdf. – ISSN 0027-8424 1091-6490

[Radepont et al. 2018] RADEPONT, M. ; LEMASSON, Q. ; PICHON, L. ; MOIGNARD, B. ; PACHECO, C.: Towards a sharpest interpretation of analytical results by assessing the uncertainty of PIXE/RBS data at the AGLAE facility. In: *Measurement* 114 (2018), pp. 501–507. – URL https://www.sciencedirect.com/science/article/pii/S0263224116303645. – ISSN 02632241

[Radhakrishnan and Murugesan 2014] RADHAKRISHNAN, A. ; MURUGESAN, V.: Calculation of The Extinction Cross Section and Lifetime of A Gold Nanoparticle using FDTD Simulations. In: *Light and Its Interactions with Matter* 1620 (2014), pp. 52–57. – URL https://aip.scitation.org/doi/abs/10.1063/1.4898219. – ISSN 0094-243x

[Rashed 1990] RASHED, R.: A Pioneer in Anaclastics: Ibn Sahl on Burning Mirrors and Lenses. In: *Isis* 81 (1990), Nr. 3, pp. 464–491. – URL https://www.journals.uchicago.edu/doi/pdfplus/10.1086/355456. – ISSN 0021-1753 1545-6994

[Rayleigh 1900] RAYLEIGH, L.: Remarks upon the law of complete radiation. In: *Philosophical Magazine* 49 (1900), Nr. 296-01, pp. 539–540. – ISSN 1478-6435

[Rayleigh 2009a] RAYLEIGH, L.: XVIII. On the passage of electric waves through tubes, or the vibrations of dielectric cylinders. In: *The London, Edinburgh, and Dublin Philosophical Magazine and Journal of Science* 43 (2009), Nr. 261, pp. 125–132. – URL https://doi.org/10.1080/14786449708620969. – ISSN 1941-5982 1941-5990

[Rayleigh 2009b] RAYLEIGH, L.: XXVIII. On the propagation of electric waves along cylindrical conductors of any section. In: *The London, Edinburgh, and Dublin Philosophical Magazine and Journal of Science* 44 (2009), Nr. 267, pp. 199–204. – URL https://doi.org/10.1080/14786449708621052. – ISSN 1941-5982 1941-5990

[Rayleigh 2009c] RAYLEIGH, L.: XXXVII. On the passage of waves through apertures in Plane screens, and allied problems. In: *The London, Edinburgh, and Dublin Philosophical Magazine and Journal of Science* 43 (2009), Nr. 263, pp. 259–272. – URL https://doi.org/10.1080/14786449708620990. – ISSN 1941-5982 1941-5990

[Rebhan 2003] REBHAN, A.: *Einführung in die Quantenelektrodynamik.* 3 Nov 2003 2003

[Regener 1933] REGENER, E.: The energy flow of the cosmic radiation. In: *Zeitschrift Fur Physik* 80 (1933), Nr. 9-10, pp. 666–669. – URL https://doi.org/10.1007/BF01335703. – ISSN 0044-3328

[Remler and Madden 1990] REMLER, D. K. ; MADDEN, P. A.: Molecular-Dynamics without Effective Potentials Via the Car-Parrinello Approach. In: *Molecular Physics* 70 (1990), Nr. 6, pp. 921–966. – ISSN 0026-8976

[Renger 2007] RENGER, G.: *Primary Processes of Photosynthesis, Part 1.* Cambridge: Royal Society of Chemistry, 2007 (Comprehensive Series in Photochemistry & Photobiology). – URL https://doi.org/10.1039/9781847558152. – ISBN 978-0-85404-369-9

[Reynolds and Lund 1988] REYNOLDS, C. S. ; LUND, J. W. G.: The Phytoplankton of an Enriched, Soft-Water Lake Subject to Intermittent Hydraulic Flushing (Grasmere, English Lake District). In: *Freshwater Biology* 19 (1988), Nr. 3, pp. 379–404. – URL https://onlinelibrary.wiley.com/doi/pdf/10.1111/j.1365-2427.1988.tb00359.x. – ISSN 0046-5070

[Riaz et al. 2017] RIAZ, M. ; EARLES, S. K. ; KADHIM, A. ; AZZAHRANI, A.: Computer analysis of microcrystalline silicon hetero-junction solar cell with lumerical FDTD/DEVICE. In: *International Journal of Computational Materials Science and Engineering* 06 (2017), Nr. 03. – ISSN 2047-6841 2047-685X

[Ridgway 1973] RIDGWAY, D.: Interview with Melvin Calvin. In: *Journal of Chemical Education* 50 (1973), Nr. 12. – URL https://pubs.acs.org/doi/pdf/10.1021/ed050p811. – ISSN 0021-9584 1938-1328

[Romer 1676] ROMER, O.: Demonstration touchant le mouvement de la lumiere trouvé par M. Romer de l'Academie royale des sciences. Le Journal des Scavans. Paris 1676. In: *Le Journal des Scavans* (1676), pp. 233–236

[Rothkopf 2014] ROTHKOPF, D.: A World disrupted: The Leading Global Thinkers of 2014. In: *Foreign Policy* 2014 (2014), Nr. 11/12

[Roy 2018] ROY, A.: Arthur Holly Compton (1892-1962). In: *Resonance-Journal of Science Education* 23 (2018), Nr. 9, pp. 943–948. – URL https://link.springer.com/content/pdf/10.1007%2Fs12045-018-0700-5.pdf. – ISSN 0971-8044

[Royant et al. 2000] ROYANT, A. ; EDMAN, K. ; URSBY, T. ; PEBAY-PEYROULA, E. ; LANDAU, E. M. ; NEUTZE, R.: Helix deformation is coupled to vectorial proton transport in the photocycle of bacteriorhodopsin. In: *Nature* 406 (2000), Nr. 6796, pp. 645–8. – URL https://www.nature.com/articles/35020599.pdf. – ISSN 0028-0836 (Print) 0028-0836 (Linking)

[Rumsfeld 2002] RUMSFELD, D. E.: *DoD News Briefing – Secretary Rumsfeld and Gen. Myers*. 12 Feb 2002 2002. – URL https://archive.defense.gov/Transcripts/Transcript.aspx?TranscriptID=2636

[Runge 1895] RUNGE, C.: Ueber die numerische Aufloesung von Differentialgleichungen. In: *Mathematische Annalen* 46 (1895), pp. 167–178. – URL https://gdz.sub.uni-goettingen.de/id/PPN235181684_0046

[Russell and Saunders 1925] RUSSELL, H. N. ; SAUNDERS, F. A.: New Regularities in the Spectra of the Alkaline Earths. In: *The Astrophysical Journal* 61 (1925). – ISSN 0004-637X 1538-4357

[Ryu et al. 2006] RYU, J. h. ; ZIERENBERG, R. A. ; DAHLGREN, R. A. ; GAO, S.: Sulfur biogeochemistry and isotopic fractionation in shallow groundwater and sediments of Owens Dry Lake, California. In: *Chemical Geology* 229 (2006), Nr. 4, pp. 257–272. – URL https://www.sciencedirect.com/science/article/pii/S000925410500450X. – ISSN 00092541

[Ryu et al. 2010] RYU, W. ; BAI, S. J. ; PARK, J. S. ; HUANG, Z. ; MOSELEY, J. ; FABIAN, T. ; FASCHING, R. J. ; GROSSMAN, A. R. ; PRINZ, F. B.: Direct extraction of photosynthetic electrons from single algal cells by nanoprobing system. In: *Nano Lett* 10 (2010), Nr. 4, pp. 1137–43. – URL https://www.ncbi.nlm.nih.gov/pubmed/20201533https://pubs.acs.org/doi/pdf/10.1021/nl903141j. – ISSN 1530-6992 (Electronic) 1530-6984 (Linking)

[Sakharov 1975] SAKHAROV, A.: *Nobel Lecture – Peace, Progress, Human Rights*. 11 December 1975 1975. – URL https://www.nobelprize.org/nobel_prizes/peace/laureates/1975/sakharov-lecture.html

[Salpeter 1952a] SALPETER, E. E.: Nuclear Reactions in the Stars .1. Proton-Proton Chain. In: *Physical Review* 88 (1952), Nr. 3, pp. 547–553. – ISSN 0031-899x

[Salpeter 1952b] SALPETER, E. E.: The Reaction Rate of the Proton-Proton Chain. In: *Astrophysical Journal* 116 (1952), Nr. 3, pp. 649–650. – ISSN 0004-637x

[Samgin and Ezin 2014] SAMGIN, A. L. ; EZIN, A. N.: Room-temperature proton-hopping transport in rutile-type oxides in the field of resonant laser radiation. In: *Technical Physics Letters* 40 (2014), Nr. 3, pp. 252–255. – ISSN 1063-7850

[Samuelson 2015] SAMUELSON, P. A.: *Rational Theory of Warrant Pricing*. Book Section Chapter 11, pp. 195–232. In: *Henry P. McKean Jr. Selecta*, Birkhäuser, 2015. – URL https://doi.org/10.1007/978-3-319-22237-0_11https://link.springer.com/content/pdf/10.1007%2F978-3-319-22237-0_11.pdf. – ISBN 978-3-319-22236-3 978-3-319-22237-0

[Santos et al. 2016] SANTOS, A. ; INFANTE, V. ; BAMSEY, M. ; SCHUBERT, D. ; IEEE: *A case study in the application of failure analysis techniques to Antarctic Systems: EDEN ISS*. New York: Ieee, 2016 (2016 Ieee International Symposium on Systems Engineering). – pp. 258–264. – ISBN 978-1-5090-0793-6

[Schiff 1864] SCHIFF, H.: Mittheilungen aus dem Universitätslaboratorium in Pisa: Eine neue Reihe organischer Basen. In: *European Journal of Organic Chemistry* 131 (1864), Nr. 1

[Schilcher 2019] SCHILCHER, K.: *Quantenelektrodynamik Kompakt*. Berlin, Boston: De Gruyter, 2019. – URL QuantenelektrodynamikKompakt. – ISBN 978-3-11-048858-6

[Schinckus 2014] SCHINCKUS, C.: *A Methodological Call for a Quantum Econophysics*. Book Section Chapter 28, pp. 308–316. In: *Quantum Interaction*, Springer, 2014 (Lecture Notes in Computer

Science). – URL https://link.springer.com/content/pdf/10.1007%2F978-3-642-54943-4_28.pdf. – ISBN 978-3-642-54942-7 978-3-642-54943-4

[Schneider 1989] SCHNEIDER, J. H.: How a Rectangular Potential in Schroedingers Equation Could Explain Some Experimental Results on Cold Nuclear Fusion. In: *Fusion Technology* 16 (1989), Nr. 3, pp. 377–378. – ISSN 0748-1896

[Schramski et al. 2015] SCHRAMSKI, J. R. ; GATTIE, D. K. ; BROWN, J. H.: Human domination of the biosphere: Rapid discharge of the earth-space battery foretells the future of humankind. In: *Proc Natl Acad Sci U S A* 112 (2015), Nr. 31, pp. 9511–7. – URL https://www.ncbi.nlm.nih.gov/pubmed/26178196. – ISSN 1091-6490 (Electronic) 0027-8424 (Linking)

[Schrodinger and Penrose 2012] SCHRODINGER, E. ; PENROSE, R.: *What is Life?* Cambridge University Press, 2012. – URL https://doi.org/10.1017/CBO9781107295629. – ISBN 9781107295629

[Schroer 2011] SCHROER, B.: Pascual Jordan's legacy and the ongoing research in quantum field theory. In: *The European Physical Journal H* 35 (2011), Nr. 4, pp. 377–434. – URL https://doi.org/10.1140/epjh/e2011-10015-8. – ISSN 2102-6459 2102-6467

[Schubert 2017] SCHUBERT, D.: Greenhouse production analysis of early mission scenarios for Moon and Mars habitats. In: *Open Agriculture* 2 (2017), Nr. 1, pp. 91–115. – URL https://www.degruyter.com/downloadpdf/j/opag.2017.2.issue-1/opag-2017-0010/opag-2017-0010.pdf. – ISSN 2391-9531

[Schubring 1996] SCHUBRING, G.: *The Cooperation between Hermann and Robert Grassmann on the Foundations of Mathematics*. Dordrecht: Springer, 1996. – ISBN 978-94-015-8753-2

[Schwartzstein 2016] SCHWARTZSTEIN, P.: Takepart features: Farming the Sahara Its population soaring, Egypt is facing a food-supply crisis. Can the government make a desert bloom? In: *Takepart* (2016), 8 January 2016. – URL http://www.takepart.com/feature/2016/01/08/desert-farming-egypt/

[Segal and Segal 1998] SEGAL, W. ; SEGAL, I. E.: The Black-Scholes pricing formula in the quantum context. In: *Proceedings of the National Academy of Sciences* 95 (1998), Nr. 7, pp. 4072–4075. – URL https://www.pnas.org/content/pnas/95/7/4072.full.pdf. – ISSN 0027-8424 1091-6490

[Shaltout 1998] SHALTOUT, M.: Solar hydrogen from Lake Nasser for 21st century in Egypt. In: *International Journal of Hydrogen Energy* 23 (1998), Nr. 4, pp. 233–238. – URL https://www.sciencedirect.com/science/article/pii/S0360319997000670. – ISSN 03603199

[Shapiro and Breit 1959] SHAPIRO, J. ; BREIT, G.: Metastability of 2s States of Hydrogenic Atoms. In: *Physical Review* 113 (1959), Nr. 1, pp. 179–181. – URL https://doi.org/10.1103/PhysRev.113.179. – ISSN 0031-899X

[Sharghi 2015] SHARGHI, K.: *One Year on Earth – Seen From 1 Million Miles*. 20 July 2016 2015. – URL https://svs.gsfc.nasa.gov/12312

[Shcheglov 2018] SHCHEGLOV, D. A.: The Mediterranean coast of Africa in Prolemy's geography and in the stadiasmus of the Great Sea. In: *Schole-Filosofskoe Antikovedenie I Klassicheskaya Traditsiya-Schole-Ancient Philosophy and the Classical Tradition* 12 (2018), Nr. 2, pp. 453–479. – ISSN 1995-4328

[Sheridan 2007] SHERIDAN, C.: Big oil's biomass play. In: *Nat Biotechnol* 25 (2007), Nr. 11, pp. 1201–3. – URL https://www.ncbi.nlm.nih.gov/pubmed/17989660. – ISSN 1087-0156 (Print) 1087-0156 (Linking)

[Sheweka 2012] SHEWEKA, S. M.: *Energy Procedia*. Vol. 18: *New Egypt with a New Hybrid Skin "A New Hybrid Architecture Vision for the Egyptians Development Corridor"*. pp. 449–457. In: *Terragreen 2012: Clean Energy Solutions for Sustainable Environment* Vol. 18. Amsterdam: Elsevier Science Bv, 2012. – URL https://www.sciencedirect.com/science/article/pii/S1876610212008260

[Shije 1999] SHIJE, A.: Statement of Amadeo Shije, Governor, Zia Pueblo. In: *Public Hearings on Official Insignia of Native American Tribes* (1999)

[Shvartsev 2018] SHVARTSEV, S. L.: The creative function of water in the formation of the world around us. In: *Geodynamics & Tectonophysics* 9 (2018), Nr. 4, pp. 1275–1291. – URL https://doi.org/10.5800/GT-2018-9-4-0395. – ISSN 2078-502X

[Sicotte et al. 2009] SICOTTE, R. ; VIZCARRA, C. ; WANDSCHNEIDER, K.: Military conquest and sovereign debt: Chile, Peru and the London bond market, 1876–1890. In: *Cliometrica* 4 (2009), Nr. 3, pp. 293–319. – URL https://link.springer.com/content/pdf/10.1007/s11698-009-0047-y.pdf. – ISSN 1863-2505 1863-2513

[Siegel 2017] SIEGEL, E.: The Sun's Energy Doesn't Come From Fusing Hydrogen Into Helium (Mostly). In: *Forbes* (2017)

[Sjoholm et al. 2012] SJOHOLM, K. H. ; RASMUSSEN, M. ; MINTEER, S. D.: Bio-Solar Cells Incorporating Catalase for Stabilization of Thylakoid Bioelectrodes during Direct Photoelectrocatalysis. In: *ECS Electrochemistry Letters* 1 (2012), Nr. 5, pp. G7–G9. – ISSN 2162-8726 2162-8734

[Slater 1925a] SLATER, J. C.: Methods for determining transition probabilities from line absorption. In: *Physical Review* 25 (1925), Nr. 6, pp. 783–790. – ISSN 0031-899X

[Slater 1925b] SLATER, J. C.: The nature of radiation. In: *Nature* 116 (1925), pp. 278–278. – ISSN 0028-0836

[Slater 1931] SLATER, J. C.: Molecular Energy Levels and Valence Bonds. In: *Physical Review* 38 (1931), Nr. 6, pp. 1109–1144. – URL https://journals.aps.org/pr/pdf/10.1103/PhysRev.38.1109. – ISSN 0031-899X

[Smith et al. 2009] SMITH, A. M. ; COUPLAND, G. ; DOLAN, L. ; HARBERD, N. ; JONES, J. ; MARTIN, C. ; SABLOWSKI, R. ; AMEY, A.: *Metabolism*. Book Section 4. In: *Plant Biology*, Garland Science, 2009. – URL http://www.garlandscience.com/res/pdf/9780815340256_ch04.pdf. – ISBN 9780815340256

[Smith 1915] SMITH, G. M.: The Development of Botanical Microtechnique. In: *Transactions of the American Microscopical Society* 34 (1915), Nr. 2. – ISSN 00030023

[Sode and Voth 2014] SODE, O. ; VOTH, G. A.: Electron transfer activation of a second water channel for proton transport in FeFe-hydrogenase. In: *Journal of Chemical Physics* 141 (2014), Nr. 22, p. 9. – URL https://aip.scitation.org/doi/pdf/10.1063/1.4902236. – ISSN 0021-9606

[Solée et al. 2013] SOLÉE, R. V. ; VALVERDE, S. ; CASALS, M. R. ; KAUFFMAN, S. A. ; FARMER, D. ; ELDREDGE, N.: The evolutionary ecology of technological innovations. In: *Complexity* 18 (2013), Nr. 4, pp. 15–27. – URL https://onlinelibrary.wiley.com/doi/full/10.1002/cplx.21436. – ISSN 10762787

[Solonec 2015] SOLONEC, T.: *Why saying 'Aborigine' isn't OK: 8 facts about Indigenous people in Australia* . 09 august 2015 2015. – URL https://www.amnesty.org/en/latest/campaigns/2015/08/why-saying-aborigine-isnt-ok-8-facts-about-indigenous-people-in-australia/

[Sommerfeld 2010] SOMMERFELD, C.: ... nach Jahr und Tag – Beobachtungen an den Trundholm-Scheiben. In: *Praehistorische Zeitschrift* 85 (2010), Nr. 2. – ISSN 0079-4848 1613-0804

[Soret 1883] SORET, J.-L.: Analyse spectrale: Sur le spectre d'absorption du sang dans la partie violette et ultra-violette. In: *Comptes rendus de l'Acadêmie des sciences (in French)* 97 (1883), pp. 1269–1270. – URL https://gallica.bnf.fr/ark:/12148/bpt6k3053w/f1269.table

[Soros 1994] SOROS, G.: *The Alchemy of Finance – Reading the Mind of the Market*. New York, Chichester, Brisbane, Toronto, Singapore: John Wiley & Sons, Inc, 1994. – ISBN 978-0-471-04313-3

[Spahr et al. 2010] SPAHR, E. J. ; WEN, L. ; STAVOLA, M. ; BOATNER, L. A. ; FELDMAN, L. C. ; TOLK, N. H. ; LUPKE, G.: Giant enhancement of hydrogen transport in rutile TiO2 at low temperatures. In: *Phys Rev Lett* 104 (2010), Nr. 20, p. 205901. – URL https://journals.aps.org/prl/pdf/10.1103/PhysRevLett.104.205901. – ISSN 1079-7114 (Electronic) 0031-9007 (Linking)

[Spano 2009] SPANO, F. C.: Analysis of the UV/Vis and CD spectral line shapes of carotenoid assemblies: spectral signatures of chiral H-aggregates. In: *J Am Chem Soc* 131 (2009), Nr. 12, pp. 4267–78. – URL http://pubs.acs.org/doi/pdfplus/10.1021/ja806853v. – ISSN 1520-5126 (Electronic) 0002-7863 (Linking)

[Sparavigna 2012] SPARAVIGNA, A. C.: Ancient bronze disks, decorations and calendars. In: *ArXiv:1203.2512* (2012). – URL https://arxiv.org/abs/1203.2512

[Sparnaay 1957] SPARNAAY, M. J.: Attractive Forces between Flat Plates. In: *Nature* 180 (1957), Nr. 4581, p. 334–335

[Sprenkle 1961] SPRENKLE, C. M.: Warrant prices as indicators of expectations and preferences. In: *Yale Economic Essays* 1 (1961), Nr. 2. – ISSN 0044-006X

[Stefan 1879] STEFAN, J.: Ueber die Beziehung zwischen der Waermestrahlung und der Temperatur. In: *Sitzungsberichte der mathematisch-naturwissenschaftlichen Classe der kaiserlichen Akademie der Wissenschaften in Wien* 79 (1879), pp. 391–428. – URL http://www.worldcat.org/title/uber-die-beziehung-zwischen-der-warmestrahlung-und-der-temperatur/oclc/934240629

[Stelmaszczuk-Górska et al. 2015] STELMASZCZUK-GÓRSKA, M. ; RODRIGUEZ-VEIGA, P. ; ACKERMANN, N. ; THIEL, C. ; BALZTER, H. ; SCHMULLIUS, C.: Non-Parametric Retrieval of Aboveground Biomass in Siberian Boreal Forests with ALOS PALSAR Interferometric Coherence and Backscatter Intensity. In: *Journal of Imaging* 2 (2015), Nr. 1. – ISSN 2313-433X

[Stewart 2007] STEWART, J. J.: Optimization of parameters for semiempirical methods V: modification of NDDO approximations and application to 70 elements. In: *J Mol Model* 13 (2007), Nr. 12, pp. 1173–213. – URL https://www.ncbi.nlm.nih.gov/pmc/articles/PMC2039871/pdf/894_2007_Article_233.pdf. – ISSN 0948-5023 (Electronic) 0948-5023 (Linking)

[Stewart 2009] STEWART, J. J.: Application of the PM6 method to modeling proteins. In: *J Mol Model* 15 (2009), Nr. 7, pp. 765–805. – URL https://link.springer.com/content/pdf/10.1007%2Fs00894-008-0420-y.pdf. – ISSN 0948-5023 (Electronic) 0948-5023 (Linking)

[Stiftung et al. 2015] STIFTUNG, V. ; 42, K. ; ZÜRICH, .: *Daylight Academy*. 2015

[Stingaciu et al. 2016] STINGACIU, L. R. ; O'NEILL, H. ; LIBERTON, M. ; URBAN, V. S. ; PAKRASI, H. B. ; OHL, M.: Revealing the Dynamics of Thylakoid Membranes in Living Cyanobacterial Cells. In: *Scientific Reports* 6 (2016). – URL https://www.nature.com/articles/srep19627.pdf. – ISSN 2045-2322

[Stöhr 1992] STÖHR, J.: *Springer Series in Surface Sciences*. Vol. 25: *NEXAFS Spectroscopy*. Berlin Heidelberg: Springer-Verlag, 1992. – ISBN 978-3-662-02853-7

[Storms 1991] STORMS, E.: Review of Experimental-Observations About the Cold Fusion Effect. In: *Fusion Technology* 20 (1991), Nr. 4, pp. 433–477. – URL https://www.tandfonline.com/doi/pdf/10.13182/FST91-A29661. – ISSN 0748-1896

[Strasser and Butler 1977a] STRASSER, R. J. ; BUTLER, W. L.: Energy transfer and the distribution of excitation energy in the photosynthetic apparatus of spinach chloroplasts. In: *Biochimica et Biophysica Acta (BBA) – Bioenergetics* 460 (1977), Nr. 2, pp. 230–238. – URL https://www.sciencedirect.com/science/article/pii/0005272877902092. – ISSN 00052728

[Strasser and Butler 1977b] STRASSER, R. J. ; BUTLER, W. L.: Fluorescence emission spectra of Photosystem I, Photosystem II and the light-harvesting chlorophyll a/b complex of higher plants. In: *Biochimica et Biophysica Acta (BBA) – Bioenergetics* 462 (1977), Nr. 2, pp. 307–313. – ISSN 00052728

[Strasser and Butler 1977c] STRASSER, R. J. ; BUTLER, W. L.: Fluorescence emission spectra of photosystem I, photosystem II and the light-harvesting chlorophyll a/b complex of higher plants. In: *Biochim Biophys Acta* 462 (1977), Nr. 2, pp. 307–13. – URL https://www.ncbi.nlm.nih.gov/pubmed/588570. – ISSN 0006-3002 (Print) 0006-3002 (Linking)

[Strippoli et al. 2005] STRIPPOLI, P. ; CANAIDER, S. ; NOFERINI, F. ; D'ADDABBO, P. ; VITALE, L. ; FACCHIN, F. ; LENZI, L. ; CASADEI, R. ; CARINCI, P. ; ZANNOTTI, M. ; FRABETTI, F.: Uncertainty principle of genetic information in a living cell. In: *Theor Biol Med Model* 2 (2005), p. 40. – URL https://www.ncbi.nlm.nih.gov/pubmed/16197549. – ISSN 1742-4682 (Electronic) 1742-4682 (Linking)

[Stuhlinger and Ordway 1994] STUHLINGER, E. ; ORDWAY, F.: *Wernher Von Braun Crusader for Space: A Biographical Memoir*. Krieger Publishing Company, 1994. – ISBN 978-0894648427

[Stumm and Sulzberger 1992] STUMM, W. ; SULZBERGER, B.: The cycling of iron in natural environments: Considerations based on laboratory studies of heterogeneous redox processes. In: *Geochimica et Cosmochimica Acta* 56 (1992), Nr. 8, pp. 3233–3257. – URL https://doi.org/10.1016/0016-7037(92)90301-X. – ISSN 00167037

[sugar 2011] SUGAR: Life cycle assessment report from Energy Biosciences Institute highlights challenges for policy makers in the biofuels sector. In: *International Sugar Journal* 113 (2011), Nr. 1347, pp. 158–158. – ISSN 0020-8841

[Sumida et al. 2013] SUMIDA, A. ; MIYAURA, T. ; TORII, H.: Relationships of tree height and diameter at breast height revisited: analyses of stem growth using 20-year data of an even-aged Chamaecyparis obtusa stand. In: *Tree Physiol* 33 (2013), Nr. 1, pp. 106–18. – URL https://www.ncbi.nlm.nih.gov/pmc/articles/PMC3556985/pdf/tps127.pdf. – ISSN 1758-4469 (Electronic) 0829-318X (Linking)

[Sun and Tomanek 1989] SUN, Z. ; TOMANEK, D.: Cold fusion: How close can deuterium atoms come inside palladium? In: *Phys Rev Lett* 63 (1989), Nr. 1, pp. 59–61. – URL https://www.ncbi.nlm.nih.gov/pubmed/10040432. – ISSN 1079-7114 (Electronic) 0031-9007 (Linking)

[Svedberg and Fåhraeus 1926] SVEDBERG, T. ; FÅHRAEUS, R.: A New Method for the Determination of the Molecular Weight of the Proteins. In: *Journal of the American Chemical Society* 48 (1926), Nr. 2, pp. 430–438. – URL https://pubs.acs.org/doi/pdf/10.1021/ja01413a019. – ISSN 0002-7863 1520-5126

[Svedberg and Rinde 1924] SVEDBERG, T. ; RINDE, H.: The Ultra-Centrifuge, a New Instrument for the Determination of Size and Distribution of Size of Particle in Amicroscopic Colloids. In: *Journal of the American Chemical Society* 46 (1924), Nr. 12, pp. 2677–2693. – URL https://pubs.acs.org/doi/pdf/10.1021/ja01677a011. – ISSN 0002-7863 1520-5126

[Swedan 2019] SWEDAN, N.: Photosynthesis as a thermodynamic cycle. In: *Heat and Mass Transfer* (2019). – URL https://doi.org/10.1007/s00231-019-02768-x. – ISSN 0947-7411 1432-1181

[Tainter 1990] TAINTER, J. A.: *Collapse of Complex Societies*. 1st edition. Cambridge, UK: Cambridge University Press, 1990 (New Studies in Archaeology). – p. 264. – URL https://archive.org/details/TheCollapseOfComplexSocieties/mode/2up. – ISBN 978-0521386739

[Tainter and Taylor 2013] TAINTER, J. A. ; TAYLOR, T. G.: Complexity, problem-solving, sustainability and resilience. In: *Building Research & Information* 42 (2013), Nr. 2, pp. 168–181. – ISSN 0961-3218 1466-4321

[Taleyarkhan et al. 2002] TALEYARKHAN, R. P. ; WEST, C. D. ; CHO, J. S. ; LAHEY, JR., R. T. ; NIGMATULIN, R. I. ; BLOCK, R. C.: Evidence for nuclear emissions during acoustic cavitation. In: *Science* 295 (2002), Nr. 5561, pp. 1868–73. – URL https://www.ncbi.nlm.nih.gov/pubmed/11884748. – ISSN 1095-9203 (Electronic) 0036-8075 (Linking)

[Tamm 1932] TAMM, I.: Ueber eine moegliche Art der Elektronenbindung an Kristalloberflaechen. In: *Zeitschrift fuer Physik* 76 (1932), Nr. 11-12, pp. 849–850. – URL https://link.springer.com/content/pdf/10.1007%2FBF01341581.pdf. – ISSN 0044-3328

[Tanabashi 2018] TANABASHI, M.: Review of Particle Physics. In: *Physical Review D* 98 (2018), Nr. 3. – URL https://journals.aps.org/prd/abstract/10.1103/PhysRevD.98.030001. – ISSN 2470-0010 2470-0029

[Tayebi and Zelevinsky 2016] Tayebi, A. ; Zelevinsky, V.: The Holstein polaron problem revisited. In: *Journal of Physics A: Mathematical and Theoretical* 49 (2016), Nr. 25. – URL https://doi.org/10.1088/1751-8113/49/25/255004. – ISSN 1751-8113 1751-8121

[Taylor and Tainter 2016] Taylor, T. G. ; Tainter, J. A.: The Nexus of Population, Energy, Innovation, and Complexity. In: *American Journal of Economics and Sociology* 75 (2016), Nr. 4, pp. 1005–1043. – ISSN 0002-9246

[Teller 1969] Teller, E.: Internal Conversion in Polyatomic Molecules. In: *Israel Journal of Chemistry* 7 (1969), Nr. 2, pp. 227–235. – ISSN 00212148

[testamentum 1994] testamentum, n.: *Novum Testamentum. Graece et Latine*. Stuttgart: Deutsche Bibelgesellschaft, 1994. – ISBN 978-3438054012

[Thiem 1979] Thiem, J.: The Great Library of Alexandria burnt – towards the history of a symbol. In: *Journal of the History of Ideas* 40 (1979), Nr. 4, pp. 507–526. – URL https://www.jstor.org/stable/2709356. – ISSN 0022-5037

[Thompson 1946] Thompson, J. E.: *Calculus for the practical man*. Bombay, Calcutta, Madras: MacMillan and Co, 1946 (Mathematics for Self-Study)

[Thorp and Kassouf 1967] Thorp, E. O. ; Kassouf, S. T.: *Beat the Market: A Scientific Stock Market System*. 1st edition. Random House, 1967. – URL https://www.amazon.com/Beat-Market-Scientific-Stock-System/dp/0394424395. – ISBN 978-0394424392

[Tien and Ottova-Leitmannova 2000] Tien, H. T. ; Ottova-Leitmannova, A.: *Membrane Science and Technology*. Vol. 5: *Chapter 9 Membrane photobiophysics and photobiology*. pp. 493–576. In: *Membrane Biophysics – Planar Lipid Bilayers and Spherical Liposomes* Vol. 5, 2000. – ISBN 9780444829306

[Tirosh 2006] Tirosh, R.: Ballistic protons and microwave-induced water solitons in bioenergetic transformations. In: *International Journal of Molecular Sciences* 7 (2006), Nr. 9, pp. 320–345. – URL https://www.mdpi.com/1422-0067/7/9/320. – ISSN 1422-0067

[Tokita et al. 2008] Tokita, Y. ; Shimura, J. ; Nakajima, H. ; Goto, Y. ; Watanabe, Y.: Mechanism of intramolecular electron transfer in the photoexcited Zn-substituted cytochrome c: theoretical and experimental perspective. In: *J Am Chem Soc* 130 (2008), Nr. 15, pp. 5302–10. – URL https://www.ncbi.nlm.nih.gov/pubmed/18348525. – ISSN 1520-5126 (Electronic) 0002-7863 (Linking)

[Torbert and Johnson 2001] Torbert, H. A. ; Johnson, H. B.: Soil of the Intensive Agriculture Biome of Biosphere 2. In: *Journal of Soil and Water Conservation* 56 (2001), Nr. 1, pp. 4–11. – URL http://www.jswconline.org/content/56/1/4.extract. – ISSN 0022-4561

[Torres et al. 2019] Torres, I. ; Sánchez, M.-T. ; Cho, B.-K. ; Garrido-Varo, A. ; Pérez-Marín, D.: Setting up a methodology to distinguish between green oranges and leaves using hyperspectral imaging. In: *Computers and Electronics in Agriculture* 167 (2019). – URL https://doi.org/10.1016/j.compag.2019.105070. – ISSN 01681699

[Toth and Jóźków 2016] Toth, C. ; Jóźków, G.: Remote sensing platforms and sensors: A survey. In: *ISPRS Journal of Photogrammetry and Remote Sensing* 115 (2016), pp. 22–36. – URL https://www.sciencedirect.com/science/article/pii/S0924271615002270. – ISSN 09242716

[Tributsch 2000] Tributsch, H.: Light Driven Proton Pumps. In: *Ionics* 6 (2000), Nr. 3-4, pp. 161–171. – ISSN 0947-7047

[Tsimilli-Michael et al. 2000] Tsimilli-Michael, M. ; Eggenberg, P. ; Biro, B. ; Köves-Pechy, K. ; Vörös, I. ; Strasser, R. J.: Synergistic and antagonistic effects of arbuscular mycorrhizal fungi and Azospirillum and Rhizobium nitrogen-fixers on the photosynthetic activity of alfalfa, probed by the polyphasic chlorophyll a fluorescence transient O-J-I-P. In: *Applied Soil Ecology* 15 (2000), Nr. 2, pp. 169–182. – URL https://doi.org/10.1016/S0929-1393(00)00093-7. – ISSN 09291393

[Tung et al. 1984] Tung, J. H. ; Salamo, X. M. ; Chan, F. T.: Two-photon decay of hydrogenic atoms. In: *Physical Review A* 30 (1984), Nr. 3, pp. 1175–1184. – URL https://journals.aps.org/pra/pdf/10.1103/PhysRevA.30.1175. – ISSN 0556-2791

[Turner 2012] Turner, S. B.: The case of the Zia: Looking beyond trademark law to protect sacred symbols. In: *Yale Law School Legal Scholarship Repository* (2012)

[Udriste et al. 2018] Udriste, C. ; Golubyatnikov, V. ; Tevy, I.: Economic Cycles of Carnot Type. In: *arXiv:1812.07960* (2018). – URL https://arxiv.org/abs/1812.07960

[Uz et al. 2019] Uz, S. S. ; Kim, G. E. ; Mannino, A. ; Werdell, P. J. ; Tzortziou, M.: Developing a Community of Practice for Applied Uses of Future PACE Data to Address Marine Food Security Challenges. In: *Frontiers in Earth Science* 7 (2019). – URL https://www.frontiersin.org/articles/10.3389/feart.2019.00283/full

[Valero 2006] Valero, F. P. J.: Keeping the DSCOVR mission alive. In: *Science* 311 (2006), Nr. 5762, pp. 775–776. – URL http://science.sciencemag.org/content/sci/311/5762/775.3.full.pdf. – ISSN 0036-8075

[van der Paauw 1935] van der Paauw, F.: Die Wirkung von Blausäure auf die Kohlensäureassimilation und Atmung von Stichococcus bacillaris. In: *Planta* 24 (1935), Nr. 2, pp. 353–360. – URL https://link.springer.com/content/pdf/10.1007/BF01910953.pdf. – ISSN 0032-0935 1432-2048

[van der Pauw 1958] van der Pauw, L. J.: A Method of Measuring Specific Resistivity and Hall Effect of Discs of Arbitrary Shapes. In: *Philips Technical Review* 20 (1958)

[van Oijen et al. 1999] van Oijen, A. M. ; Ketelaars, M. ; Kohler, J. ; Aartsma, T. J. ; Schmidt, J.: Unraveling the electronic structure of individual photosynthetic pigment-protein complexes. In: *Science* 285 (1999), Nr. 5426, pp. 400–402. – URL http://science.sciencemag.org/content/sci/285/5426/400.full.pdf. – ISSN 0036-8075

[van Oijen et al. 2001] van Oijen, A. M. ; Ketelaars, M. ; Matsushita, M. ; Kohler, J. ; Aartsma, T. J. ; Schmidt, J.: Unraveling the electronic structure of individual photosynthetic pigment-protein complexes. In: *Biophysical Journal* 80 (2001), Nr. 1, pp. 151A–151A. – ISSN 0006-3495

[Van Siclen and Jones 1986] Van Siclen, C. D. W. ; Jones, S. E.: Piezonuclear fusion in isotopic hydrogen molecules. In: *Journal of Physics G: Nuclear Physics* 12 (1986), Nr. 3, pp. 213–221. – ISSN 0305-4616

[Varo et al. 1992] Varo, G. ; Zimanyi, L. ; Chang, M. ; Ni, B. ; Needleman, R. ; Lanyi, J. K.: A residue substitution near the beta-ionone ring of the retinal affects the M substates of bacteriorhodopsin. In: *Biophys J* 61 (1992), Nr. 3, pp. 820–6. – URL https://www.ncbi.nlm.nih.gov/pubmed/1504253. – ISSN 0006-3495 (Print) 0006-3495 (Linking)

[Venturi et al. 2005] Venturi, M. ; Balzani, V. ; Gandolfi, M. T.: Fuels from solar energy: A dream of Giacomo Ciamician, the father of photochemistry. In: *Proceedings of the 2005 Solar World Congress* (2005). – URL https://www.gses.it/pub/Ciamician.pdf

[Verhoeven 1996] Verhoeven, J. W.: Glossary of terms used in photochemistry (IUPAC Recommendations 1996). In: *Pure and Applied Chemistry* 68 (1996), Nr. 12, pp. 2223–2286. – URL https://www.degruyter.com/view/journals/pac/68/12/article-p2223.xml. – ISSN 1365-3075 0033-4545

[Vinson et al. 2011] Vinson, J. ; Rehr, J. J. ; Kas, J. J. ; Shirley, E. L.: Bethe-Salpeter equation calculations of core excitation spectra. In: *Phys Rev B Condens Matter Mater Phys* 83 (2011). – URL https://www.ncbi.nlm.nih.gov/pmc/articles/PMC6508618/pdf/nihms977147.pdf. – ISSN 1098-0121 (Print) 1098-0121 (Linking)

[Voelkel 2008] Voelkel, J. R.: Publish or Perish: Legal Contingencies and the Publication of Kepler's Astronomia nova. In: *Science in Context* 12 (2008), Nr. 1, pp. 33–59. – ISSN 0269-8897 1474-0664

[Voityuk et al. 1998] VOITYUK, A. A. ; MICHEL-BEYERLE, M. E. ; ROSCH, N.: Quantum chemical modeling of structure and absorption spectra of the chromophore in green fluorescent proteins. In: *Chemical Physics* 231 (1998), Nr. 1, pp. 13–25. – URL https://www.sciencedirect.com/science/article/pii/S0301010498000809. – ISSN 0301-0104

[Voityuk and Rosch 2004] VOITYUK, A. A. ; ROSCH, N.: *INDO/S*. 2004. – URL https://doi.org/10.1002/0470845015.cu0004

[von Ditfurth 2015] VON DITFURTH, H.: *Im Anfang war der Wasserstoff*. 20. Auflage. dtv Wissen, 2015. – ISBN 978-3-423-33015-2

[von Goethe 1810] VON GOETHE, J. W.: *Zur Farbenlehre*. Tuebingen: J.G. Cottasche Buchhandlung, 1810

[von Helmholtz 1867] VON HELMHOLTZ, H.: *Handbuch der physiologischen Optik*. Leipzig: Leopold Voss, 1867. – URL https://archive.org/details/handbuchderphysi00helm

[von Weizsäcker 1937] VON WEIZSÄCKER, C. F.: Über Elementumwandlungen im Innern der Sterne (On element conversions in the inside of stars. I.). In: *Physikalische Zeitschrift der Sowjetunion* 38 (1937)

[von Weizsäcker 1938] VON WEIZSÄCKER, C. F.: Über Elementumwandlungen im Innern der Sterne (Element conversions in the interior of stars. II). In: *Physikalische Zeitschrift der Sowjetunion* 39 (1938)

[Vredenberg and Bulychev 2002] VREDENBERG, W. J. ; BULYCHEV, A. A.: Photo-electrochemical control of photosystem II chlorophyll fluorescence in vivo. In: *Bioelectrochemistry* 57 (2002), Nr. 2, pp. 123–128. – URL https://www.sciencedirect.com/science/article/pii/S1567539402000622. – ISSN 1567-5394

[Wacker 2016] WACKER, A.: *Fermi's golden rule*. 2016. – URL http://www.matfys.lth.se/staff/Andreas.Wacker/Scripts/fermiGR.pdf

[Wallenhorst 1981] WALLENHORST, S. G.: The Drake Equation Reexamined. In: *Quarterly Journal of the Royal Astronomical Society* 22 (1981), Nr. 4, pp. 380–387. – URL https://ui.adsabs.harvard.edu/link_gateway/1981QJRAS..22..380W/ADS_PDF. – ISSN 0035-8738

[Walters et al. 1980] WALTERS, C. ; HOOVER, R. A. ; KOTRA, R. K.: Interstellar colonization: A new parameter for the Drake equation? In: *Icarus* 41 (1980), Nr. 2, pp. 193–197. – ISSN 00191035

[Wang et al. 2017] WANG, P. ; CHANG, A. Y. ; NOVOSAD, V. ; CHUPIN, V. V. ; SCHALLER, R. D. ; ROZHKOVA, E. A.: Cell-Free Synthetic Biology Chassis for Nanocatalytic Photon-to-Hydrogen Conversion. In: *ACS Nano*. 11 (2017), pp. 6739–6745

[Wang et al. 2016] WANG, Y. ; QU, T. F. ; ZHAO, X. Y. ; TANG, X. H. ; XIAO, H. ; TANG, X. X.: A comparative study of the photosynthetic capacity in two green tide macroalgae using chlorophyll fluorescence. In: *Springerplus* 5 (2016), p. 12. – URL https://springerplus.springeropen.com/track/pdf/10.1186/s40064-016-2488-7. – ISSN 2193-1801

[Warner 2013] WARNER, J.: The Toshka mirage in the Egyptian desert – River diversion as political diversion. In: *Environmental Science & Policy* 30 (2013), pp. 102–112. – URL https://www.sciencedirect.com/science/article/pii/S146290111200202X. – ISSN 1462-9011

[Warshel and Chu 2001] WARSHEL, A. ; CHU, Z. T.: Nature of the surface crossing process in bacteriorhodopsin: Computer simulations of the quantum dynamics of the primary photochemical event. In: *Journal of Physical Chemistry B* 105 (2001), Nr. 40, pp. 9857–9871. – URL https://pubs.acs.org/doi/pdf/10.1021/jp010704a. – ISSN 1520-6106

[Watt 2003] WATT, J. G.: A brief history of the Chilean nitrates industry. In: *Cim Bulletin* 96 (2003), Nr. 1073, pp. 84–88. – ISSN 0317-0926

[Webb and Bustin 1988] WEBB, J. ; BUSTIN, R.: Eratosthenes revisited. In: *The Physics Teacher* 26 (1988), Nr. 3, pp. 154–155. – URL https://aapt.scitation.org/doi/pdf/10.1119/1.2342460. – ISSN 0031-921X

[Weber and Cabras 2019] WEBER, G. ; CABRAS, I.: *The Ecological Economy of Georgescu-Roegen*. Book Section Chapter 10, pp. 221–238. In: *Economic Theory and Globalization*, Palgrave Macmillan, 2019. – URL https://doi.org/10.1007/978-3-030-23824-7_10. – ISBN 978-3-030-23823-0 978-3-030-23824-7

[Webster 1997] WEBSTER, R.: Regression and functional relations. In: *European Journal of Soil Science* 48 (1997), Nr. 3, pp. 557–566. – URL https://onlinelibrary.wiley.com/doi/abs/10.1111/j.1365-2389.1997.tb00222.x. – ISSN 1351-0754

[Weisskopf and Wigner 1930a] WEISSKOPF, V. ; WIGNER, E.: Berechnung der natuerlichen Linienbreite auf Grund der Diracschen Lichttheorie. In: *Zeitschrift fuer Physik* 63 (1930), Nr. 1-2, pp. 54–73. – URL https://link.springer.com/content/pdf/10.1007%2FBF01336768.pdf. – ISSN 1434-6001 1434-601X

[Weisskopf and Wigner 1930b] WEISSKOPF, V. ; WIGNER, E.: Über die natuerliche Linienbreite in der Strahlung des harmonischen Oszillators. In: *Zeitschrift fuer Physik* 65 (1930), Nr. 1-2, pp. 18–29. – URL https://link.springer.com/content/pdf/10.1007%2FBF01397406.pdf. – ISSN 1434-6001 1434-601X

[Wells 1895] WELLS, H. G.: *The Time Machine – An Invention*. New York: Henry Holt And Company, 1895. – p. 232. – URL https://archive.org/download/timemachineinven00well/timemachineinven00well.pdf

[Werdell et al. 2019] WERDELL, P. J. ; BEHRENFELD, M. J. ; BONTEMPI, P. S. ; BOSS, E. ; CAIRNS, B. ; DAVIS, G. T. ; FRANZ, B. A. ; GLIESE, U. B. ; GORMAN, E. T. ; HASEKAMP, O. ; KNOBELSPIESSE, K. D. ; MANNINO, A. ; MARTINS, J. V. ; MCCLAIN, C. R. ; MEISTER, G. ; REMER, L. A.: The Plankton, Aerosol, Cloud, Ocean Ecosystem Mission: Status, Science, Advances. In: *Bulletin of the American Meteorological Society* 100 (2019), Nr. 9, pp. 1775–1794. – URL https://journals.ametsoc.org/doi/pdf/10.1175/BAMS-D-18-0056.1. – ISSN 0003-0007

[Wick 1954] WICK, G. C.: Properties of Bethe-Salpeter Wave Functions. In: *Physical Review* 96 (1954), Nr. 4, pp. 1124–1134. – ISSN 0031-899X

[Wien 1896] WIEN, W.: Ueber die Energievertheilung im Emissionsspectrum eines schwarzen Koerpers. In: *Annalen der Physik und Chemie* 294 (1896), Nr. 8, pp. 662–669. – ISSN 00033804 15213889

[Wien and Lummer 1895] WIEN, W. ; LUMMER, O.: Methode zur Pruefung des Strahlungsgesetzes absolut schwarzer Koerper. In: *Annalen der Physik und Chemie* 292 (1895), Nr. 11, pp. 451–456. – URL https://onlinelibrary.wiley.com/doi/pdf/10.1002/andp.18952921103. – ISSN 00033804 15213889

[Wigner 1993] WIGNER, E. P.: *Über die Operation der Zeitumkehr in der Quantenmechanik*. Book Section Chapter 15, pp. 213–226. In: *The Collected Works of Eugene Paul Wigner*, Springer, 1993. – URL https://doi.org/10.1007/978-3-662-02781-3_15https://eudml.org/doc/59401. – ISBN 978-3-642-08154-5 978-3-662-02781-3

[Willen 1991] WILLEN, E.: Planktonic Diatoms – an Ecological Review. In: *Archiv Fur Hydrobiologie* (1991), pp. 69–106. – URL https://www.researchgate.net/publication/29721360_Planktonic_diatoms_-_An_ecological_review/link/0a85e534e9770ad30c000000/download. – ISSN 0003-9136

[Wissner and Tainter 2012] WISSNER, A. ; TAINTER, J. A.: *The Collapse of Complex Societies*. 18 November 2012 2012. – URL https://www.youtube.com/watch?v=ddmQhliVM48https://www.youtube.com/watch?v=G0R09YzyuCI

[Wolff 2018] WOLFF, E.: Die Krise am Horizont. Zehn Jahre nach der Finanzkrise – Keines der Probleme geloest / Arbeitskreis Finanzkrise und Alternativen. attac Augsburg, 23 Oktober 2018. – Report. – URL https://www.youtube.com/watch?v=NKhbD-WKA6k

[Woo et al. 2017] Woo, K. C. ; Kang, D. H. ; Kim, S. K.: Real-Time Observation of Nonadiabatic Bifurcation Dynamics at a Conical Intersection. In: *J Am Chem Soc* 139 (2017), Nr. 47, pp. 17152–17158. – URL https://pubs.acs.org/doi/pdf/10.1021/jacs.7b09677. – ISSN 1520-5126 (Electronic) 0002-7863 (Linking)

[Wood et al. 2018] Wood, W. H. J. ; MacGregor-Chatwin, C. ; Barnett, S. F. H. ; Mayneord, G. E. ; Huang, X. ; Hobbs, J. K. ; Hunter, C. N. ; Johnson, M. P.: Dynamic thylakoid stacking regulates the balance between linear and cyclic photosynthetic electron transfer. In: *Nature Plants* 4 (2018), Nr. 2, pp. 116–127. – URL https://www.nature.com/articles/s41477-017-0092-7.pdf. – ISSN 2055-026X

[Wuersten 2016] Wuersten, F.: Agrovet-Strickhof innovative Cooperation. In: *Schweizer Archiv fuer Tierheilkunde* 158 (2016), Nr. 5, p. 320. – URL https://sat.gstsvs.ch/de/sat/archiv/2016/052016.html

[Xiao et al. 2006] Xiao, Y. M. ; Fisher, K. ; Smith, M. C. ; Newton, W. E. ; Case, D. A. ; George, S. J. ; Wang, H. X. ; Sturhahn, W. ; Alp, E. E. ; Zhao, J. Y. ; Yoda, Y. ; Cramer, S. P.: How nitrogenase shakes – Initial information about P-cluster and FeMo-cofactor normal modes from nuclear resonance vibrational Spectroscopy (NRVS). In: *Journal of the American Chemical Society* 128 (2006), Nr. 23, pp. 7608–7612. – URL https://pubs.acs.org/doi/pdfplus/10.1021/ja0603655. – ISSN 0002-7863

[Yang et al. 2017] Yang, B. ; Knyazikhin, Y. ; Mottus, M. ; Rautiainen, M. ; Stenberg, P. ; Yan, L. ; Chen, C. ; Yan, K. ; Choi, S. ; Park, T. ; Myneni, R. B.: Estimation of leaf area index and its sunlit portion from DSCOVR EPIC data: Theoretical basis. In: *Remote Sens Environ* 198 (2017), pp. 69–84. – URL https://www.ncbi.nlm.nih.gov/pubmed/28867834. – ISSN 0034-4257 (Print) 0034-4257 (Linking)

[Yang et al. 1996] Yang, F. ; Moss, L. G. ; Phillips, Jr., G. N.: The molecular structure of green fluorescent protein. In: *Nat Biotechnol* 14 (1996), Nr. 10, pp. 1246–51. – URL https://www.ncbi.nlm.nih.gov/pubmed/9631087. – ISSN 1087-0156 (Print) 1087-0156 (Linking)

[Yariv 1977] Yariv, A.: The application of time evolution operators and Feynman diagrams to nonlinear optics. In: *IEEE Journal of Quantum Electronics* 13 (1977), Nr. 12, pp. 943–950. – ISSN 0018-9197

[Yarkony 1999] Yarkony, D. R.: S1–S0 Internal Conversion in Ketene. 1. The Role of Conical Intersections. In: *The Journal of Physical Chemistry A* 103 (1999), Nr. 33, pp. 6658–6668. – URL https://pubs.acs.org/doi/pdf/10.1021/jp9910136. – ISSN 1089-5639 1520-5215

[Ye et al. 2011] Ye, N. h. ; Zhang, X. w. ; Mao, Y. z. ; Liang, C. w. ; Xu, D. ; Zou, J. ; Zhuang, Z. m. ; Wang, Q. y.: 'Green tides' are overwhelming the coastline of our blue planet: taking the world's largest example. In: *Ecological Research* 26 (2011), Nr. 3, pp. 477–485. – URL https://doi.org/10.1007/s11284-011-0821-8. – ISSN 0912-3814 1440-1703

[Youngs and Somerville 2017] Youngs, H. ; Somerville, C.: Implementing industrial-academic partnerships to advance bioenergy research: the Energy Biosciences Institute. In: *Curr Opin Biotechnol* 45 (2017), pp. 184–190. – URL https://www.ncbi.nlm.nih.gov/pubmed/28458111. – ISSN 1879-0429 (Electronic) 0958-1669 (Linking)

[Zhang and Liu 2020] Zhang, L. ; Liu, Y.: Potential interventions for novel coronavirus in China: A systematic review. In: *J Med Virol* 92 (2020), Nr. 5, pp. 479–490. – URL https://www.ncbi.nlm.nih.gov/pubmed/32052466https://onlinelibrary.wiley.com/doi/full/10.1002/jmv.25707. – ISSN 1096-9071 (Electronic) 0146-6615 (Linking)

[Zheng et al. 1991] Zheng, C. ; McCammon, J. A. ; Wolynes, P. G.: Quantum simulations of conformation reorganization in the electron transfer reactions of tuna cytochrome c. In: *Chemical Physics* 158 (1991), Nr. 2-3, pp. 261–270. – URL https://www.sciencedirect.com/science/article/pii/030101049187070C. – ISSN 03010104

Index

https://doi.org/10.1515/9783110629941-013